TRENDS IN ERGONOMICS/ HUMAN FACTORS II

TRENDS IN ERGONOMICS/ HUMAN FACTORS II

Edited by

Ray E. EBERTS and Cindelyn G. EBERTS

School of Industrial Engineering
Purdue University
West Lafayette
Indiana
U.S.A.

1985

NORTH-HOLLAND
AMSTERDAM • NEW YORK • OXFORD

ISBN: 0 444 87751 7

Published by:
ELSEVIER SCIENCE PUBLISHERS B.V.
P.O. Box 1991
1000 BZ Amsterdam
The Netherlands

Sole distributors for the U.S.A. and Canada:
ELSEVIER SCIENCE PUBLISHING COMPANY, INC.
52 Vanderbilt Avenue
New York, N.Y. 10017
U.S.A.

Library of Congress Cataloging in Publication Data

Main entry under title:

Trends in ergonomics/human factors II.

Papers presented at the Second Annual Mid-Central Ergonomics/Human Factors Conference held at Purdue Univ., June 13-15, 1985.
Includes bibliographies and index.
1. Human engineering--Congresses. I. Eberts, Ray E., 1954- . II. Eberts, Cindelyn G. III. Mid-central Ergonomics/Human Factors Conference (2nd : 1985 : Purdue University)
TA166.T743 1985 620.8'2 85-4592
ISBN 0-444-87751-7

PRINTED IN THE NETHERLANDS

PREFACE

This book consists of the papers presented at the Second Annual Mid-Central Ergonomics/Human Factors Conference held at Purdue University in West Lafayette, Indiana 13–15 June, 1985. The papers are the result of recent research in human factors and ergonomics laboratories from the U.S. plus work from Europe and Canada. The paper topics in the conference were encouraged to be comprehensive and, thus, all the major areas of current research in human factors and ergonomics have been included in this volume. To organize this wide range of material, the book has been divided into nine chapters. The objective of the conference was to bring together researchers and practitioners to share and exchange information, ideas, and theory in applying human factors and ergonomics research.

Ray E. Eberts
Cindelyn G. Eberts

School of Industrial Engineering
Purdue University
West Lafayette, Indiana

CONTENTS

I

METHODOLOGY

Trends in Ergonomics/Human Factors II
R.E. Eberts and C.G. Eberts (Editors)

THE TWO WORLDS OF HUMAN FACTORS

David Meister

U.S. Navy Personnel Research and Development Center
San Diego, California

The major problem facing Human Factors is the gap between what the researcher provides as research products and what the practitioner working in system development needs. Human Factors research does not satisfy the needs of the practitioner because it is not oriented toward system-relevant questions and because no effort is made to develop a human performance database that can be useful in system development. Underlying these inadequacies is the over-emphasis on hypothesis-testing and the researcher's lack of interest in collecting usable data. Remedies must be achieved in the university and government.

THE PROBLEM:

The major problem facing the Human Factors (HF) discipline today--one that it will continue to face in the near if not long term--is not a technical one (although it has technical antecedents and consequences). It is the communications gap between the HF researcher and the HF practitioner, the specialist working in the applications area. The gap between what the researcher provides as research outputs and what the practitioner needs and what he can do with the researcher's outputs.

My thesis is that the practitioner can do very little with these outputs and that as a consequence he is left very much in limbo. The survey of HF system developers that I reported on a few years ago (Meister, 1979) indicated that the HF specialist working in development is highly dependent on his own resources, his intuition and his personal and communal experience. Perhaps that should be enough for him to do his job, but he does not feel so. He expects that research will supply an ingredient he lacks--substantive data and design principles. He feels however that much of the research being performed is of little use in his work, the work of analyzing system requirements, making design recommendations and evaluating design products. This lack of utility applies just as much to the analyses and evaluations performed in selection and training of system personnel.

Very little of what the practitioner does is assisted by relevant, substantive data. This is not to say that there are no relevant data but that in comparison with the number of system development questions that arise, these data are pitifully few. In consequence the practitioner is unable to prove to Engineering and to the development community at large the value of his efforts. This inability is a major contributor to the engineer's negative attitude toward HF. Efforts have been made (eg, Price, Fiorello, Lowery,

Smith and Kidd, 1980) to demonstrate the value of HF in system development by anecdotes, case histories and logic, but these lack the one conclusive factor: data.

Those of us who have worked in system development feel that our analyses and evaluations should be founded on a bedrock of quantitative data and principles derived from systematic research. If researchers also feel the same way - and it is hard to conceive of their not feeling thus - it is obligatory that they supply practitioners with the answers to the latters' questions.

Researchers may feel that their outputs are indeed relevant and that the gap lies in the practitioner's inability to make use of those outputs. This question can be debated, of course. If, for example, a researcher does a study involving the Theory of Signal Detection, should his conclusions from that study bear a direct or indirect relationship to the optimal design of new types of detection displays? And who has the responsibility for translating his research conclusions into practically applicable guidance for HF specialists working with displays? My own hypothesis is that if application mechanisms are not built into a research study to begin with, the conclusions of that study will not be very usable when someone wishes to apply them.

Some researchers assume that over some unspecified period of time their research will be translated by some unspecified process into applicable products. If one believes this, one merely has to wait and, like sunlight being transformed by photosynthesis into plant food, behavioral research conclusions will eventually become practical design guidance. There may indeed be some truth in this but it is very difficult to wait for the completion of the translation process. It seems to me that the guidance now available is not very much advanced over what it was when I started my professional work 34 years ago. If my experience is a common one, then it seems unreasonable to have to wait more than a single generation for behavioral research to bear fruit.

This is not a plea for more applied research as opposed to basic research. The choice between basic and applied research - however one delimits the two - is essentially a cul-de-sac because both are necessary. In a paper some years ago (Meister, 1980) I suggested that purely applied problems (ie, those stemming from system development) also require a foundation of basic research ("macro-basic" research).

For research to be practically valuable the HF researcher must pursue that research in the same domain as that in which the practitioner works. For the latter that domain is the human-machine system, which is hardly a novel concept these days. The essence of that concept is the commonsense assumption that there is an interaction between the physical characteristics of the machine and operator/maintainer performance. Consequently, one affects the other. It is therefore necessary for the HF practitioner to know as precisely, as quantitatively, as possible what the effect of system physical characteristics is on human performance, as reflected in error and successful task accomplishment measures. For example, he evaluates control panel adequacy by the likelihood of the operator's errorless performance. If research outputs are to be useful to him they must be framed in the same conceptual structure.

What are these system physical characteristics that affect operator performance? I cannot give you a complete list because not enough is known about them, but a few suggest themselves. Most obvious of course is the question of the relationship between number of controls and displays, their control panel arrangement, and operator error. Other factors almost certainly influence operator performance: the speed with which the operator must respond; the fluidity with which system events change and the amount of change; the amount and type of operator involvement with the machine; the precision with which analog displays must be read; the amount of system feedback provided to the operator; the concreteness of the task outputs required of him; the relationship between operator outputs and ultimate system outputs. And there are many others, doubtless. If HF researchers are not studying relationships such as these, their research products, however well they describe some types of human performance, are of little consequence to the practitioner. Unfortunately, very little HF research seems to concern itself with questions such as these. One need only examine the papers published in <u>Human Factors</u> or <u>Ergonomics</u> and ask oneself, how can I apply the conclusions reached in these studies to the relationship questions that have just been listed? The papers deal with human performance and to that extent they are <u>indirectly</u> relevant to the practitioner's concerns, but they are not directly relevant. They serve primarily as background information but they do not help to provide specific answers to system questions.

Anyone who has tried to solve specific HF problems has "gone to the literature" (a sort of intellectual journey to Mecca) and experienced the frustration of finding no specific answers. It may be too much to expect of any body of literature that it provide specific answers, because, to be generalizable, research must be general rather than specific. It is my contention however that even if HF research must be general, it is still possible for it to be useful if it is relevant to system parameters. The practitioner can make useful deductions from relevant research even if the studies he peruses do not deal specifically with his immediate problem.

System development begins with an analysis of the requirements laid upon the system: to have such and such speed or altitude or range; to have certain attributes like high reliability and maintainability. These requirement examples are physicalistic in nature; that is because almost never are system requirements of a behavioral nature. I will address the reason for this later. In any event, the HF practitioner must analyze these physical requirements to deduce from them their impact on personnel performance; and then he must suggest a physical mechanism to incorporate his behavioral conclusions into the human-machine interface or some method of cushioning the effect of the physical requirements on the operator. What is involved is a transformation from the physical to the behavioral and vice-versa.

Because this transformation is at the heart of system-relevant HF research, the researcher must directly confront that transformation in his work. He can do this only by conceptualizing how his anticipated research conclusions might be utilized in application to development. This requires of the researcher much imagination. He must ask himself, if I get the results I hope to get from my study, will it answer questions relevant to the transformation process? If the answer to his question is no, it is unlikely that his study will be of substantive value to the practitioner.

The necessity for maintaining this point of view in his research will undoubtedly limit the freedom the researcher has in conducting his studies. No researcher enjoys limitations. The ultimate utility of his work depends, however, on working within those limitations. If his research is essentially irrelevant to the practitioner it will wind up in dusty journals on library shelves. Only he and a few others concerned with the questions he addresses will be at all interested in that research.

That may of course be sufficient for those researchers who disclaim any interest in application. One wonders though at the value of the many studies performed over the years that result in serried ranks of journal volumes in libraries. It is instructive to consider that, when Payne and Altman (1962) undertook their effort to develop a quantitative probabilistic human performance data base dealing primarily with controls and displays, they found only 164 studies out of several thousands they examined that could satisfy their criteria for data selection. What was the value of the other thousands of studies they rejected?

Some may feel that the picture I paint is too sombre. Are there not design guides like Van Cott and Kinkade (1972), textbooks like McCormick and Sanders (1982), military specifications like MIL-STD 1472C (Department of Defense, 1981), and compilations of design principles like those of Smith and Aucella (1983) and Williges and Williges (1984)? Unfortunately, these provide little usable design guidance. They are valuable to the practitioner when he peruses them to tell him that in fact there is no information available on the particular problem he is facing. With this reassurance he can go ahead and use his intuition and experience without feeling that he may have overlooked some substantive data somewhere. These documents therefore provide negative reinforcement, as it were, the value of which should not however be overlooked.

THE PRACTITIONER'S NEEDS:

One cannot ask the researcher to provide certain practical outputs without specifying what the practitioner would like to have. We can then match research outputs against research needs, recognizing of course that certain practitioner desires may be unreasonable.

(1) Design standards. These I have defined as consisting of three elements: (a) A relationship between a system characteristic (e.g., number of controls and displays) and human performance, expressed ideally in the form of a task probability-of-accomplishment metric. This describes how well operators will perform if a particular system characteristic is included in design. (b) He should have data describing what the effect of a non-optimal relationship would be, again in terms of task failure or error consequences. For example, what would the effect be of including an excessive number of controls and displays on a control panel? Obviously, if element (a) in the design standard is complete, (b) automatically follows; but the data in element (a) may not be complete and then it is necessary to develop element (b) specifically. Element (b) is needed to determine the relative importance of behavioral principles in design, and to persuade equipment engineers that if design is non-optimal from a behavioral standpoint, the operator and the system will pay a price. If a non-optimal design leads to no known behavioral consequences, the designer is likely to feel that any design solution he accepts is as good as any other in terms of behavioral effects. (c) There should also be some principles for incorporating a

desired system characteristic into design, but this may be an engineering question rather than a behavioral one.

Admittedly the development of such design standards would be a difficult task, particularly the determination of the consequences on performance of a non-optimal design. The latter might require development of a number of design alternatives and the testing of each one. However, the major part of the standard, consisting of the quantitative relationship between a system characteristic and resultant operator performance seems - in theory, at any rate - perfectly possible to determine, using conventional methods of experimentation.

(2) Human performance data. The reader cannot have overlooked the assumption that the heart of the design standard is a mass of human performance data specifically related to system characteristics. The term "specifically" must be emphasized. This is because what is needed is a database, not just data (in the conventional sense of data as a whole). A database is a set of data compiled in accordance with certain taxonomic categories derived from the kinds of questions one wishes to answer. Data satisfying taxonomic criteria are accepted into the database; other data are rejected as useless for database purposes (although they may be useful for other purposes). The reader is referred to Topmiller, Eckel and Kozinsky, (1982) which is a compilation of all extant (published) databases that might be used for human reliability analyses. These databases are all phrased in either a probability of error or probability of task accomplishment metric, meaning that the studies from which they came provided data that were already or could be transformed into the probability metric. Information must also be available about how the raw data were collected and any factors that might affect their application to real world problems. For example, most data are collected in the laboratory environment which may be either too limiting or too free as compared with the real world. In this case the data must be modified to take account of the environmental discrepancy. The point of all this is simply that a database is not merely a compilation of all data bearing more or less on certain parameters; the data must satisfy certain rigid standards to be acceptable.

Most important, the development of a usable human performance database is an effort of some magnitude. Studies must be analyzed; data extracted; certain taxonomic and other criteria must be applied; data must be accepted or rejected; if data are accepted, they may have to be modified to put them into the probabilistic form for combination with other data. This last is most important. In a database data are melded into each other; each individual set of data does not stand alone. The point is to incorporate data from a number of studies into a single value.

One cannot therefore point at the literature in general and say, "there is a database", because it is nothing of the sort. Because of the effort involved in developing a database, there have been almost no attempts since the original prototype was published (Munger, Smith and Payne, 1962). This despite the urgings of a number of people including myself. Unfortunately those in government who support research efforts have felt that the task of extracting usable material from accumulated literature is not sufficiently "original", as if the data refinement and compilation effort were only a task that a drudge would attempt.

SOME EXPLANATIONS:

I have tried to find some explanation for the state of affairs in HF that I have described. Others may have other explanations.

To me the largest part of our problem results from the fact that the HF research paradigm taken over from Psychology is oriented around hypothesis-testing. The popular concept of science is that it begins with theory from which hypotheses are derived; and that these hypotheses must be tested under highly controlled (i.e., usually the laboratory but occasionally also real world) conditions.

Although data are involved in hypothesis-testing, data are not a primary concern in hypothesis-testing. The essential thing is the statistical test of significance which determines whether an hypothesis can be rejected (as unlikely) or accepted. The hypothesis-testing arranges the conditions of stimulus presentation and subject responses so that usually only the extremes (and perhaps one or two other points) of a distribution of conditions are tested. This leaves most of the continuum of data conditions untested. For example, if there are 10 conditions under which data may be gathered, the experimenter may collect his data only under conditions 1, 5 and 10 (as representing the totality of a relationship), because this is sufficient in his mind to test his hypothesis. However, the data from the limited number of conditions is rarely sufficient for HF work which, because it is performed in an engineering orientation, requires engineering precision and completeness.

If we look further to the outputs of hypothesis-testing studies, we find a set of conclusions which are phrased in qualitative generalities. There are two reasons for this. Study results should be maximally generalizable and so their conclusions must be geared to the widest possible range of conditions (hence general); and in any event the researcher is not concerned about including actual numerical values in his conclusions because it is the hypothesis being tested that is of interest. So the conclusions are almost always qualitative. The researcher is not interested in assembling the study data and certainly not in combining those data with other data attacking the same question. He is not concerned with performance prediction in the same way the HF practitioner is; for the latter, the prediction of performance to be achieved with a certain type of design is the major measure of the effectiveness of that design (from a behavioral standpoint).

What can one do with qualitative conclusions such as, for example, performance is optimal when workload is moderate? The statement in the conclusion may be entirely correct and just as unusable because it does not contain enough specifics to apply to real world situations. The common design principle found in MIL STD 1472C (Department of Defense, 1981) that less important controls and displays should be placed toward the periphery of a control panel tells us very little more than common sense would tell us because there are no human performance data associated with the principle. What, for example, is the effect on operator performance if the principle is violated?

One consequence of this lack of substantive relevant data is that it is almost impossible to establish human performance requirements for new systems. Or, lacking requirements data, it is almost impossible to present "normative" standards to the developer: given such and such a situation, empirical data tell us that humans will ordinarily perform in such and such a manner.

The system design process begins with and stands or falls on the requirements imposed initially on the system, but almost never is there a behavioral requirement. HF specialists have difficulty imposing such behavioral requirements because, lacking data, any requirements they impose on the system may demand more of system personnel than those personnel can perform. There are exceptions, of course. Psychophysical data tell us that if a symbol is less than 10-12 seconds of arc in size, it will not usually be perceived. Other data tell us that one cannot demand a response of the operator faster than 200-400 milliseconds. But these are extreme limiting factors for quite simple, discrete actions. Data are lacking about how people ordinarily perform in more complex circumstances. The lack of such normative data also results in part from an unwillingness of HF researchers to perform their studies in the real world of system use. Normative data are descriptive data, which are usually gathered outside of the laboratory (although not necessarily so). Most researchers have had little experience with actual system development and usage so that they are unaware of the kinds of problems encountered in that environment.

The consequence of the lack of substantive relevant data is that if by chance behavioral requirements are imposed on new system development--for example, the requirement to design a system so that a 10th grade high school student can operate it--we cannot satisfy the requirement because we do not know what 10th graders can do with certain kinds of systems.

What I have said may sound impossibly pessimistic, as if the whole HF field is blithely marching down a cul-de-sac, the road of physicalistic science which has only limited value for behavioral scientists even if it is valuable for physicists. However, if this is what I am suggesting, I reflect the viewpoint of many if not most practitioners as they look at HF research vis a vis themselves.

PROBLEM SOLUTION

The solution to such a problem is not an easy one. The problem may be inherent in the contradictory aims that researchers and practitioners have and hence a total solution may not be in the cards. Nevertheless, a partial solution may be achievable.

I am tempted to believe that any fundamental change in research outlook must start in the universities which train embryo HF specialists. Most of our people (at least the older generation) have had at least some of their training in Psychology, where the orientation is overwhelmingly on theory and hypothesis-testing. The indoctrination these students receive in their graduate work is pervasive and long lasting.

Although one would not expect the academic community to change its research principles completely, it might be possible for it to make its students more aware of the needs of the practitioner; to suggest that there is perhaps more to research than hypothesis-testing; to place more specific emphasis on systems and system development.

Government too has a role to play in terms of the research it will support. Far more research needs to be done on the system-operator relationships I have mentioned and how engineers design; the latter is the context for the incorporation of HF principles in new system development (see Meister,1971).

Government support is also necessary for the extraction of data from the literature to generate a true human performance data base. Again I come back to the point: what good are data that are never utilized except by a small band of scholars (see Meister, 1982).

I am somewhat pessimistic about the possibility of a significant revolution occurring in the HF research outlook. The odds are at best 1 in 10. We must however at least examine the problem, even if it is only to determine whether or not a problem exists. Otherwise, we may find ourselves with a HF discipline that is highly scientific (by conventional research criteria) but of only limited use either to the HF practitioner or to the world as a whole.

REFERENCES:

[1] Department of Defense. Military standard 1472C. Human engineering design criteria for military systems, equipment and facilities. Washington, D.C. 2 May 1981.

[2] McCormick, E.J. and Sanders, M.S. Human Factors in Engineering Design, 5th Edition. (McGraw-Hill, New York, 1982).

[3] Meister, D. Where and what are the data in Human Factors? Proceedings of the Human Factors Society Annual Meeting, 1982, 722-726.

[4] Meister, D. The concept of macro-basic research in Human Factors. Proceedings of the Human Factors Society Annual Meeting, 1980, 458-461.

[5] Meister, D. The influence of government on human factors research and development. Proceedings of the Human Factors Society Annual Meeting, 1979, 5-13.

[6] Meister, D. Human Factors: Theory and Practice. (Wiley, New York, 1971)

[7] Munger, S.J., Smith, R.W. and Payne, D. An index of electronic equipment operability: Data Store. Report AIR-C43-1/62-RP(1), American Institute for Research, Pittsburgh, PA (January 1962). (AD 607 161)

[8] Payne, D. and Altman, J.W. An index of electronic equipment operability: report of development. Report AIR-C-43-1/62-FR, American Institute for Research, Pittsburgh, PA (January, 1962).

[9] Price, H.E., Fiorello, M., Lowery, J.C., Smith, M.G., and Kidd, J.S. The contribution of human factors in military system development: methodological considerations. Technical Report 476, Army Research Institute, Alexandria, VA (July 1980).

[10] Smith, S.L. and Aucella, A.F. Design guidelines for the user interface in computer-based information systems. Report ESD-TR-83-122, Mitre Corporation, Bedford, MA (March 1983).

[11] Topmiller, D.A., Eckel, J.S. and Kozinsky, E.D. Human reliability databank for nuclear power plant operators. Vol. I: Review of existing human error reliability databanks. Report NUREG/CR-2744/1 of 2, SAND82-7057/1 of 2, General Physics Corporation, Dayton, OH (December 1982) (NUREG 2744-V1)

[12] Van Cott, H.P. and Kinkade, R.G. (eds.) Human Engineering Guide to Equipment Design, Revised Edition. (Govt. Printing Office, Wash.1972)

[13] Williges, B. and Williges, R.C. Dialogue design considerations for interactive computer systems, in: F.A. Muckler (ed). Human Factors Review 1984 (Human Factors Society, Santa Monica, CA, 1984).

The opinions expressed are those of the author alone and do not represent the Department of Defense.

Trends in Ergonomics/Human Factors II
R.E. Eberts and C.G. Eberts (Editors)

AVERAGED CROSS-CORRELATIONS: A METHODOLOGY FOR VALIDITY ASSESSMENT[1] IN SMALL SAMPLES

William P. Dunlap, Ned C. Silver, Robert E. Hunter
Tulane University
and
Alvah C. Bittner, Jr.
Naval Biodynamics Laboratory
New Orleans, Louisiana, U.S.A.

Precision in estimating validity coefficients can be gained by measuring the predictor and criterion more than once, then averaging all possible cross-correlations. In studying this procedure approximate equations were developed for the standard errors of averaged cross-correlations, and tested for accuracy by Monte-Carlo methods. The precision of averaged cross-correlations was examined over a variety of conditions. Greatest increases in precision were found when the correlations involved were small.

INTRODUCTION

In research, there often exists a tradeoff between the number of subjects used and the number of pieces of data collected per subject. Repeated measures research is often characterized by small sample sizes. This is particularly true if data collection takes more than one day; even if subjects are paid, attrition is often great. The result of this problem is that many data sets with multiple observations have rather limited numbers of subjects. If the planned method of analysis is correlational (e.g. multiple correlation or factor analysis), then conventional wisdom requires that large sample sizes be used. For example, sample sizes suggested for meaningful factor analysis are an absolute minimum of five subjects per measure, but never less than 100 subjects in all (Gorsuch, 1974). The present study demonstrates that one can estimate validities with precision by averaging cross-correlations. Such estimates, as will be seen, show stability equivalent to single estimates of the same correlations obtained with much larger samples. Under some conditions, the standard errors of the resulting estimates are reduced to the point that the average cross-correlations obtained from as few as 20 individuals are as precise as single estimates obtained from samples of more than 100.

Dunlap, Jones, and Bittner (1983) investigated the question of averaging correlations across blocks of trials. Assume for example, that the same variable is measured 25 times. If the sample size is small, estimates of the intercorrelations from trial to trial will be quite unstable (as will the means and standard deviations from trial to trial). More stable estimates of the means could be obtained by blocking the data into five trial blocks. Although it is somewhat counterintuitive, such blocking of the data does not improve the accuracy of cross-correlations estimated from the blocks of data. Instead, the correlations simply increase in size following the Spearman-Brown formula, but show the same standard errors, which are determined by the number of subjects involved. Dunlap, Jones, and Bittner (1983) showed that accuracy of the estimated correlations could be substantially improved by averaging correlation coefficients across trials. They also showed that empirical standard errors obtained by Monte Carlo simulation agreed quite well with theoretical predictions based on asymptotic formulae originally derived by Karl Pearson, and presented by Steiger (1980). It turns out that the gain in precision obtained by averaging correlations is dependent on the number of coefficients averaged, the size of the population correlation being estimated, and yet is fairly independent of the sample size involved. Furthermore, the greatest increase in accuracy occurs where it is most needed, when the population correlation is small.

It should be clear that the above research addressed a specific case, averaging correlations in blocks where the same variable is measured repeatedly. Bittner (1982) provided a similar analysis of the case where one might estimate reliability by averaging m(m-1)/2 intercorrelations of a variable measured m times. Bittner, Dunlap, and Jones (1982) made preliminary progress toward analyzing the case of estimating cross-correlations using averaged correlations in small samples. Unfortunately, the latter two studies made only limited use of Monte Carlo simulation, therefore, the accuracy of the predictions based on Pearson's asymptotic equations were not adequately established. Also, the Bittner *et al.* (1982) study considered only the case where the reliabilities among measures of the first variable, X, equaled the reliabilities among measures of the second variable, Y; (i.e., $r_{xx} = r_{yy}$). In this study, the Steiger (1980) equations are solved for the more complicated case where r_{xx}, r_{yy}, and r_{xy} are free to assume any value, given the Cauchy-Schwarz boundary restriction (Kendall & Stuart, 1977) that r_{xy} is less than or equal to $(r_{xx} \cdot r_{yy})^{1/2}$.

METHOD

Mathematical Derivation: Let there be n measures on variable X and m measures on variable Y. Furthermore, let the X's be intercorrelated by the common population correlation r_{xx}, the Y's by r_{yy}, and the cross-correlations of each X with each Y have the common population correlation, r_{xy}. Because it is intended to average all sample cross-correlations, r_{xy}, we first need to consider the variance-covariance matrix of the r_{xy}'s of which there are nm in all; therefore, the resulting variance-covariance matrix is an nm x nm array.

Asymptotic expressions for the variances and covariances involved were originally derived by Pearson for use with large sample sizes (N approaching infinity). These expressions are given in a mathematically convenient form by Steiger (1980). Replacing the Greek symbols sigma and rho by s and r, respectively, the equations are as follows:

$$Ns_{jk}^2 = (1-r_{jk}^2)^2 \quad (1)$$

$$Ns_{jk,jh} = r_{kh}(1-r_{jk}^2-r_{jh}^2)-1/2(r_{jk}r_{jh})(1-r_{jk}^2-r_{jh}^2-r_{kh}^2) \quad (2)$$

$$Ns_{jk,hm} = 1/2[(r_{jh}-r_{jk}r_{kh})(r_{km}-r_{kh}r_{hm})+(r_{jm}-r_{jh}r_{hm})(r_{kh}-r_{kj}r_{jh})$$
$$+(r_{jh}-r_{jm}r_{mh})(r_{km}-r_{kj}r_{jm})+(r_{jm}-r_{jk}r_{km})(r_{kh}-r_{km}r_{mh})] \quad (3)$$

Equation 1 represents the variance of a correlation coefficient, Equation 2 expresses the covariance between two correlations sharing the common variable, j, and Equation 3 gives the covariance between two correlations sharing no common variables.

For the present problem, there are nm variances (see Equation 4), along the major diagonal of the variance-covariance matrix. There are two types of covariance sharing a common variable, since this common variable can be either an X (see Equation 5) or a Y (see Equation 6). There are nm(m-1) of the former and nm(n-1) of the latter. Covariances sharing no common variable (see Equation 7) are all the same, and there are nm(m-1)(n-1) of these.

$$Ns_{xy}^2 = (1-r_{xy}^2)^2 \quad (4)$$

$$Ns_{xy1,xy2} = r_{yy}(1-2r_{xy}^2)-1/2r_{xy}^2(1-2r_{xy}^2-r_{yy}^2) \quad (5)$$

$$Ns_{x1y,x2y} = r_{xx}(1-2r_{xy}^2)-1/2r_{xy}^2(1-2r_{xy}^2-r_{xx}^2) \quad (6)$$

$$Ns_{x1y1,x2y2} = (r_{xx}-r_{xy}^2)(r_{yy}-r_{xy}^2)+r_{xy}^2/2[(1-r_{xx})^2+(1-r_{yy})^2] \quad (7)$$

The Fisher-z transformation is recommended when averaging correlation coefficients. The rationale for this recommendation follows from the observation that the sampling distribution of the untransformed r's is skewed, which, with positive correlations, produces a bias toward underestimation of the true population value (Kendall & Stuart, 1977). The Fisher-z transformation, $z=1/2\mathrm{Ln}[(1+r)/(1-r)]$, almost entirely corrects this skew prior to averaging. More importantly, the average of the Fisher-z transformed correlations provides for an asymptotically maximum likelihood estimate of the population correlation, which is near maximum for even moderate sample sizes ($N \geq 10$). Steiger gives the necessary formulae for converting the variance and covariance expressions to a form appropriate for Fisher-z transformations. The conversion to Fisher-z form simply requires dividing the variance or covariance terms by $(N-3)(1-r_{xy}^2)^2$, because in the present problem the indices jk, jh, and hm each involve an XY pair.

The variance of the average Fisher-z converted cross-correlations can now be computed by pre- and postmultiplying the variance-covariance matrix by vectors containing the constant 1/(nm). This is equivalent to multiplying the four terms represented by Equations 4-7 by the number of each type, summing these products, dividing by $(N-3)(1-r_{xy}^2)^2$, then dividing again by

n^2m^2. The standard error is, of course, the square root of the above computation. The standard error (SE) will decrease as n and m increase, although the extent to which SE decreases will be dependent on r_{xx}, r_{yy}, and r_{xy}. Using the back solution from the standard error of a single Fisher-z, one can compute the effective sample size (EN) from the following expression:

$$EN=1/SE^2+3 \qquad (8)$$

The quantity EN is an estimate of the number of subjects who would be required to get an equivalently accurate estimate of the cross-correlation between X and Y with only single tests of X and Y.

Monte Carlo Simulation: The above derivations are based on asymptotic (N approaching infinity) expressions and their small sample characteristics remain to be determined. Due to the complexity of these expressions, their small sample behavior appeared to be best explored by Monte Carlo methods. The first stage of the simulation was to derive an algorithm for generating random samples of data from a population with a known intercorrelation matrix. Without loss of generality, the means and standard deviations of these random variables respectively were set to zero and one. Independent random normal variates were generated by the method of Box and Muller (1958), using seeds from the DEC-20 internal random number generation function RAN (see Edgell, 1979, for the characteristics of this function). What was needed was a matrix of weightings to be applied to the independent random variables to form linear composites that had the desired population correlation matrix. These weightings were derived by what amounts to the reverse of a factor analysis. Factor analysis takes intercorrelated variables and finds sets of weightings that form linear composites (factor scores) that are independent. These weightings used in reverse will produce intercorrelated variables from originally independent variables. One problem encountered is that many of the correlation matrices of research interest produce mathematical difficulties for the computer algorithms used by standard factor analysis software. This problem was avoided by using an earlier and more primitive method of factor analysis, called square root factor analysis or diagonal factor analysis (see Gorsuch, 1974). A Monte Carlo simulation program was written to generate data with the given r_{xx}, r_{yy}, and r_{xy} population correlations under a variety of n, m, and sample size values.

The program was written in FORTRAN and was run on a DEC-20 computer. For each empirical value presented in the following tables, the program generated 5000 data sets. For each data set the average of the Fisher z-transformed cross-correlations was computed. The standard error of the latter estimate of the cross-correlation was then computed across the 5000 determinations.

RESULTS

The findings of the Monte Carlo study are summarized in Table 1. In this table, theoretical values for the standard error, calculated as described under "mathematical derivation", are compared to empirical values of the standard error, estimated by the Monte Carlo simulation. The reliabilities are assumed to be equal, $r_{yy} = r_{xx}$; therefore the values under the column labeled r_{xx} actually are the values of the population correlation among both X's or Y's. Also, Table 1 treats only the case where X's and Y's were measured equal numbers of times, labeled "Repetitions." In general, it can be seen that the theoretical and empirical values of the standard errors are

Table 1
Theoretical (T) and Empirical (E) Standard Errors with Equal Reliabilities (r_{xx}=r_{yy}) and Various Cross-correlations, r_{xy}, with Equal Numbers of Repeated Measures (1,2,3,5, or 10) of X and Y for Sample Sizes of 10, 20, and 30.

Repetitions		1		2		3		5		10	
r_{xy}	r_{xx}	T	E	T	E	T	E	T	E	T	E
						N=10					
.0	.2	.378	.381	.227	.227	.176	.173	.136	.129	.106	.104
.1	.2	.378	.379	.226	.226	.176	.172	.136	.133	.107	.103
.0	.4	.378	.370	.265	.259	.227	.221	.197	.191	.174	.167
.1	.4	.378	.384	.264	.260	.226	.217	.196	.187	.173	.167
.2	.4	.378	.379	.262	.259	.224	.218	.194	.183	.171	.162
.0	.8	.378	.377	.340	.341	.328	.323	.317	.308	.310	.304
.2	.8	.378	.375	.339	.333	.326	.321	.315	.309	.307	.303
.4	.8	.378	.375	.334	.324	.319	.309	.308	.299	.299	.290
.6	.8	.378	.370	.323	.317	.305	.300	.290	.281	.280	.270
						N=20					
.0	.2	.243	.237	.146	.144	.113	.112	.087	.086	.068	.066
.1	.2	.243	.246	.145	.145	.113	.113	.088	.086	.069	.067
.0	.4	.243	.239	.170	.172	.146	.143	.126	.125	.112	.109
.1	.4	.243	.241	.169	.168	.145	.143	.126	.122	.111	.109
.2	.4	.243	.243	.168	.165	.144	.143	.124	.122	.110	.109
.0	.8	.243	.242	.218	.212	.210	.212	.204	.205	.199	.201
.2	.8	.243	.242	.217	.215	.209	.204	.202	.199	.197	.197
.4	.8	.243	.241	.214	.212	.205	.205	.197	.195	.192	.191
.6	.8	.243	.241	.207	.206	.196	.192	.186	.187	.180	.179

in fairly close agreement; where disagreement exists, the tendency is for the theoretical calculation to overestimate the empirical value, which is to err in the conservative direction. It is clear that as the sample size increases, this tendency for overestimation decreases markedly. The average absolute difference between the theoretical and empirical standard errors were .0049, .0015, and .0005, for sample sizes of 10, 20, and 30, respectively; therefore, the errors are at most in the third decimal place and of acceptable magnitude with N as small as 20. By examining the discrepancies between theoretical and empirical values for N = 10, certain trends become apparent: overestimation is greater the greater the number of repetitions; overestimation is less the larger the value of the cross-correlation. The error is so small with N's of 20 that the above trends were not apparent in the latter data.

Similar comparisons of theoretical and empirical standard errors were made for cases where r_{xx} did not equal r_{yy}, and where the repetitions of X are not always equal to the number for Y. Again, it was seen that theoretical values for the standard error of the cross-correlations averaged in Fisher-z form fall reasonably close to the empirical values. The discrepancies that did exist were for the theoretical standard errors to exceed the empirical. An examination of trends in the differences between theoretical an empirical values revealed the same tendencies noted for Table 1; greater overestimation for larger numbers of replications, and less overestimation with larger population cross-correlations.

Table 2
Effective Sample Sizes for Equal Reliabilities ($r_{xx}=r_{yy}$) and Various Cross-Correlations, r_{xy}, with Equal Numbers of Repeated Measures (2,3,4,5,6,7,8,9,10, or 20) of X and Y for Sample Sizes of 10, 20, and 30.

Repetitions		2	3	4	5	6	7	8	9	10	20
r_{xy}	r_{xx}										
						N=10					
.0	.2	22	35	47	57	66	74	81	87	92	125
.1	.2	23	35	47	57	65	73	79	85	90	119
.0	.4	17	22	26	29	31	33	34	35	36	41
.1	.4	17	23	26	29	31	33	34	35	36	41
.2	.4	18	23	27	30	32	34	35	36	37	42
.0	.8	12	12	13	13	13	13	13	13	13	14
.2	.8	12	12	13	13	13	13	13	14	14	14
.4	.8	12	13	13	14	14	14	14	14	14	15
.6	.8	13	14	14	15	15	15	16	16	16	16
						N=20					
.0	.2	50	81	109	134	156	175	192	207	220	298
.1	.2	50	81	109	134	155	173	189	203	215	285
.0	.4	37	50	59	66	71	75	78	81	83	95
.1	.4	38	51	60	66	72	76	79	82	84	96
.2	.4	38	51	61	68	73	77	81	83	86	98
.0	.8	24	26	27	27	27	28	28	28	28	29
.2	.8	24	26	27	27	28	28	28	29	29	29
.4	.8	25	27	28	29	29	30	30	30	30	31
.6	.8	26	29	31	32	32	33	33	34	34	35

Because the Monte Carlo studies confirmed the prediction that theoretical equations for the standard errors would suffice even for relatively small sample sizes, Tables 2 and 3 used those equations to depict the improvement in accuracy that can be obtained by averaging cross-correlations after Fisher-z transformation. The values in these tables are effective sample sizes which were computed by Formula 8. These values estimate the number of subjects that would be required to provide a sample correlation of equivalent accuracy with only one measurement of X and Y, therefore only a single estimate of the cross-correlation, rather than the multiple cross-correlations averaged under the present procedure. For example, in Table 2,

Table 3
Effective Sample Sizes for Equal and Unequal Reliabilities ($r_{xx} \geq r_{yy}$) and Various Cross-Correlations, r_{xy}, with Different Allocations of Repeated Measures of X and Y for N=10.

		Reps of X	9	8	7	6	5	4	3	2	1
		Reps of Y	1	2	3	4	5	6	7	8	9
r_{xy}	r_{xx}	r_{yy}									
.0	.8	.8	11.5	12.4	12.8	12.9	12.9	12.9	12.8	12.4	11.5
.0	.6	.6	13.9	16.5	17.5	18.0	18.1	18.0	17.5	16.5	13.9
.0	.8	.6	11.5	13.6	14.5	15.0	15.3	15.4	15.3	15.0	13.9
.0	.6	.4	13.9	18.4	20.8	22.1	22.8	23.0	22.7	21.4	18.0
.0	.4	.2	18.0	27.6	33.9	38.0	40.4	41.2	40.1	36.3	27.2
.0	.8	.4	11.5	15.1	17.1	18.3	19.0	19.5	19.6	19.4	18.0
.0	.6	.2	13.9	21.0	25.8	29.3	31.6	33.0	33.4	32.2	27.2
.0	.8	.2	11.5	17.1	21.1	24.0	26.1	27.7	28.7	28.9	27.2
.0	.9	.1	10.7	17.0	22.1	26.5	30.2	33.3	35.8	37.7	38.0
.2	.8	.8	11.6	12.5	12.9	13.0	13.1	13.0	12.9	12.5	11.6
.2	.6	.6	14.1	16.8	17.9	18.4	18.6	18.4	17.9	16.8	14.1
.2	.8	.6	11.6	13.8	14.8	15.3	15.5	15.6	15.5	15.2	14.0
.2	.6	.4	14.1	18.8	21.3	22.7	23.4	23.7	23.3	22.0	18.4
.2	.4	.2	18.6	28.3	34.6	38.6	41.0	41.9	41.0	37.4	28.6
.2	.6	.2	14.2	21.5	26.6	30.2	32.7	34.1	34.6	33.3	28.2
.2	.8	.4	11.7	15.4	17.5	18.7	19.5	20.0	20.1	21.0	18.6
.2	.8	.2	11.7	17.6	21.8	24.9	27.2	28.8	29.8	29.9	27.7
.2	.9	.1	10.9	17.4	23.0	27.7	31.8	35.1	37.8	39.4	38.4
.4	.8	.8	11.8	12.9	13.3	13.5	13.6	13.5	13.3	12.9	11.8
.4	.6	.6	14.8	17.9	19.2	19.8	20.0	19.8	19.2	17.9	14.8
.4	.8	.4	12.4	16.7	19.0	20.4	21.3	21.7	21.7	21.0	18.6
.4	.8	.6	12.2	14.6	15.7	16.2	16.5	16.6	16.4	15.9	14.3
.4	.6	.4	15.2	20.4	23.1	24.6	25.4	25.6	25.2	23.7	19.6
.4	.8	.2	12.6	19.3	24.2	27.9	30.5	32.3	33.1	32.4	28.1
.6	.8	.8	12.5	14.0	14.5	14.8	14.9	14.8	14.5	14.0	12.5
.6	.6	.6	16.8	20.1	21.5	22.1	22.3	22.1	21.5	20.1	16.8
.6	.8	.6	14.1	16.9	18.0	18.6	18.7	18.6	18.1	17.0	14.4

the entry "22" in the column headed by 2 repetitions indicates that if variables X and Y were both measured twice on 10 subjects, yielding four estimates of the cross-correlation to be averaged, that resulting average would have the accuracy that could be obtained by running 22 subjects on whom X and Y were measured only once. It should be noted that these effective sample sizes are actually underestimated, particularly with N = 10, because the standard errors in this case are overestimated slightly. As the number of repetitions increases, the effective sample size increases monotonically. Furthermore, the reliabilities of the two measures have a profound effect on the accuracy of the estimated cross-correlation, with greatest accuracy or largest effective sample size when the reliabilities are small. With N = 20 it is clear that under some conditions with only as many as four repetitions on the two measures, effective sample sizes of over 100 may be obtained. Surprisingly, the size of the cross-correlation has only slight and somewhat inconsistent effects on the effective sample size.

The case where reliabilities differ is considered in Table 3, and here a somewhat different but interesting question arises. If one planned to run 10 measures in all on the subjects, what distribution or allocation of those measures among variables X and Y results in optimal precision of the final estimated cross-correlation. The maximum effective sample size for each row has been underlined in Table 3, and it is clear that more measures should be allocated to the variable with the lowest reliability. This finding is strongest when the cross-correlation is smallest, and the optimum allocation is dependent on the discrepancy between the two reliabilities.

DISCUSSION

Many of the findings of the present study ran parallel to those of the earlier and less complex investigation of Dunlap _et al_. (1983). To summarize, theoretical predictions of the standard errors of cross-correlations averaged in Fisher-z form, were fairly close to empirical values generated by Monte Carlo simulation, even for quite small sample sizes. Where deviations between the two were seen, they tended to be in the conservative direction; the theoretical formulae overestimated the standard error, therefore, underestimated the actual increase in precision to be gained by averaging correlation. It was concluded that the theoretical predictions were of sufficient accuracy to permit their use in tabulating effective sample sizes under a variety of conditions, rather than the more exact but considerably more expensive process of simulation.

Examination of the effective sample sizes showed that precision increased as a function of the numbers of repetitions of X and Y, and, therefore, as a function of the number of correlations averaged. Furthermore, the greatest increases in precision occurred when the reliabilities of the measures were low; precisely the situation where accuracy is most needed. These two findings are entirely consistent with the findings of Bittner (1982), Bittner, _et al_. (1982), and Dunlap, _et al_. (1983). The somewhat surprising result was that the effective sample size was almost completely controlled by the size of the intracorrelations among X's and among Y's, the reliabilities, and minimally affected by the magnitude of the cross-correlation, the quantity being estimated. It appears quite feasible to use the tables and formulae provided by the present study to aid experimental planning in the study of cross-correlation or validity. Not only should it be possible to use smaller sample sizes than previously thought necessary, but it is also possible to use the above information to optimize the allocation of the various times a subject is tested on measures X and Y, depending upon their relative reliabilities, to maximize the precision of the resulting estimated cross-correlation, as is done in Table 3. In a more general sense it is hoped that attention to the problem of averaging correlations across repeated measures will also contribute to increasing attention to more general problems of stability of validities and reliabilities across repeated measures. Clearly, the above formulation assumes stability in both which may or may not be true in practice.

FOOTNOTES

[1]This research was sponsored by the Office of Naval Research under Contract No. N00205-84-M-1689. The views expressed are those of the authors and do not necessarily represent those of the Department of the Navy.

REFERENCES

[1] Bittner, A. C., Jr. Averaged correlations between parallel measures: Reliability estimation. (Proceedings of the Eighth Psychology in DoD Symposium, USAF Academy, Colorado Springs, Colorado (1982).
[2] Bittner, A. C., Dunlap, W. P., & Jones, M. B. Averaged cross-correlations with differentially stable variances: Fewer subjects required with repeated measures. Proceedings of the 26th annual meeting of the Human Factors Society, Seattle, Washington. (1982).
[3] Box, G. E. P., & Muller, N. E. A note on the generation of random normal deviates. Annals of Mathematical Statistics 29 (1958) 610-611.
[4] Dunlap, W. P., Jones, M. B., & Bittner, A. C. Average correlations vs. correlated averages. Bulletin of the Psychonomic Society 21 (1983) 213-216.
[5] Edgell, S. E. A statistical check of the DECsystem-10 FORTRAN pseudorandom number generator. Behavior Research Methods & Instrumentation, 11 (1979) 529-530.
[6] Gorsuch, R. L. Factor Analysis (W. B. Saunders, Philadelphia, 1974).
[7] Kendall, M., & Stuart, A. The Advanced Theory of Statistics (Vol. 1) (Macmillan, New York, 1977).
[8] Steiger, J. H. Tests for comparing elements of a correlation matrix. Psychological Bulletin 87 (1980) 245-251.

II

PERCEPTION

Trends in Ergonomics/Human Factors II
R.E. Eberts and C.G. Eberts (Editors)

PROCESS CONTROL MONITORING: THE EFFECTS OF SPATIAL AND VERBAL ABILITY AND CONCURRENT TASK DEMAND

Christopher D. Wickens and Annette Weingartner

Department of Psychology
University of Illinois at Urbana-Champaign
Champaign, Illinois
U.S.A.

Eighteen subjects of either high spatial (and low verbal) or high verbal (and low spatial) ability performed a simulated process monitoring and failure detection task concurrently with either a spatial or a verbal secondary task. The results indicated that: (1) the spatial group performed better on the spatial task and the verbal group performed better on the verbal task. (2) Both groups showed most interference between the primary process monitoring task and the spatial side task. (3) The two groups showed identical patterns of interference across the verbal and spatial tasks.

A number of researchers have recently invoked the concept of an "internal model" as essential to the process controller's task (Umbers (1979); Landeweerd (1979); Rasmussen (1981)). Such a model can be expected to serve at least two functions (Wickens (1984a)). It may allow the controller to predict the process response to control input and therefore serve as the basis for open loop control (Kragt and Landeweerd (1974)), and it may create a set of expectancies for how the process variables should respond to inputs and covary with each other. When these expectancies are violated, the model may serve as the basis for detection and diagnosis of process failures (Rasmussen (1981)).

While the internal model is a useful hypothetical construct, it is not easy to operationally define and investigate (Gentner and Stevens (1983)). One important defining characteristic of model, however, is the extent to which it is represented in terms of verbal propositions as opposed to analog spatial images. While the process itself <u>is</u> clearly analog, there is no reason why the representation might not be in terms of a series of verbal propositions of the form: "if the pressure is high at X, the temperature should decline at Y". In one investigation of this sort, Landeweerd (1979) examined the process controller's tasks from the perspective of individual differences in ability along the verbal-spatial continuum and has concluded that those with good verbal-causal understandings of the process make better controllers, while those with a better visual-spatial image are better at fault diagnosis.

A second approach to investigating the code of representation is through dual task methodology (Wickens (1984a,b)). The assumption is made that process monitoring depends upon active comparisons of values in working memory. The research of Baddeley and his colleagues (Baddeley and Hitch (1974)); Baddeley and Lieberman (1980)) indicates that there appear to be two working memory systems, one involving a verbal phonetic rehearsal loop and the other described by the metaphor of a visual spatial scratchpad. From the present perspective, the important characteristic of these two systems is that they appear to depend upon different processing resources (Wickens (1984b)). Therefore tasks depending upon verbal working memory will be more likely to be disrupted by concurrent phonetic and verbal activity, while tasks employing spatial working memory will be more disrupted by concurrent spatial and manual tasks (Baddeley and Lieberman (1980); Brooks (1968)). The dual task methodology thus offers another means for assessing the code of representation of the internal model of the process. If the model is spatial, concurrent performance of the primary monitoring task with a secondary spatial task should be more difficult than concurrent performance with a secondary verbal task. If the model is verbal, then the converse results should be obtained.

The objective of the present study is to combine the individual differences and dual task methodology to infer the code of representation of the internal model of a complex dynamic system for individuals of high spatial or high verbal ability. This objective is to be accomplished by examining the level of performance on a multivariate process monitoring task obtained by the two groups, and the pattern of dual task interference of each group, as each was required to time-share system monitoring with verbal and with spatial secondary tasks. Three alternative patterns of data can be predicted, each corresponding to different characteristics of the internal model that might be exhibited by the two groups.

(1) Flexible model. Each ability group (verbal-spatial) employs an underlying model of the dynamic system that capitalizes upon that group's respective strength. Hence, performance of each group is expected to be most disrupted by the concurrent task that corresponds to the group's label (i.e., verbal task for the verbal group, spatial for the spatial). In addition, there might potentially be a disadvantage for the verbal group on primary task performance because they deploy a model code that is less compatible with the inherently spatial analog system that is to be monitored.

(2) Fixed model-fixed capacity. Both ability groups possess the same underlying internal model code. Because the system is inherently a spatial analog one, it is logical to assume that the internal model code would also be spatial. As a consequence, the spatial secondary task should disrupt primary task performance (or be disrupted by the primary task) to a greater extent than the verbal. In addition to this main effect however, there are two nested predictions of how the spatial and verbal groups might differ in their interference patterns. These are based upon different assumptions of how spatial subjects differ from verbal in terms of their working memory capacity. According to a fixed capacity model, both groups possess the same spatial memory capacity, but the spatial group deploys this capacity to spatial tasks with greater efficiency. Hence, the performance resource function relating performance on the task (vertical axis) to resources invested in the task (horizontal axis) would appear as shown in Figure 1a. The spatial group indicated by the dashed line requires fewer resources to perform the primary task effectively, and hence would be less disrupted by allocating a given amount of those resources to the spatial side task than would be the verbal group. Such a model might also assume that the spatial group would produce somewhat better primary task performance.

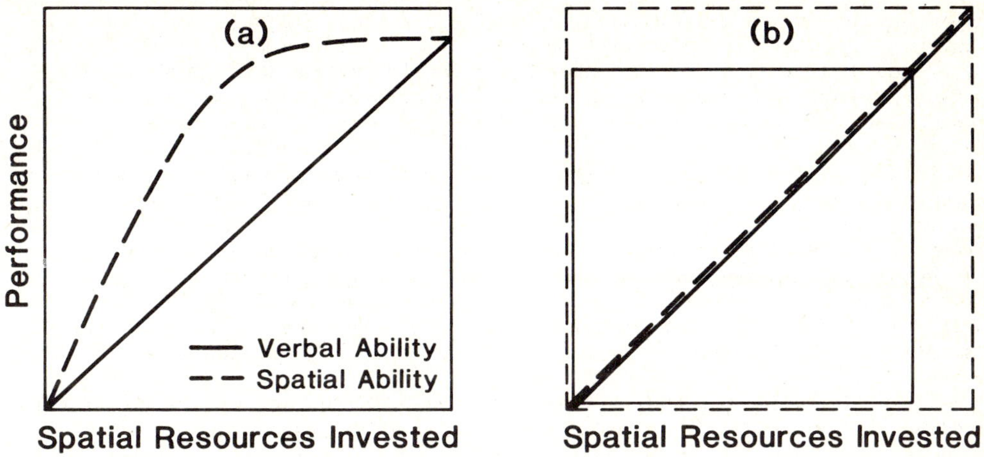

Figure 1: The performance resource function of two models: (a) fixed capacity for both ability groups, (b) variable capacity between ability groups.

(3) Fixed model-variable capacity. This model also assumes that both ability groups deploy a spatial model, and so again predicts that both will be more disrupted by performing the spatial secondary task. However, it also predicts that the spatial group differs from the verbal by having more total spatial resources to utilize, even though these might not be deployed in a more efficient manner as in model (2). Such a difference, shown in the hypothetical performance resource functions of Figure 1b, would suggest that the difference between groups is a general one related to the resources available and is not task specific related to the resources consumed. It would predict an equal level of performance loss of the two groups as the same amount of resources are withdrawn, but both the baseline and the dual task level of performance would presumably be higher for the spatial ability group.

In the present experiment subjects were given the primary task of monitoring 5 continuously varying indicators of the state of a slowly changing closed loop dynamic system. Concurrently each of two secondary tasks were imposed, one requiring spatial working memory and the other verbal. The two groups of subjects were selected from a larger population to have either high spatial and low verbal abilities (relative to the population mean), or low spatial and high verbal abilities. These groups will be referred to as "spatial" and "verbal," respectively.

METHOD

A more detailed description of the methods is reported in Wickens and Weingartner (1985). Eighteen subjects were selected from a larger pool of 77 subjects who had taken a battery of verbal and spatial tasks. These 18 were divided into a verbal and spatial group, each selected as nearly as possible to score in the upper quartile of one ability and the lower quartile of the other. All subjects (9 males, 9 females distributed between groups) were paid $3.00/hr for participation.

For the primary process monitoring task, subjects viewed a display with five time-varying vertical bargraphs. Each bargraph represented the value of one variable in a simulated nuclear process. The simulation was of a complex, negative feedback, stable system "driven" by a slowly changing random input signal. When failures occurred, they appeared as gradual ramp changes in either one of two gain values from normal values to terminal steady state values. The ramps were ten seconds in length. From the subject's point of view, this produced gradual changes in the amplitude and frequency response of different variables, and in the degree of coupling or covariation between pairs of variables within the system. After each failure was initiated the subject was given ten seconds to detect it. If a response was not made within this interval, the sequence was scored as a MISS and the system was automatically reset to normal operation. If the failure was detected, the sequence was scored as a HIT and the system again, automatically reset itself at the end of the ten second interval. Failures in the system occurred at random intervals, averaging four per three minute trial. Prior to extensive training on monitoring and understanding the system, the causal model underlying the relation between its variables was carefully explained to subjects using both verbal and graphical representations.

On two-thirds of the detection trials subjects also performed one of two kinds of concurrent tasks. Both were presented auditorily to prevent peripheral visual interference with the primary task. The verbal task required subjects to retain lists of abstract words in working memory, occasionally responding with specific words to probe stimuli. The spatial task, adopted from Brooks (1967) required subjects to "fill in" an imagined 4 x 4 matrix with different digital values. The two secondary tasks were calibrated for equal difficulty as measured by percent correct. Subjects performed these three task combinations, along with single task trials on the secondary tasks alone, over four days. Data from the third and fourth days (six replications of each trial type) were used for analysis.

RESULTS

Data from both the primary and secondary task were subjected to repeated measures analyses of variance. For the secondary tasks, only accuracy (% correct) was employed, whereas separate analyses were conducted on primary task latency, correct detection, and false alarm rate. Rather than describing each ANOVA in turn, the effects will be discussed in terms of certain fundamental underlying hypotheses.

It was necessary first to insure that the two ability groups that had been selected on the basis of paper-and-pencil tests continued to be differentiated on performance of the dynamic tasks employed in the main experiment. Simultaneously, it was essential for the hypotheses under consideration to guarantee that the two tasks were in fact spatial and verbal in their underlying processing demands. Both effects could be simultaneously validated by the presence of a strong "crossover" interaction between task type and ability group.

As indicated in Table 1, which plots the accuracy of each secondary task as a function of ability groups, such a crossover interaction was clearly and unequivocally obtained ($F(1,16) = 4.26$, $p = .05$). Both tasks and ability groups showed roughly the same level of overall performance (both main effects showed $F < 1$), but each group performed best with the task compatible with their own abilities.

		Task Type	
		Verbal	Spatial
Ability	Verbal	81	77
	Spatial	75	82

Table 1. Accuracy (% correct) on secondary task.

		Secondary Task		
		None	Verbal	Spatial
Ability	Verbal	96	95	91
	Spatial	97	95	94

Table 2. % correct detection on primary task.

Having confirmed then that the secondary tasks were in fact "spatial" and "verbal" (as operationally defined by group ability differences), it was then possible to examine their influences on the primary task. Table 2 shows the measure of primary task failure detection accuracy for the single and the two dual task conditions. These detection data were submitted to a 3 (condition) x 2 (group) repeated measure ANOVA. From the table it is apparent that there was a dual task decrement--detection accuracy was lower with the secondary tasks present than when they were absent. Secondly, this decrement appears to be greater with the spatial than with the verbal task for both groups. These effects were substantiated statistically by the main effect of task ($F(2,32) = 4.19$, $p < .05$), and by the subsequent HSD planned comparisons test between the means which indicated a reliable loss of performance accompanying the introduction of the spatial side task ($HSD(32) = 4.02$, $p < .05$), but not the verbal side task. There were no significant effects or interactions involving group, in the primary task data. That is, both groups showing the same (statistically) pattern of interference effects. Neither of the other primary task dependent variables, detection latency, and false alarm rate showed statistically reliable effects of any of the independent variables.

The secondary task data as a source of information regarding dual task interference are shown in Figure 2. For both ability groups, the data show a pattern of interference supporting the conclusion drawn from the primary task data, that the failure detection task requires spatial resources. Under single task conditions, the spatial tasks are performed as well or better than the verbal, while under dual task conditions the pattern reverses. Stated in other terms, the spatial task shows a substantial dual task decrement, while the verbal task shows almost none. This effect is substantiated by the significant task x condition interaction ($F(1,16) = 6.39$, $p < .05$). The most interesting aspect of the data however is that the pattern of task load by task type interaction is identical for both ability groups, and the 3-way interaction of group x task x load failed to approach significance ($F(1,16) = 0.32$). The main difference between the two groups is the improved performance of the spatial group on the spatial task, an interaction already noted above.

Finally, it should be noted that there were reliable effects of practice (across the six trials in each condition) on primary task response latency ($F(5,80) = 2.78$, $p < .05$) and secondary task performance ($F(5,80) = 3.18$, $p < .05$). The latter measure also showed a significant task load (single-dual) by practice interaction ($F(5,80) = 3.45$, $p < .05$). This interaction reflected a large increase in dual task performance from trial 1 to trial 2, which was not manifest in performance of the single task. Neither task type nor ability type entered into any reliable interactions with practice.

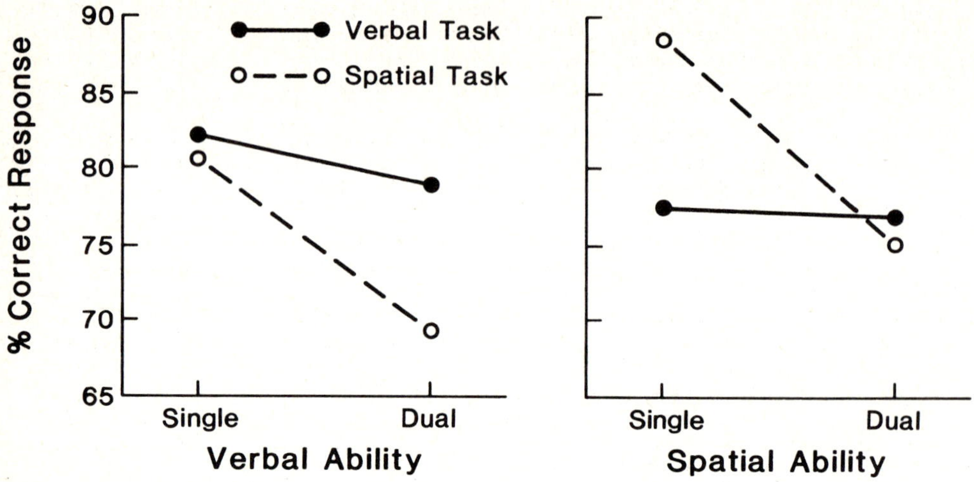

Figure 2: Effects of task type, task load, and ability group on secondary task accuracy.

DISCUSSION

The results confirm that the demands of the primary task are spatial for both ability groups. Therefore the code of the internal model appears to be task-driven rather than ability-driven. Both groups showed substantial interference on both the primary and the spatial secondary task, while both also showed nearly perfect time-sharing with the verbal secondary task. The results also clearly indicated that the ability and task type manipulations were successful in the direction intended. The interesting question then is how the two ability groups differed in terms of their time-sharing of verbal and spatial material. The answer to this question can be considered in the framework of the fixed and varied capacity models described in Figure 1.

The data in Figure 2 indicate that the fixed capacity model must be rejected. This model predicts that the spatial ability group uses its fixed quantity of spatial resources more efficiently than does the verbal group. Were this the case, then dividing resources between tasks should produce a lesser decrease in performance on the primary and/or the secondary task for the spatial group. According to the data presented in Table 2 and Figure 2, this was clearly not the case.

The varied capacity model then seems to be more plausible. According to this model the spatial group differs from the verbal simply by having more total capacity available; for example, a "larger" visual-spatial scratchpad, or one from which information is lost at a slower rate. This interpretation assumes that all spatial resources are deployed by both groups when performing the spatial secondary task. Hence, secondary task performance will be better for the spatial group, who can place more resources in service (e.g., 4 units versus 3 for the verbal group). It assumes secondly that an equivalent amount of resources (e.g., 3 units) are deployed to the single task failure detection task by each group (this is the maximum of 3 units for the verbal group and one unit less than maximum for the spatial group). Hence, sharing resources between the two tasks will leave the same "shortfall" between demanded and available resources in dual task conditions

(7 demanded - 4 available = 3, for the spatial group; 6 - 3 = 3 for the verbal group). Since this shortfall predicts the magnitude of the dual task decrement, such an accounting can readily explain the equivalence in time-sharing performance between groups shown in Figure 2.

This accounting of the results is less than fully satisfying however in one respect. If the spatial group has more spatial resources available, why don't they deploy more of them to improve performance on the primary task, relative to the verbal group? One possibility is that such improvements were in fact present, but they were distributed across the three performance measures of the primary task (hits, false alarms, and latency) in such a way that none of the measures alone showed reliable group differences. In fact, all three measures did "favor" the spatial group in single as well as dual task performance. Single task response latency was slightly faster for the spatial group (3.08 vs. 3.12 sec). Accuracy was higher (97% vs. 96%) and false alarm rate was lower (17% vs. 22%).

A second possible accounting of the results is that the pattern of task interference was attributable to factors that were quite independent from the source of group differences. According to this view, the spatial and verbal groups may indeed have differed in terms of their working memory capacities; however, the primary task was considerably more visual than spatial in its demands and the two groups did not differ in terms of their basic visual capacity (Baddeley and Lieberman (1980)). Hence, the interference that was obtained was simply the consequence of visual interference resulting perhaps from a greater degree of unwanted (and unnecessary) visual scanning that occurred when the spatial side task was performed. This scanning would of course induce a loss in performance because of the heavy visual demands of the primary task.

In conclusion, the results stand in contrast with those of Landeweerd (1979) who found ability differences in performance on process control monitoring kinds of tasks. Such differences as noted were not observed here. The results do appear to support the interpretation that the processes used to perform the failure detection task used here were visual-spatial and not verbal phonetic. As noted above, whether the "spatial" or the "visual" is the more important component of demands cannot be easily determined. The former refers to characteristics of an internal model described by the capacity based first hypothesis; the latter to the peripheral scanning interference of the second hypothesis. The data do suggest the feasibility of imposing concurrent tasks of verbal phonetic nature on the process monitor, with little fear of interference. Hence, the data are compatible with the emerging distinction between verbal and spatial resources as a useful dichotomy for describing and predicting dual task performance (Wickens (1984b)). The precise way in which this performance is influenced by ability differences clearly require further research.

ACKNOWLEDGMENTS

This research was supported by a grant MDA903-83-K00255 from the Army Research Institute. The authors would like to thank Dr. Amir Mane for his efforts in programming the failure detection task, and Dr. David Dupree for providing the test data from which the sample of subjects were chosen.

REFERENCES

[1] Baddeley, A.D. and Hitch, G., in: Bower, G. (ed.), Recent Advances in Learning and Motivation (Vol. 8) (Academic Press, New York, 1974).

[2] Baddeley, A.D. and Lieberman, K., Spatial working memory, in: Nickerson, R.S. (ed.), Attention and Performance VIII (Erlbaum, Hillsdale, NJ, 1980).

[3] Brooks, L.R., Spatial and verbal components in the act of recall, Cand. Jrnl. of Psych. 22 (1968) 349-368.

[4] Genter, D. and Stevens, A., Mental Models (Erlbaum, Hillsdale, NJ, 1983).

[5] Kragt, H. and Landeweerd, J.A., Mental skills in process control, in: Edwards, E. and Lees, F.P. (eds.), The Human Operator in Process Control (Taylor & Francis, London, 1974).

[6] Landeweerd, J.A., Internal representation of a process fault diagnosis and fault correction, Ergonomics 22 (1979) 1343-1351.

[7] Rasmussen, J., Models of mental strategies in process plant diagnosis, in: Rasmussen, J. and Rouse, W.B. (eds.), Human Detection and Diagnosis of System Failures (Plenum Press, New York, 1981).

[8] Umbers, I.G., Models of the process operator, Intl. Jrnl. of Man-Machine Studies 11 (1979) 263-284.

[9] Wickens, C.D., Engineering Psychology and Human Performance (Charles Merrill, Columbus, OH, 1984a).

[10] Wickens, C.D., Processing resources in attention, in: Parasuraman, R. and Davies, R. (eds.), Varieties of Attention (Academic Press, New York, 1984b).

[11] Wickens, C.D. and Weingartner, A., Characteristics of the internal model of the process monitor: Use of dual task and individual differences methodology, Cognitive Psychophysiology Laboratory Technical Report CPL-85-1/ARL-85-1, Dept. of Psych., Univ. of Illinois at Urbana-Champaign (1985).

[12] Brooks, L.R., The suppression of visualization in reading, Quart. Jrnl. of Exptl. Psych. 19 (1967) 289-299.

Trends in Ergonomics/Human Factors II
R.E. Eberts and C.G. Eberts (Editors)

MISPERCEPTION OF SMALL CARS

Ray E. Eberts and Allen G. MacMillan

Purdue University
School of Industrial Engineering
West Lafayette, IN 47907

Small cars are overrepresented in accidents. As the size of cars decrease, the problem of small car accidents will increase. Several reasons exist why small cars are in more accidents. In the experimentation reported, a possible perceptual reason for small car accidents is investigated. If the visual angle of the car on the eye's retina is used for distance judgment, then small cars will look like large cars farther away and, thus, other drivers would tend to take more chances in front of a small car or follow a small car closer. The data from the experiment indicate that visual angle is a salient distance cue for judging the distance of cars.

INTRODUCTION

We know that the size of the average car is decreasing. We also know that small cars are in more accidents. This increase in the proportion of small cars implies that we can expect more accidents. To help prevent this possibility, we must try to determine why small cars are in more accidents. Once this is determined, we can try to change drivers' behaviors or change the design of small cars to lessen the probability of accidents.

Cars in accidents can be divided into two basic categories -- those cars that cause the accidents and those cars which are hit by others. For the cars that cause accidents, analysis of the data has shown that small cars are over represented because they are driven more by young drivers; these young drivers are in more accidents anyway. When the driver's age is taken into account, small cars do not necessarily cause more accidents than others. For the cars that are hit by other drivers, such a ready explanation is not available. The following experiment concerns itself with one possibility why small cars could be hit more by other cars; the possibility exists that small cars are misperceived.

There are two conditions where a small car can be hit by another car. An accident can occur if the small car is followed too closely or if someone pulls in front of an oncoming small car either to make a turn or to pass another car. In one study that looked at following behavior on highways, Evans and Wasielewski (1983) found that small cars were followed more closely than larger cars. We have found no data to indicate that people take more chances in front of oncoming small cars but the possibility does exist that this may be the case as well.

To investigate the problem of small car collisions, we did an experimental study of one possible cause of these accidents; the misperception of distances of small cars. Perception problems have been found to be a factor in many accidents.

MISPERCEPTION OF DISTANCES

In an analysis of accident data, Treat and Shinar (1976) concluded that human errors account for 60% of all accidents; environmental and vehicle factors only account for 30% and 10% of the accidents, respectively. Well over 90% of the information drivers need for safe driving is obtained visually (Gioia & Morphew, 1968). In addition, Puchonok (1974) found that 25% to 60% of all accidents are caused by some kind of information failure. Putting these facts together, we can conclude that when an information failure causes an accident, the information failure is due to a lack or misperception of visual information.

One of the most important driving decisions which has to be made is the decision about how far away another vehicle is; this is the basis for making decisions on following distance, for turning left in front of an oncoming car, and for passing another vehicle in front of an oncoming car. This decision is based entirely on visual information. The following discussion will show how small car distance can be misperceived so that other drivers would follow closer to small cars and would take more chances in front of small cars.

Several visual cues are available for the perception of distance. These cues include texture (the texture is less distinct on farther objects), interposition (an object which obscures the view of another is closer), clarity (similar to texture), convergence (parallel lines converge as they get farther away), and binocular disparity (three -dimensional or stereoscopic vision). The first four cues are all two dimensional; the last cue is the only one which is three dimensional. The binocular disparity cue is useful only in a very limited range; only from about 0-10 feet. The cues that are important for making driving decisions are all two dimensional and, thus, could be studied with pictures and slides with little or no loss of important dimensional information.

One of the most salient distance cues available is the visual angle of an object in the visual scene. The visual angle refers to the angle formed by the object as it passes through the lens of the eye (see Figure 1). Generally, for similarly sized objects, the larger the visual angle the closer is the perceived object. The visual angle can be misleading as a cue of distance if a class of objects, such as a and b in Figure 1, is of widely varying size. If this is the case, then the other visual cues must be used to

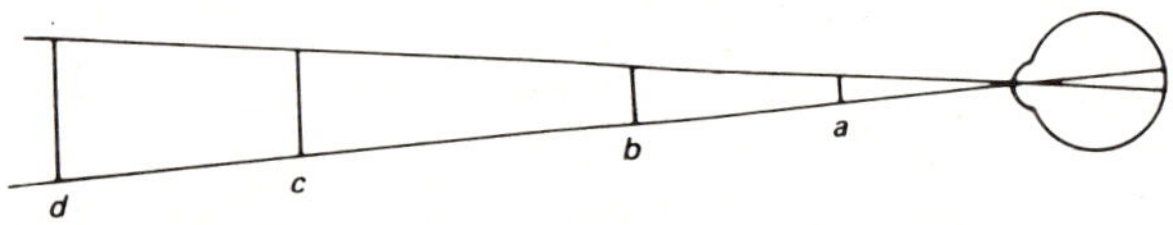

FIGURE 1
Representation of the Visual Angle

correctly judge the distance. For a class of objects that is fairly uniform in size, such as automobiles have been in the past, visual angle will be a good and consistent cue for distance perception. In the last couple years, however, the size of cars has been becoming more variable and, thus, visual angle is becoming less and less of a consistent cue for distance perception. If people use visual angle as a distance cue (due to their past experience with uniformly sized cars), then distance misperceptions for small cars could occur; these misperceptions could be potentially fatal.

As an example of how car distance could be misperceived, consider the following. If a driver is driving on a freeway and is following another car by maintaining a constant visual angle regardless of the width of the leading car, the driver would have to follow closer to a small car than to a larger car. If a Honda is misperceived as a Cadillac, then the Honda would be follwed 13.5 feet closer than it should be followed if the Cadillac's visual angle is maintained for the Honda. In general, using this argument, a small car will be perceived as a larger car farther away. Thus, other drivers would have a tendency to follow closer to a small car and also take more chances in front of a small car because they would think that they would have more time than they actually had to carry out the maneuver.

PURPOSE OF THE EXPERIMENT

The purpose of this experiment is to determine if drivers use the visual angle of a car, to the exclusion of other important visual cues, to make distance perceptions. With a widely varying size of car, as used in the experiment and as are available on the roads, then visual angle could be a misleading cue and cause distance misperceptions. These distance misperceptions could then be the basis for accidents due to the failure of visual information processing. Two distinct models of how people perceive distance are tested in the experiment. The one model (an ideal observer model) states that drivers optimally use the visual distance cues available to make distance perceptions. Under this model, distance is not misperceived if all these visual cues are available. The second model (the visual angle model) states that drivers will predominately base their distance decisions on the visual angle formed by the car. Under this model, even with other visual distance cues available, drivers will misperceive distance for the small cars. They will think that the small car is farther away than it actually is.

In the following experiment, slides of varying sized cars were taken at several distances and visual angles. Subjects were asked to estimate the distance of the car in feet. This task is not normally done when driving, but since we are not necessarily interested in absolute distance perception but, rather, relative distances to differentiate between the two models, this kind of task should not be problematic.

To help compare the models, the slides were divided into two groups. The first slide group had a clear background with all visual cues, except the unimportant binocular disparity, available. For the second slide group, the dark background, other distance cues surrounding the car were blacked out so that subjects only had visual angle to base their decisions on. By necessity, since no other cues are available, this group would have to adhere to the visual angle model for distance perception. Thus, this control group, the dark background slides, can be compared to performance with the clear background slides to help differentiate between the models.

METHOD

Stimuli. Nine cars were chosen to represent the range from the smallest car to the largest car. The smallest car, a Subaru, is classified as a minicar and is not, at this time, available in this country. The largest car, a Mercury Marquis, can be classified as a large luxury car.

The slides were taken on a country road. No other traffic was in the picture. In doing the task so that subjects would not focus on any one background cue to estimate distance, the camera was randomly placed at various positions along the road. To take the pictures from the perspective of a following car, a 35 mm camera was placed on a tripod at driver height. A 49 mm lens was used so that the picture distortion could be minimized and the picture would be similar to that seen by a driver. Black -and-white film was used to eliminate any effects of distance perception which may have occurred due to the color of the car.

The nine cars were placed at various distances along the road. These distances were chosen so that each car would have a picture taken at a constant visual angle. The ten visual angles chosen were from 2 degrees (large distances) to 20 degrees (short distances) in increments of 2 degrees. The nine cars taken at 10 visual angles thus produced 90 slides. Two sets of slides were taken. On one set of slides, the periphery (everything except the car) was blacked out to eliminate other distance cues. For the other set of slides, the background was kept clear.

Subjects. Ten Purdue students, both male and female, participated in the study. All were seniors or graduate students, and all had normal or corrected-to-normal vision.

Procedure. The subjects were seated in a darkened classroom. The slides were projected on a screen in the front of the room. The size of the image was adjusted so that the visual angle of the projected image would be the same as the angle presented to the camera when the slides were taken. The subjects were given a pencil and a booklet to record their answers.

Before the experiment actually began, the subjects were shown a series of ten slides to help them calibrate their distance judgments. On the same country road as the cars, an object was shown at varying distances from 20 to 200 feet in increments of 20 feet. For each slide, the experimenter informed the subjects of the distance of the object. Next, subjects were given 40 practice trials which were not analyzed. For the 40 slides, 30 were from the clear background set. The practice slides ranged from the closest distance to the farthest distance; the intermediate slides were randomly chosen. Subjects wrote down responses for these practice slides, but no feedback on the correctness of their responses was given.

In the experimental session, subjects were shown 360 slides altogether (two replications of the 180 slides). Each slide was presented for 1-1.5 seconds. This time corresponds to the time drivers usually spend to process visual information. After this time period, the slide was blacked out and the subjects were given 3-5 seconds to write an estimate of the distance, in feet, in their booklet. After a group of 40 slides, the subjects were given a 30 second rest break.

In the presentation, the 90 clear background slides were presented first, then the 90 dark background slides. The order of presentation was reversed for the next replication of 180 slides. The experiment lasted approximately 50 minutes.

RESULTS

The data of the actual distances are plotted in Figure 2. This figure represents what the data would look like if the subjects adhered to the ideal observer model and used all the visual distance cues to accurately estimate the car distances. If the subjects adhered to the visual angle model using only visual angle to estimate distance and, thus, in some cases misperceiving the distance, the slopes of the visual angles as a function of cars would all be flat. For the ideal observer model the 10 lines of the visual angles over the nine cars would fan out; for the visual angle model these lines would all be flat and parallel with each other. The reported distances from the subjects are represented in Figures 3 and 4 for the clear and dark backgrounds, respectively. The slopes of the ten lines are less than what could be expected from the ideal observer model but slightly more than what could be expected from the visual angle model. The slopes of the best-fitting regression lines for the ten visual angles are listed in Table 1 for the two models and the reported distances. For the ideal observer model, these slopes would increase from .69 to 7.00 as the visual angle is decreased and for the visual angle model these slopes would all be 0. The slopes of the best-fitting regression lines for the reported

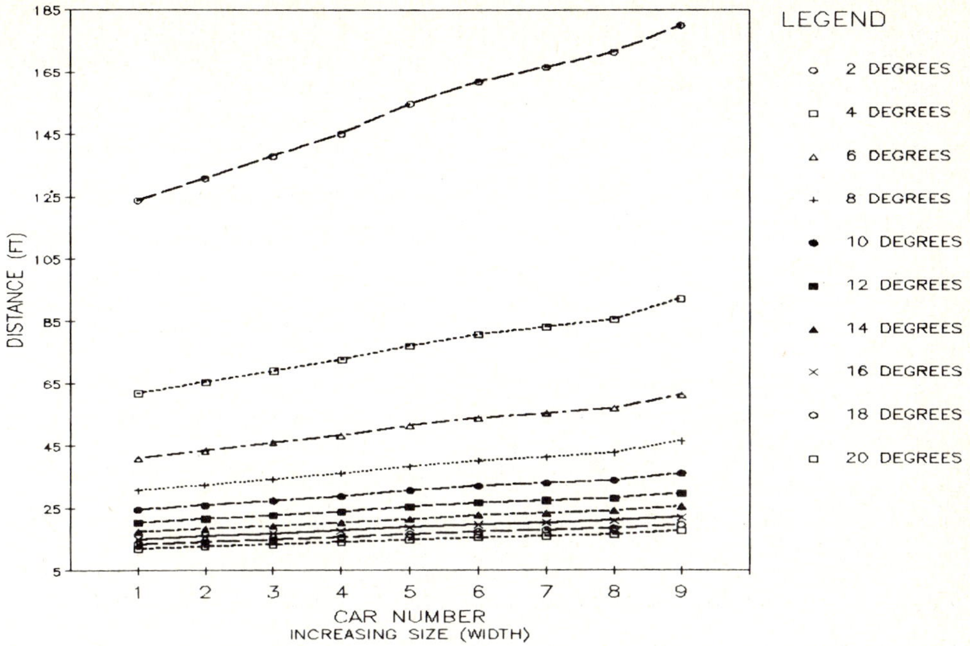

FIGURE 2
Ideal Observer Model

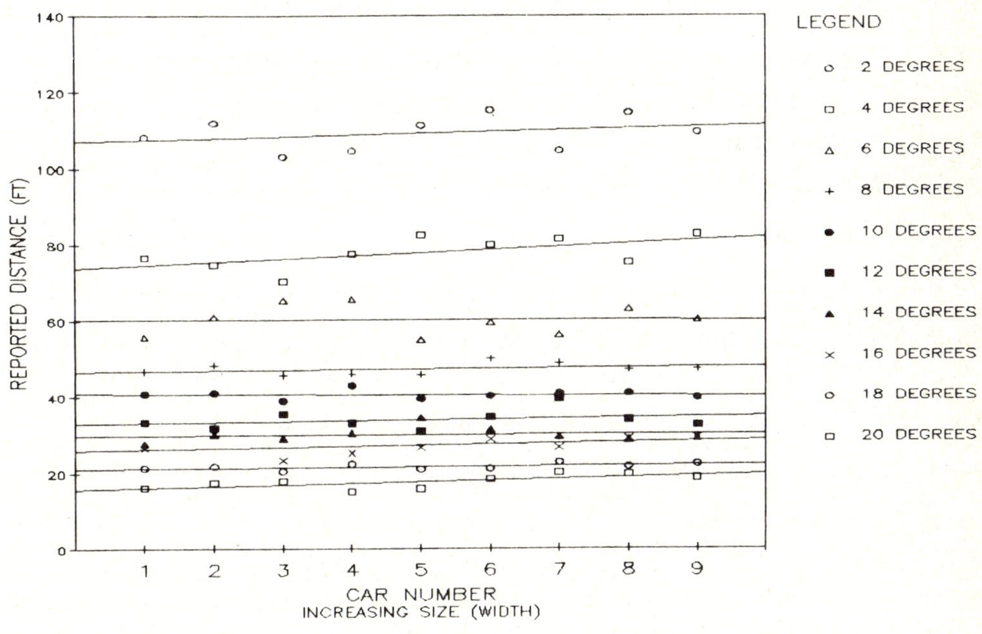

FIGURE 3
Reported Distance For the Clear Background

distances of the subjects were all below what could be expected for the ideal observer model. In fact, only one of the slopes, .83 for the 4 degree clear background, falls within the range of the ideal observer model and this is much less than expected for that model.

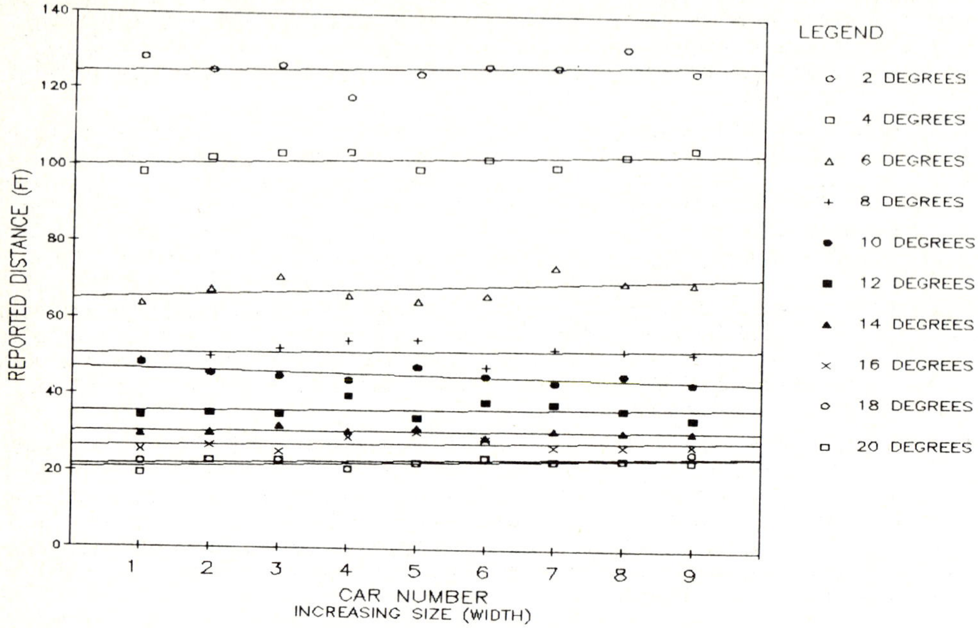

FIGURE 4
Reported Distance For the Dark Background

Visual Angle	Ideal Observer Model	Visual Angle Model	Reported Clear	Reported Dark
2	7.00	0	.43	.28
4	3.65	0	.83	.42
6	2.42	0	-.01	.66
8	1.87	0	.13	.23
10	1.43	0	-.08	-.30
12	1.16	0	.20	.18
14	1.00	0	.04	.11
16	.87	0	.27	.22
18	.76	0	.11	.31
20	.69	0	.40	.38

TABLE 1
Slopes of Best-Fitting Regression Lines for Visual Angles

DISCUSSION

Which of the two models, the ideal observer or visual angle model, did the data support? Some support exists for both models. A discussion of the support for the two models follows. With the ideal observer model, we would have expected the slopes of Table 1 to be high for the low visual angles and low for the high visual angles. With the visual angle model, we would expect low slopes for each of the ten best-fitting regression lines from Table 1. Again, partial fit to this model was found by the low slopes for the regression lines.

It appears that drivers do predominately use visual angle as a source for distance judgment, although the positive slopes seen in the results do not entirely support that conclusion. Instead, the subjects did make some correction for the actual distance of the car; this correction was not enough to accurately predict distance nor was it enough to alleviate the problem of misperception of the small car's distance. This causes many problems due to misperception of small cars. For the cars and distances tested in this study, at one extreme small cars can be perceived to be 40 feet farther away than they actually are. This misperception means that other drivers would be more likely to follow small cars closer and to take more chances in front of small cars thinking that they have more time to maneuver than they actually do have. This experiment shows, therefore, that small cars will be in more accidents due to this misperception.

Is there any way to correct this problem? It is possible that if drivers were made aware of the kinds of cues that they automatically use then they could possibly make better estimates when the deficiencies of the processing are pointed out. The existence of the correction in the second stage of the information processing model indicates that they can make such corrections; they must be made to realize that the correction they make is not adequate. The research to address this question could be carried out in the future.

REFERENCES

1. Evans, L., and Wasielewski, P., Risky driving related to driver and vehicle characteristics, Accident Analysis & Prevention 15 (1983) 121-136.
2. Gioia, A. J., and Morphew, C. E., Evaluation of driver vision (General Motors Technical Report) (General Motors Corporation, Detroit, 1968).
3. Puchonok, K., Accident cause analysis (National Highway Safety Administration Report DOT-HS-800-716) (Cornell Aeronautical Laboratory, Ithaca, NY, 1974).
4. Treat, J. R., and Shinar, D. A., Methodology for assessing and classifying traffic accident causes (National Highway Safety Administration Report DOT-HS-801-979) (Department of Transportation, Washington, 1976).

ACKNOWLEDGEMENT

This research was supported by a grant from the AAA Foundation. Mr. Sam Yaksich was the contract monitor.

Trends in Ergonomics/Human Factors II
R.E. Eberts and C.G. Eberts (Editors)

SEEING TONES AND HEARING RECTANGLES: ATTENDING TO SIMULTANEOUS AUDITORY AND VISUAL EVENTS

Patricia A. Casper and Barry H. Kantowitz

Department of Psychological Sciences
Purdue University
West Lafayette, Indiana 47907
U.S.A.

The allocation of attention in dual-task situations depends on both the overall and the momentary demands associated with both tasks. Subjects in an inclusive-or reaction-time task responded to changes in simultaneous sequences of discrete auditory and visual stimuli. Performance on individual trials was affected by (1) the ratio of stimuli in the two tasks, (2) response demands of the two tasks, and (3) patterns inherent in the demands of one task.

It is not unusual for humans to be engaged in two or more tasks concurrently at any given moment in a typical day. However, humans, unlike machines, are limited in their ability to process sequences of simultaneous events and subsequently perform one task in isolation much better than two tasks at once. Common as this divided attention, or timesharing situation is, researchers disagree when the issue of the locus of the information processing bottleneck arises. Do the limitations lie in the perceptual phase of processing, or in the final stages of response preparation? Recently, Klapp, Hill and Tyler (1983) have argued that two (or even many) tasks are easy to perform simultaneously if they have identical or harmonically-related rhythms. Earlier experiments by Klapp (1979) cited interference at the motor level as the cause of poorer performance in non-related rhythmic tasks. In an experiment designed to distinguish perceptual from motor-based interpretations of dual-task performance, Klapp et al. instructed subjects to monitor a tone and a light that cycled on and off in rhythmically-repeating patterns, and to press a response key when they noticed that either the tone or the light was no longer being presented. No overt rhythmic responses were required by the subjects. The stimuli either had identically-synchronized on and off phases, were shifted for the tone stimulus by a short lag, or had completely different on and off periods. Klapp et al. regarded the former two cases as being temporally compatible, while the latter was not compatible. The performance measure used was the dual-task decrement--the difference between performance on the dual-task and a single-task control. They found that performance in the compatible conditions was better than that in the incompatible condition. The explanation Klapp et al. gave for these results pertained to the "Gestalt" of the two tasks, in that harmonically-related tasks are easier to organize into a "perceptual whole" than unrelated tasks. According to Klapp et al. , this is the only plausible explanation of how people are able to master such difficult tasks as, for example, flying a helicopter, or other tasks requiring the processing of many different inputs simultaneously.

An alternative account for behavior demanding multiple execution of different responses may be found in the attention literature (Kantowitz,

1974) which also predicts better performance for two simultaneous compatible tasks than for simultaneous incompatible tasks (Greenwald, 1972). However, an attentional theory is a more parsimonious account of dual-task behavior than a Gestalt interpretation since it is able to predict behavior on specific trials during the dual-task situation. Eijkman and Vendrik (1965) have shown that it is possible for humans to detect increments in the intensity of stimuli presented to two different sensory channels simultaneously without interference between the two. Subjects detected increments in intensity of auditory and visual stimuli that were presented continuously for a total duration of 2 seconds, with approximately a 10 second inter-stimulus interval. In Eijkmann and Vendrik's relatively easy task there was no interference between the two modalities, but it is possible that tasks of a more difficult nature will be more susceptible to degradations in performance. In another signal-detection experiment Sorkin, Pohlmann, and Woods (1976) reported that detection of events in one channel (in a two-channel auditory task) is dependent on events occurring in the other channel. These results are in line with an attentional theory that would predict fluctuating relationships between performance in the two simultaneous tasks depending on the trial-to-trial demands placed on the central processing system. More specifically, in situations where one task demands twice the amount of processing as the other, performance should be better for the more frequently-performed task on those trials when it is executed in isolation than when two responses must be executed simultaneously. Likewise, if two tasks are performed that require equal and concurrent processing, performance in one task will be better on those trials where no response is required to the other channel.

A "new and improved" methodology. The present experiment was designed to test the attentional view's prediction of behavior at the trial level. Subjects are presented sequential stimuli in both the auditory and the visual channels simultaneously, and are asked to respond to changes in the pitch of the former and the hue of the latter. In one condition, twice as many visual (2:1) as auditory stimuli are presented. In another condition, there is a 1:1 relationship between the occurrence of the stimuli. Both the 2:1 and the 1:1 conditions may be regarded as temporally compatible, in Klapp et al.'s sense. It is predicted that in the 2:1 condition performance for the visual stimuli will be better on those trials where an auditory stimulus does not occur. Similarly, in the 1:1 condition it is expected that performance on the visual stimuli will be worse when a response to a change in the auditory channel is required.

Another condition was included in the present experiment in an attempt to separate two variables--the occurrence of a stimulus versus the value of that stimulus--that were confounded in Klapp et al.'s study. In Klapp et al.'s experiment only one light and one tone were used--both of which were either on or off at any given moment. The present experiment, by including two types each of auditory and visual stimuli, manipulated not only the temporal structure of the stimulus sequences, but the patterns of change of the stimuli within a sequence as well. Subjects were instructed to respond when a stimulus in one modality appeared that was different from the previously-presented stimulus in that modality. Sequences could either have a fixed pattern of changes (requiring regular and predictable responses), or they could have a random pattern of change.

In addition to including conditions designed to specifically test the microstructure of an attentional explanation of behavior, the present experiment represents an improvement over the methodology used by Klapp et al. (1983). First, due to equipment limitations, Klapp et al. were forced to terminate their stimulus sequences at the point when a response had occurred. The present study allowed indefinite stimulus sequence durations. Second, since Klapp et al. only examined subjects' responses to the termination of a stimulus, it was not possible to measure the number of missed responses, since eventually the subject would realize that a response was required. The present study measured responses to qualitative changes in the stimuli, which allowed observations of both hits and misses because a "same-different" recognition task was substituted for Klapp et al.'s "on-off" detection task. Third, in Klapp et al.'s experiment there was only one response key available for the subject to use for both sets of stimuli, making it impossible to determine exactly which stimulus had elicited a response from the subject. This method precluded Klapp et al. from determining false alarm rates--in fact, the only response measure they used was reaction time. The present experiment had separate response keys for both the auditory and the visual channel, allowing us to determine exactly which responses occurred for each stimulus modality. Further, both d' and reaction time measures were combined in an attempt to more clearly describe the amount of information being transmitted by the subject.

METHOD

Subjects. Sixteen male undergraduate students from Purdue University participated in the experiment. Each subject was paid three dollars an hour for his participation.

Materials and Design. A within-subjects design was used, with all of the subjects receiving both double- and single-stimulation conditions. Four double-stimulation sequences were given, then four single-stimulation sequences, followed by four more double-stimulation sequences. A Latin square design was used to counterbalance the single-stimulation conditions, while a modified version of the switchback design was used to counterbalance the four double-stimulation conditions.

A 2 x 2 design was used for the double-stimulation conditions, with sequences varying both in temporal structure and in the pattern of change within a sequence. Sequences could have either a 1:1 temporal relationship between the visual and auditory stimuli, or a 2:1 ratio of visual to auditory stimuli. In the 1:1 case, both an auditory and a visual stimulus appeared simultaneously on each stimulus presentation, or trial. In the 2:1 case, a visual and an auditory stimulus were presented, followed by a visual stimulus alone, followed by another visual-auditory stimulus pair, etc. throughout the sequence. Within a sequence, there were either fixed or random patterns in which the stimuli changed. In the fixed pattern condition, the color value of the visual stimuli changed on every other presentation, and in the random condition a table of random numbers was used to predetermine the color of stimuli within a sequence. In both fixed and random conditions the overall probability of change within a sequence was .5. Figure 1 shows the temporal relationships between the stimulus sequences.

The auditory stimuli used in the experiment consisted of high- and low-pitched tones of 3615 Hertz and 1420 Hertz at 75 dB (A) SPL which were presented to the subjects over Grason Stadler (model TDH39-300Z) headphones. The visual stimuli consisted of 1.25 cm by 2.5 cm vertical red and blue rectangles presented on a Sony Trinitron 12-inch color television at 2.41 vertical degrees of visual angle. The television was situated 60 cm in front of the subjects. An Apple II computer was used to generate the tones and produce the rectangles on the television screen.

Both the tones and the rectangles were presented for a duration of 100 msec, with an inter-stimulus interval (ISI) of 1600 msec. A sequence consisted of 100 stimulus presentations resulting in a total sequence duration of 2 minutes and 40 seconds.

Subjects responded by pressing one of two response keys located on the table directly in front of them. The key for the tones was on the left and the key for the rectangles was on the right. The static force required to depress the keys was 60 grams. A Cyborg model 91A ISAAC computer interface was used to read responses and to time intervals to the nearest msec.

Single-stimulation training. Training was first given on the single-stimulation task. Subjects were instructed to attend to a sequence of tones or rectangles and to respond by pressing the appropriately-labelled key when they noticed that a change occurred in the stimuli during a sequence. A change was defined as any stimulus presentation that was different from the previous stimulus in a particular sequence; for example, if a red rectangle is presented, followed by a blue rectangle on the next presentation, the subject should press the key labelled "rectangles". If another red rectangle is presented after the blue rectangle, another "rectangle" response is required. The subjects were informed that a change may or may not occur each time that a stimulus is presented, but they were not told that a pattern may exist in the stimuli. Subjects were told to use only one hand to respond. Two sequences were presented during single-stimulation training--a randomly-changing sequence of tones, and a randomly-changing sequence of rectangles. Both sequences used in training consisted of 100 stimulus presentations.

Double-stimulation training. Next the subjects were trained on two double-stimulation tasks--a randomly-changing 1:1 sequence (both a tone and a rectangle occur on each stimulus presentation), and a randomly-changing 2:1 sequence (where a tone only occurs on every other presentation). Subjects were told that the stimuli in either modality may or may not change on any given presentation, and that a change may occur in none, one, or both sequences on any given presentation. Following a sequence their mean reaction times to the tones and the rectangles (in msec) were displayed on the screen in front of them.

RESULTS

Effects of Temporal Structure. D' scores were calculated separately for all subjects in all conditions from the hit (the subject responds "yes" given that a change has occurred) and false alarm (the subject responds "yes" given that no stimulus change has occurred) data. Figure 2 shows the results of the analyses of the tone and rectangle data for the effects of temporal structure on d' scores (rectangles, $F[1,15] = 6.21$, $p < .05$; tones $F[1,15] = 73.0$, $p < .001$). There was a significant block effect for both the tones and the rectangles (tones, $F[1,15] = 4.87$, $p < .05$; rectangles, $F[1,15] = 8.91$, $p < .05$).

Figures 3 and 4 show the results of an analysis performed on a dependent variable that combined d' scores with reaction times--$(d')^2/RT$ (Moray & Fitter, 1973). Figures 3 and 4 are for double-stimulation data. Figure 3 shows the combined effects of temporal structure, block, and pattern of change for the rectangle data. There was a main effect of temporal structure, $F(1,15) = 29.10$, $p < .001$, pattern of change, $F(1,15) = 16.86$, $p < .005$, and block, $F(1,15) = 9.75$, $p < .01$. In addition, significant interactions were found between block and pattern of change, $F(1,15) = 5.78$, $p < .05$, temporal structure and pattern of change, $F(1,15) = 16.0$, $p < .005$, and block, temporal structure, and pattern of change,

F(1,15) = 5.21, p < .05.

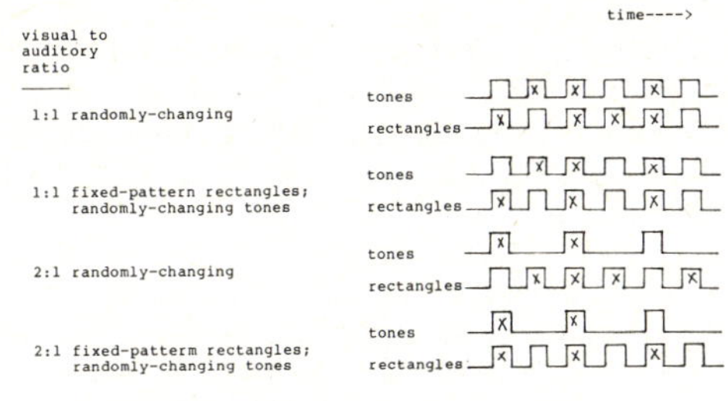

Figure 1. Temporal relationships among treatment conditions (An "X" represents a stimulus change)

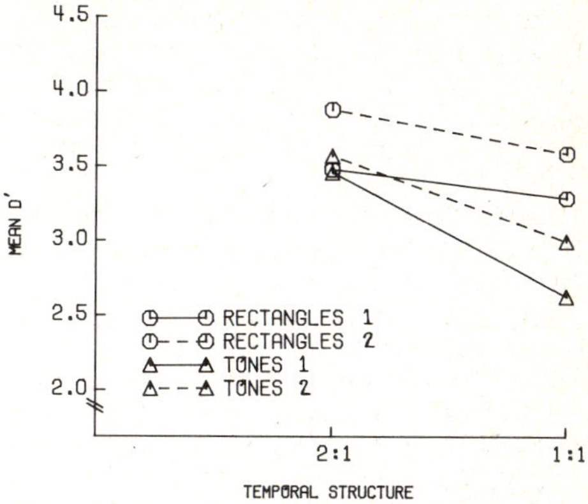

Figure 2. The effect of temporal structure on rectangle and tone d' scores.

The results of the analysis on the tone data are presented in Figure 4. There was a main effect of temporal structure, $F(1,15) = 28.0$, $p < .001$, of pattern of change, $F(1,15) = 8.78$, $p < .05$, and of block, $F(1,15) = 5.0$, $p < .05$. A significant interaction was found between temporal structure and the pattern of change, $F(1,15) = 11.13$, $p < .005$.

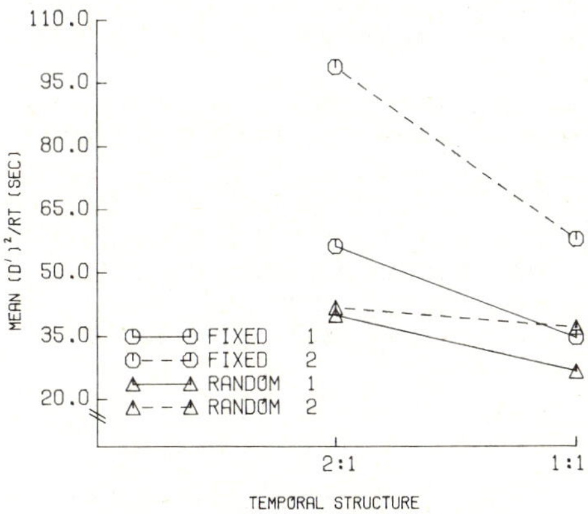

Figure 3. Rectangle $(d')^2/RT$ as a function of temporal structure, block, and pattern of change.

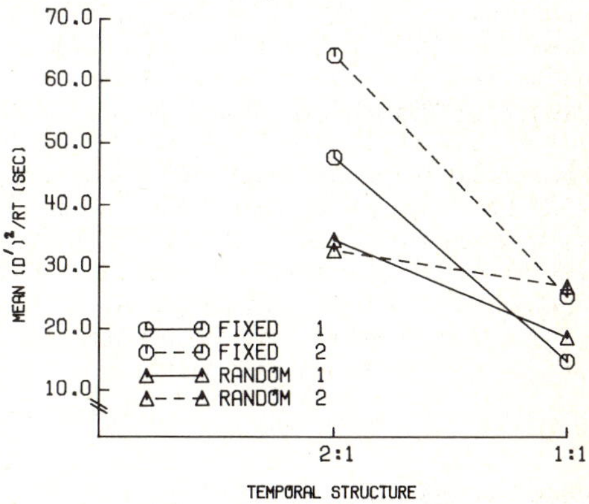

Figure 4. Tone $(d')^2/RT$ as a function of temporal structure, block, and pattern of change.

Figure 5 presents the combined results for the single- and double-stimulation tasks. The data for this analysis were taken from block 2, and the means are given in Table 1. There was a main effect of the type of task (single-stimulation, 2:1, 1:1 rectangles to tones) for both fixed and random pattern tones and rectangles (Table 2). T-tests of the differences between the single-stimulation, 2:1, and 1:1 conditions were significant at the .01 level for all differences except those between the 2:1 and 1:1 random pattern tones and rectangles, which were significant at the .05 level (Table 3).

Single-Stimulation analyses. The effects of pattern of change on $(d')^2/RT$ scores for both tones and rectangles in the single-stimulation conditions are shown in Figure 6. There was no statistical difference, $F(1,15) = 1.41$, $p > .05$, between the two patterns of change for the rectangle stimuli. However, for the tones, $F(1,15) = 14.0$, $p < .005$, the difference was significant.

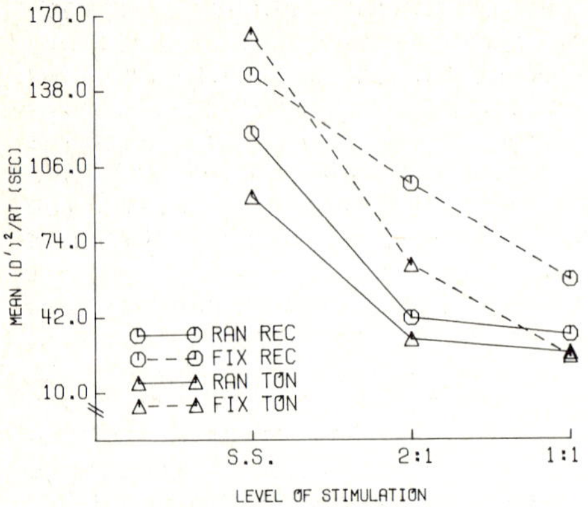

Figure 5. Combined single- and double-stimulation data for both fixed- and random-pattern tones and rectangles. Data is from block 2 of double-stimulation.

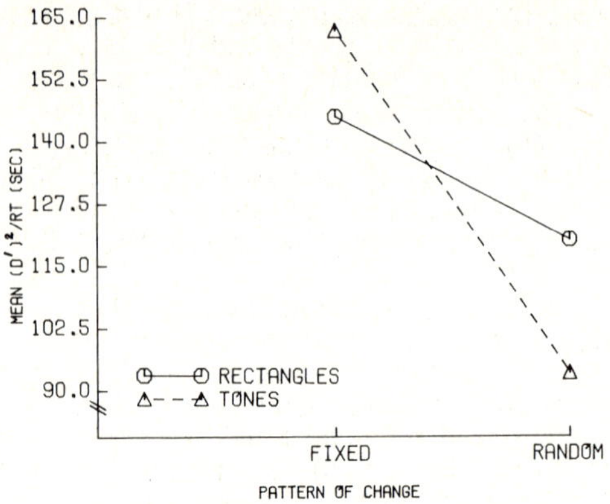

Figure 6. The effect of pattern of change on $(d')^2/RT$ for both tones and rectangles in single-stimulation.

Table 1

Combined single- and double-stimulation means

		S.S	2:1	1:1
Random pattern of change	Tones	93.51	32.67	26.89
	Rectangles	120.15	41.84	34.57
Fixed pattern of change	Tones	162.38	64.21	25.30
	Rectangles	145.00	98.78	57.93

Table 2

F and p values for Figure 5

		F(1,15)	p
Fixed pattern of change	Tones	33.78	< .001
	Rectangles	8.72	= .001
Random pattern of change	Tones	48.50	< .001
	Rectangles	46.45	< .001

Table 3

t-values for Figure 5

Condition		SS - 2:1 t(15)	SS - 2:1 p	2:1 - 1:1 t(15)	2:1 - 1:1 p
Fixed pattern of change	Tones	11.4	< .01	4.5	< .01
	Rectangles	4.4	< .01	3.9	< .01
Random pattern of change	Tones	16.4	< .01	1.6	> .05
	Rectangles	16.0	< .01	1.5	> .05

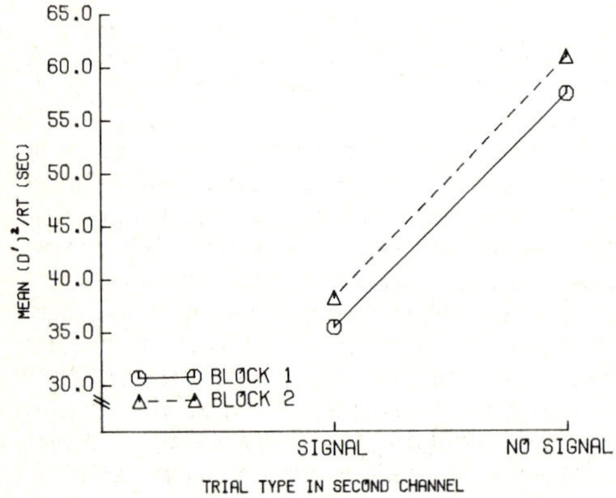

Figure 7. Rectangle $(d')^2/RT$ as a function of the presence of an auditory stimulus in the 2:1 random-pattern of change condition.

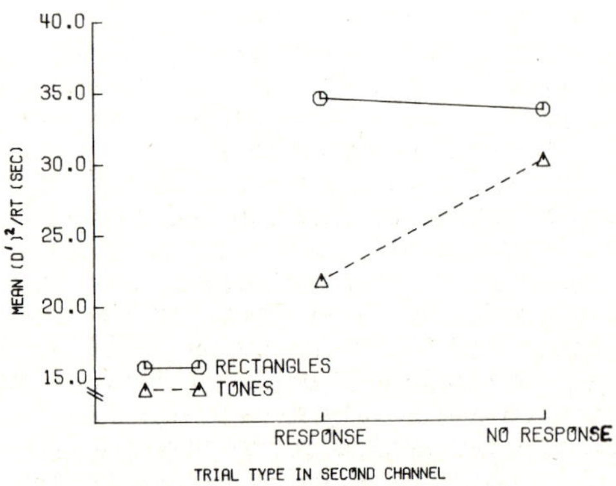

Figure 8. Rectangle and tone $(d')^2/RT$ as a function of the type of trial in the second modality in the 1:1 random-pattern of change condition.

Trial-by-trial analyses. The data from the 2:1 random-pattern of change conditions were analyzed for the effects of the presence of an auditory stimulus (on any given trial) on the performance for the rectanges on those trials. Figure 7 shows $(d')^2/RT$ scores for the rectangles in both blocks on trials with and without corresponding auditory stimuli. The analysis yielded a significant main effect of the presence of an auditory stimulus on rectangle performance, $F(1,15) = 4.97$, $p < .05$, but no interaction between block and the presence of auditory stimulus, $F(1,15) = .007$, nor any difference between the two blocks, $F(1,15) = .624$.

An analysis similar to the one described in the last paragraph was done on the 1:1 random-pattern data. Rectangle performance on trials where responses were required to the other channel (that is, a change in the pitch of the tones occurred) was compared to performance on those trials where no tone response was required (Figure 8). Analysis of variance found no effect of a change in the pitch of the tones on rectangle performance, $F(1,15) = .09$, $p > .05$, nor any block effect, $F(1,15) = 1.80$, $p > .05$. There was no interaction between block and tone change, $F(1,15) = .02$.

Performance was also measured for the tones on trials with and without concurrent rectangle changes (Figure 8). Means for tones with rectangle changes were 16.99 and 26.58 in the first and second blocks, while the means for tones without corresponding rectangle changes were 25.87 and 34.35 in blocks 1 and 2. Concurrent rectangle changes had a significant effect on tone performance, $F(1,15) = 6.80$, $p < .05$, as did the block, $F(1,15) = 5.22$, $p < .05$. There was no interaction between block and rectangle change, $F(1,15) = .08$.

DISCUSSION

Trial-by-trial results. The most important hypotheses proposed in this paper concern performance on specific trials during a sequence of stimuli. In the introduction to this paper, we predicted that it would be easier to detect changes in one modality when there was either 1) no stimulus in the other modality, or 2) no response required to a stimulus in the other modality. As was expected, performance (as measured by both d' and $[d']^2/RT$ scores) on the rectangles was better in the 2:1 condition on those trials when a rectangle occurred alone. Since there were no events to attend to in the auditory channel on those trials, the resources previously allocated to that channel (in previous trials where both a tone and a rectangle were presented) were switched over to the visual channel. Likewise, when events occurred simultaneously in both channels, processing resources available for the rectangles were less than those available when no tones occurred and performance declined.

The data from the 1:1 condition did not lend themselves to such a straightforward interpretation as those in the 2:1 condition. Rectangle performance was unaffected by the nature of the events occurring in the auditory channel; however, tone performance improved or declined depending on the trial-to-trial demands in the rectangle modality. These results could be taken as support for the idea of visual dominance (Posner, Nissen, & Klein, 1976), or as evidence of a bias towards attending to one modality over the other (Tanner & Swets, 1954). Despite instructions to the subjects explicitly directing them to pay equal attention to both the tones and the rectangles performance was better for rectangles (Figure 8). Thus, the greater overall frequency of rectangles taken over the entire experiment may have induced a response bias favoring the visual modality. We are repeating the experiment using only the 1:1 condition to evaluate

this hypothesis.
Single-stimulation results. In the single-stimulation control condition, a main effect of pattern of change was found only for the tone data. Since there were no differences in the d' scores for both the tones and the rectangles it can be concluded that the reaction times contributed the significant part of the difference for the tones when the $(d')^2/RT$ scores were figured. Inspection of the percent correct and false alarm data revealed that the majority of the subjects had reached a ceiling for correct responding (95 - 100%), thus indicating that the task was too easy to reveal any differences between the two patterns of change.
Single- versus double-stimulation. Crucial to an attentional explanation of dual-task behavior is the prediction that single tasks in isolation are easier to perform than dual tasks. The analyses combining data from both single- and double-stimulation conditions showed a significant effect of the level of stimulation, with single-tasks easier than dual-tasks, and the 2:1 ratio easier than the 1:1 ratio. These results were in line with our proposal that single-tasks are easier than dual-tasks, and that a 2:1 relationship between tasks yields better performance than a 1:1 relationship in a dual-task environment.

It is unclear from the results of this experiment whether the performance advantage of the 2:1 over the 1:1 condition was due to perceptual or response factors. Is the 2:1 condition easier because there are fewer tone stimuli to perceive or because there are fewer responses required to the tones? There were no conditions included in the present experiment to separate these interpretations, although Klapp et al. (1983) implicated perceptual interference in their results of better performance for compatible rhythms over incompatible ones. An experiment currently in progress was designed to isolate any response conflict factors affecting performance in the 1:1 condition. Subjects are given the same task as in the previous experiment under two different ISI and speed-emphasis conditions in an attempt to increase the overall level of false alarms. In the 1:1 condition, if the trials with false alarms to the tones have a lower d' for the corresponding rectangles from the same trial than the trials where no false alarms occur to the tones, a response factor may be the cause of poorer performance overall in the 1:1 condition as compared with the 2:1 condition, and dual-task performance could be explained by a response conflict model (Kantowitz, 1974). If rectangle d' is unaffected by simultaneous tone false alarms it would appear that the bottleneck is located in the front end of the human information-processing system.

Acknowledgment

This research was supported by Cooperative Agreement NCC 2-228 from the National Aeronautics and Space Administration, Ames Research Center; S.G. Hart was the Technical Monitor.

References

[1] Eijkman, E., and Vendrik, A. J. H. Can a sensory system be specified by its internal noise? The Journal of the Acoustical Society of America 37(6) (1965) 1102-1109.
[2] Greenwald, A. G. On doing two things at once: Time sharing as a function of ideomotor compatibility. Journal of Experimental Psychology 94 (1972) 52-57.

[3] Kantowitz, B. H. Double Stimulation: In B.H. Kantowitz (Ed.), Human Information Processing: Tutorials in Performance and Cognition (Potomac, MD.: Erlbaum, 1974).
[4] Kantowitz, B. H., and Sorkin, R. D. Human Factors: Understanding People-System Relationships (New York: Wiley, 1983).
[5] Klapp, S. T. Doing two things at once: The role of temporal compatibility. Memory and Cognition 7(5) (1979) 375-381.
[6] Klapp, S. T. Temporal compatibility in dual motor tasks II: Simultaneous articulation and hand movements. Memory and Cognition 9(4) (1981) 398-401.
[7] Klapp, S. T., Hill, M. D., and Tyler, J. G. Temporal compatibility in dual motor tasks: A perceptual rather than motor interpretation. Paper presented at the 24th annual meeting of the Psychonomic Society, San Diego, CA (1983).
[8] Moray, N., and Fitter, M. A theory and the measurement of attention. In S. Kornblum (Ed.), Attention and Performance IV. New York: Academic Press (1973).
[9] Posner, M. I., Nissen, M., and Klein, R. Visual Dominance: an information-processing account of its origins and significance. Psychological Review 83(2) (1976) 157-171.
[10] Sorkin, R. D., Pohlman, L. D., and Woods, D. D. Decision interaction between auditory channels. Perception and Psychophysics 19(4) (1976) 290-295.
[11] Tanner, W. P. Jr., and Swets, J. A. A decision-making theory of visual detection. Psychological Review 61 (1954) 401-409.

Trends in Ergonomics/Human Factors II
R.E. Eberts and C.G. Eberts (Editors)
Elsevier Science Publishers B.V. (North-Holland), 1985

EYES/EARS; SPATIAL/VERBAL: HOW SHOULD A COCKPIT TALK TO A PILOT?

Christopher P. Robinson
Aeronautical Systems Division
Wright-Patterson AFB, OH, 45433

Stephen J. Base
Ray E. Eberts
Industrial Engineering, Purdue University
West Lafayette, IN 47907

Multiple resource theory suggests that since the modality/code combination used for encoding primary flight information by a pilot is visual/spatial, secondary information should use auditory/verbal displays. However, direct mapping between visual/spatial displays and responses presents and advantage which must be considered. A dual task tracking and firing combat simulator was developed to test the relative effectiveness of presenting secondary information using visual/spatial, visual/verbal, auditory/spatial, and auditory/verbal modes. Results indicated that the auditory/verbal display was superior to the others in terms of the number kills and the number of errors. This mode was also rated superior on a questionnaire.

INTRODUCTION

The interface between the aircraft and its ultimate controller - a human pilot - must be designed to maximize the efficiency and accuracy of the human-computer interaction. The piloting of a modern tactical aircraft involves the accurate performance of a difficult tracking task while simultaneously making decisions; the outcome of which will often determine the pilot's survival. While the information presented to the pilot regarding the tracking task is practically limited to the visual spatial mode (i.e. visual contact with the enemy aircraft), on-board computers can make use of other input modalities to present decision-aiding information to the pilot. These include auditory/verbal (generated speech) and auditory/spatial (source direction and pitch). The other two alternatives, one of which has been used already and the other being implemented currently, are visual/-verbal (alphanumeric) and visual/spatial (pictorial). The question of which mode is most effective for which circumstances is currently commanding much debate with clear advantages and disadvantages on all sides (Robinson, 1984). This study used a dual task track and kill paradigm to compare these four methods of presenting secondary information to a pilot.

Hawkins, Reising, Lizza, and Beachy (1983) compared the effectiveness of speech, alphanumeric, and pictorial modes of presenting emergency information to the pilot. Their study showed no significant performance differences between the three modes, though preference ratings showed that the subjects would rather have speech presentation.

In a study limited to the visual input modality, Aretz and Calhoun (1982) designed an experiment which compares different aspects of pictorial and alphanumeric displays and their integration with each other. In a fixed base simulator, subjects were required to maintain flight control while retrieving weapons stores information. This information was presented alphanumerically, color pictorially, black and white pictorially, and with a mixture of alphanumerics and color pictorial. Results indicated that alphanumeric displays had a shorter task completion time. However, this method was not statistically better than the color pictorial or combination methods.

Moroze and Koonce (1983) tested performance differences between traditional round-dial displays, digital Heads-Up-Displays (HUD), and spatial HUD displays. The "spatial" HUD consisted of a linear-tape altimeter, and was not purely spatial since it requires numbers on the tape for reference. The hypothesis was that the linear tape method would result in better performance than the round-dials and digital HUD since it provided information in a method more consistent with the subjects' internal model of the aircraft's dynamics. Without this consistency, the subject would have to go through more transmission coding processes, with corresponding increases in workload and probability of error. In the study, subjects "flew" a microcomputer simulator program, performing various maneuvers while responding to a recognition task (picking out odd-even-odd sequences in a random series of digits presented auditorially). The results showed that the traditional displays were more effective while the subjects were still learning; however, after criterion levels had been met there were no significant differences between the display types.

Wickens, Sandry, and Vidulich (1983) suggest that the use of speech synthesis is an obvious way to reduce the load on the visual perception modality. This application cannot be made, however, without consideration of the stimulus-central processing-response (S-C-R) compatibility, i.e. the concept that visual inputs and manual responses, along with auditory input and vocal responses, result in faster responses than when the stimulus and response are not similar.

PURPOSE

According to Wickens (1984), the primary stimulus inputs can be described by two main axes; modality, which includes visual and auditory inputs, and code, which includes spatial and verbal forms of information. These divisions separate stimulus information into four modality/code combinations, with each type of information drawing from its respective resource "pools". A pilot engaged in flying a fighter aircraft in combat draws heavily upon her visual and spatial encoding and central processing resource pools for tracking the enemy aircraft.

The purpose of the proposed research was to help determine which of the four encoding resource pool combinations should be utilized to display necessary information to a pilot in a high visual tasking environment. Multiple resource theory might suggest that transmission modes other than visual/spatial would be appropriate for such a situation. However, as Wickens points out, the other transmission modes do not share the same degree of stimulus-central processing-response compatibility as the visual-spatial mode when the response is limited to a manual mode. To what extent is this potential disadvantage compensated for by designing displays which take advantage of under-utilized resource pools? What is the interaction between modality and code for information displays in such a scenario? The experiment was designed to explore these questions.

METHOD

The experiment used a pursue and kill paradigm in which the subjects were tasked with chasing and destroying as many enemy aircraft as possible in the allotted time. Based on simulated range information generated by an on-board computer, the subject had to select an appropriate range-specific weapon to attack the enemy aircraft. The range information was presented via four different modality/code combinations (visual/spatial, visual/verbal, auditory/spatial, and auditory/verbal) to test the relative effectiveness of each.

The paradigm consisted of two basic tasks which were not entirely independent. The primary task consisted of a dual-axis tracking problem while the secondary task consisted of the weapon decision firing problem. The tasks were not independent because the enemy aircraft must be successfully tracked AND the correct weapon must be chosen in order to achieve a kill.

The dual-axis tracking problem simulated actual aircraft flight dynamics and control. The pitch axis gave first-order responses to control inputs while the yaw axis gave second-order responses. The subject had an "inside out" view of the target. The stick was pulled back to "pitch up" the simulator and pushed forward to "dive," while a left input yawed the simulator to the left and a right input yawed it to the right. To successfully track, the enemy airplane had to be kept inside a window sight generated by the CRT. To simulate evasion, random "gust" inputs were applied to the enemy aircraft necessitating constant correction inputs.

To destroy an enemy aircraft the subject had to fire the appropriate range-specific weapon while the aircraft was in the sight. The range numbers represent a distance of one to six miles from the target. The subject had to decide which of three weapon types, listed in Table 1, to use in attacking the target.

Table 1. Weapons Types

WEAPON TYPE	EFFECTIVE RANGE
Cannon	1 or 2 Miles
Missile Type-1	3 or 4 Miles
Missile Type-2	5 or 6 Miles

The range was a randomly generated set of numbers between one and six inclusive. Once notified of the distance to the target, the subject decided which weapon to use and fired by pushing the corresponding function key. The subjects received feedback on hits and out-of-range shots. Only by using the correct weapon for the current range (which updated approximately every three seconds), with the target in the sight, could a kill be achieved. The total number of shots possible was limited by time. As described before, the four different transmission modes depicted in Table 2 were used to test which mode best allowed the subject to integrate the range information with the tracking task to solve the firing problem.

Table 2. Four Range Transmission Modality/Code Combinations

VISUAL SPATIAL (V/SP)	AUDITORY SPATIAL (A/SP)
VISUAL VERBAL (V/Ve)	AUDITORY VERBAL (A/Ve)

The simulation mockup consisted of an IBM Personal Computer with color graphics, speech synthesis boards, and a joystick. The color CRT was used to simulate a cockpit forward view with a HUD installed, and the joystick was the aircraft control stick. For the V/Sp display, identical vertical analog bargraphs were presented on both sides of the CRT. This symmetry was incorporated to avoid having the subjects' view directed to just one half of the screen. The relative height of the graphs corresponded to the distance of the target. The vertical arrangement was selected to help insure stimulus-response (S-R) compatibility with the weapons choice key layout; on the IBM PC the function keys are arranged in vertical columns. Thus, there was a direct mapping between display and response. The V/Ve display consisted of numerals displayed in the same areas on the CRT as the bargraphs. The A/Sp display consisted of six different tones one octave apart generated by the PC soundboard; the lowest tone corresponding with the shortest range on up to the highest tone corresponding to the furthest range. Again, this provided a direct spatial mapping from the display to the response. The A/Ve range display used speech synthesis boards which verbally stated the numeric range.

A crossover experimental design was used in which all test subjects were exposed to all four transmission mode combinations. To separate the effects of order from mode in the statistical analysis, eight experimental mode order sequences were applied. In these eight sequences each mode was represented twice in each of the four "time" positions, and each of these different mode sequences was used twice in a randomized ordering to accommodate sixteen test subjects. All of the test subjects were right handed nonpilot adults with a background in engineering.

Each subject was given a briefing on the nature and purpose of the experiment, followed by three two-minute primary task tracking runs designed to teach the subject the tracking task and reduce the effect of learning on the rest of the experiment. This was followed by a two-minute primary single-task tracking run for data. Then the subject completed four dual task test runs, experiencing each of the four transmission modes. For each mode the subject was first given a one-minute practice session with the enemy aircraft "fixed" in the sight so that the subject could practice firing his weapons with generated range values for the specific transmission mode being evaluated. This was followed by a two-minute data run with full dual task tracking and firing problem. After all the transmission modes had been tested, a final two-minute primary task tracking data run was accomplished to determine the degree of tracking learning which had taken place during the experimental session. Finally, the subjects were asked to respond to a questionnaire. The experiment took about 50 minutes for each subject.

A number of measurements were taken during the sessions. On tracking and dual task data runs the tracking performance was measured by the Root Mean Square (RMS) of the distance that the target aircraft was away from the center of the sight. For the dual task data runs, the number of kills, out-of-range shots, misses, and potential shots (target was in the sights but no shot was taken) were recorded. Furthermore, when the subject made a weapon selection error, the wrong selection along with the correct choice was recorded. In the questionnaire, each subject rated the effectiveness of each of the modes on a scale ranging from negative three to three, rank-ordered the mode preferences, and gave written explanations for these preferences.

RESULTS

The four treatment combinations had a significant effect (see Figure 1) on the number of kills made by subjects ($F(3,24)=4.495$, $p<.012$). A

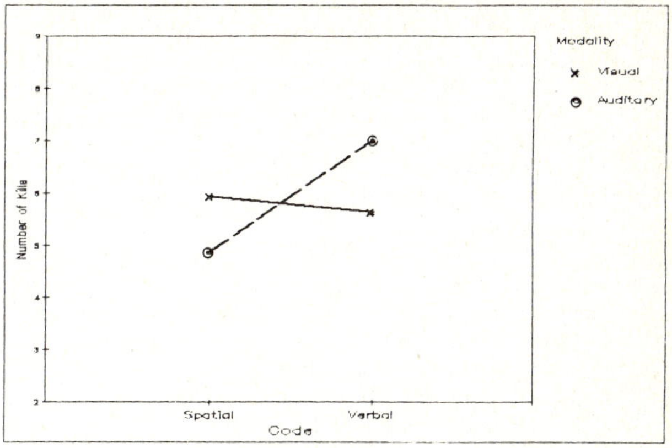

Figure 1. Effects of Modality and Code on Number of Kills.

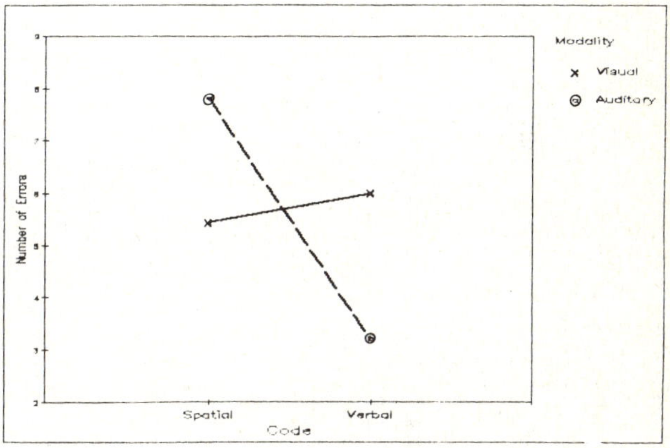

Figure 2. Effects of Modality and Code on Number of Errors.

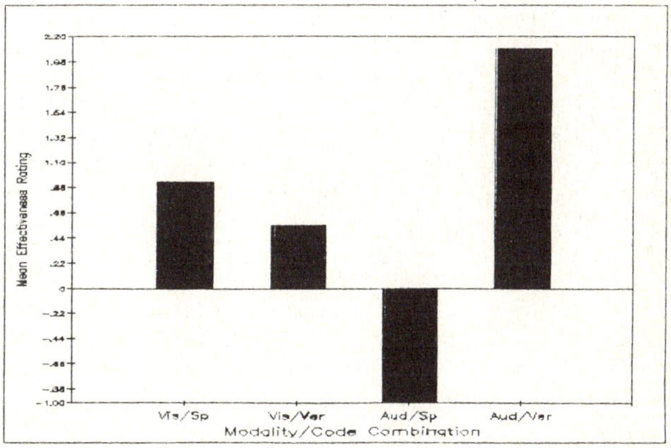

Figure 3. Effectiveness Ratings by Modality and Code Combination.

Newman-Keuls test for paired comparisons was performed on the means (Anderson and McLean, 1974). This test showed that at the 95 percent confidence level, kills with the A/Ve combination were significantly greater than both the A/Sp and V/Ve combinations. No other pairwise comparisons were statistically significant. The experimental design allowed indirect testing of the effects of a treatment combination by time interaction on kills; the analysis showed this interaction to be nonsignificant.

The treatment combinations also significantly affected (see Figure 2) the number of errors made in weapon selection ($F(3,24)=17.635$, $p<.0001$). A Newman-Keuls test on these means showed that subjects made significantly more errors using the A/Sp display than in all the other modes. In addition, both the V/Sp and the V/Ve modes resulted in more errors than in the A/Ve mode. The difference between V/Sp and V/Ve was not significant. Treatment combinations did not have a significant effect on the tracking RMS performance.

The subjective effectiveness ratings (see Figure 3) were significantly affected by the information modes, ($F(3,24)=13.69$, $p<.0005$). The A/Ve display mode was rated higher than the other three modes, and the V/Sp and V/Ve displays were rated higher than the A/Sp display.

Finally, the modality by code interaction had a statistically significant effect on number of kills ($F(1,8)=13.0$, $p<.007$, see Figure 1). This interaction also affected the number of errors, ($F(1,8)=36.07$, $p<.007$, see Figure 2), and subjective ratings ($F(1,8)=25.7$, $p<.001$).

DISCUSSION

The results of this experiment tend to confirm the viewpoint presented by Wickens, et al. (1983), that voice synthesis may be a good means of reducing the load on the visual perception modality. When the secondary information was presented using the A/Ve combination, the best performance was elicited in terms of errors. However, none of the modes tested resulted greater tracking degradation than any other mode; all of the effects were with the firing problem itself. Also, the subjective ratings tend to agree with those found by Hawkins, et al. (1983).

As might be predicted by multiple resource theory, a subject engaged in a two-axis visual tracking task of sufficient difficulty would have the capacity of the visual/spatial pool nearly saturated. The most underutilized pool in such a situation would be on the opposite ends of the mode-code axes, or the auditory/verbal pools. Thus information which is presented utilizing this combination of resources should be processed and responded to more efficiently and accurately than when the other combinations are utilized.

In the paradigm used in this study, however, the responses were made manually, i.e. the weapon was fired by pushing the proper function key. In the A/Ve mode, then, there was not a direct S-R relationship between the information and the response keys. In the V/Sp, and, to a lesser extent in the A/Sp modes, this S-R relationship did indeed exist which may explain why the number of kills made in the A/Ve mode was not significantly greater than in the V/Sp mode. Why did this advantage not play more of a role in the comparison with the verbal modes in regards to the number of errors made? A possible explanation might be in the experimental paradigm. In this experiment, it was not always essential for the subjects to respond immediately to new information; the firing task was not response-time intensive. Based on the "model human processor" proposed by Card, Moran, and Newell (1983), a single task processing cycle should take on the order of 300 msec. In this experiment, the range information updated every three seconds. Therefore, even in the dual-task setting, the subjects had plenty of time to perceive, process, and respond to the new range information, i.e. select the proper weapon, while they tried to center the target in their sight. The longer response time resulting from more processing steps required to translate verbal information to a spatial manual response did not incur a significant degradation in performance, as the subjects had sufficient time between range changes.

Another interesting finding comes from the analysis of the modality by code interaction effects. While the A/Ve mode was very effective, the A/Sp was very ineffective. In fact, both the spatial and verbal codes in conjunction with the visual modality elicited better performance than the A/Sp combination. These results suggest that if different resource pool combinations are to be used to present secondary information, then both the modality AND the code should be different. One aspect of the experiment which may have influenced on the data is that six tones may have been more than the subjects should be expected to distinguish consistently (Miller, 1956).

CONCLUSIONS

The results of the experiment provide support for the multiple resource theory-based idea that the speech synthesis display may be appropriate for presenting secondary information to pilots whose visual and spatial resource pools are saturated. To the extent that the simulation actually does replicate the flying experience, a case can be made for the consideration of the integration of such displays into military aircraft. The experiment may not, however, have brought out all of the positive aspects of direct mapping between stimulus and response, i.e. high S-R compatibility; this aspect should be given more attention before speech displays are incorporated too freely.

REFERENCES

1. Anderson, Virgil L. and McLean, Robert A. Design of Experiments: A Realistic Approach. New York: Marcel Dekker, Inc. 1974.

2. Aretz, A.J. and Calhoun, G.L. Computer Generated Pictorial Stores Management Displays for Fighter Aircraft. Proceedings of the Human Factors Society 26th Annual Meeting, Seattle, Wa., Oct 1982, 455-459.

3. Card, Stuart K.; Moran, Thomas P.; and Newell, Allen. The Psychology of Human-Computer Interaction. Hillsdale, N.J.: Erlbaum. 1983.

4. Hawkins, James; Reising, John; Lizza, Gretchen; and Beachy, Keith. Is a Picture Worth 1000 Words--Written or Spoken? Proceedings of the Human Factors Society 27th Annual Meeting, Norfolk, Va., Oct 1983, 970-972.

5. Miller, George A. The Magical Number Seven, Plus or Minus Two: Some Limits on Our Capacity for Processing Information. The Psychological Review, 1956, 63, 81-97.

6. Moroze, Michael L. and Koonce, Jefferson M. A Comparison of Analog and Digital Scales for Use in Heads-Up Displays. Proceedings of the Human Factors Society 27th Annual Meeting, Norfolk, Va., Oct 1983, 938-940.

7. Robinson, Christopher P. A Comparison of Pictorial and Speech Warning Messages in the Modern Cockpit. M.S. Thesis, Dept. of Industrial Engineering, Purdue University, 1984.

8. Wickens, Christopher D. Engineering Psychology and Human Performance. Columbus, OH: Charles E. Merrill. 1984.

9. Wickens, Christopher; Sandry, Diane; and Vidulich, Michael. Compatibility and Resource Competition between Modalities of Input, Central Processing, and Output. Human Factors, 1983, 25(2), 227-248.

Trends in Ergonomics/Human Factors II
R.E. Eberts and C.G. Eberts (Editors)

DETECTION AND IDENTIFICATION OF ABNORMALITIES IN CHEST X-RAYS: EFFECTS OF READER SKILL, DISEASE PREVALENCE, AND REPORTING STANDARDS

Raja Parasuraman
The Catholic University of America, Washington D.C. 20064

The effects of disease prevalence on diagnostic accuracy and reporting criteria in reading chest x-rays were examined for staff and resident radiologists. Staff radiologists, but not residents, adjusted reporting criteria appropriately with a change in disease prevalence. The staff were also more accurate in both disease detection and identification, but the residents were not significantly worse if both first and second choice diagnoses were evaluated. Measures of diagnostic accuracy based on the ROC were independent of disease prevalence. The results confirm the utility of ROC analysis in evaluating diagnostic performance in chest radiography, and suggest that optimising reporting standards, and allowing readers to report at different levels of confidence, are possible techniques for reducing error.

INTRODUCTION

Observer errors of perception and interpretation can severely limit the utility of diagnostic tests in medicine. Surveys show that 10-40% of the significant abnormal findings in samples of radiographs are not detected initially (Garland, 1949; Herman & Hessel, 1975; Rhea et al., 1979; Tuddenham, 1962), an error rate that amounts to several million errors annually (Berlin, 1977). Not all of these errors will affect patient management and health, but in general errors are costly and time-consuming, and the associated legal and social costs can be considerable. In 1971, for example, many coal miners with pneumoconiosis who sought compensation under the Federal Coal Miners Health and Safety Act (1969: PL-91-173) had chest radiographs that were initially classified as normal. This led to lengthy appeal and review procedures (Morgan et al., 1973). The effectiveness of chest radiography in preventing mortality and in monitoring the environment of workers exposed to hazardous chemicals and dust is also critically dependent on the level of reader performance (Liddell & McDonald, 1980). Furthermore, a recent survey found that approximately 50% of the malpractice suits filed against radiologists in Cook Co., Ill. are based on allegations of radiological omissions or misinterpretations (Berlin, 1980).

Errors in radiology have traditionally been attributed to technical limitations (e.g., poor image quality) or to "human error" (e.g., faulty search) (Boyington, 1980; Tuddenham, 1962). From the perspective of statistical decision theory, errors are a joint product of both detectability and the criterion or reporting standard used to choose one diagnostic alternative over the other (Lusted, 1971; Metz, 1978). Diagnostic performance can be evaluated by analysing the ROC, which is a plot of the proportion of true-positive versus false-positive reports at different reporting criteria (Swets & Pickett, 1982). This approach suggests two possible ways in which errors may be reduced. Image-processing techniques aimed at improving disease detectability may be useful (Swets, 1979; Todd-Pokropek, 1980). Once the limit of improvement with image-enhancement has been reached, however, or if the existing system cannot be changed, manipulating reporting criteria may be more effective in reducing error.

At a given level of detectability, there is a tradeoff between the two types of error possible, false-positives and false-negatives. The relative importance of these errors varies with the diagnostic setting, as does the appropriate reporting criterion. This suggests that training methods should aim to optimise the reporting criterion associated with the decision goal appropriate to the setting. Disease prevalence typically varies with diagnostic setting. Thus, the present study examined the effects of changes in disease prevalence on the reporting standards and diagnostic accuracy of two groups of skilled chest x-ray readers, staff and resident radiologists. The goal of the study was to examine whether a shift in disease prevalence is effective in inducing changes in reporting criteria, and whether such changes are optimal. When disease prevalence is relatively low, as in screening tests, a relatively strict reporting criterion is optimal for maximising expected value, assuming the costs and benefits of positive and negative outcomes are equal (Green & Swets, 1966). On the other hand, when disease prevalence is relatively high, as in a referral setting, a lax criterion is optimal. A related point of interest in the study was to examine the effects of disease prevalence on ROC-based measures of diagnostic accuracy (detectability). The utility of such measures in the field would be considerably enhanced if their theoretical invariance with such decision-related variables as disease prevalence, decision goals, and the costs and benefits of outcomes, holds in practice. Finally, this study used the recent extension of ROC analysis to include indentification (Green & Birdsall, 1978; Starr et al., 1975) to examine both disease detection and disease classification.

METHOD

Four first and second year radiology residents and four staff radiologists participated. A total of 120 standing posterior-anterior chest films having equal numbers of normal and abnormal findings were selected, each of which had been confirmed by previous history or by follow-up. There were 8 possible types of abnormality. Two 90-film sets differing in the relative proportion of abnormal cases (disease prevalence) were compiled, a low disease prevalence (LDP) set (33.3%) and a high disease prevalence (HDP) set (66.7%) These prevalence rates, chosen to correspond roughly to those likely in "screening" and "referral" situations respectively, are meant to be suggestive only. It would be preferable to use a much lower prevalence rate, say in the range 3-5%, to be representative of screening, but this would entail a great many trials to estimate error probabilities accurately. Actual prevalence rates may be lower or higher in particular situations, but disease prevalence will generally be higher for referral than for screening. Andersson et al. (1979), reported malignancy rates of 3.2% and 52.9% for breast-cancer screening and referral respectively.

A standard reporting form consisting of a 5-category confidence rating scale (1-definitely abnormal; 2-probably abnormal; 3-possibly abnormal; 4-probably normal; 5-definitely normal) and a list of the 8 disease types was used. If an abnormality was reported (ratings 1, 2, and 3), first and second choice disease types had to be identified. Readers attended for several sessions over a 5-week period and read both the LDP and HDP sets of films twice in counterbalanced order. Films were viewed one at a time on a standard viewbox at a self-paced rate. Readers were told that abnormal cases could be expected to occur "about one-third" or "about two-thirds" of the time in the LDP and HDP sets respectively. They were asked to give equal weight to avoiding false-positive and false-negative errors and to assume symmetrical costs and benefits of outcomes.

RESULTS

For disease detection, ROCs were fitted to each reader's data using a maximum likelihood solution (Gray & Morgan, 1972). χ^2 tests of goodness of fit showed that individual ROCs were fitted by straight lines on normal-deviate axes for 14 of the 16 ROCs analysed ($\chi^2 < 1.94$, $P > .37$). A number of measures of disease detectability and reporting criteria were then derived from the ROCs. Table 1 gives the mean values of two detectability measures, A_z and d_a. A_z is an estimate of the area under the normal ROC and varies from 0.5 to 1.0 (Swets & Pickett, 1982); d_a is a robust analog of d' (Simpson & Fitter, 1973). Staff radiologists had significantly higher A_z values than the residents for both the LDP and HDP conditions ($F(1,6)=16.1$, $P < .01$). Disease prevalence had no effect on A_z in both groups of readers ($F(1,6)=0.01$), indicating that A_z was a criterion-free measure of detectability. It was also found that ROC slope was not affected by reader skill ($F(1,6)=0.00$) or by disease prevalence ($F(1,6)=0.16$). Analysis of variance also confirmed the effect of reader skill on the d_a measure of detectability ($F(1,6)=19.69$, $P < .01$), and the invariance of d_a with disease prevalence ($F(1,6)=0.27$).

Table 1 gives the maximum-likelihood estimates of the response criterion measure Z_k, which is the normal deviate of the estimated false-positive probability at criterion. The staff radiologists adopted lower, more liberal criteria when reading in the HDP set as compared to the LDP set. However, disease prevalence had no effect on the reporting criteria of the residents, as was confirmed by the significant interaction between disease prevalence and groups ($F(1,6)=6.1$, $P < .05$). Furthermore, the residents used more liberal criteria than the staff radiologists ($F(1,6)=8.2$, $P < .05$).

The Z_k measure of the criterion is relatively robust. The more conventional measure, the likelihood ratio β, is less robust, but is useful because it can be compared with the optimal values of β for maximising expected value or the proportion of correct responses. Given a symmetrical cost-benefit matrix, the value of β that maximises expected value is 1-p/p, where p is the a priori probability of abnormality (disease prevalence). Thus, the optimal values of β for the LDP and HDP conditions are 2.0 and 0.5 respectively. Figure 1.A plots the obtained values of β against the optimal values for each reader. There was considerable deviation from the optimum, but the criterion values of the staff were correctly rank-ordered with the optimal values, except for one reader. The criteria of the residents showed no consistent relationship with the optimal values.

The results obtained with two measures of the criterion, Z_k and β, indicate that the residents showed little change in their reporting criteria with a change in disease prevalence, whereas the staff adopted more liberal criteria

	Staff LDP	Staff HDP	Residents LDP	Residents HDP
	Disease Detection			
A_z	0.92	0.91	0.85	0.87
d_a	2.01	1.86	1.50	1.58
Z_k	1.08	0.70	0.60	0.68
β	1.17	0.60	0.57	0.83
P(R)	0.37	0.66	0.45	0.62
	Disease Identification			
$P(C)_1$	0.60	0.54	0.54	0.58
$P(C)_2$	0.68	0.72	0.69	0.71
	Detection and Identification			
$P(A)_1$	0.74	0.70	0.61	0.66
$P(A)_2$	0.82	0.78	0.79	0.82

TABLE 1. Mean values of detectability and reprting criteria for staff and resident radiologists for low (LDP) and high (HDP) disease prevalence. P(R) is the overall probability disease report. P(C) is the overall probability of correct disease identification. P(A) is the area under the joint detection and identification ROC. The subscripts for P(C) and P(A) refer to whether only first (1) or both first and second (2) choice diagnoses were scored as correct.

when disease prevalence was high. The residents did not adjust their reporting rates in the optimal direction; if anything, their criteria changed in the opposite direction to that predicted (see Table 1 and Figure 1.A). It is possible that readers used some decision goal other than maximising expected value. Figure 1.B shows "probability-matching" functions (Craig, 1976) for each reader. Each plot shows the relationship between the overall probability of positive reports (both true and false) and the a priori probability of disease for the LDP (0.33) and HDP (0.67) conditions. As Figure 1.B shows, the suboptimal behavior of the staff radiologists vis a vis the expected-value goal can be viewed as a form of probability matching. The reporting behavior of the residents also conformed to probability matching, although to a lesser extent.

The degree of agreement between readers was examined by computing the kappa statistic for all possible pairs of readers within each of the two groups. kappa compensates for agreement arising by chance alone, and varies from +1 (perfect agreement) to -1 (perfect disagreement), with 0 representing chance agreement (Cohen, 1980). The computing formula is (Po-Pc)/(1-Pc), where Po is the observed probability of agreement between two judges, and Pc is the agreement attributable to chance (obtained from the marginal distributions). Table 2 shows the intra- and inter-reader values of kappa. Among the staff, kappa values were moderately high and significantly greater than 0 ($z > 3.9$, $P < .001$ for each pair). Intra- and inter-reader agreement among the residents was relatively poor, and could mostly be attributed to chance.

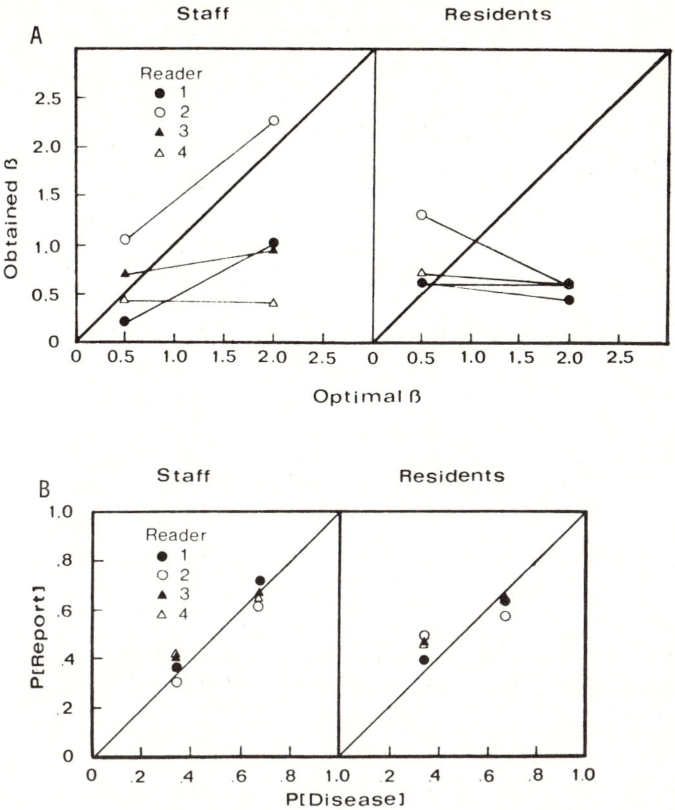

FIGURE 1. (A) Observed and optimal values of β for individual staff and resident readers. (B) Probability matching functions for individual readers. P[Report] is the overall probability of both true and false reports.

Finally, Table 1 gives the mean disease identification values, P(C), and the mean values of the area under the joint detection and identification ROC, P(A). Values are shown for first choice diagnoses only (1) as well as for either first or second choices (2). Analysis of variance showed that the staff were more accurate in differential diagnosis when only first choices were scored ($F(1,6)=7.83$, $P < .05$). Staff and residents were not significantly different if both first and second choices were scored ($F(1,6)=0.00$). Disease prevalence had no influence on accuracy in either case.

DISCUSSION

The finding that the radiology residents did not adjust their reporting criteria appropriately in response to a change in disease prevalence has implications for training. Radiology residents are usually the first readers of a chest radiograph, often under conditions of severe time pressure when on the "firing line" in an emergency ward. These initial reports are verified by a staff radiologist, who may have to make a significant number of corrections. Rhea et al. (1979) carried out a quality audit of the initial reports made in an emergency ward, and found that over a 20-day period 326 changes in the residents' reports out of a total of 3,300 reports had to be made. These figures point to the need for improved training programs to enhance the performance of radiology residents. The present results suggest that programs shouldinclude training to adopt appropriate reporting criteria and maintain them consistently over time, and training to shift criteria appropriately in response to expected changes in disease prevalence. These two requirements may pose conflicting demands on readers; nevertheless, it is important that all readers, regardless of skill, are aware of the effects of variations in reporting criteria on performance. Garland (1949) first showed that a reader's "decision attitude" affects diagnostic performance. He asked radiologists to read a group of chest films using either a liberal or a conservative criterion, and found that a liberal reporting policy increased the number of true-positives while also increasing the number of false-positives. The extent to which readers' criteria fluctuate over time is not known, but it seems clear that radiology residents, as well as more experienced readers would benefit from feedback regarding individual and appropriate criterion levels, especially as new imaging techniques are introduced (Cuaron et al., 1980). This may also be effective reducing the inter-reader variation in performance found in this and in previous studies (Fletcher & Oldham, 1949; Slovic et al., 1971).

	Staff					Residents			
	1	2	3	4		1	2	3	4
1	.44	.34	.42	.19	1	.20	.18	.06	-.06
2		.40	.46	.30	2		.01	-.27	.07
3			.59	.29	3			.27	.30
4				.33	4				.09

TABLE 2. Intra-reader and inter-reader agreement (kappa) in reports of disease presence and absence (collapsed over confidence ratings and over disease prevalence conditions) for staff and resident readers. Intra-reader (diagonal entries) and inter-reader (off-diagonal entries) values of kappa are based on N's of 180 and 360 respectively.

The expected changes in β with an increase in disease prevalence were not found in the present study, although the staff radiologists' criteria did change in the predicted direction. Reporting criteria in the high disease prevalence condition were not as stringent as the optimal criterion. It is possible to explain this type of "conservatism," which is quite common in psychophysical studies (Green & Swets, 1966), if one assumes that costs and benefits vary with disease prevalence. As Craig (1976) has pointed out, however, such post-hoc suppositions are indicative of a predictive flexibility that is both a strength and a weakness of the expected-value model of decision making. One possibility for further research would be to examine the effects of an explicit cost-benefit matrix (cf. Mobley & Goldstein, 1978). The alternative model suggested by Craig (1976), that observers use a probability matching rule, is consistent with the reporting behavior of the staff, but less so of the residents. Since readers were informed of the disease prevalence, the tendency to match reporting rates to expected disease rates is perhaps not surprising. Given that such information is often not available in practice, it would be interesting to examine whether skilled and unskilled readers adjust their reporting rates to maximise expected value, match expected disease rate, or use some other response strategy when the disease prevalence is unknown.

The staff radiologists were superior to the radiology residents in disease detection for both conditions of disease prevalence. Further, the staff were more accurate in disease identification, if, as is usually the case, only one choice of diagnosis was called for. An unexpected result, however, was that residents were no worse than their more experienced colleagues if second choices were also scored. Second choices were correct more often than chance and thus carried significantly more information that pure guesses. These results could possibly be used in enhancing the diagnostic performance of residents. The performance gain noted here may not, however, generalize to other imaging techniques such as computed tomography, which are generally more informative as far as differential diagnosis is concerned. Nevertheless, allowing readers to report at different levels of confidence may be a useful technique for reducing error in all types of medical imaging.

REFERENCES

[1] Andersson, I., Andren, L., Hildell, J., et al. Breast cancer screening with mammography. Radiology 132 (1979) 273-276.
[2] Berlin, L. Does the "missed" radiographic diagnosis constitute malpractice? Radiology 123 (1977) 523-527.
[3] Berlin, L. Malpractice and radiologists. Amer. Jrnl. Roentgen. 13 (1980) 587-591.
[4] Boyington, P. Perception. Radiography 46 (1980) 11-13.
[5] Cohen, J. A coefficient of agreement for nominal scales. Educ. Psychol. Meas. 20 (1960) 37-46.
[6] Craig, A. Signal recognition and the probabillity-matching rule. Percept. Psychophys. 20 (1976) 157-162.
[7] Cuaron, A., Acero, A.P., Cardenas, M., et al. Interobserver variability in the interpretation of myocardial images with Tc-99m-labeled disphosphonate and pyrophosponate. Jrnl. Nuc. Med. 21 (1980) 1-9.
[8] Fletcher, C.M. & Oldham, P.D. The problem of consistent radiological diagnosis in coal miners' pneumoconiosis. Brit. Jrnl. Industr. Med. 6 (1949) 168-183.
[9] Garland, L.H. On the scientific evaluation of diagnostic procedures. Radiology 52 (1949) 309-327.
[10] Green, D.M. & Birdsall, T.G. Detection and recognition. Psychol. Rev. 85 (1978) 192-206.
[11] Green, D.M. & Swets, J.A. Signal Detection Theory and Psychophysics. (Wiley, New York, 1966).
[12] Grey, D.R. & Morgan, B.J.T. Some aspects of ROC curve fitting: Normal and logistic models. Jrnl. Math. Psychol. 9 (1972) 128-139.
[13] Herman, P.G. & Hessel, S.J. Accuracy and its relationship to experience in the interpretation of chest radiographs. Invest. Rad. 10 (1975) 62-67.
[14] Liddell, F.D.K. & McDonald, J.C. Radiological findings as predictors of mortality in Quebec asbestos workers. Brit. Jrnl. Industr. Med. 37 (1980) 257-267.
[15] Lusted, L.B. Signal detectability and medical decision-making. Science

171 (1971) 1217-1219.
[16] Metz, C.E. Basic principles of ROC analysis. Sem. Nuc. Med. 8 (1978) 283-298.
[17] Mobley, W.H. & Goldstein, I.L. The effects of payoffs on the visual processing of dental radiographs. Hum. Factors 20 (1978) 385-390.
[18] Morgan, R.H., Donner, M.W., Gayler, B.W., et al. Decision processes and observer error in the diagnosis of pneumoconiosis by chest radiography. Amer. Jrnl. Roentgen. 117 (1973) 757-764.
[19] Rhea, J.T., Potsaid, M.S., & DeLuca, S.A. Errors of interpretation as elicited by a quality audit of an emergency radiology facility. Radiology 132 (1979) 277-280.
[20] Simpson, A.J. & Fitter, M.J. What is the best measure of detectability? Psychol. Bull. 80 (1973) 481-488.
[21] Slovic, P., Rorer, L.G., & Hoffman, P.J. Analysing use of diagnostic signs. Invest. Rad. 6 (1971) 18-26.
[22] Starr, S.J., Metz, C.E., Lusted, L.B., & Goodenough, D.J. Visual detection and localization of radiographic images. Radiology 116 (1975) 533-538.
[23] Swets, J.A. ROC analysis applied to the evaluation of medical imaging techniques. Invest. Rad. 14 (1979) 109-121.
[24] Swets, J.A. & Pickett, R.M. Evaluation of Diagnostic Systems. (Academic, New York, 1982).
[25] Todd-Pokropek, A. Image processing in nuclear medicine. IEEE Trans. Nuc. Sci. NS-27 (1980) 1080-1094.
[26] Tuddenham, W.J. Visual search, image organization, and reader error in roentgen diagnosis. Radiology 78 (1962) 694-704.

Trends in Ergonomics/Human Factors II
R.E. Eberts and C.G. Eberts (Editors)

APPLICATION OF FEATURE ANALYSIS TO VISUAL SEARCH TIME OF INSPECTION TASKS

Stuart Martin Weiss, James W. Barany, and James G. Sturdevant

School of Industrial Engineering
Purdue University
West Lafayette, Indiana 47907 U.S.A.

AT&T Telecommunications

Transforming the concepts of pattern recognition to an individual inspection task would indicate that an inspector recognizes a general set of attributes (or visual features) which implies the presence of a defect as opposed to having a unique mental representation (an exact matching mental template) for recognizing each of type defect, or different levels of a given type of defect.

INTRODUCTION

Two current theoretical explanations of psychological pattern recognition are those that specify some form of feature analyzing mechanism within the pattern recognition process and those that involve the concept of template matching. Most current psychological models employ the feature analysis approach (Forgus & Melamed, 1976) because:

1) research in sensory neurophysiology (e.g., Hubel & Wiesel, 1959) has demonstrated that there are feature analyzing cells in the cortical and sub-cortical visual areas in cats and other mammals; and

2) there are serious difficulties inherent in the template matching approach.

The template matching approach postulates that a visual pattern is recognized by matching it with a stored (memorized) template or general model at some cognitive information processing stage. Problems become apparent, however, when one considers that the pattern to be recognized can assume several variations. For example, the pattern may be of any size; may be placed at any retinal location; may be rotated to any orientation; or be distorted if the observer is subject to astigmatisms, cataracts, or other visual aberrations. How could a sufficient number of templates be stored in memory to match all of the possible variations of the myriad of patterns that an individual might encounter? How could a particular template be selected for use? It would be necessary to postulate some mechanism that selectively changes the stored template to sequentially match them to some input for recognition. According to Forgus & Melamed (1976), parallel application of a class of templates is unlikely because it would be prohibitively complex. The feature analysis approach postulates that collections of feature or pattern elements that uniquely identify a pattern, such as a letter, are stored in memory. It is assumed that at the retinal (or some other early) level of information processing, the visual system is sensitive to a number of these elements, and this information would be evaluated at some higher level of processing. Here, the information concerning the frequency and identity of the perceived pattern elements would be used to estimate the similarity of the perceived pattern to the members of a repetoire of pattern descriptions. This approach is more efficient than template matching because it assumes that more patterns are associated with unique features that can be analyzed regardless of pattern size, retinal location, or orientation.

Neisser (1966) stated that the ability to recognize meaningful patterns involves the synthesis of separable attributes and characteristics (or visual features) that describe a given figure. These attributes may be measurable (objective - such as the length of lines or the particular angles between lines which form the figure), or they may be non-measurable (subjective - such as appearance or color where the attributes are; too much, too little, too light, or too dark). In a series of experiments, which were designed to study the cognitive processes involved in visual search, Neisser (1964) timed subjects as they scanned lists of fifty letters in search of specific targets. He found that the time required to detect a target was an increasing linear function of its location within the list. Over the course of several days in the experiments, the speed with which the subjects scanned increased dramatically. This finding implies that, with practice, subjects develop efficient search strategies. Neisser (1964) concluded that subjects "may find techniques that enable them to examine several lines at once, whereas at first they had fixated each item separately and successively."

The context of the task also affects the speed of visual search. For example, the letter "Z" is detected more easily against a background of letters containing round features (e.g., DGOQRU) than in a background of letters with angular features.(e.g., EIMWVX) Subjects were found by Neisser (1964) to detect the "Z" twice as fast in the round background than in the angular background. According to Neisser (1964), the subject does not need to identify each letter in order to determine if it is a "Z". In the analysis of each line, the subject needs only to detect some of the unique features that identify the particular letters. However, when the defect and background contain similar visual features, the subject must search through more features to distinguish a "Z" from a background of angular letters (which may resemble a "Z") than a background of round letters.

The present application of the concepts of psychological pattern recognition to an industrial inspection tasks assumes that an inspector recognizes a set of attributes (or visual features) that indicates the presence of a defect, rather than maintaining a unique mental representation (referred to in the psychological literature as an exact matching mental template) for recognizing either each type of defect or the different levels of a given type of defect. Several specific predictions concerning the behavior of inspectors follow if, as assumed above, defect recognition involves the detection of visual features such as lines, angles, and curves. Suppose there is, for example, a self-terminating inspection task where a part is immediately rejected when an error is detected. When the inspector is presented with a part that has similar inspection points (e.g., soldered printed wiring areas or terminals on circuit boards), each composed of similar visual features, the inspector will have greater difficulty distinguishing between them than when the inspection points have only a few visual features in common. Here, an inspection point is defined in terms of a set of visual features that represent a recognizable pattern which is characteristic of either a defect or non-defect. The time required to detect a defect should increase when the inspection points are similar because more features of each inspection point have to be identified (checked) before the distinguishing features of a defect (or defect pattern) are detected. If no error is found, the search will continue exhaustively until all inspection points of the part are scanned, or an error is detected. Therefore, a reasonable expectation is that it will take longer to search a part that has no error because every visual feature of each inspection point must be identified before passing a part as acceptable.

Another aspect involves the frequency of errors that can be made in the inspection task. An error may occur when the inspector:

1) acknowledge the presence of a defect when no defect exists; or

2) fails to detect the presence of an existing defect.

The former is termed a type I error, while the latter is termed a type II error. When an inspector fails to detect the presence of the defect, a defect pattern composed of features similar to those of a non-defective item is probable. Thus, the inspector fails to recognize the distinguishing feature of the defect (or defect pattern) because the similarity of defects and non-defects produces difficult conditions for defect discrimination.

Regardless of whether or not a defect is present, the nature of search itself depends, at least to some extent, upon the physical characteristics of the background of visual elements containing no defects. When the background is composed of complex visual elements (i.e., each element is defined by a large number of visual features), the search will be slower relative to the search that is employed for a simple background. Thus reaction times for the detection of a defect against a complex background will be large, relative to those observed when the background is simple.

The present research concerns the nature of the search for a defect and how search is affected by the visual context in which it is embedded. The approach taken was to:

1) Analyze separately the affect of defect conspicuity on the time required for defect detection and visual search;

2) Determine the affect of defect conspicuity on inspection accuracy; and

3) Study the affect of defect position on the time required for defect detection.

METHODOLOGY

Subjects. Five undergraduate engineering students and five industrial inspectors volunteered to serve as subjects. All of the subjects had normal (20/20) or corrected to normal vision. None of the student subjects has previous inspection experience. The inspectors were compensated by their employer for their participation in this experiment. They had from two to five years of experience with visual search tasks in the inspection of solder connections on complex circuit boards.

Apparatus. Stimulus image generation and data recordings were performed by an IBM Model 5150 Personal Computer. A graphics monitor (Amdek-Color II, screen size 10.50 x 7.50 inches), located at eye level and at a viewing distance of approximately 25 inches, was used for the presentation of stimuli. Subject responses were recorded by means of a light pen (FTG Data Systems, Model FT-156). During each experimental session, the illumination in the laboratory was reduced to prevent glare on the screen surface of the graphics monitor.

Task. The experimental task was to inspect a representation of a printed circuit board and to indicate whether the board was acceptable or defective. The "board" pattern was formed by a 20 x 20 matrix of yellow letters superimposed upon a green field.

Procedure. Prior to the experimental session, all subjects were informed that they were going to participate in an experiment involving the measurement of the "types of demands" placed on individuals, such as visual inspectors, who perform complex visual search tasks. Subjects were told that during the experiment they would be asked to determine whether or not there were any defects present in a series of parts that would be displayed by a computer. In addition, each subject was given instructions concerning the nature of the inspection task and the operation of the light pen.

Subjects were told that the purpose of the study was to determine how accurately and how quickly they could perform the inspection task. Each subject was given ten practice trials to become familiar with the apparatus and task requirements. Student subjects were not provided with any instructions concerning how to "scan" the part for defects. Inspectors were asked to employ the same scanning pattrns that they would perform with real circuit boards.

At the beginning of a trial, the phrase "READY-LOCATE PEN TO HOME POSITION" appeared at the top of the screen. The subject initiated a trial by touching the pen to a target square located below this phrase. Following this signal, the computer randomly selected a part configuration and constructed the appropriate stimulus matrix. As each matrix was displayed to the subject, he/she scanned the matrix to see if a defect was present. If a defect was found, the subject pointed to the error using the light pen. If no defect was found, the subject was instructed to point the light pen to a zone outside the matrix labelled "PASS." When the subject failed to correctly identify a defective part, an error signal in the form of a tone was sounded for a period of five seconds and the defect blinked on the screen. Following the entry of the subject's response, the "READY" message was returned to the screen in preparation for the next trial.

Performance Measures. On each trial, the subject's response (screen location indicated by light pen), inspection time (measured from stimulus onset to response), and stimulus information (stimulus identification number, position in the stimulus presentation sequence, letters that defined the stimulus, and location of a defect when one was present) were recorded. These data were utilized to determine the mean and standard deviation of the time required for the pass/reject decision and the proportion of items correctly identified as acceptable and defective.

Stimulus Materials. Four different types of "circuit boards" were constructed in which the context letters used to form the matrix contained either "round" or "angular" visual features, and the letter "Z" was the target (defect) letter. In condition one, the context letters, used to construct the matrix, contained many visual features that were not found in the target letter. In order words, the contrast between the defect and the background was relatively high. Conversely, in condition two, the context letters used to construct the matrix contained many visual features that were similar to those found in the target letter. In this case, the contrast between the defect and the background was relatively low. Finally, in conditions three and four, each "board" type (previously discussed) were present with no error embedded in the matrix.

The four computer generated part configurations were each individually displayed in a random order sixty (60) times during the experiment. For each of the part configurations (either easy or difficult) that contained a defect, sixty trials were presented in which the defect appeared five times in one of twelve possible matrix positions. An additional sixty trials were presented for each part configuration in which the defect position was randomly selected. These presentations served to prevent subjects from associating particular positions with a higher frequency of defect occurrence and were not used in any of the analyses. Computer programs used in this experiment were translated into assembly language using the IBM Basic Compiler. The interstimulus delay was four seconds. The experimental session lasted approximately two and a half hours. A ten-minute rest period was provided for each subject after completion of 180 of the 360 trials.

RESULTS

Analysis of Defect Detection Time and Detetion Errors. The mean values of defect detection time for easy and difficult part configurations were, respectiely, 6.030 and 8.799 seconds for student subjects, and 5.211 and 7.852 seconds for inspectors. These values are plotted in Figure 1. The ratio of the mean defect detection time with a difficult part configuration to the mean defect detection time with an easy part configuration was 1.462 and 1.507, respectively, for student subjects and inspectors.

A three-way (Part Configuration x Defect Position x Subjects) analysis of variance of defect detection time for student subjects on the simulated visual inspection task indicated that the main effects of part configuration, defect position, and subjects were significant. $F(1,4) = 6.59$, $p<.10$; and $F(11,44) = 2.37$, $p<.05$; and $F(4,451) = 71.10$, $p<.001$ respectively. The analysis of variance of defect detection time for inspectors indicated that all main effects were significant: part configuration, $F(1,4) = 129.49$, $p<.001$; defect position $F(11,44) = 2.32$, $p<.05$; subjects, $F(4,449) = 11.65$, $p<.001$.

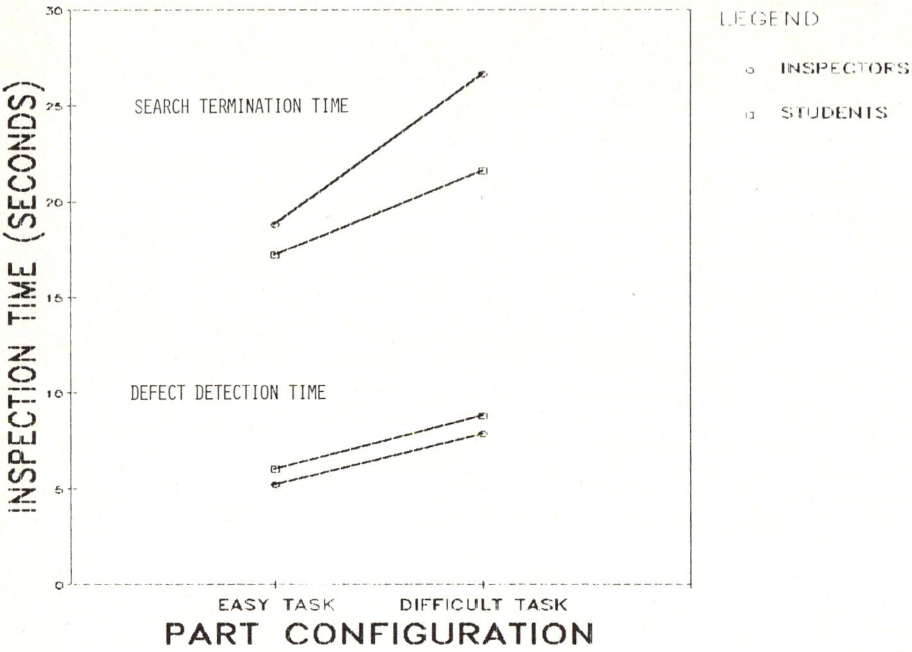

FIGURE 1
Mean Search Termination Time and Defect Detection Time Per Part Configuration For Student Subjects and Inspector

The student subjects incorrectly identified 31 of the 600 matrices that actually contained defects as being "defect-free." The breakdown of errors for easy and difficult part configurations were six (or 1.00 percent) and twenty-five (or 4.17 percent), respectively, for an average performance error of 5.17 percent. Thus, the students were able to perform at an average accuracy level of 94.83 percent. The inspectors incurrently identified 29 of the 600 matrices which actually contained defects as being "defect-free." The breakdown of errors for easy and difficult part configurations were seven (or 1.17 percent) and twenty-two (3.67 percent), respectively, for an average performance level of 4.84 percent. Thus, the inspectors were able to perform at an average accuracy level of 95.16 percent.

Analysis of Search Termination. The mean values of search termination time (accept a part as "defect free") for easy and difficult part configurations were, respectively, 17.205 and 21.603 seconds for student subjects, and 18.802 and 26.649 seconds for inspectors. These values are plotted in Figure 1 (shown previously). The ratio of the mean search termination time with a difficult part configuration to the mean search termination time with an easy part configuration was 1.256 and 1.417, respectively, for student subjects and inspectors.

A two-way (Part Configuration x Subjects) analysis of variance of search termination time from student subjects indicated than the main effects of part configuration and subjects were significant, F (1,4) = 17.41, $p<.05$; and F (4,590) = 77.46, $p<.001$, respectively. The analysis of variance of search search termination time for inspectors indicated that all main effects were significant: part configuration, F (1,4) = 35.56, $p<.01$; subjects, F (4,590) = 133.52, $p<.001$.

Discussion. The inspection time elapsed before defect detection on a difficult task was determined to be 46.2 percent greater for student subjects and 50.7 percent greater for inspectors than the inspection time elapsed before defect detection in an easy task. Similarly, the inspection time elapsed before search termination on a difficult task was determined to be 25.6 percent greater for student subjects and 41.7 percent greater for inspectors than the inspection time elapsed before search termination on an easy task. This suggests that subtle or obscure defects require additional time for visual search and detection than defects of a gross nature. In addition, inspectors have a greater chance of failing to detect a defect when the visual features contained in the defect are masked by similar acceptable features contained in the other inspection points of the part. These results indicate that defect conspicuity has a significant negative effect on the accuracy of visual inspection, which apparently can not be overcome by providing inspectors with an unlimited amount of inspection time. Inspection accuracy deteriorates as the defects become less obvious or conspicuous. The data also support a finding by Morawski (as cited by Drury, 1982) that the optimum time per item for a search task ranges from two to three times the mean time needed to locate a single defect.

The main effect of defect position and the defect position x subject interaction term accounted for approximately seven percent of the total variation in inspection time for the student population and approximately nine percent of the total variation for the inspector population. In the context of the application of the concepts of psychological pattern recognition to an inspection task, it is apparent that the time required to detect a defect should depend upon the number of defect-free inspection points that must be searched before a point containing a defect is encountered. In the present experiment, a stimulus consisted of a two dimensional matrix in which a defect was embedded (i.e., the defect occurred in a particular cell). Several researchers (Heron, 1957; Harcum & Dyer, 1962; Harcum, 1967) have shown that subjects typically search stimulus arrays, such as those that were employed here, in a top-to-bottom, left-to-right manner. If it is assumed that the subjects in the present experiment typically employed such systematic search behavior and that defect detection time is related to the number of defect-free inspection points preceding a defect, then it is reasonable to expect that defect detection time should be an increasing function of the position of the cell in which the defect occurs.

CONCLUSIONS

Frequent visual inspection is required during many manufacturing operations to maintain quality assurance. Reducing the amount of time allowed for visual search as product quality improves is common practice. This philosophy makes the assumption that the defects that remain at later manufacturing steps will be more obvious and therefore easier to detect.

The present study concerning an investigation of human performance on visual inspection tasks has shown that defect types differ with regard to their complexity and the amount of time required for detection.

The results of this research suggest that the time required for the cognitive components of inspection of an item will be a function of the following:

1) Presence or Absence of a Defect; visual fields that do not contain a defect require more time for visual search,

2) Defect Type; the contrast between the visual features of a defect and its backround, and

3) Defect Position; the time required to detect a defect is a function of the number of elements containing no defects that must be scanned prior to the occurrence of an element containing a defect.

In addition, the influence of multiple defects on the time required for inspection needs to be ascertained in future research.

Therefore, the results of these experiments indicate that even in relatively simple tasks, the visual search and decision making processes utilized by the inspector are subject to appreciable error. Thus, gaining a better understanding of the dynamics of the inspection process and visual search behavior of an inspector may serve as the basis for equating the difficulty (in terms of the under mental processes) of different inspection tasks, determining the time required for these tasks, determining the time required for these tasks, and developing a method for training inspectors.

REFERENCES

[1] Drury, C. G., Improving Inspection Performance, in: Salvendy, G. (ed.), Handbook of Industrial Engineering (John Wiley & Sons, New York, 1982).

[2] Forgus, R. and Melamed, L., Perception: A Cognitive Stage Approach (McGraw-Hill, New York, 1976).

[3] Harcum, E. R. and Dyer, D. W., Monucular and Binocular Reproduction of Binary Stimuli Appearing Right and Left of Fixation, American Journal of Psychology, 75 (1962) 56-65.

[4] Harcum, E. R., Two Possible Mechanisms of Differential Set in Tachistoscopic Perception of Multiple Targets, Perceptual and Motor Skills, 25 (1967) 289-304.

[5] Heron, W., Perception as a Function of Retinal Locus and Attention American Journal of Psychology, 70 (1957) 38-48.

[6] Hubel, D. and Wiesel, T., Receptive Fields of Single Neurons in the Cat's Striate Cortex, Journal of Physiology, 148 (1959) 574-591.

[7] Neisser, V., Cognitive Psychology (Appleton-Century Croft, New York, 1966).

[8] Neisser, V., Visual Search, Scientific American, 210 (1964) 94-103.

Trends in Ergonomics/Human Factors II
R.E. Eberts and C.G. Eberts (Editors)

A Contingent Criterion Model of Computer Assisted Detection

Donald E. Robinson
Department of Psychology
Indiana University
Bloomington, Indiana

Robert D. Sorkin
Department of Psychological Sciences
Purdue University
West Lafayette, Indiana

In this paper we consider a person-machine system consisting of an automated alarm system and a human monitor. A Contingent Criterion Model of such a system is developed using basic concepts of statistical decision theory. According to the model, the human should establish two criteria for responding: one contingent on an alarm from the automated detector and one, on no-alarm. The model shows large gains in performance compared to either detector alone. The degree to which human subjects perform in the manner suggested by the model is evaluated in two experiments.

INTRODUCTION

Sorkin and Woods [1] have described an alerted monitor system having two components: an automated alerting subsystem and a human operator. The automated system continuously monitors ongoing processes and makes decisions about system status. Occasionally, the automated system will alert the human operator to some special condition. When alerted, the human operator may analyze the available data, confirm or disconfirm the actions of the automated system, and, perhaps, take further action. Air traffic control centers and the flight-decks of commercial aircraft are examples of alerted monitor systems.

The ability to predict and control operator performance in alerted monitor systems is important. Problems are known to exist in systems of this type, e.g. aircraft flight-decks [2] and nuclear power plant control rooms [3]. However, there has been very little theoretical or experimental work relevant to the analysis of combined (human plus automated alarm) detection systems. Two exceptions are an early paper of Pollack and Madans [4] and the recent work of Sorkin and Woods. There are currently no general principles or guidelines for the design of alerted monitor systems or for the setting of subsystem parameters. A general model of the alerted monitor system and determination of typical operator behavior would be an important step toward the improved design of such systems.

In this paper we briefly review a general approach to the evaluation of detection systems, present a model of the performance of human operators responding to alarms, and evaluate the success of the model in describing the performance of two different such person-machine systems.

EVALUATION OF COMBINED DETECTION SYSTEMS. The Theory of Signal Detectability (TSD) provides the means for evaluating the performance of detection systems. This theory is well known [5, 6], and we will assume that the reader is familiar with it. The TSD postulates that a detector makes a measurement, which may be described as a multi-dimensional vector, from which is computed a uni-dimensional statistic, x. Due to the presence of noise (N), x has associated with it a probability density, f(x), which is usually assumed to be the normal density function. There are two densities: f(x|N) and f(x|SN), depending on whether the signal (S) was present or not. The detector makes a decision about the presence or absence of the signal by comparing the value of x which arises from a particular observation to a criterion value of x, C. The detector says "yes" if x is greater than C, "no" otherwise. For an ideal detector, C is a function of the a priori probability that the signal will occur as well as of the values and costs associated with the possible outcomes of the decision: 'hit,' 'miss,' 'false alarm,' or 'correct rejection.' Another way in which the criterion is specified is in terms of Beta (B), the ratio of the ordinates at C, e.g. f(xc|SN)/f(xc|N). The performance of the detector will depend on the amount of overlap of the N and SN distributions as well as on the placement of the criterion. When the distributions are normal, the normalized distance between the means (d') is frequently chosen as a measure of detectability.

THE CONTINGENT CRITERION MODEL. The detectabilities of the human detector and automated detector are defined as d'h and d'a, respectively. If these detectors make independent observations on the same noisy channel, the optimal combination of the observations would result in a system detectability equal to d'opt, the square root of the sum of squares of the d's of the two detectors. In the case of an automated alarm system monitored by a human detector, the observation made by the automated detector is not available to the human: only a binary response (alarm or no alarm) is provided. In this case, the performance of the combined detectors ,d'ah, cannot reach d'opt. How should the human detector operate to maximize system performance when only the binary response of the alarm system is available? For a single detector, the value of B which maximizes the expected value is a function of the a priori probabilities, as well as of the values and costs. The a priori probabilities of 'signal' and 'no signal' may be altered by the human's knowledge of the first detector's response. If the second detector has knowledge of the original a priori probabilities and of the 'hit' and 'false alarm' rates of the first detector, the a posteriori probability of the presence of the signal may be computed using Bayes' theorem. There will be two estimates of the a posteriori probability: one conditional upon the first detector responding 'yes' and one upon the first detector responding 'no.' These two estimates of the a posteriori probabilities may then used as a priori probabilities to establish two values of B, By and Bn, as well as their associated criteria, Cy and Cn. For the second detector to establish the two criteria optimally, the second detector must have knowledge of the 'hit' and 'false alarm' proportions of the first detector, of the relevant values and cost, and of its own sensitivity.

Since the second detector uses two criteria, Cy and Cn, there will be two pairs of hit and false alarm probabilities: [Py(y|SN), Py(y|N)] and [Pn(y|SN), Pn(y|N)]. These four probabilities, along with the hit and false alarm probabilities of the first detector, Pa(y|SN) and Pa(y|N), may be used to obtain the hit and false alarm probabilities of the combined detection system:

$$Pah(y|SN) = Pa(y|SN)[Py(y|SN) - Pn(y|SN)] + Pn(y|SN) \quad \text{Eq. 1}$$

$$Pah(y|N) = Pa(y|N)[Py(y|N) - Pn(y|N)] + Pn(y|N) \quad \text{Eq. 2}$$

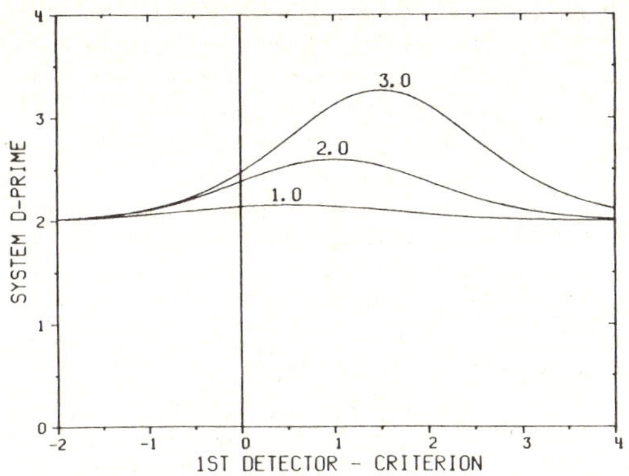

Fig. 1 d'ah as a function of Ca. d'h=2.0. Parameter is d'a.

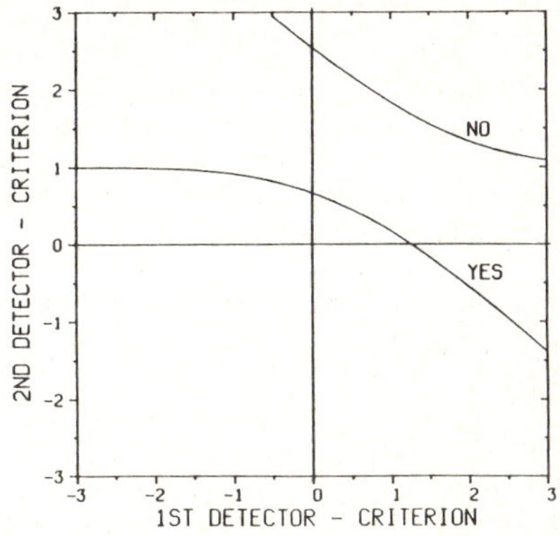

Fig. 2 Cn and Cy as a function of Ca. Predictions of the model.

The Contingent Criterion (CC) Model, as defined by the Equations 1 and 2, predicts that a combined detection system can achieve large improvements compared to either detector alone. For example, when Ba=1.0 and d'a=d'h=1.0, d'opt is 1.41, while the system d', d'ah, is 1.30. Thus, the CC model achieves an efficiency of 0.92; if d'a=d'h=2.0, d'opt=2.83 while d'ah=2.60: an efficiency of 0.92. If the two detectors are not equally sensitive, the efficiency is still high, but the order is important. If d'a=2.0 and d'h=1.0, d'ah =2.05 while d'opt=2.24. If the order is reversed, (e.g. d'a=1.0 and d'h=2.0) d'opt is unchanged, but d'ah increases to 2.16: an increase in efficiency of 4% (0.92 to 0.96). If Ba is not equal to 1.0, smaller gains result. Fig. 1 makes it clear that the model provides efficient use of the available information. This figure shows how performance of the system varies as a function of the criterion of the first detector (Ca) for three values of d'a: 1.0, 2.0, and 3.0. For all three curves, d'h=2.0.

The two criteria of the second detector, Cy and Cn, will vary with changes in Ca. Fig. 2 show how this occurs when d'a=d'h=2.0. Note that, except for extreme values of Pa(y|N), the difference between Cy and Cn changes only slightly with changes in Ca. The general shape of the two functions in Fig. 1 is changed little if the two detectors are not equally sensitive. However, as the automated detector becomes more sensitive relative to the human, the separation between the two criteria increases. This is consistent with the intuitively reasonable notion that as the sensitivity of the automated detector increases relative to that of the human, the human should rely more and more on the decision made by the alarm system. The converse is true if d'h>d'a. As the human detector becomes more sensitive relative to the automated one, the two criteria move closer together.

The Contingent Criterion Model is a model for the optimum performance of a combined detection system. That is, the model leads to the specification of the criteria for the second detector that will maximize the expected value for the system. Although it is not expected that human operators will function optimally, the model may be used to describe and evaluate the performance of the human.

The experiments reported here are an initial effort toward an evaluation of the CC Model. Neither experiment attempts to simulate a realistic work-place environment. The two experiments differ in the tasks required and in the sense modality through which information is conveyed.

EXPERIMENT I

In this experiment, subjects were required to detect a tonal signal which was occasionally presented during a brief noise-burst. In some blocks of trials, a simulated alarm system was present.

METHOD. Four undergraduate students were paid for their participation in the experiment. Each subject received several thousand practice trials before any of the data reported here were collected. Signal and noise stimuli were generated by a computer and reproduced through 14-bit digital-to-analog converters at a rate of 20,000 samples per second. The noise channel was band-pass filtered from 100 to 3000 Hz. The signal and noise waveforms were then added and presented to the subjects through headphones. Within the 100- to 3000-Hz passband, the long term power spectrum of the noise was white, and the distribution of instantaneous pressures was approximately Gaussian. The average spectrum level of the noise (No) was 50 dB SPL. The duration of each noise burst was 148-ms. The signal was a 108-ms sample of a 500-Hz sinusoid, centered in the noise burst.

The data were collected using a single-interval, four category rating procedure. Indicator lights on the subject's console marked the beginning of each trial, the observation period, and provided feedback. The signal was presented on 50 percent of the trials. The subjects were instructed to use one of four response buttons to indicate their level of confidence that a signal had been presented on a particular trial. Trial-by-trial feedback was provided to the subjects at the end of each trial.

In each two-hour experimental session, eight blocks of 100 trials were presented. On one-half of the blocks during each daily session, the subjects were unaided by the automated detector. On the other one-half, the subjects were told that an automated detector was operating and that the detector would indicate its decision by illuminating one of two lights on the subjects' console. These lights were turned on at the beginning of the trial and turned off at the end of the feedback period.

The value of d' for the automated detector was selected to match, as nearly as possible, that of the listeners. The chosen value of d'a was 1.0. It was anticipated that a value of 10 log(E/No) of 8.0 would result in d'h of approximately 1.0. Three alarm criteria for the automated detector were selected to yield probabilities of the "yes" response from the automated detector, Pa(y), of 0.25, 0.50, and 0.75. The values of d'a actually obtained were 0.91, 0.95 and 0.95 with associated values of Ca of -0.24, 0.48, and 1.21, respectively.

The automated detector was defined on a statistical basis only. That is, when d'a=1.0 and Ca=0.50 (Ba=1.0), for example, the automated detector said "yes" on 0.69 percent of the signal-plus-noise trials and on 0.31 percent of the noise-alone trials independent of the waveform presented on each trial. This is equivalent to a zero correlation between the noise samples presented to the listeners and those presented to the automated detector.

RESULTS. The data from the four-category rating task result in three estimates of d'. For the analyses which follow, the average of these three estimates is used. Since the CC Model maximizes P(C), not d', it will be desirable to describe the data in terms of P(C). Values of d' have been converted to P(C)max [6]. These values of P(C) will be referred to as Ph(C) and Pah(C) for the human-alone and the human-plus-alarm, respectively. Figure 3 shows the values of Pa(C), Ph(C) and Pah(C) for one subject at each of the three values of Ca. It can be seen that the combined detectors perform better than either of the two alone. The data for this subject are representative of those for the other subjects. The average values of Ph(C) for the four subjects, over the three criteria, are: 0.62, 0.65, 0.65, and 0.73. The average values of Pah(C) are 0.67, 0.68, 0.69, and 0.77. A comparison of these values to those predicted by the model reveals that the CC Model does a good job of describing the performance of the combined detectors. The predicted values of Pah(C) are 0.69, 0.70, 0.70, and 0.76. Except for the middle criterion, Ca=0.48, the subjects perform slightly worse than does the model: the observed values of Pah(C) are approximately 97% of the predicted values. In the one exception, two of the four subjects perform better than does the model.

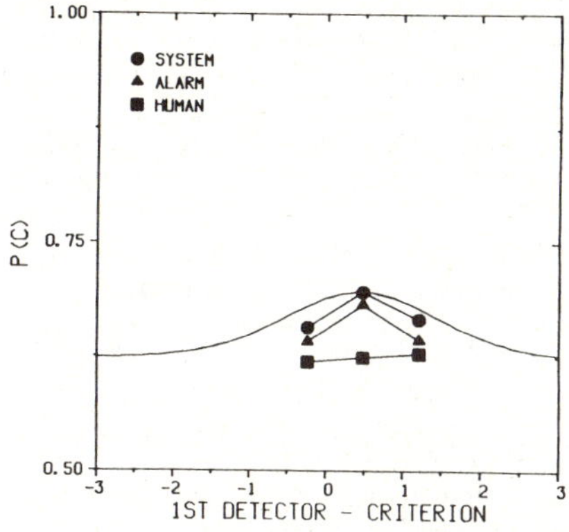

Fig. 3. P(C) as a function of Ca for Sub. 1. Smooth curve: model predictions.

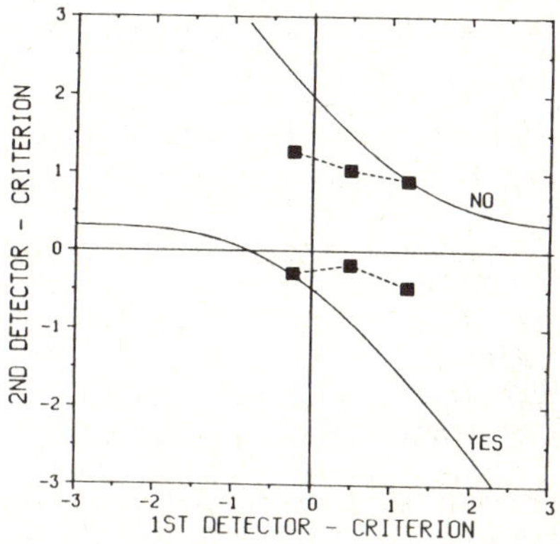

Fig. 4. Cy and Cn as a function of Ca. Smooth curves: model predictions. Data: Sub. 1.

The CC Model suggests that human subjects should use different criteria for making their decisions depending upon which response was made by the automated detector. Figure 4 shows the criteria used by one subject as well as those predicted by the model. Note that this subject established two criteria. The other three subject's data are similar to those shown. Over all criteria and subjects, the average separation between the criteria is 1.25 z-score units. Observe that the separation of the criteria used by this subject is less than that of the model. The average distance between the criteria predicted by the model is 2.13. Also observe that the change in the criteria of this subject with changes in Ca is less than that seen in the model. This is true of the data for the remaining three subjects.

EXPERIMENT II

In this experiment subjects performed two tasks concurrently: a scrolling letter detection task and a diagnostic decision task aided by a simulated automated detector. The stimuli for both tasks were presented on a video display terminal (VDT).

METHOD. The subjects were three undergraduate students who were paid for their participation at an hourly rate plus an incentive bonus contingent upon good performance. Each subject received three weeks of practice before the data reported here were collected.

The purpose of the scrolling detection task was to prevent the subjects from giving their full attention to the diagnostic decision task. The subjects were paid for correct responses and penalized for incorrect responses on both tasks. Since performance on the scrolling task was approximately constant over experimental conditions, data from the scrolling task will not be considered in this paper.

The diagnostic decision task consisted of an array of four, two-digit numbers (e.g. +1.2 -0.3 -0.6 -2.1) scrolling up the VDT. The numbers were generated from normal distributions having a unit variance and either a zero mean for a no-signal state or a mean of 1.0 for the signal state. (Note that, for an ideal observer, d' based on four, independent samples would be 2.0) The subject pressed a key on a response box if he or she believed that the number array indicated the presence of the signal, otherwise no response was made. If a response was made, a "D" appeared to the right of the numbers if that trial had actually been a signal (or dangerous) trial; otherwise, an "S" (for safe) was displayed. On some trials, a 55 dBA auditory alarm (500 Hz square wave) sounded simultaneously with the appearance of the number array on the VDT. This alarm indicated to the subject that the automated detector had sensed the presence of the signal. This alerting signal remained on until either the numeric array had scrolled off the screen or the subject responded. In order to determine when to activate the alarm, a fifth random normal deviate was generated (but not displayed) by the computer. This deviate was generated from a distribution having a unit standard deviation with a mean of zero on no-signal trials and a mean of 2.0 on signal trials. Thus, d'a was 2.0. If this fifth value exceeded a specified value, Ca, the alarm would sound. Three values of Ca were employed: 0.5, 1.0 and 1.5.

Each of the three values of Ca was factorially combined with three signal probability conditions [P(SN)=0.33, 0.50, and 0.67]. Points were added or subtracted for all task outcomes to determine the monetary payoff to the subject. Each value of P(SN) had a different payoff structure associated with it. In order to maximize their points, the subjects should have set conservative criteria when P(SN)=0.33; liberal criteria when P(SN)=0.67; and intermediate criteria when P(SN)=0.50. Each block of trials consisted of 100 lines of the number arrays. On a given day, each subject received only one value of Ca and all values of P(SN). The subjects had three sessions with each criterion condition before receiving the next condition in the sequence.

RESULTS. Although the criteria used by the subjects were influenced in the expected manner by the probability-payoff condition, this variable had no effect on the subjects' sensitivity (d'h) in the decision task. Thus, the data are collapsed over the various probability-payoff conditions. Figure 5 shows values of Pa(C), Ph(C) and Pah(C) for one subject at each of the three values of Ca. It can be seen that the combined detectors perform better than either of the two alone. The data for this subject are representative of those for the other subjects. A comparison of the mean values of P(C) over subjects within each condition reveals that in all cases, the performance of the human-plus-alarm is better than the alarm-alone or the human-alone. The average values of Ph(C) for the three subjects, over the three values of Ca, were 0.77, 0.80, and 0.77. The average values of Pah(C) were 0.85, 0.87, and 0.87. As was the case with Experiment I, the model describes the data well. The predicted values of Pah(C) are 0.86, 0.88, and 0.87.

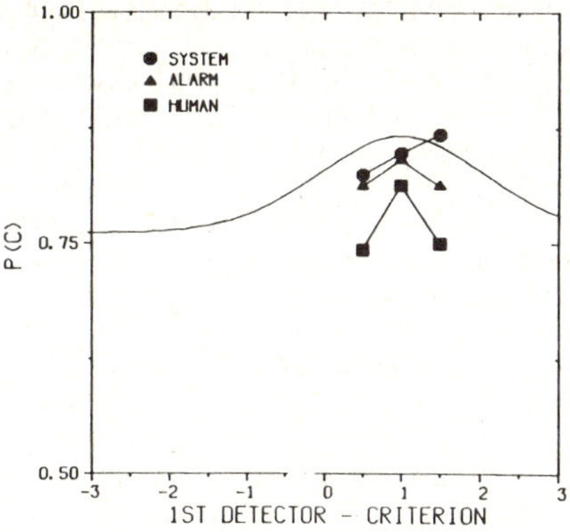

Fig. 5. P(C) as a function of Ca for Sub. 1. Smooth curve: model predictions.

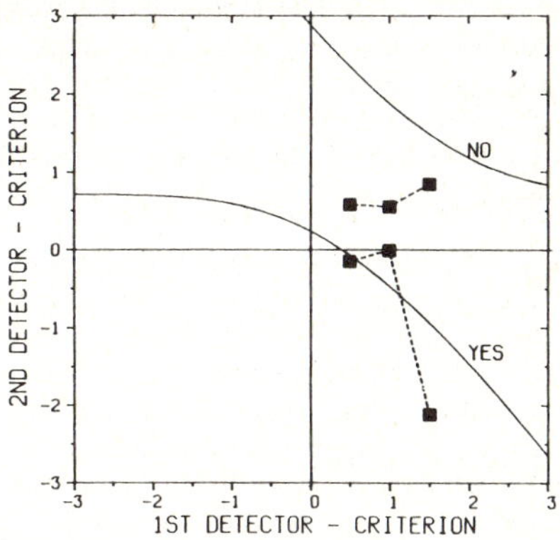

Fig. 6. Cy and Cn as a function of Ca. Smooth curves: model predictions. Data: Sub. 1.

Figure 6 shows the criteria used by one subject as well as those predicted by the model. First, observe that this subject clearly established two criteria. The other subjects' data are similar to the those shown. Over all criteria and subjects, the average separation between the criteria is 1.66 z-score units. Second, note that the separation of the criteria used by this subject is less than that of the model. The average distance between the criteria predicted by the model is 2.20. Thus, as was true for Experiment I, the subjects do not separate the two criteria sufficiently, nor change them appropriately with changes in Ca, to maximize performance.

DISCUSSION

The two experiments lead to the same general conclusions. The combined person-machine system is always better than the poorer of the two detectors. In most cases the combined detectors perform better than the more sensitive detector. The fact that the Contingent Criterion Model does a reasonably good job of describing the data is encouraging. However, it is clear that the human subjects do not do as well as is possible given the information available. This is not surprising. The model is not intended to predict behavior. Rather, it is ideal. The model indicates what should be done to maximize the proportion of correct responses of the combined person-machine system. We may inquire, however, what aspect of the subjects' behavior results in less than optimal performance. One possibility is that the sensitivity of the human is decreased due to the presence of the automated detector. It appears that, if anything, the opposite is true. There is evidence that some subjects showed increased sensitivity when the automated detector was present.

Data from both experiments indicate that the subjects did not choose the two criteria properly. In all cases, the criteria chosen by the subjects are closer together than required to maximize system performance. Nor do they move the criteria as much as they should when Ca is changed. The separation between the two criteria should decrease as the sensitivity of the automated detector decreases. If the subjects underestimate the sensitivity of the automated detector, the two criteria would be placed too close together. Another possibility is that the subjects assumed that the values and costs associated with their decisions were not the same on trials on which the automated detector responded "yes" as they were when it responded "no." If such was the case, the two criteria could be placed too close together. In order to establish the two criteria optimally, a human detector must have knowledge of the sensitivity and criterion of the automated detector, of their own sensitivity, and of the relevant values and costs. If this knowledge is not available or is incorrectly assessed, the performance of the combined system will suffer.

It is possible that human operators can be trained to use the available information more efficiently and to set criteria which will lead to improved system performance. In the experiments reported here, no effort was made to train the subjects in the placement of their criteria. The subjects were merely told to use the response of the automated detector in any way they chose in order to do the best they could. No mention was made of establishing different criteria based on the response of the automated detector. It seems likely that with proper training human subjects can approach the predictions of the Contingent Criterion Model.

An important consideration which follows from the general Contingent Criterion Model is that system performance is dependent not only upon the behavior (sensitivity and criterion placement) of the human operator, but also upon the criterion (threshold or alarm-setpoint) of the automated alarm system. System performance may be improved by changing this parameter of the alarm system.

REFERENCES

[1] Sorkin, R. D. and Woods, D. D. Systems with human monitors: A signal detection analysis. Human-Computer Interaction, In press. (1985).

[2] Patterson, R. D. Guidelines for auditory warning systems on civil aircraft. CAA Paper 82017. (London: Civil Aviation Authority, November, 1982).

[3] Seminara, J. L., Gonzalez, W. R., and Parsons, S. O. Human Factors Review of Nuclear Power Plant Control Room Design. (Palo Alto, CA: Electric Power Research Institute, NP-309 1977).

[4] Pollack, I. and Madans, A. B. On the performance of a combination of detectors. Human Factors, 6 (1964) 523-531.

[5] Green, D. M. and Swets, J. A. Signal Detection Theory and Psychophysics. (New York: Wiley, 1966).

[6] Egan, J. Signal detection theory and ROC analysis. (New York: Academic Press, 1975).

ACKNOWLEDGMENTS

The authors wish to thank Dr. Earl S. Stein of the Federal Aviation Administration for his technical assistance and encouragement. Also contributing to this work were Bruce G. Berg of Indiana University and E. MacDonald of Purdue University. This research was sponsored by the U. S. Department of Transportation under contract DTRS 5683-C-00047. This report represents the scientific position of the authors and not necessarily that of the DOT or the FAA.

Trends in Ergonomics/Human Factors II
R.E. Eberts and C.G. Eberts (Editors)
Elsevier Science Publishers B.V. (North-Holland), 1985

AN EVALUATION OF VISUAL CONTRAST SENSITIVITY MEASURES

David L. Kohfeld

Department of Psychology
Southern Illinois University
Edwardsville, Illinois 62026
U.S.A.

Visual contrast sensitivity measures were obtained using a video display that generated vertical sine-wave gratings. The results indicated that visual resolving power is best at intermediate spatial frequencies and progressively less keen at relatively extreme frequencies. A systems-identification method of deconvolution suggested that distributions of contrast sensitivity measures are comprised of at least two component processes, only one of which reflects "true" sensory sensitivity. The results also indicated that the psychophysical method of increasing contrast generates the most criterion-free measures of contrast sensitivity.

INTRODUCTION

An important aspect of visual perception is the resolving power of the visual system, that is, its ability to determine very small spatial patterns, or to discriminate fine detail in a visual display. One way to measure spatial resolution is to present a grating of vertical black test bars on a white background and to reduce the difference between the luminance of the light bars and the luminance of the dark bars until a person can no longer detect the dark bars on the white background; the resulting adjustment is known as the _contrast threshold_. Contrast is usually defined as ($L_{max} - L_{min} / L_{max} + L_{min}$), where L_{max} is the luminance of the brightest part of the grating and L_{min} is the luminance of the darkest part. The reciprocal of this contrast value is usually computed and is plotted as _contrast sensitivity_ (Thomas, 1975). Contrast threshold measurements are usually obtained for a number of sinusoidal gratings of increasing _spatial frequency_. The resulting data are presented in the form of _contrast sensitivity functions_ (CSFs), in which either contrast threshold (Sekuler & Tynan, 1977) or contrast sensitivity (Campbell & Robson, 1968; Ginsburg & Cannon, 1983) is plotted on a logarithmic ordinate against spatial frequency on a logarithmic abscissa.

Recent research by Ginsburg and his associates (Ginsburg, 1981; Ginsburg & Cannon, 1983; Ginsburg, Evans, Sekuler, & Harp, 1982) has explored several psychophysical methods for obtaining contrast sensitivity measures; i.e., Bekesy tracking, the method of adjustment, and the method of increasing contrast. According to Ginsburg and Cannon, the method of increasing contrast was the best of the three methods because the instructions to subjects were easiest to follow, less time was needed for data collection, the measures were most consistent (lowest variability), and perhaps most importantly in the present context, the threshold values were least influenced by the response criterion bias that obscures the "pure" relation between CSFs and sensory processes only.

In the present investigation, entire distributions of contrast sensitivity measures were evaluated. In the first experiment, a systems-identification method of deconvolution was used to decompose the distributions into a component density that presumably reflected the separate influence of "pure" contrast sensitivity and response bias on threshold detections. The general research objective was the same as Ginsburg's: to develop a valid predictor of contrast sensitivity in other tasks, such as performance in a visual simulator, in which precise visual resolution is essential.

EXPERIMENT ONE

Method. The operators were three pilot trainees at Williams AFB, Arizona. The apparatus was an Optronix Model 200 Vision Tester, which has been described elsewhere (Ginsburg & Cannon, 1983). In the present experiment, the display screen had an average luminance of 100 cd/m^2. Seven spatial frequencies were employed: 0.5, 1.0, 2.0, 4.0, 8.0, 11.4, and 22.8 cycles per degree. The operators viewed the screen from a distance of three meters. The psychophysical method employed in Experiment 1 is known as Bekesy tracking. Specifically, the contrast on the visual display either increased or decreased at a fixed rate of 4 dB per second, as determined by the position of the operator's hand-held switch (see Ginsburg & Cannon, 1983, for procedural details involving the Bekesy tracking procedure).

Results and Discussion. Figure 1 presents log threshold contrast sensitivity as a function of log spatial frequency, in cycles per degree, for the Bekesy tracking method. The data means were pooled across the three operators, thus yielding 150 measures per data point. The top function (dashed line) is for the descending trials (decreasing contrast), and the bottom function (solid line) is for the ascending trials (increasing contrast). The data form contrast sensitivity functions (CSFs) across the seven spatial frequencies that are plotted on the logarithmic abscissa.

At least two conclusions may be drawn from the data shown in Figure 1. First, the CSFs reveal the characteristic inverted-U shape found in previous research. This indicates that the resolving power of the visual system is best at intermediate frequencies (2-8 cycles per degree) and progressively less keen at relatively extreme frequencies (0.5 and 22.8 cycles per degree). Second, the CSFs for the descending trials suggest considerably greater contrast sensitivity than those for the ascending trials. This difference was also noted by Ginsburg and Cannon, and is probably attributable to the exaggerated response bias that occurs when descending trials are coupled with ascending trials, as required by the Bekesy procedure. This particular consideration will be evaluated subsequently more fully.

Distributional Analysis. The next step in this analysis involved the preparation of histograms for each operator (3) at each spatial frequency (7) for the ascending and descending series (2), yielding a total of 42 distributions, each comprised of 50 measures. Although these histograms cannot be presented here because of space limitations, it was apparent that the shapes of the ascending distributions were quite similar across operators and spatial frequencies, whereas the descending measures formed a distinctly

FIGURE 1
CONTRAST SENSITIVITY FUNCTIONS FOR THE BEKESY TRACKING METHOD

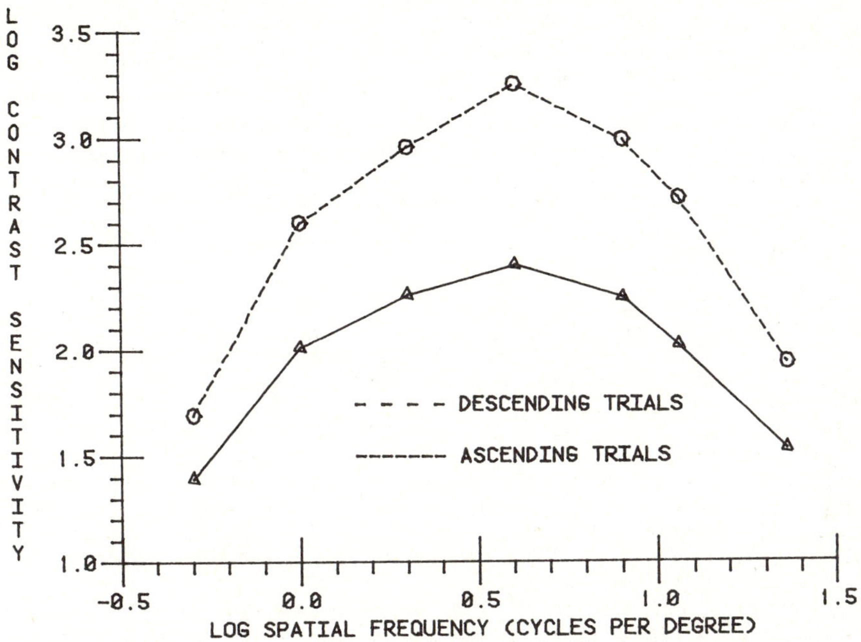

different group of common distributions. Accordingly, the two categories of histograms were collapsed across spatial frequencies by adjusting the 50 raw scores in each distribution to a common origin of 1.0 on the log sensitivity scale, and then collecting these values across the three operators and the seven spatial frequencies. Although this procedure eradicates information about the actual positions of individual spatial frequency distributions on the contrast sensitivity axis, it does not sacrifice vital information about the nature of log sensitivity as a continuous random variable. The outcome is shown in Figure 2, where proportion of threshold detections is plotted as a function of log threshold contrast sensitivity (bin widths of 0.02 log sensitivity). The ascending and descending distributions (solid lines) include 1,050 measures each. The dashed-line function in Figure 2 will be discussed subsequently.

In general, the ascending and descending histograms displayed in Figure 2 are most likely from the same Pearson "family" of distributions. That is, both distributions appear gamma in form; their means, standard deviations, and positive skewness are correlated. These properties suggest that both distributions were generated by common underlying processes that almost certainly have to do with the Bekesy tracking procedure for obtaining contrast sensitivity thresholds. But it is equally apparent that when descending thresholds are obtained, one or more of the component processes is behaving quite differently (larger mean, variance, and skewness), or perhaps another process emerges in addition to those required for ascending threshold estimations. In short, the shapes of the histograms in Figure 2 pointed to the feasibility of a deconvolution procedure which could reveal additional information about the processes that underlie these distributions.

FIGURE 2
DISTRIBUTIONS OF THRESHOLD DETECTIONS FOR THE BEKESY TRACKING METHOD

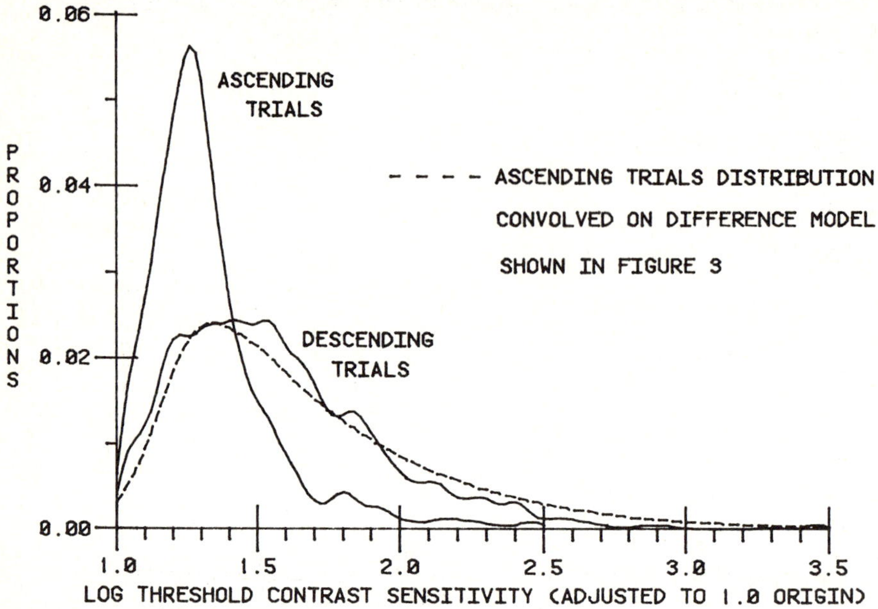

The convolution method employed in this investigation was the Z transform method (Jury, 1964), which is often associated with linear systems identification in engineering (Hsia, 1977; Sage & Melsea, 1971). This approach incorporates the idea that an observed distribution (e.g., the descending series distribution) is often the convolution of a known distribution (e.g., the ascending series distribution) and another hypothetical process (e.g., a response bias component) for which one seeks to identify a model. Conceptually, convolution methods are similar to ordinary subtractive procedures except that the former involve the deconvolution (or "subtraction") of entire distributions rather than just measures of central tendency.

Using the Z transform method of convolution, a model of the difference between the ascending and the descending distributions in Figure 2 was obtained by deconvolving the former distribution from the latter one. The resulting difference model is shown in Figure 3. The difference model selected by the convolution method was also the one whose reconvolution with the ascending distribution gave the closest approximation to the descending distribution from which it was identified. This reconvolution is the distribution composed of the dashed lines, as seen in Figure 2. Although a test for goodness-of-fit was not made, the fit is obviously quite good, which supports the idea that the model describes a linear, additive component that accounts for the difference between the ascending and descending thresholds obtained in this research.

The difference model presented in Figure 3 is most likely exponential in form. An exponential model of the component process which separates the

FIGURE 3
DIFFERENCE MODEL OBTAINED BY DECONVOLVING THE BEKESY ASCENDING DISTRIBUTION FROM THE BEKESY DESCENDING DISTRIBUTION

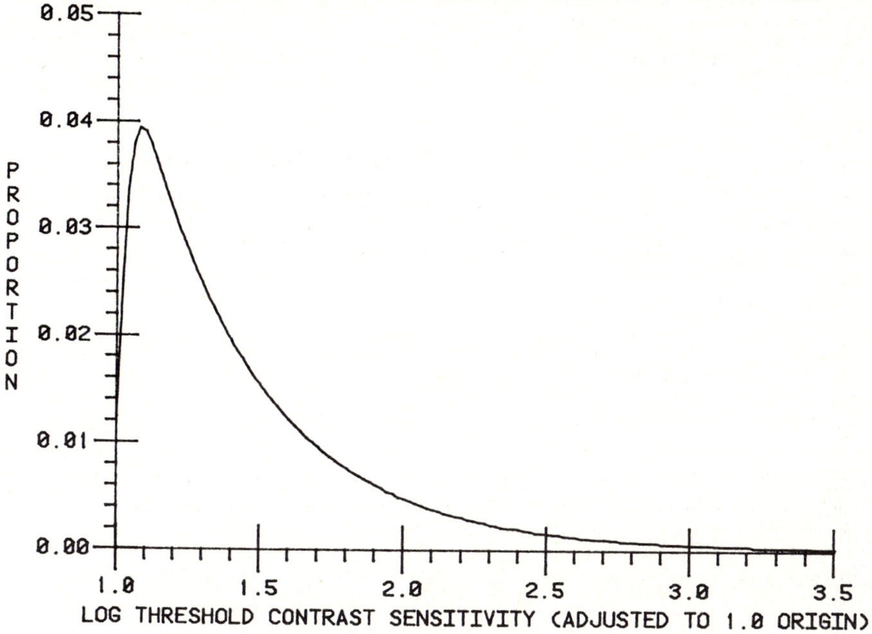

descending distribution from the ascending distribution implies that this underlying event operates much like a conditional, and constant probability mechanism. Although this model is sharply bounded from below, the plotted location of this lower bound (1.0 log unit) on the contrast sensitivity axis is arbitrary, since the distributions from which it was derived were adjusted to a common 1.0 origin. The exponential form of this model, however, suggests a process that generates increasingly larger contrast sensitivity measures over a range of about 2.0 log sensitivity units. Moreover, the conditional probability that this process will influence the contrast threshold measures in the descending series becomes progressively lower according to a fixed probability rule. Finally, an exponential model provides a description of a single underlying event or process. As applied to the present analysis, this means that the higher and more variable contrast sensitivity thresholds in the descending series might be attributable to one categorical process.

EXPERIMENT TWO

The purpose of the second experiment was to obtain contrast threshold data for the method of increasing contrast and the method of adjustment. The goal was to compare these results with those obtained from the Bekesy tracking procedure, as was reported in Experiment 1.

The operators were three Research Psychologists employed at the Human Resources Laboratory, Williams Air Force Base, Arizona. All had normal or fully-corrected vision. The apparatus and general procedures were the same as in Experiment 1, except that only six spatial frequencies were utilized (0.5, 1.0, 3.0, 6.0, 11.4, and 22.8 cycles per degree). The method of increasing contrast was utilized in the first experimental session. This method is essentially the same as the ascending series part of Bekesy tracking, except that no descending series are included. Briefly, the contrast increased at a fixed rate of 4 dB per second from below threshold to where the operator could first detect the presence of the test grating. The method of adjustment was employed in a subsequent experimental session. The operator was given a hand-held switch with an adjustable knob that allowed him to either increase or decrease the contrast between the dark bars and the light bars on the screen until he could just barely detect the presence of the grating. The operator was instructed to turn the knob as far as it would go below the threshold before a trial began. He then increased the contrast and "bracketed" the threshold point, using as much time as was needed to achieve the desired setting.

Results and Discussion. Figure 4 shows the mean threshold values, plotted in the same manner as in Figure 1. Because of the obvious similarity of the CSFs for each individual, the data were collapsed across operators, thus yielding 160 measures in each CSF data point for the adjustment method and increasing contrast method. The third CSF in Figure 4, labeled "Bekesy ascending trials," is the same as the one in Figure 1. This latter CSF is comprised of data points with 150 measures each, and is presented for comparison with the two CSFs obtained in the present experiment.

The most compelling feature of Figure 4 is the overall similarity of the CSFs for the three methods shown. This degree of similarity is somewhat

FIGURE 4
CONTRAST SENSITIVITY FUNCTIONS FOR THE ADJUSTMENT METHOD, INCREASING CONTRAST METHOD, AND THE BEKESY ASCENDING TRIALS

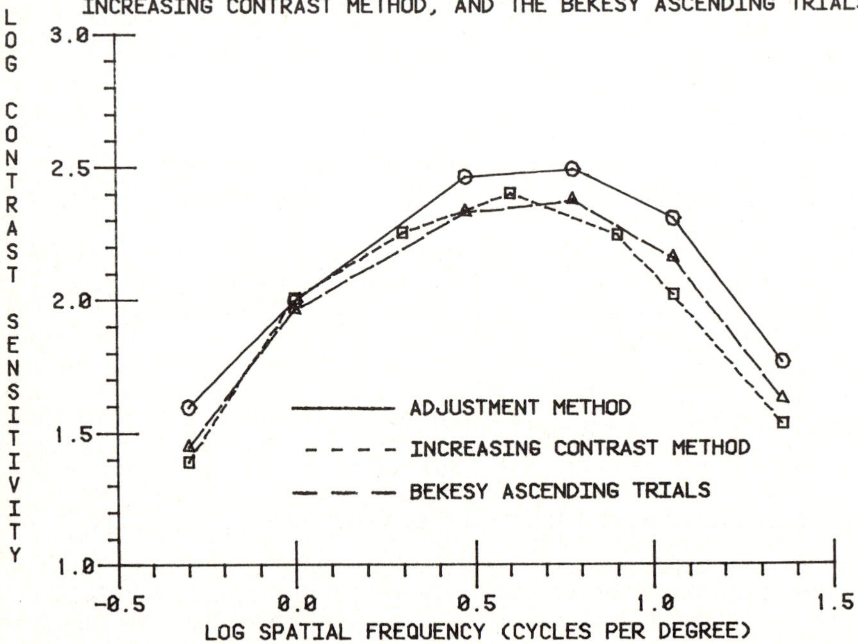

surprising when one considers that each CSF was composed of data taken from only three operators, and indeed, a separate trio of operators in the Bekesy ascending trials. These results suggest at least three important conclusions. First, the three different psychophysical methods generated such similar CSFs that one's confidence in the validity of these measures is improved considerably. Second, the elevated CSF for the Bekesy descending trials in Experiment 1 is probably based on properties of the method itself rather than "true" differences in contrast sensitivity. That is, the descending trials appear to be the "outliers" relative to three other procedures. Finally, it now seems possible that representative CSFs can be generated from relatively few operators, at least if these individuals are familiarized with the testing procedures and can generate a reasonably large sample of contrast threshold measures at each spatial frequency.

Figure 5 provides a comparison of the distributions of data for the same three methods that are depicted in Figure 4. The threshold values were adjusted to the 1.0 origin on the log sensitivity axis, just as they were in Figure 2. The distribution composed of Bekesy ascending trials (1,050 measures) is the same as the one plotted in Figure 2. Each of the distributions for the increasing contrast method and the adjustment method contain 960 values. The similarities between the general forms of the distributions shown in Figures 2 and 5 are quite apparent.

The results shown in Figure 5 allow two important conclusions. First, all three distributions are gamma in form, with characteristic positive skewness. As noted previously, this implies that the processes or events which underlie each distribution are similar for the three psychophysical methods, as

FIGURE 5
DISTRIBUTIONS OF THRESHOLD DETECTIONS FOR THE ADJUSTMENT METHOD, INCREASING CONTRAST METHOD, AND THE BEKESY ASCENDING TRIALS

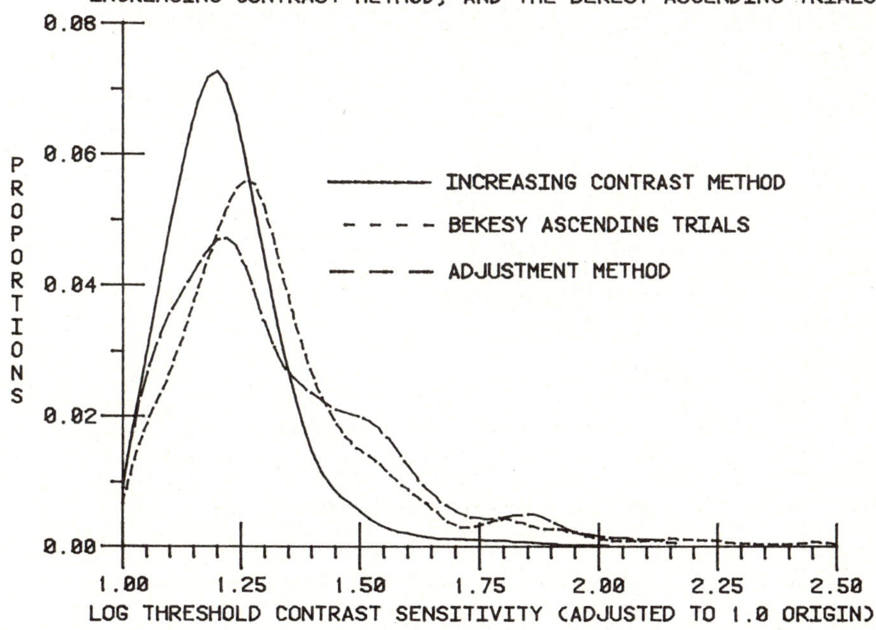

one might expect. Second, if one accepts the view that a distribution of measures with the lowest possible variability is the most desirable in terms of measuremental precision, the method of increasing contrast could serve as the most predictive method for determining contrast sensitivity skills. Indeed, Ginsburg and Cannon offered this conclusion for reasons quite similar to those presented in this paper.

GENERAL DISCUSSION

In 1963, McGill proposed that the form of some distributions would provide a kind of "signature" that reflects the nature of the processes that comprised these distributions. Since then, a number of investigators have demonstrated that simple reaction time is the convolution of two main components: sensory detection and response execution (Burbeck & Luce, 1983; Green & Luce, 1971; Kohfeld, Santee, & Wallace, 1981; Kohfeld, 1982, 1984). The present analysis of contrast sensitivity thresholds was influenced by this previous reaction time work in two important ways. First, the linear systems identification, and the associated Z transform method of convolution provided the analytic framework for decomposing the distributions of contrast sensitivity measures. Second, the reaction time process is sometimes discussed in terms of a response threshold that triggers the reaction; this is conceptually similar to the idea of the decision threshold that is determined with contrast sensitivity measures. In this vein, one can explore the parallels between the two reaction time stages, sensory detection and response execution, and the corresponding two stages in contrast sensitivity decisions.

The theory of signal detectability (TSD) provides one way to conceptualize the behavior of individuals in both reaction time and contrast sensitivity tasks. According to TSD, the operator is both a sensor and a decision maker, and therefore the evaluation of threshold measurements would require an independent assessment of the operator's sensory sensitivity, d', and response criterion, β (Dember & Warm, 1979; Green & Swets, 1966). TSD assumes further that d' and β are independent aspects of task performance, and are selectively influenced by different factors. With respect to contrast sensitivity, the visual contrast at the video screen determines d', whereas β is characterized by the operator's strategy in responding to task demands. As such, the logic of TSD may prove useful in describing the forms of the distributions of threshold contrast sensitivity. Because of the nature of this threshold task, one can suppose that these distributions reflect the behavior of both sensory mechanisms (d') and a response process (β). In other words, the distributions shown in Figures 2 and 5 are a composite of at least two component densities, d' and β, each of which is an independent random variable.

It should be noted that no prior predictions were made about the shapes ot the distributions actually obtained in this research. That is, no theoretical formulation dictated that particular forms of histograms should be generated, e.g., exponential, gamma, or normal. Thus, when the convolution method resulted in a model that was exponential in form (as seen in Figure 3), no a priori logic suggested that this model was part of the sensory component (d'), the response component (β), or perhaps a third component altogether. However, both TSD logic and common intuition provide guidance in this matter. When one serves as an operator in the descending series part of the Bekesy tracking task, it is at once apparent that deciding when the grating has disappeared is considerably more difficult than when ascending threshold decisions are required. The consensus of the operators can be offered in the quote: "I held the switch down until I was sure that the bars were gone." This sounds very much like response bias, or in TSD terminology, the operators adopted a more lenient criterion in the descending trials, thus yielding inflated contrast thresholds.

These considerations lead one to conclude that deconvolving the ascending distribution from the descending distribution resulted in a difference model that reflects a response bias component. As noted previously, its exponential form implies that it represents a single process or categorical event. This interpretation does not imply that the ascending trials are free of response bias; rather, it supports the view that descending thresholds are influenced by at least two sources of undesired response bias. The separation of the d' and β components from the distributions generated with Bekesy ascending trials, the method of adjustment, and the most reliable method of increasing contrast (see Figure 5) is a matter for future work with deconvolution methods and hazard function estimation.

ACKNOWLEDGEMENTS

The data collection, analysis, and preparation of this paper were supported by the U.S. Air Force Systems Command, the Air Force Office of Scientific Research, and the Southeastern Center for Electrical Engineering Education during the author's work as a Summer Research Fellow at the Human Resources Laboratory, Williams Air Force Base, Arizona, in 1984. Computer time and routines, including graphics, were funded by a research grant to the author from the Office of Research and Projects in the Graduate School at Southern Illinois University, Edwardsville. I am indebted to George Geri, Marc Hunter, Robert Nullmeyer, and Robert Ruhl for their assistance and collaboration during this research work.

REFERENCES

(1) Burbeck, S. L., and Luce, R. D. (1982). Evidence from auditory simple reaction times for both change and level detectors. Perception and Psychophysics, 32, 117-123.

(2) Campbell, F. W., and Robson, J. G. (1968). Application of Fourier analysis to the visibility of gratings. Journal of Physiology, 197, 551-566.

(3) Dember, W. N., and Warm, J. S. (1979). Psychology of perception (2nd Edition). New York: Holt, Rinehart and Winston.

(4) Ginsburg, A. P. (1981). Spatial filtering and vision: Implications for normal and abnormal vision. In L. M. Proenza, J. M. Enoch, and A. Jampolsky (Eds.). Clinical applications of visual psychophysics. New York: Cambridge University Press.

(5) Ginsburg, A. P., and Cannon, M. W. (1983). Comparison of three methods for rapid determination of threshold contrast sensitivity. Investigative Ophthalmology & Visual Science, 24, 798-802.

(6) Ginsburg, A. P., Evans, D. W., Sekuler, R., and Harp, S. A. (1982). Contrast sensitivity predicts pilots' performance in aircraft simulators. American Journal of Optometry and Physiological Optics, 59, 105-109.

(7) Green, D. M., and Luce, R. D. (1971). Detection of auditory signals presented at random times: III. Perception and Psychophysics, 9 257-268.

(8) Green, D. M., and Swets, J. A. (1966). Signal detection theory and psychophysics. New York: John Wiley & Sons.

(9) Hsia, T. C. (1977). Systems identification: Least squares methods. Lexington, Massachusetts: D. C. Heath.

(10) Jury, E. I. (1964). Theory and application of the Z-transform method. Harrington, New York: Krieger.

(11) Kohfeld, D. L. (1982). Reaction time stages of auditory intensity. Paper presented at the 23rd annual meeting of the Psychonomic Society, Minneapolis, Minnesota.

(12) Kohfeld, D. L. (1984). Sensory-detection models of auditory and visual reaction time processes. Paper presented at the 17th annual meeting of the Society for Mathematical Psychology, Chicago, Illinois.

(13) Kohfeld, D. L., Santee, J. L., and Wallace, N. D. (1981). Loudness and reaction time: II. Identification of detection components at different intensities and frequencies. Perception and Psychophysics, 29, 550-562.

(14) McGill, W. J. (1963). Stochastic latency mechanisms. In R. D. Luce, R. R. Bush, and E. Galanter (Eds.). Handbook of mathematical psychology, Vol. 2. New York: John Wiley & Sons.

(15) Sage, A. P., and Melsea, J. L. (1971). System identification. New York: Academic Press.

(16) Sekuler, R., and Tynan, P. (1977). Rapid measurement of contrast-sensitivity functions. American Journal of Optometry & Physiological Optics, 54, 573-575.

(17) Thomas, J. P. (1975). Spatial resolution and spatial interaction. In E. C. Carterette and M. P. Friedman (Eds.). Handbook of perception V. New York: Academic Press.

Trends in Ergonomics/Human Factors II
R.E. Eberts and C.G. Eberts (Editors)

INSPECTOR ACCURACY AS EFFECTED BY INCOMING QUALITY, COMPLEXITY OF INSPECTION TASK AND TIME ALLOWED FOR INSPECTION

PROFESSOR ABDUL RAOUF

UNIV. OF PETROLEUM & MINERALS
DHAHRAN, SAUDI ARABIA

FARAG ELFEITURI

UNIV. OF WINDSOR
WINDSOR, ONTARIO, CANADA

Accuracy of visual inspection is influenced by three major sets of factors: inspector related factors, task related factors, and factors related to the environment. It is generally agreed that inspectors make two types of errors namely: reporting a defective component when it is not defective (type I error), and failing to report a defect (type II error).

This paper presents the findings of an experimental study which examined potential inspection accuracy differences, in terms of two types of errors when varying the following factors:

(1) different incoming quality levels having poisson distribution (5 levels)

(2) different levels of task complexity (2 levels)

(3) different times allowed to complete inspection (2 levels)

Findings are based on the results obtained using fifteen subjects while performing a computer controlled simulated inspection task.

INTRODUCTION

The output of production systems invariably contain some defective units caused by workers, machines, material, or a combination of all of them. The inspection process is the basic source of information on these errors. Inspection tasks may be classified into three basic categories, these are: tasks involving visual scanning, tasks involving measurements, and tasks involving monitoring of a process.

The accuracy of inspection process is the key element to better production quality. However, the inspection accuracy is influenced by a number of factors. These factors can be categorized into three major groups:

(1) Inspector related factors, such as the age, experience, sex, etc.

(2) Task related factors, such as task pacing, task perception, etc.

(3) Environmental and organizational factors, such as, illumination, temperature, motivation, etc.

The inspector is an important component to consider in evaluating a quality control system. The inspector accuracy is the degree to which the inspector correctly makes decision on product quality, i.e., accepts products which meet specification and rejects products which fail to meet specifications. Every component in the processing system has certain reliability and inevitable variations due to many factors. The inspector, like the other components in the system, has less than perfect performance. In fact, inspectors make two types of errors. One is the type I error, in which a conforming quality characteristics is classified as nonconforming quality characteristics is classified as nonconforming; the other is the type II error, in which a nonconforming characteristic is passed as conforming. The efficiency of the quality control systems is dependent upon the ability of the inspectors to detect defects. The ability to do this is affected by many factors such as incoming quality, time allowed for inspection and so on. The effects of some of these factors on the inspection accuracy have been the subject of many studies. Most of the studies did not consider the distribution of defects being poisson. The findings of some of the studies are contradicting.

Consequently this study was undertaken to determine the effects of incoming quality (defect rates, having poisson distribution), time allowed for inspection and task complexity on inspection accuracy in a simulated paced visual inspection task. There are a number of factors in the inspection tasks which, under certain circumstances, may affect the accuracy of the resulting inspection information. These factors, which are mentioned in different researches, are summarized in Table 1.

Table 1: Factors Influencing Inspection Accuracy

A. TASK RELATED FACTORS:

1. Paced vs. unpaced
2. Defect conspicuity
3. Task complexity
4. Density of items
5. Rate of defects
6. Importance of the characteristic being inspected
7. The need of usage of aids such as:
   (a) magnification
   (b) overlays
   (c) viewing screen
   (d) automatic scanner
8. Design of work place
9. Time of the day
10. Number of faults occuring simultaneously
11. Method of inspection
12. Inspection rate
13. Display characteristics
14. Task perception

B. INSPECTOR RELATED FACTORS:

1. Visual acuity
2. Age
3. Experience
4. Sex

5. Intelligence
6. Social factors
7. Psychological factors
8. Number of inspectors
9. Eye movement
10. Sleep deprivation
11. Personality
12. Level of training

C. <u>ENVIRONMENTAL AND ORGANIZATIONAL FACTORS:</u>

1. Illumination
2. Noise
3. Temperature
4. Humidity
5. Motivation
6. Incentives
7. Music while working
8. Management standards
9. Job rotation
10. Feedback and feedforward

Table 1 does give an impression of the wide range of factors that can affect the inspection process. These factors were grouped into three main categories. However, the classification is somewhat arbitrary, because there is a strong inter-relation among the factors in the different classes.

Jacobson (18) reported that some physical characteristics of inspectors might affect their performance. He concluded that age can affect the accuracy of the inspectors. He found a slight relationship between visual acuity and accuracy.

Harris (14) has studied the effect of defect rate on inspection accuracy. He concluded that inspection accuracy decreased with reduction in defect rate. However, McCornack (22) reported that "usually the inspectors are better at identifying good product rather than bad product by quite a margin". This means that inspection accuracy increases with reduction in defect rate.

Tiffin and Rogers (34) reported that there is very little relationship between inspector accuracy and visual acuity, height, weight, age or experience of the inspectors. Contrary to expectation, experienced inspectors are not the better ones.

A study of the effects of certain visual factors on the efficiency of textile workers was made by Ayers (1). The objectives of his study were to discover if inspectors with deficient vision are unsatisfactory in quality and/or quantity of production and to discover if visual factors vary in their influence upon different phases of jobs performance. He found very low correlations between inspector accuracy and vision tests, age and job experience.

Williges and Streeter (37) were concerned with the characteristics of the display of inspection items and their subsequent effects. Their study dealt with simple visual scanning inspection in which the main purpose was to investigate the effects of display arrangement which could structure the inspection task either by providing a pattern of scanning or by pacing the rate of inspection. Two types of displays were used, they are dynamic display and static display. It was found that structuring according to display arrangements seems to have little effect on inspectors' performance. The major result of the study suggested that inspectors pace their inspections according to the display presentation rate, but the overall result may be inferior inspection performance when compared to alternative inspector-paced displays.

The effect of the complexity of the inspection task were studies in many papers. Complexity was defined by Harris (16) as a function of the number or parts which make up an equipment item and the way the parts are interrelated or arranged. McFarling and Heimstra (23) defined the complexity as the number of items, components, or potential targets per product to be inspected. Complexity was measured by simply counting the number of major parts making up each item and the way the parts were arranged. However, one can extend the definition of complexity to include the number of good and defective parts contained in a product, the arrangement and distribution of the parts in the product, the number of types of defects per product and the degree of similarity between good and defective components since all these factors can be related to product complexity rather than any other variable.

One of the earliest studies on the effect of product complexity was conducted by Harris (16) who investigated the effect of the complexity of electronic equipment on inspection performance. Complexity was considered to increase as the number of parts increased and as the arrangement of parts become less orderly. It was hypothesized that the percentage of defects detected would be inversely related to the complexity of the task even when inspectors are given an unlimited amount of time in which to inspect. The inspection accuracy was measured in terms of the percentage of defects detected. No considerations were given to the false alarm reported. The results of the study indicated that inspection accuracy has an almost perfect inverse relationship with the product complexity has a significant determinal effect on inspection accuracy, an effect that can not be overcome simply by extending the amount of inspection time allowed. However, the effect of the amount of time on inspection accuracy was found to have a significant affect when other factors were considered. For example, Drury (6) has described the effects of speed of working on inspection accuracy in seven different studies and showed that similar results were obtained irrespective of the type of product examined. Moreover, the results showed that as more time is allowed to inspect each item, the probability of defect detection increases.

Another interesting result was obtained by McFarling and Heimstra (23) who investigated the influence of three factors namely, task pacing, product complexity and task perception, in inspection accuracy. The task consisted of inspecting simulated printed circuits varying in circuit complexity. They reported that decision time is seen to increase with increasing circuit complexity, type II error increases also, while type I error decreases.

There is a great number of papers and research work which have been done in the area of inspection accuracy under different single or combination of factors which affect the accuracy of the inspection process. A brief summary of selected papers is present here.

Fox and Haslegrave (12) who studied the effects of the probability of a defect occurring on the efficiency of the inspector and have found the same results of Harris (14) which was discussed above.

Drury and Addison (8) investigated the importance of feedback and fault density and found that feedback does increase the performance of the inspector. Also, a drop in the fault detection with increased quality was reported to show that the inspection group were attempted to keep the outgoing percentage defective constant.

Drury and Sheehan (10) reported that a small negative correlation between inspector efficiency and age exists. In the same study they did not find a close relation between the detection accuracy and visual acuity. Another study by Sheehan and Drury (31) reported a decrease in inspection accuracy as a function of increasing age.

Raouf and Elfeituri (30) have studied the effect of defect rate on inspection error rate. In that study the task used was a visual paced task developed at 5 levels of defect rates. The subjects were asked to count the number of D's in arrays consisted of letters O's and D's displayed on a CRT terminal connected to a PDP-11 Computer. The number of D's (defects) in each array was developed as a poisson random number generated as a part of a computer program. Each array was 15x15, the D's in each array were distributed randomly. The time allowed to inspect each array (TAI) was 20 sec. The distinguishing between the letters O and D is difficult as they appear on the CRT terminal, and this difficulty is used as a measure of task complexity.

The performance of the subjects was measured in terms of probability of making type I and type II errors.

The results of that study concluded that type II error increases with increase in defect rate, the relationship can be given by:

Prob. of making type II error = $0.057 + 0.008\,X$ where X is the defect rate.

Type I error decreases insignificantly with the increase in defect rate, the relationship can be written as:

Prob. of making type I error = $0.00093 - 1.23 \times 10^{-5} X$.

As a result of this literature search it is clear that conflicting thoughts can be recognized. As an example, there are different schools or theories in the literature concerning the effects of the incoming quality, product complexity and inspection rate on inspection accuracy. These are illustrated in Table 2.

As a result of this conflicting results, it is clear that, considering the importance of the inspection to the modern industry, the need for an adequate study in this area is overshelming. "In a word, the area of inspector accuracy is wide open for some sophisticated experimentation" (22).

Another important factor to be realized is that the experience with the variations in the number of defects per unit indicate that the distribution of the number of defects per unit follow very closely the form of a poisson distribution with mean equal to C. (is used as well). Since the theoretical conditions for the applicability of the poisson distribution call for the count of the number of occurrences of an event that has an infinite number of opportunities to occur, and a small constant probability of occurrence at any opportunity (for practical purposes infinite may be

Table 2: Summary of Some Conflicting Theories

Factor	Conflicting Theories	Supporting Authors
Incoming quality	1) A decrease in the incoming quality causes a decrease in inspection accuracy because of the increased number of chances for error.	McCornack (22).
	2) A decrease in the incoming quality causes an increase in inspection accuracy because the work for the inspector in this case is boring.	Harris (14).
Task complexity	1) As the task complexity increases, the inspectors' performance decreases	McFarling & Heimstra (23); Harris (16).
	2) Inspectors' performance do not change with the change in task complexity.	Williges & Streeter(37); Kochhar & Jaisingh(21); Drury (6).
Time Allowed for inspection	1) As more time is allowed to inspect each item, the performance of the inspectors increases.	Smith & Barany (33); Drury (6).
	2) The decrease in the inspection accuracy can not be simply overcomed by extending the amount of inspection time allowed.	Harris (16).

interpreted as meaning very large); whenever the number of defects per unit can be transformed into fraction defective and the value of this fraction is less than or equal to 0.10, the number of defects per unit follows very closely to the poisson distribution (11). Yet, except for the study of Jain (19), all the studies cited earlier have not considered defects having poisson distribution.

The review of the literature leads to the fact that more efforts should be devoted to clarify some of the points which are still unsettled and to bring some attention to other subjects which have not been studied as of yet.

METHOD

The Experimental Conditions

The independent variables of this study were investigated at the following levels:

1. Time Allowed for Inspection (TAI): The inspection task was presented at two levels. These are of 20 seconds per array and 25 seconds per array.

2. The Task Complexity (TC): In this study task complexity represented by the degree of disciminability between the defects and the good characteristics. Products contain two or three types of defects. (the case of products containing one type of defects was discussed in 30).

3. The Incoming Quality (Defect Rate) (DR): This factor was studied at five levels which were C = 2, 5, 8, 10 and 12.

The Task

The task consisted of arrays of O's, C's, D's, and G's displayed on a CRT display connected to a PDP 11 computer. The number of C and D and C,D and G in each array was a poisson random number. The number of each letter in the arrays was chosen randomly, also, the distribution of the different letters in an array was random as well. The method of choosing the number and the distribution of each defect "letter" randomly because it was suggested that this method match the real inspection situation in real life. Details of generating the displays are given elsewhere (Raouf and ElFeituri, 30).

The Subjects

The subjects selected for this investigation are fifteen subjects, ten of them are male subjects (average 26, Std. Dev. 1) and five females (average 25.8, Std. Dev. 1), from the students of the University of Windsor. None of the subjects had any physical disabilities. The only restriction applied on the selection of the subjects was that they must have normal vision and be able to maintain constant level of vigilance.

The experiments lasted for approximately three hours for each subject. The subjects were paid a nominal amount for performing the task.

Experimental Routine and Randomization

Before the beginning of experimentation, each subject was given an oral briefing on how the experiment was to be conducted. In order to minimize the effect of learning, the subjects were allowed adequate time to become accustomed to the task. To eliminate the effects of fatigue, each subject was asked to take a break whenever he/she began to feel tired. Moreover, they were given a break time between each run, the subjects were allowed to relax and walk around. The break time was between 20 - 30 minutes. The sequence of the experimental conditions were randomized for each subject.

Design of the Experiment

The design of this experiment is based on the techniques given by Mandenhall (26) and Hicks (17) for factorial experiments. This particular model is a 15x2x2x5 factorial (mixed) with subjects as a random factor and inspection rate, task complexity, and defect rates as fixed factors. The same design is used for each of the two measures of inspector accuracy, namely, type I and type II errors.

Data Collection

Each subject performed the experiment at each level of defect rate, task complexity, and inspection rate. The performance was measured in terms of the probability of making type I and type II errors. Each subject had performed the experiment twice.

The Analysis of Variance (ANOVA) precedures were conducted for each type of error. All means and interactions were tested. The Duncan Multiple Range Tests were performed on the means of those factors and interactions which were significant.

The Probability of Making Type II Error

The probability of making type II error under all experimental conditions are shown graphically in Figure (1).

The results of the analysis of variance for the probability of type II error as a measure of the inspector accuracy was carried out.

All the main effects SUB, DR, TAI and TC are significant at 0.0001 level, three of the first order interaction terms are significant (SUB x TC, SUB x TAI and TC x TAI) at different levels of significance.

It was noticed that the DR of 2 (on the average, the samples contain 2 defects) results in the best inspector accuracy of 2.5% type II error, while the DR of 12 results in the lowest inspector accuracy of 8.26% when accuracy is measured by probability of making type II error. Duncan Multiple Range Test showed that inspection accuracy was different at each of the DR's.

The analysis of the TC showed that level III results in the lowest type II error of 4.1%, level II results in a 6.8% type II error and level I results in a 11.6% type II error. Duncan tests showed that each level of TC produces an inspector accuracy (in terms of type II error) significantly different from the other levels.

It was observed that when TAI is 25 sec. the probability of making type II error is lower, at 4.5% while the 20 sec. rate gives a 6.4% type II error. Each level of TAI is significantly different from the other.

The SUB x TAI interaction is a significant factor based on the ANOVA results. Further analysis showed that for each subject probability of making a type II error was higher at lower levels of TAI. Analysis of TC x TAI interaction had shown that probability of making type II errors varied as TC and TAI varied. The probability of making type II errors was observed to be higher for higher level of TC and lower TAI.

The Probability of Making Type I Error

The probability of making type I error under all experimental conditions are shown graphically in Figure 2.

Analysis of variance for this measure of inspector accuracy showed that SUB and DR were found to be significant factors. Duncan's Multiple Range tests indicated the different levels of the TAI, TC and SUB were not significantly different from each other, while the first and second level of DR were significantly different from all other levels.

Probability of making type I errors at different DR levels was observed to be the inverse of the effects of defect rate on type II errors.

The mean percentage of type I error for the entire experiment is 0.00172, while the mean percentage of type II error is 0.055. Therefore, the mean ratio of type II inspector error to type I is approximately 32 to 1. This result substantiates the findings of Davis (4) and McKnight (24) who assumed a significantly large ratio of type II error to type I error, in the formulation of their sampling plan formulas.

Conclusions & Applications

Following conclusions can be drawn.

DR: The effect of DR on type II error, which is the significant measure of performance in this study, substantiates the first theory as subscribed by McCornack (22) that a decrease in the incoming quality (increase in DR) of a lot causes a decrease in inspector accuracy. One interpretation of this result could be that, increasing the defect rate causes the increase in the number of defects in each unit or sample and this, in turn, causes an increase in the possibility of not detecting the defects.

The effect of DR on type I error, as a second measure of performance, is not significant in this particular study. This is in agreement with the findings of (24). However, a slight decrease in probability of making type I error is observed with the increase in DR.

The study substantiates another phenomena, that was observed by many reseachers, for example, Drury (5) and (6), which indicates that the probabilities of type I and type II errors are related. As one probability increases the other decreases. Limiting conditions for this relationship need to be investigated further.

TAI: Inspector accuracy is higher in this study when TAI is at the higher level (25 sec.), and inspector accuracy (in terms of the probability of making type II error) decreases steadily as the TAI decreases. Limiting conditions for this factor need to be investigated further.

No significant effect of TAI on type I is observed. However, from the results of Duncan's Multiple Range test, it was noticed that type I error decreases with the increase in TAI. That is, as the subject is allowed longer time to inspect each unit, both type I and type II errors decrease. The steady decrease in inspector accuracy observed as the TAI is decreased gives an indication of the accuracy to expect for levels beyond the range of this study. But further study is needed to validate the continuance of this trend. At both levels of TAI, inspector accuracy is not significantly affected by changes in the TAI, while, on the other hand, inspector accuracy affected significantly by changes in TC, which means that there is a significant interaction between TAI and TC.

TC: The results of this study indicates that TC has a significant determinal effect on inspector accuracy. As task complexitydecreases, inspector accuracy (measured in terms of type II error) increases. These results confirm those by McFarling and Heimstra (23) and Harris (16).

At the second level of task complexity the probability of type II error is 6.85%. When TC is at the lower level (Level III), the inspector accuracy is the highest (at 4.14% type II error). On the other hand, task complexity does not significantly affect the probability of making type I error. However, a very slight increase in probability of type I error with the decrease in TC can be observed. This result confirms the findings of McFarling and Heimstra (23) who reported a significant increase in probability of making type I error with decreasing task complexity. The effects of task complexity on inspection accuracy in terms of making type II error, can be reduced significantly by increasing TAI.

Inspector Accuracy: Within the limits of this investigation and for tasks similar to the task used in this study, the probability of making type II error seems to be a more realistic criterion for measuring inspector accuracy than the probability of type I error. Since in visual inspection tasks, the possibility of counting a good characteristic as a defect seems to be very small (the mean of probability of type I of the entire experiment is 0.172% while the mean of probability of type II error is 5.5%).

APPLICATIONS

The results of this study may be applied to paced visual scanning inspection task where the inspector has to search for defects by making a point by point examination of a unit in a certain period of time. The results can be applied particularly to tasks that required the detection of defects in units having an array shape such as electronic circuity on circuit boards, improper connections on a chasis or scratches on an areas of finished surface.

BIBLIOGRAPHY

1. Ayers, A.W., (1942), "A Comparison of Certain Visual Factors with the Efficiency of Textile Inspectors", J. of Applied Psychology, Vol.26, pp.22.
2. Bloomfield, J.R., (1972), Visual Search in Complex Fields: Size Differences Between Target Disc and Surrounding Discs", Human Factors, Vol. 14, No. 2, pp. 139.
3. Case, K.E., (1980), "The P Control Chart Under Inspection Error", J. of Quality Technology, Vol. 12, No. 1, pp. 1.
4. Davis, A.S., (1966), "The Sensitivity of Sampling Inspection to Inspection Error", Master's Thesis, Texas Tech. College.
5. Drury, C.G., (1974), "The Human Factor in Industrial Inspection", Quality Progress, Vol. 7, No. 12, pp. 14.
6. Drury, C.G., (1973), "The Effect of Speed of Working on Industrial Inspection Accuracy", Applied Ergonomics, Vol. 4, No. 1, pp. 2.
7. Drury, C.G. and Fox, J.G., (1969), "Human Reliability in Quality Control, Taylor and Francis, London.
8. Drury, C.G. and Addison, J.L., (1973), "An Industrial Study of the Effect of Feedback and Faulty Density on Inspection Performance:, Ergonomics, Vol. 16, No. 2, pp. 156.

9. Drury, C.G. and Clement, M.R., (1978), "The Effect of Area, Density and Number of Background Characters on Visual Search", Human Factors, Vol. 20, No. 5, pp. 597.

10. Drury, C.G. and Sheehan, J.J., (1969), "Ergonomic and Economic Factors in an Industrial Inspection Task", The Int. J. of Production Research, Vol. 7, No. 4, pp. 333.

11. Duncan, A.J., (1974), "Quality Control and Industrial Statistics, Richard D. Irwin Inc., Homewood, Ill., 4th Edition.

12. Fox, J.G., and Haslegrave, C.M., (1969), "Industrial Inspection Efficiency and the Probability of a Defect Occurring", Ergonomics, Vol. 12, No. 5, pp. 713.

13. Fishman, G.S., (1973), "Concepts and Methods in Discrete Event Digital Simulation, John Wiley & Sons Ltd.

14. Harris, D.H., (1968), "Effect of Defect Rate on Inspection Accuracy", H. of Applied Psychology, Vol. 52, No. 5, pp. 377.

15. Harris, D.H. and Chaney, F.B., (1969), "Human Factors in Quality Assurance, John Wiley & Sons Ltd.

16. Harris, D.H., (1966), "Effect of Equipment Complexity on Inspection Performance", J. of Applied Psychology, Vol. 50, No. 3, pp. 236.

17. Hicks, C.R., (1973), "Fundamental Concepts in the Design of Experiments, Holt, Renehart and Winston, 2nd Edition.

18. Jacobson, H.J., (1952), "A Study of Inspector Accuracy", Industrial Quality Control, Vol. 9, pp. 16.

19. Jain, J.K., (1977), "A Model for Determining the Optimal Number of Inspectors Minimizing the Inspection Cost", M.A.Sc. Thesis, University of Windsor.

20. Johnson, S.L., and Funke, D.J.,(1980), "An Analysis of Human Reliability Measures in Visual Inspection", J. of Quality Tech., Vol. 21, No. 2, pp. 71.

21. Kochhar, D.S. and Jaisingh, S.C., (1980), "Contemporary Approaches to Paced Visual Inspection", AIIE Trans., Vol.12, No. 1, pp. 38.

22. McCornack, R.K., (1961), "Inspector Accuracy: A Study of the Literature:, Sandia Corporation Tech. Memo, pp. 53.

23. McFarling, L.H. and Heimstra, N.W., (1975), "Pacing, Product Complexity, and Task Perception in Simulated Inspection", Human Factors, Vol. 17, No. 4, pp. 361.

24. McKnight, K.A., (1967), "An Investigation of the Effects of the Two Types of Errors on Sampling Inspection Plans", M.S. Thesis, Texas Tech. College.

25. Megaw, E.D., (1979), "Factors Affecting Visual Inspection Accuracy", Applied Ergonomics, Vol. 10, No. 1, pp. 27.

26. Mendenhall, W., (1968), "The Design and Analysis of Experiments", Duxbury Press, 1968.

27. Monk, T.H. and Brown, B., (1975), "The Effect of Target Surround Density on Visual Search Performance", Human Factors, Vol. 17, No. 4, pp. 356.

28. Nelson, J.B. (1977), "Human Performance on Paced-Inspection Tasks: A Survey of Past and Future Research", Proceedings of Spring Conference, AIIE.

29. Raouf, A., (1975), "Predicting Human Performance in a Quality Control System: Some Experimental Findings", Quality Assurance, Vol. 1, No. 1, pp.3.

30. Raouf, A. and Elfeituri, F., (1982), "Effect of Incoming Quality on Inspection Error Rate: An Experimental Study", ASQC Forum 28, Toronto.

31. Sheehan, J.J. and Drury, C.G., (1971), "The Analysis of Industrial Inspection", Applied Ergonomics, Vol. 2, No. 2, pp. 74.

32. Smith, G.L., "Inspector Performance on Microminiature Tasks", A Research Paper in Ref. 7, pp. 149.

33. Smith, L.A. and Barany, T.W., (1970), "An Elementary Model of Human Performance on Paced Visual Inspection Tasks", AIIE Trans., Vol. 2, No. 4, pp. 298.

34. Tiffin, J. and Rogers, H.B., (1941), "The Selection and Training of Inspectors", Personel, Vol. 18, pp. 171.

35. Wallack, P.M. and Adams, S.K., (1970), "A Comparison of Inspector Performance Measures", AIIE Trans., Vol. 2, No. 2., pp. 97.

36. Wang, S.C., (1975), "Human Reliability in Visual Inspection", Quality, Vol. 14, No. 9, pp. 34.

37. Williges, R.C. and Streeter, H., (1971), "Display Characteristics in Inspection Tasks", J. of Applied Psychology, Vol.55, No.2, pp.123.

38. Zunzanyika, X.K. et al., "Effects of Information on Industrial Inspection Performance", A Research Paper in Ref. 7, pp. 189.

TYPE II ERROR VS DR.TC.TAI (Fig.1)

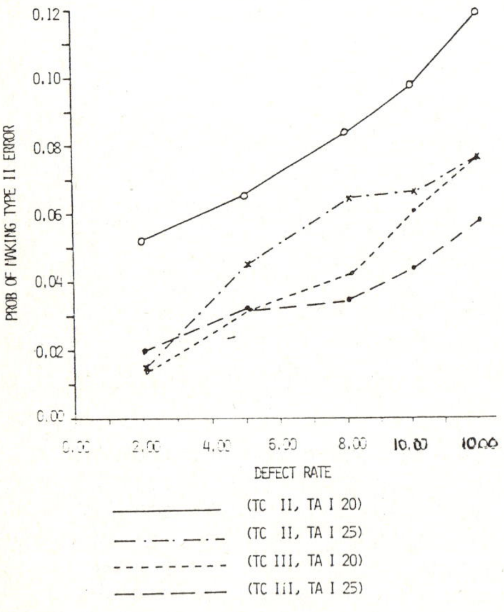

TYPE I ERROR VS DR.TC.TAI (Fig.2)

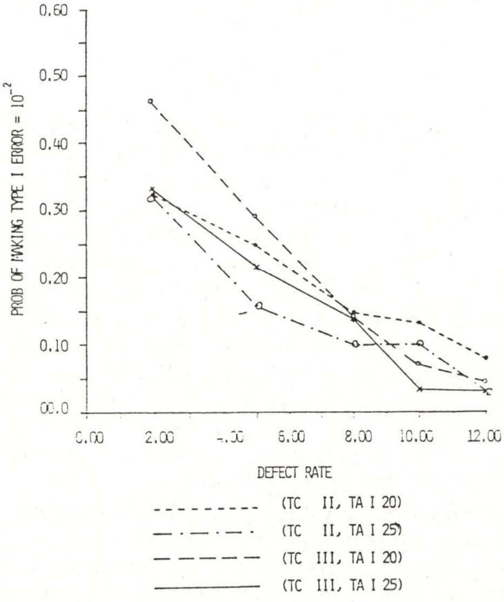

Trends in Ergonomics/Human Factors II
R.E. Eberts and C.G. Eberts (Editors)

ACCURACY OF CHECK-READING DIALS

Anil Mital and Sukumar Ramanan
Ergonomics Research Laboratory
University of Cincinnati (ML 72)
Cincinnati, Ohio 45221

Check reading tasks associated with intermittent and accurate monitoring of qualitative displays were simulated to determine the effects of several task related factors (exposure time, pointer position, percentage of deviant dials, and background color) and operator sex on human performance. Twenty-eight male and twenty-two female subjects performed the simulated tasks. The results of this simulation helped in resolving the long standing controversy about the position of pointers in check reading dials; no difference was found between the 9 o'clock and the 12 o'clock positions. For shorter exposure times (0.25 and 0.50s) dials with white background were found superior. However, performance was superior with dials which had black background when the exposure time was 0.75s. Overall, the performance of males was better than the performance of females. The human performance was also found to be independent of the number of deviant dials in the simulated tasks.

INTRODUCTION AND BACKGROUND

"Check-Reading" is the intermittent and accurate monitoring of a display of instruments to simply determine if a condition is normal or not. Situations requiring frequent, rapid, and accurate reading of instrument dials are frequently encountered in power plant control rooms, aircrafts, and many industries and are becoming even more pervasive due to increasing automation in industry. In these situations, several instrument dials are clustered together and the purpose of check reading is to assure normal working and to detect an abnormal condition or error, if one exists. The normal is generally defined by the pointer taking an orientation of 9 o'clock, 12 o'clock, etc. The normal operation is indicated by all pointers taking the same alignment.

Situations which generally require check-reading are critical in nature. Therefore, it is crucial that any error or errors be detected quickly and accurately. Several researchers have studied the problem of arranging the cluster of instrument dials over the past year. The bulk of work was, however, done between 1948 and 1964. Although, since 1964, no scientific investigation has been reported, the final solution to the problem is yet to be determined. The conclusions reported by many earlier studies are also contradictory in nature.

The factors that are likely to have an effect on the accuracy of check-reading can be divided into three main groups:

1. Task factors such as exposure time, duration of the check reading task, percentage of deviant dials at any given time, the number of dials in the cluster, the dial diameter, pointer orientation, and background color of the dial.

2. Individual factors such as visual acuity, experience, training, sex and age.

3. Environmental factors such as temperature, humidity, lighting and noise.

While the first group of factors deal with the design of dials and configuration of clusters, factors in the other two groups influence, or are expected to influence, human performance. Conclusive information about all these factors should lead to more accurate check-reading.

Several of the factors listed above have been investigated in the past by various researchers (Johnsgard (1953), Ross et al. (1955), Spragg and Rock (1952), Spragg and Wulfeck (1955), Warrick and Grether (1948), White (1951), Dashevsky (1964), Oatman (1964)). The most critical of all the design factors appears to be the orientation of the instrument dial pointer. To date there is no conclusive evidence as to which configuration, from among the several studied from time to time, is best. Johnsgard (1953) has shown that the type of pointer pattern facilitates efficient check-reading. Pointer positions of 12 o'clock and 9 o'clock have been found to be superior compared to other alignments. However, there is no clear agreement on which of these two is better. Dashevsky (1964) did not find any consistent difference between the two positions, although in some studies prior to his (Warrick and Grether (1948), White (1951)), the 9 o'clock position was found to be superior.

The effect of many other design factors such as exposure time, background color of the dial, percentage of deviant dials present, etc. has never even been investigated. It was hypothesized in this study that the simultaneous variation of task factors will influence the check reading accuracy. The purpose of this study was, therefore, to determine the effect of simultaneously varying several task factors on the check reading accuracy. How does the sex of the operator affect performance was also of interest. The specific factors investigated were: (1) operator sex, (2) exposure time, (3) pointer position, (4) percentage of deviant dials, and (5) background color of dials. Extended pointer dials were used in the simulation. It has been reported that these lead to superior performance (Dashevsky (1964), Oatman (1964)).

METHOD

Twenty-eight males and twenty-two females, ranging in age between 20 to 25 years, voluntarily participated in the experimant. All subjects had normal vision, with or without correction lenses. Only one subject performed at one time.

Experimental Design

A randomized complete block factorial design was used. Subjects were treated as blocks. As mentioned earlier, five different independent variables were used: operator sex (both male and female subjects participated), exposure time (0.25, 0.50, and 0.75s), pointer position (both 9 o'clock and 12 o'clock positions were used to determine which of the two was superior), percentage of deviant dials (1%, 2%, and 3% - to determine if the performance changes with the number of deviant dials), and background color of dials (black dials on white background and white dials on black background - Figures 1 and 2). There were a total of thirty-six treatment combinations (3 exposure times x 2 pointer positions x 3 percentages of deviant dials x 2 backgrounds) per subject. Subjects performed treatments in a random order. The number of errors made by subjects was the response variable. Several variables, such as illumination, size of the projected dial, etc., were controlled. Details are given in the experimental setup sub-section.

Apparatus

A standard 35 mm slide projector, with a standard lens, was used to project clusters of deviant dials on the screen. The exposure time was controlled with the help of a Hunter Decade Interval Timer (Model 111-B, Series D) which was connected to a Gerbrands Tachistascopic shutter and driver (Model G 1166) (Figure 3).

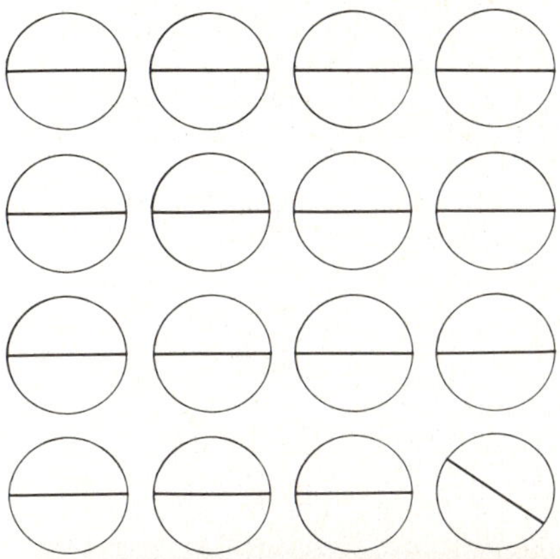

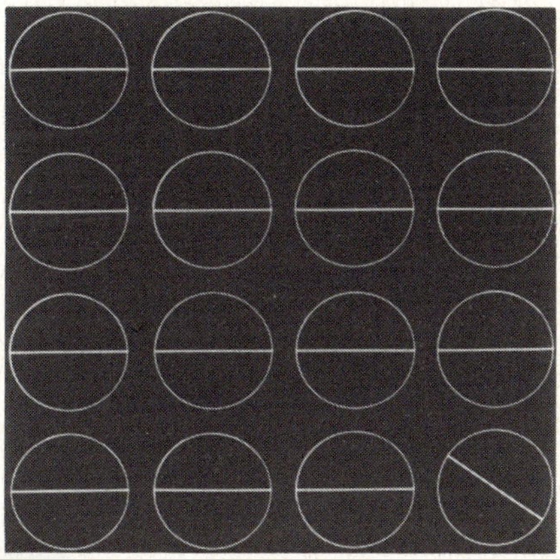

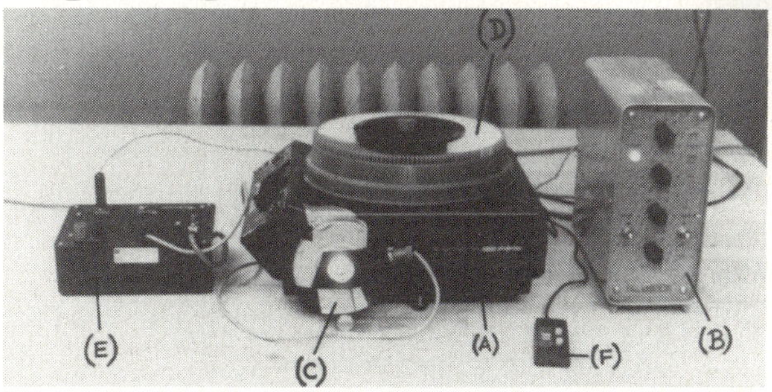

Figures 1, 2, 3 (in clockwise order)

(A) Kodak 35 mm Slide Projector. (B) Hunter Decade Interval Timer (Model 111-B,Series D).
(C) Gerbrands Tachistoscopic Shutter (Model G1166). (D) Kodak 140 Slide Tray.
(E) Gerbrands Tachistoscopic Shutter Driver (Model G1166). (F) Remote Control Switch.

Experimental Setup

Thirty-six combinations of 90 slides/combination, with 16 dials per slide in a 4x4 matrix (Dashevsky (1964), Oatman (1964)), were projected on a screen. The exposure time, percentage of deviant dials, the pointer position, and background color varied from one combination to the next in a random order. Each combination had 1440 dials (90 slides x 16 dials/slide). The randomization was done by generating random numbers with the help of a FORTRAN program.

The projector was placed 7 feet from the screen at a height of 34 inches. The projected dial diameter was maintained at 3 inches (Johnsgard (1953), Warrick and Grether 1948), Oatman (1964)); the projected pointer width was 3/32 inch. All slides were photographed and processed together to control background intensity. The screen illumination was maintained at approximately 9 foot lamberts (Spragg and Rock (1952), Spragg and Wulfeck (1955)).

Experimental Procedure

The subject was seated 50 inches from the screen (Johnsgard (1953), Warrick and Grether (1948), Oatman (1964)). Only one combination was screened at a time. The value of exposure time, percentage of deviant dials, pointer position, and background color changed only between the combinations, but not within the combination. Slides were screened one after another in sequence. Each slide had only one deviant dial present, if at all it had any. There were a total of 15 slides with deviant dials for 1% deviant dials, 29 slides with deviant dials for 2% deviant dials, and 44 slides with deviant dials for 3% deviant dials. The deviant dial could be present in any of the sixteen positions.

Only one slide, out of 90 for a given combination, was projected at a time. The slide was exposed for a certain period of time, as dictated by the combination (0.25, 0.50, or 0.75s), and the subject was required to respond by identifying the location of the deviant dial, if one was present. This response was recorded by the experimenter on the data sheet affixed to an illuminated clipboard. The subject was also required to indicate non-existence of a deviant dial if none existed. All ninety slides were presented in the above manner. The protocol was the same for each combination and for each of the fifty subjects.

Each subject was also given a practice session, for familiarization, prior to actual collection of data for the experiment.

RESULTS AND DISCUSSION

The essential requirements of a good information display are: (i) to convey the necessary information and (ii) to do so without any ambiguity and error. Additionally, in case of check-reading tasks, the consequences of making an error can be extremely serious. Keeping this in mind, the number of errors made was chosen as the performance (evaluation) criterion. Subjective opinion of the subject was also recorded.

The data were statistically analyzed to determine the effects of various factors, described earlier, on performance. The Statistical Analysis System (SAS) software was used. Data analysis was carried out in two steps. In the first step, all data were included and the effect of operator sex on performance was determined. It was found that males were 36% more efficient than females. In other words, they made thirty-six percent fewer errors compared to females. This difference in performance was significant at less than one percent level.

Since there were large differences in performance between males and females, their data were analyzed separately. For both males and females, the experimental data wre subjected to an analysis of variance (ANOVA). Tables I and II show ANOVAs for males and females, rspectively. In both cases, the results were very similar. It appears that simultaneous variation of some of the task variables, from among those investigated, does indeed influence the accuracy of check-reading dials.

Exposure Time

Three different exposure times were investigated: 0.25, 0.50, and 0.75s. As the exposure time increased, fewer errors were made. The percentage of errors for each exposure time were significantly different from each other, both for males and females ($p < .001$). The average percengage error for each exposure time is shown in Table III, separately for males and females.

In previous studies, the exposure time has ranged from 0.04 s (Oatman (1964)) to 0.25s (Dashevsky (1964)). It is clear that the accuracy of check reading is a function of viewing time duration. Fewer errors will be made with a careful and relaxed look rather than with a rapid cursory look.

Background Color

In this study, two different background colors were studies: white background, with black dials and pointers, and black background, with white dials and pointers. The color effect was significant at less than one perent level for both males and females. In general, fewer errors were made with black dials and pointers on white background thatn with white dials and pointers on black background. However, this trend was not consistent. When the exposure time increased to 0.75s, performace of males and females was better with the black background color compared to the white background color. The interaction between color and exposure time was highly significant ($p < .0001$). Figure 4 shows the interaction between the two factors.

This finding is not really surprising. Several previous studies have also reported contradictory findings with regard to black and white background colors. While Holmes (1931) and Taylor (1934) have stated that black color on white background is better, Craik (1941) and Berger (1944) contradicted them by stating exactly the opposite.

TABLE I. Analysis of Variance for Male Data.

Source	Degrees of Freedom	Sum of Squares	F-Value	Level of Significance (%)
Subjects	27	0.082	4.58	0.01
Exposure Time (EX)	2	0.033	24.88	0.01
Background Color (BA)	1	0.024	36.32	0.01
Pointer Position (PP)	1	0.0003	0.43	51.14
Percentage of Deviant Dials (PD)	2	0.0001	0.11	90.01
EX*BA	2	0.19	14.28	0.01

Note: Only those statistically significant interactions, which were meaningful, are included.

TABLE II. Analysis of Variance for Female Data.

Source	Degrees of Freedom	Sum of Squares	F-Value	Level of Significance (%)
Subjects	21	0.061	4.27	0.01
Exposure Time (EX)	2	0.087	63.62	0.01
Background Color (BA)	1	0.036	51.93	0.01
Pointer Position (PP)	1	0.00003	0.04	83.90
Percentage of Deviant Dials (PD)	2	0.002	1.36	25.64
EX*BA	2	0.038	28.16	0.01

Note: Only those statistically significant interactions, which were meaningful, are included.

TABLE III. Standard Deviation (S) and Average ($\bar{x}$) of Percentage Error for Various Task Factors

Factor	Males $\bar{x}$	Males S	Females $\bar{x}$	Females S
Exposure Time				
0.25 second	2.18	3.55	3.29	4.07
0.50 second	1.28	2.70	1.70	2.88
0.75 second	0.82	2.13	0.79	2.04
Background Color				
White	0.95	2.15	1.27	2.37
Black	1.93	3.47	2.61	3.91
Pointer Position				
9 o'clock	1.36 |	2.73	1.88	3.08
12 o'clock	1.47 |	3.05	1.92	3.44
Percent of Deviant Dials				
1%	1.36 |	2.90	1.82	3.38
2%	1.42 |	3.11	2.09	3.50
3%	1.45 |	2.60	1.75	2.81

Note: Vertical line indicats that these values are not different from each other ($p \geq .10$).

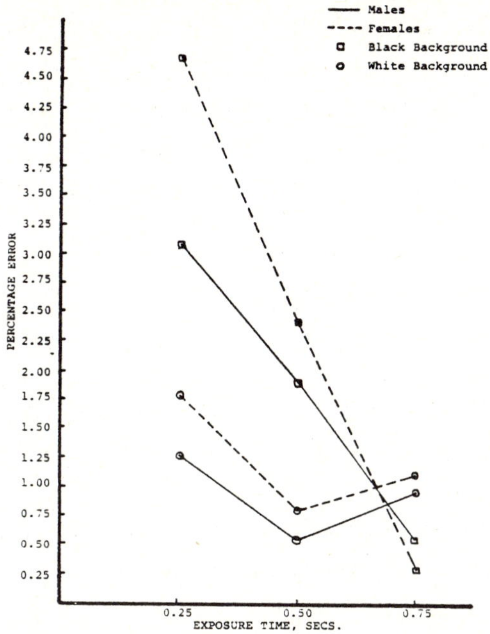

Figure 4

Theoretically, both combinations have the same contrast. Therefore, for the same exposure time there should be no diffrence between black on white and white on black. However, this was not the case. It appears that selection of background color is important in check-reading tasks and depending upon the time routinely available, either a black or a white background should be selected.

Pointer Position

There was no difference in performance of males and females when dials with pointers in either 9 o'clock or 12 o'clock position were used. The difference in percentage error (Table III) was insignificant ($p \geq .10$). This finding is consistent this those of Dashevsky (1964). Thus from the performance viewpoint, either of the two positions is acceptable.

Subjects, however, preferred the 12 o'clock positoin. Therefore, from the design standpoint the 12 o'clock position is recommended.

Percentage of Deviant Dials

One of the objectives of this study was to determine the effect of varying the percentage of deviant dials on performance. Three different percentages were used: 1, 2, and 3 percent. No significant difference in performance was found ($p < .0001$). Subjects responded to increased number of deviant dials without compromising their performance (Table III). Presence of more deviant dials, therefore, is not an alarming concern as far as human performance is concerned.

CONCLUSIONS

The findings of this study lead to the following conclusions:

1. Accurate performance in check-reading tasks is significantly influenced by the sex of the operator. On the average, in this study, male subjects performed their tasks 36% more efficiently than females. However, as the exposure time increases, the performance differences disappear.

2. For exposure times up to 0.50s, black dials and pointer on a white background are desirable. With 0.75s, or greater, exposure time, the black background is preferable. In general, larger exposure times (more than 0.75 s) are desirable.

3. There is no difference between the 9 o'clock and 12 o'clock pointer positions. On the basis of subjective preference, however, the 12 o'clock position is recommended.

4. Human performance is not affected by the number of deviant dials encountered during check reading tasks.

REFERENCES

[1] Berger, C.J., Stroke-Width, Form and Horizontal Spacing of Numerals as Determinants of Threshold of Recognition, Journal of Applied Psychology 28 (1944), 208-231.

[2] Craik, K.J.W., Instrument Lighting for Night Use, Air Ministry Flying Personnel Defense Comm. Rep. No. FPRC 342 (1941).

[3] Dashevsky, S.G., Check Reading Accuracy as a Function of Pointer Alignments, Patterning, and Viewing Angle, Journal of Applied Psychology 48 (1964), 344-347.

[4] Holmes, G., The Relative Legibility of Black Print and White Print, Journal of Applied Psychology 15 (1931), 248-251.

[5] Johnsgard, K.W., Check-Reading as a Function of Pointer Symmetry and Uniform Alignment, Journal of Applied Psychology 37 (1953), 407-411.

[6] Oatman, L.C., Check-Reading Accuracy Using an Extended-Pointer Dial Display, Journal of Engineering Psychology 3 (1964), 123-131.

[7] Ross, S., Katchmar, L.T., and Bell, H., Multiple Dial Check Reading: Pointer Symmetry Compared with Uniform Alignment, Journal of Applied Psychology 39 (1955), 215-218.

[8] Spragg, S.D. and Rock, M.L., Dial Reading Performance as a Function of Brightness, Journal of Applied Psychology 36 (1952), 128-137.

[9] Spragg, S.D. and Wulfeck, J.W., The Effect of Immediately Preceding Task Brightness of Visual Performance, Journal of Applied Psychology 39 (1955), 237-243.

[10] Taylor, C.G., The Relative Legibility of Black and White Print, Journal of Educational Psychology 25 (1934), 561-578.

[11] Warrick, M.J. and Grether, W.F., The Effect of Pointer Alignment on Check Reading of Engine Instrument Panels, Memo. Report MCREXD 694-17, USAF Air Material Command (1948).

[12] White, W.J., Effect of Pointer Design and Pointer Alignment on the Speed and Accuracy of Reading Groups of Simulated Engine Instruments, USAF AML Tech. Rep. No. 6014 (1951).

[13] White, W.J., Warrick, M.J. and Grether, W.F., Instrument Reading: III. Check Reading of Instrument Groups, Journal of Applied Psychology 37 (1953), 302-307.

III

STRESS, FATIGUE, WORKLOAD AND VIGILANCE

Trends in Ergonomics/Human Factors II
R.E. Eberts and C.G. Eberts (Editors)

THE PRINCIPLE OF MAXIMAL ADAPTABILITY IN SETTING STRESS TOLERANCE STANDARDS

P.A. Hancock

Department of Safety Science and
Human Factors Department
Institute of Safety and Systems Management

and

M.H. Chignell

Department of Industrial and Systems Engineering

University of Southern California
Los Angeles, California
U.S.A.

The present paper explores the impact of stress on human behavior at several operational levels. Stress is linked to current notions of attention. An isomorphic relationship between actions at a physiological and psychological level is proposed and made explicit in the case of one environmental stressor. This relationship is generalized into a theory of stress and capability and is used to examine practical problems of stressor interactions.

INTRODUCTION

In this paper we intend to present a different perspective from which to view the action of stress upon human behavior. It is built upon earlier work which described preliminary proposals that are more explicitly developed here (e.g., Hancock, 1984a; Hancock & Pierce, 1984). We intend to relate the action of stress to current notions of human attention (e.g., Kahneman, 1973; Wickens, 1980) and to show that there is a form of isomorphic relationship between the modes of behavior at a physiological and psychological level. Particular examples are drawn from performance under thermal stress for two reasons. First, we have more familiarity with such effects than those under a wide variety of alternate stressors and second, thermal stress and extremes of heat in particular, represent one of the most simple stress cases. This latter point is developed below.

STRESS

Generically, stress may be regarded as being composed of three co-existing, non-exclusive entities. The first of these is an environmental value or an input element which reflects a specific metric of the physical surround. Typically, this is viewed as some adverse circumstance that the human is obliged to respond to, but is better thought of as a dynamic change in condition from the current state. The second element in the three part definition is this response undertaken by the human usually in mitigation of the perturbing input. The third and final entity is some form of output which reflects resultant individual response. Each of these elements may be distinguished at a physical, physiological, psychological and social level depending upon the focus of interest. Stress at any level has widespread effects at each alternate level, but exactly how these diverse responses are "mapped" to each other remains largely unknown. In the present paper, we show for one example how such responses at a physiological and psychological level may be mapped and thus what isomorphic relationship exists. An immediate problem is the quantitative identification of stress within each of the three elements named and this is achievable using a heat stress example.

SIMPLICITY OF THERMAL STRESS

Quite simply, for thermal stress it is possible to establish quantifiable components in each of the above three elements. Input can be represented by a series of physical measures which can be combined in differing ways to form heat stress scales. To improve such scales for use in conjunction with exposed individuals it is also feasible to include human reactions to heat and thereby produce a more practical scale. This results from the combination of the input stress element and the response. This response, the acute and chronic adaptation process, is designed to dissipate the impact of the thermal stress and may be quantified simply. Amongst such adaptation processes the most obvious examples are thermoregulatory actions generated by exposure to relatively short term, extreme conditions. Finally, physiological output may be viewed as change in either peripheral, or peripheral and deep, body temperatures and can be measured accordingly. Output on a psychological level is measured through efficiency of behavioral response assessed most often through the use of mental or neuromuscular tasks. While this overall identification is also possible for a wide variety of alternate stressors and indeed is necessary for the elaboration of a generalized theory, it is often less straightforward than the thermal stress example.

There is one further benefit from using thermal stress and extreme heat in particular. One problem that can beset research on stress effects is the variation in reaction of the individual subjects involved. Among others, Rohles (1971) has indicated that such differences covary with the stress condition such that differences are minimized at extremes (see also Webb, 1963). Therefore, at thermal extremes the human response to stress represents one of the more clear situations, unencumbered by wide interpersonal variation and overall trends of interest may be more clearly captured.

STRESS AND ATTENTION

Under extreme heat stress, Hancock (1981) has indicated that highly simple mental performance tasks are degraded only marginally before the subject reaches the physiological tolerance limit. Further analysis has indicated that more complex tasks, e.g., dual compared to single task performance, were degraded by smaller perturbations to deep body temperature (Hancock, 1982). A recent assessment of vigilance indicated that any non-compensable change in deep body temperature is sufficient to reduce the long-term ability to sustain attention (Hancock, 1984b). These results have been explained typically through the use of the arousal construct, as has the action of a variety of state and environmental stressors (Poulton, 1977).

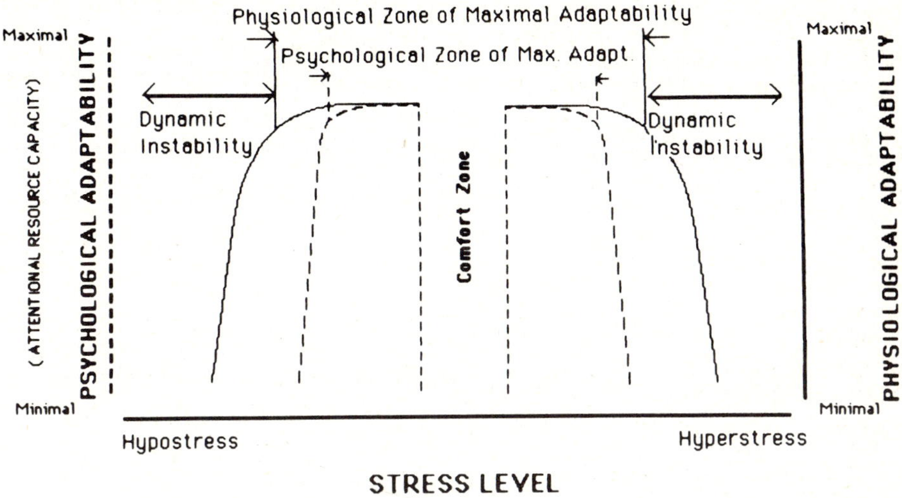

Figure 1

Change in Physiological Adaptability (Solid Lines) and Psychological Adaptability or Attentional Resource Capacity (Outer Dashed Lines) as a function of Stress Level. Embedded is a Comfort Zone (Inner Dashed Lines). Within the Zones of Maximal Adaptability, negative feedback control predominates. Beyond Zonal Limits, dynamic instability predominates where positive feedback gives rise to progressive System failure.

In contrast, recent models of attentional capacity may be used to assess the effect of stress on performance. Kahneman (1973) proposed that this attentional capacity was not a fixed, or static, quantity. Indeed, such capacity was viewed as dynamic and could vary according to several factors, each of which contributed to arousal state. Similarly, various environmental and task demands might act to reduce capacity. Cohen (1980) has proposed that stress may compete for or reduce the level of capacity. More recently, Hancock and Pierce (1984) have stated that stress which overcomes efficient homeostatic action acts to drain attention. Their summary, which describes performance breakdown under the sequential reduction of attention was derived from examples of vigilance capability in thermal extremes. However, it should be noted that as capacity is reduced, the operator compensates by narrowing attention to focus on the informational cues of greatest salience (Easterbrook, 1959; Hancock & Dirkin, 1983).

Kahneman's model of attention assumed a global pool of capacity, where attentional resources could be allocated to tasks as required. In Wickens' (1980) hybrid conceptualization it is assumed that there are separate and independent pools of attentional resources which will each be responsive to a unique type of task demand (e.g., auditory vs. visual input, manual vs. vocal output). Both views of attention appear to be applicable to the analysis of human performance. Task demands may require the use of individual resource pools depending on specific characteristics of that task whereas, in our view, environmental stressors act in a more global manner and reduce capacity across all pools of attentional resources.

A related view of attention is that some tasks demand attentional resources due to the unpredictable nature of the stimulus-response relationship. Other tasks are more predictable because of their consistency, allowing automated processing to occur (Schneider & Shiffrin, 1977). One of the characteristics of these latter tasks is that they may be performed in a relatively "attention free" fashion. In the present context, such a task would be less vulnerable to the attention draining effects of stress. This would result in an increase of the effective zone of psychological maximal adaptability, see Figure 1. The greater the task automation the closer the former zone comes to the limits of physiological adaptability.

MAXIMAL ADAPTABILITY

Small amounts of stress can be adapted to, but increasing levels of stress will eventually produce a breakdown in task performance to be followed by physiological incapacitation. The amount of input stress which can be tolerated by a human performer defines a zone of maximal adaptability, where physiological and psychological mechanisms can achieve compensation to the environmental disturbance. The proposed relationship between stress and adaptability is shown in Figure 1. A stressor (represented on the abscissa) such as thermal stress, will vary between extreme values, with a zone of comfort occurring at some intermediate region. We recognize that any specific input stress may exhibit unipolarity or bipolarity, although Figure 1 illustrates the bipolar case which is more pertinent to the thermal stress examples given. The comfort zone will be included within a larger region where psychological adaptability remains viable. As the level of input stress becomes more extreme, however, increasing discomfort will be followed by a rapid decrease in psychological adaptability (the dotted line). For the specific case of heat stress, genesis of the failure in task efficiency occurs at that point where complete physiological compensation to the environment is no longer possible and change in body temperature (the output stress entity) becomes manifest. Eventually, loss of psychological adaptability is followed by a similar decrease in physiological adaptability. In this event, homeostatic mechanisms are overcome by the stressor and the regulatory system changes mode of operation from the typical negative feedback to a positive feedback function. This is represented for both physiological and psychological processes as a region of dynamic instability in Figure 1. The end point of the psychological adaptability is the absence of all attentional resources and this we take to be a reasonable definition of unconsciousness. The end point of physiological adaptability is the failure to maintain life sustaining functions.

The boundaries of the zones of maximal adaptability of physiological and psychological response may be continuous, requiring a statistical definition as in the concept of a psychophysical threshold. Alternatively, the boundaries may themselves represent discontinuities. In this latter case, the point of singularity can be detected using some form of trend analysis for the case of a single stressor. However, in many industrial environments, the human operator is exposed to the impact of multiple stressors. In this multidimensional case where interaction of several stressors can occur, the point of discontinuity can be represented as a cusp on a topological manifold (Figure 3). The type of cusp will depend on the nature of the interacting stressors and may be described using the tenets of catastrophe theory (Zeeman, 1977).

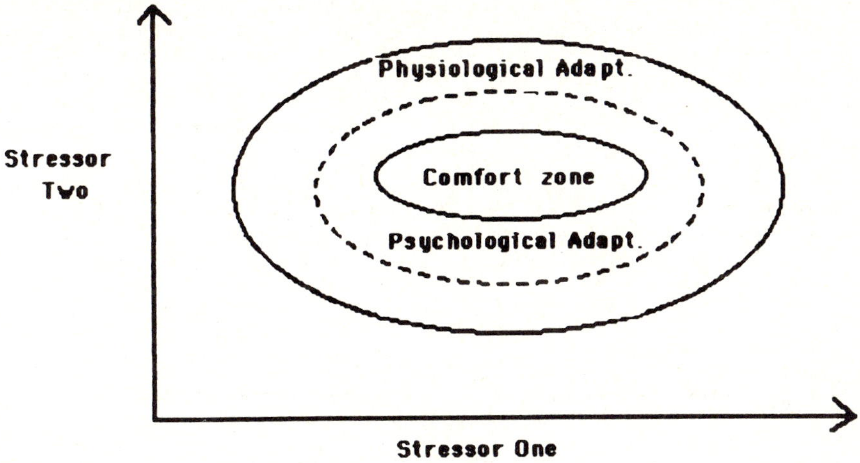

Figure 2

Zones of Comfort, Psychological Adaptability and Physiological Adaptability within a Spatial Representation of the Interaction between Two Stressors.

The basic model describing changing adaptability due to stress level can be also applied to situations where there are multiple stressors with interacting effects on both physiological and psychological response. Interactions between stressors can be defined using a spatial model. Figure 2 shows zones of comfort, psychological adaptability and physiological adaptability, respectively, in a bivariate space of stressors. The shape of these zones will depend on the nature of their interaction. Pure additivity (no interaction) will result in circular zones, whereas interactions will produce elliptical or more irregular shapes. The pattern of nesting for the three zones reflects the increasingly strict requirements for physiological adaptability, psychological adaptability, and comfort, with associated restrictions in the appropriate ranges of environmental variation. Figures 1 and 2 can be combined to show the rapid decrease in adaptability which will occur outside the zones of moderate stress level, as shown in Figure 3. The level of psychological adaptability for a particular combination of stressors is represented as a point on the surface of a three-dimensional manifold (dotted lines). This manifold is nested inside another (continuous lines) which shows a similar response for physiological adaptability. The larger size of the manifold for physiological adaptability reflects the assumption (also illustrated in Figure 2) that the zone of maximal adaptability is larger for physiological than for psychological response. The same descriptive model can be used for multivariate combinations of three or more stressors. The region of physiological or psychological adaptability would then be represented by an n+1 dimensional manifold with adaptability being equivalent to the height over an n-dimensional plane of stressors. The purpose of this topographic model is to provide an intuitive understanding of the relationship between stress and adaptability. Knowledge about where maximal psychological adaptability occurs can form the basis for early detection of stress loadings which will eventually lead to systemic breakdown caused by dynamic instability of physiological response.

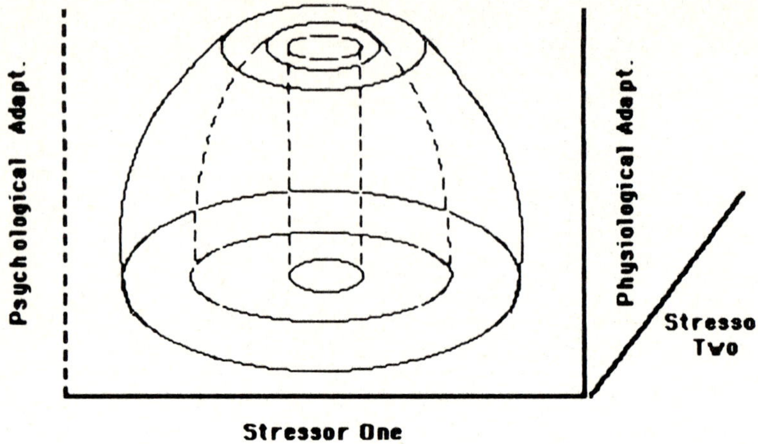

Figure 3

The Effect of Two Stressors on Physiological and Psychological Adaptability Comfort is represented as an all-or-none concept (Central Dotted Cylinder). Extreme levels of Stress lead to a rapid decline in Psychological (Dotted Line) and Physiological (Continuous Line) Adaptability. This results in nested three-dimensional Manifolds as shown. Note that Figure 2 is a plan view of the present Figure.

STRESSOR INTERACTIONS

Together, Figures 2 and 3 imply that stressors can be located as orthogonal axes in the spatial model. In practice, two stressors may be synergistic, as appears to be the case with noise and vibration, or the stressors may be antagonistic as in the example of noise and sleep loss, where the presence of one stressor tends to offset the effect of the other. Interactions between stressors necessitates a vector representation where the stressors can be located in oblique as well as orthogonal relationships, as shown in Figure 3. Each stressor can be represented as a vector within a multi-dimensional space. A region centered on the origin of this space will represent the comfort zone. The orientation of the vector will reflect the qualitative nature of the stressor. The length (magnitude) of the vector will vary with the level of the stressor. Vector summation can be used to predict interactions between stressors. Ekman and Lindman (1961) developed a vector model of multidimensional scaling and similarity judgments. In a close analogy with the present model, the length of the vector represented its intensity while the orientation of the vector referred to its qualitative character. The description of stressors within a vector (spatial) model allows quantitative assessment of stressor interactions and the prediction of the location of points of maximal adaptability for multiple stressor combinations. A multivariate array of stressors can be replaced with a single vector which represents their summation. This vector will then define a point on a multidimensional manifold (as illustrated in Figure 3) which will in turn specify the levels of psychological and physiological adaptability.

The vector model is a convenient way of thinking about stressor interactions but quantification of stressors in terms of vector orientations and magnitudes is difficult. Scaling of stressors is not yet possible as insufficient quantitative data are available. Much of the relevant research has focused on stressors taken singly rather than in multivariate combinations. Factor analysis or multidimensional scaling (Kruskal, 1977; Davison, 1983) can be used to transform physical specifications of stressors into a reduced set of latent factors (dimensions), providing that sufficient data about similarities or correlations in the effects of different stressors are collected. Alternatively, canonical correlation analysis (Harris, 1975) might be used, although the resulting canonical variates are often difficult to interpret without further analysis (Bentler & Huba, 1982). Initially, exploratory investigations of the relationship between stress and adaptability are likely to be most useful (Skinner, 1978). In the absence of detailed information on how stressors interact, even an approximate scaling of stressors would be helpful in developing a model of the relationship between stress and adaptability. Such a model would indicate the progress of an exposed individual towards regions of dynamic instability. Intuitively, an episode of exposure to environmental stressors would trace a path on the surface of a multidimensional manifold. Breakdown in physiological or psychological adaptability would correspond to leaving the plateau of the manifold (cf. Figure 3) where adaptability is not impaired significantly by the environmental stressors.

SUMMARY

In many human endeavors, a major concern is the permissible level of stress to which an individual may be exposed. While limits have been described for individual stressors, a coherent overall theory has yet to emerge. Although the arousal position served a useful function in this context, its lack of predictive power is a serious drawback. This paper has outlined a different perspective. The current position is based on the concept of adaptability in both physiological and psychological terms, where the latter construct is tied to recent theories of human attention. Research findings concerning performance in the face of heat stress suggest that significant breakdown in capability is associated with the point at which thermoregulatory action is no longer capable of maintaining a state of dynamic stability. We expect this isomorphism between physiological action and psychological response to hold for multiple stress conditions.

Knowledge of this zone of maximal adaptability will allow safe stressor limits to be specified. The concept of maximal adaptability can be extended to combinations of multiple stressors and it is hypothesized that the principle of maximal adaptability will also apply to stressor interactions (Figure 3). Spatial models of stressor interaction will be useful in describing the relationship between stress and human adaptability, but considerable empirical work is required. The effect of stress is seen in a reduction of available attentional capacity (psychological adaptability) and an increase in physiological strain which has to be compensated for by regulatory systems (physiological adaptability). This concept allows for the development of more precise tolerance limits which may be based not only upon input values of the physical stressor, as is the case for most of the present exposure standards, but also allows for the designation of permissible tolerances for each individual worker.

REFERENCES

[1] Bentler, P.M. and Huba, G.J., Rotation in canonical correlation analysis, in: Hirschberg, N. and Humphreys, L.G. (eds.), Multivariate Applications in the Social Sciences (Erlbaum, N.J., 1982).

[2] Cohen, S., The aftereffects of stress on human performance and social behavior: a review of research and theory. Psych Bull 88 (1980) 82-108.

[3] Davison, M.L., Multidimensional Scaling (Wiley, New York, 1983).

[4] Easterbrook, J.A., The effect of emotion on cue utilization and the organization of behavior. Psych Rev. 66 (1959) 183-201.

[5] Ekman, G. and Lindman, R., Multidimensional ratio scaling and multidimensional similarity. Reports from the Psychological Laboratories, University of Stockholm. (1961) 103.

[6] Hancock, P.A., Heat stress impairment of mental performance: a revision of tolerance limits, Aviation, Space and Environmental Medicine, 52 (1981) 177-180.

[7] Hancock, P.A., Task categorization and the limits of human performance in extreme heat, Aviation, Space and Environmental Medicine, 53 (1982) 778-784.

[8] Hancock, P.A., Environmental stressors, in: Warm, J.S. (ed.), Sustained Attention in Human Performance (Wiley, New York, 1984) (a).

[9] Hancock, P.A., Effect of environmental temperature on display monitoring: an overview with practical implications, Am Ind Hygiene Assoc J. 45 (1984) 122-126 (b).

[10] Hancock, P.A. and Dirkin, G.R., Stressor induced attentional narrowing: implications for design and operation of person-machine systems. Proc Hum Fac Assoc Canada. 16 (1983) 19-21.

[11] Hancock, P.A. and Pierce, J.O., Toward an attentional theory of performance under stress: evidence from studies of vigilance in heat and cold, in: Mital, A. (ed.), Trends in Ergonomics/Human Factors I (North-Holland, Amsterdam, 1984).

[12] Harris, R.J., A Primer of Multivariate Statistics (Academic Press, New York, 1975).

[13] Kahneman, D., Attention and Effort (Prentice-Hall, N.J., 1973).

[14] Kruskal, J.B., Multidimensional scaling and other methods for discovering structure, in: Enslein, K., Ralston, A. and Wilf, H.S. (eds.), Statistical Methods for Digital Computers Vol III (Wiley, New York, 1977).

[15] Poulton, E.C., Arousing stresses increase vigilance, in: Mackie, R.R. (ed.), Vigilance, Operational Performance and Physiological

Correlates (Plenum, New York, 1977).

[16] Rohles, F.H. Psychological aspects of thermal comfort, ASHRAE Jrnl. (1971) 86-90.

[17] Schneider, W. and Shiffrin, R.M., Controlled and automatic human information processing I: detection, search, and attention. Psych Rev 84 (1977) 1-66.

[18] Skinner, H.A., The art of exploring predictor-criterion relationships, Psych Bull. 85 (1978) 327-337.

[19] Webb, P., Pain limited heat exposures, in: Hardy, J.D. (ed.), Temperature, its Measurement and Control in Science and Industry (Reinhold, New York, 1963).

[20] Wickens, C.D., The structure of attentional resources, in: Nickerson, R. & Pew, R. Attention and Performance VIII (Erlbaum, N.J., 1980).

[21] Zeeman, E.C., Catastrophe Theory: selected papers 1972-1977 (Addison-Wesley, Massachusetts, 1977).

Trends in Ergonomics/Human Factors II
R.E. Eberts and C.G. Eberts (Editors)

AN INVESTIGATION OF MENTAL FATIGUE MEASUREMENTS DURING A READING TASK

Thomas R. Huston
Richard L. Shell

Department of Mechanical and Industrial Engineering
University of Cincinnati
Cincinnati, Ohio 45221

This study investigated the relationship between two mental fatigue measurements--EEG brain wave signals and subjective ratings of mental fatigue. Eighteen males engaged in a five hour reading task. During this task, the subjects were monitored for their EEG signals and completed a subjective mental fatigue rating form. Correlation analysis was the statistical technique used to compare the two types of mental fatigue measurements. The results reveal some moderate correlations between the EEG signals and the subjective ratings.

INTRODUCTION

In recent years, the United States' workforce has undergone a significant change. The percentage of workers involved in manufacturing have declined while the percentage in the service sector has grown. This growth in the service sector has resulted in a greater number of white collar or knowlege workers. This increased number of knowledge workers presents a host of new human factors problems.

Perhaps the least understood of these is the problem of mental fatigue. The topic of mental fatigue is hardly new. There has been a proliferation of mental fatigue studies since the 1940s. However, these studies have typically focused on psycho-physical tasks. As an example, there are numerous studies concerning pilot mental fatigue. While there exists a paucity of information regarding knowledge worker mental fatigue, most of these studies have been too short in duration and have failed to model the type of activity in while the knowledge worker engages.

The objectives of this study were threefold. First of all, there is not an universally accepted measurement for detecting mental fatigue although many measurements have been suggested. Hence, the relationship between two of the more common mental fatigue measurements were studied. They were the EEG signals and the subjective mental fatigue rating measurements. Secondly, a mental task was selected to model an activity in which all knowledge workers engage--reading. Thirdly, the time duration or the mental task was chosen to have a sufficient duration to simulate the time a knowledge worker might work without an extended restbreak.

METHODS

Experimental Design

Eighteen male subjects voluntarily participated in the experiment. The subjects were Engineering students from the University of Cincinnati and were all in good health. The subjects engaged twice in a five hour mental task.

The mental task involved completing standardized reading tests. The reading tests had a difficulty of education level 16. The reading tests consisted of a mixture of the Graduate Record Exam (GRE), Legal Scholastic Aptitude Test (LSAT), and Graduate Management Aptitude Test (GMAT) segments.

Throughout the mental task, the subjects were monitored for their EEG signals. An array of equipment was used in the monitoring process. A Narco-system DMP-4 four channel desk physiograph was the key instrument in the array. The physiograph employed silver plated electrodes to detect the EEG signal. A Narco Biosystem MKIV preamplifier was used to magnify the weak EEG signal output by the body. The EEG signal was also sent thorugh a Khron-Hite 3550 low pass filter and then on to a HP Fourier 5466A analog to digital signal converter.

Every hour, the subjects were given a short restbreak during which they completed a subjective mental fatigue rating form. The subjective rating form design followed the recommendations given by Granjean (1981) and Cooper and Hall (1969). The rating form was bipolar in design consisting of four categories with descriptors. The four categories were the subjects' feelings of "fresh/weary", "awake/sleepy", "strong/weak", and "interest/boredom". The rating form was readily interpretive by the use of a five point scale. An example of the subjective form is outlined in Figure 1.

Figure 1
Subjective Rating Form

Fresh	1	2	3	4	5	Weary
Awake	1	2	3	4	5	Sleepy
Strong	1	2	3	4	5	Weak
Interested	1	2	3	4	5	Bored

The experimental sessions were conducted in a closed room to minimize noise disturbance. The room was climate controlled to eliminate any thermal environmental impact on the results. Moreover, the experiments were administered at the same daily time to obviate any circadian rhythm interference.

Data Analysis

One of the problems associated with the data analysis is that there is not an universally accepted means of evaluating the EEG signal. The means for evaluating the EEG range from a simple visual analysis to the sophisticated computer signal analysis. In this investigation, power spectral analysis, a Fourier Analysis technique was chosen as the means to analyze the EEG signal.

The power spectrum was calculated as follows:

$$Gxx(f) = Sx(f)\, S^{*}x(f)$$

where $Gxx(f)$ = Power spectrum
$Sx(f)$ = Fourier transform of the EEG time signal
$S^{*}x(f)$ = Complex conjugate of $Sx(f)$

Integration of the power spectrum over the frequency bandwidth yielded the energy contained within a frequency bandwidth. In this research, the bandwidths were the delta, theta, alpha, and beta bandwidths. Where the frequency bandwidths were defined as follows:

$0 < f < 4$ hz delta wave bandwidth
$4 < f < 8$ hz theta wave bandwidth
$8 < f < 13$ hz alpha wave bandwidth
$13 < f < 30$ hz beta wave bandwidth

The integral over the frequency bandwidths revealed the energy content in these bandwidths. However, to present the energy contents in a more easily understood fashion, a percent energy content index was determined for the four EEG bandwidths. This was computed by dividing the energy within each bandwidth by the total energy content of the power spectrum. This percent energy content of each bandwidth was then averaged on an hourly basis to facilitate comparison with the subjective rating results.

The subjective rating data simply was collected and did not require any transformation.

Correlation analysis was the technique selected to compare the two mental fatigue measurements. The correlation analysis was executed via the CORR procedure of the SAS statistical package.

RESULTS

The results of the correlation analysis are easily visualized in a matrix format. Examine Table 1. This analysis revealed a series of ineresting points. First, although none of the correlations were exceedingly strong, there were a number of weak correlations that were significantly different than zero. Also, there were significant correlations for the "awake/sleepy" category and each of the four EEG wavebands. This was interesting since it suggested that a subject's inner feelings about being awake or sleepy were moderately correlated with the percent energy content EEG wavebands. Somewhat similar observations were made for the "fresh/weary", and "strong/weak" categories. In each instance, the categories had significant correlations with three of the four wavebands.

A third observation involved the theta waveband. The correlations between theta and the four subjective categories were all nonzero and negative. These negative values implied that as the theta percentage values increased the subjects exhibited somehwat higher feelings of weariness, sleepiness, weakness, and boredom. Conversely, the alpha and beta wavebands each contained two very small correlation values. The correlations of the "interest/boredom" category were generally not significant. The exception was the theta waveband. This seemed to suggest that there was not any relationship between the delta, alpha, and beta wavebands and a subjective feeling of "interest/boredom".

CONCLUSIONS

Based upon the results of this study, the following conclusions were drawn. The theta waveband was the most promising for detecting changes of the subjects' inner feelings of mental fatigue within the four rating areas. While the alpha and beta wavebands were the least promising. Also, there appears to be a moderate relationship between the energy contents of the EEG wavebands and the subjects' feelings of being "awake/sleepy", "fresh/weary", and "strong/weak". Finally, the subjects' feelings of "interest/boredom" was only moderately related to the theta waveband and to none of the other wavebands.

Table 1
Correlation Matrix*

	Delta	Theta	Alpha	Beta
Fresh-weary	0.194	-0.243	0.152	0.083
	0.009	0.001	0.040	0.267
Awake-sleepy	0.295	-0.276	0.219	0.161
	0.001	0.002	0.003	0.030
Strong-weak	0.158	-0.147	0.058	0.145
	0.033	0.048	0.432	0.051
Interest-boredom	0.085	-0.206	0.047	0.044
	0.253	0.005	0.523	0.555

*First value represents the correlation between the two areas whereas the second value represents the probability that the correlation value is different than zero.

REFERENCES

[1] Borg, G., Subjective Aspects of Physical & Mental Load, Ergonomics, 1978, Vol. 21, No. 3, pp. 187-194.

[2] Cameron, C., A theory of Fatigue, Ergonomics, 1973, Vol. 16, No. 5, pp. 633-648.

[3] Cooper, G. E., and Hall, R. P., The Use of Pilot Rating in the Evaluation of Aircraft Handling Qualities, Moffet Field, California, NASA Ames Research Center, NASA TN-D-5133, April 1969.

[4] Granjean, E., Fiting the Task to the Man, London: Taylor and Francis Ltd., 1981.

[5] Muscio, B., Is A Fatigue Test Possible?, British Journal of Psychology, 1921, Vol. 12, pp. 31-46.

[6] Okogbaa, O. G., An Empirical Model for Mental Work Output and Fatigue, Unpublished Ph.D. Dissertation, University of Cincinnati, 1983.

[7] Shell, R. L., The Impact of Automation on Work Measurement, Proceedings 1982 Fall Industrial Engineering Conference, Institute of Industrial Engineering, 1982.

[8] Shell, R. L., and Okogbaa, O. G., The Effect of Mental Fatigue on Knowledge Worker Performance, Proceedings, 1983 Fall Industrial Engineering Conference, Institute of Industrial Engineering, 1983.

[9] Wierwille, W. W., Physiological Measures of Aircrew Mental Workload, Human Factors, 1979, Vol. 21, No. 5, pp. 575-593.

Trends in Ergonomics/Human Factors II
R.E. Eberts and C.G. Eberts (Editors)

EFFECTS OF COGNITIVE FACTORS ON SUBJECTIVE FATIGUE AND PERFORMANCE

Gayle Wisdom and Raja Parasuraman

Catholic University of America
Washington, D.C.
U.S.A.

The effects of knowledge of end of session and work set (amount or time) on subjective fatigue and performance were investigated. Subjects performed an addition task and periodically completed a subjective fatigue checklist. ANOVA revealed two significant main effects for fatigue ratings and a complex three way interaction on problems completed between Knowledge, Workset and Time. Results supported the validity of The School of Aerospace Medicine Subjective Fatigue Checklist and were discussed in relation to previous research.

INTRODUCTION

Subjective fatigue scales are frequently employed in operator workload assessment (Stave, 1977) and measurement of monotony and boredom (Davies, Shackelton, & Parasuraman 1983; Thackray, 1981) during continuous work. The objective of such evaluations is to measure the perception of fatigue in response to various workload demands and to assess its relationship to performance. However, cognitive factors as well as task demands may influence responses on subjective measures. Consequently, the valid interpretation of subjective fatigue or workload measures rests on identifying those factors that affect the perception of fatigue independent of actual workload demands. In addition, cognitive factors may differentially affect subjective fatigue ratings and performance.

The effects of one cognitive variable, prior knowledge of work session length have been investigated in a number of early studies (Barmack, 1939; Bills and Brown, 1929; Krueger, 1937). Prior knowledge of work session length was manipulated by informing subjects of the amount of work to be done with no time limits imposed (Amount Set), or of the time alloted for the task with no amount of work designated (Time Set), or both. The results of these studies, which use primarily mental tasks, (e.g. addition of series of digits), were inconsistent. When Time Set (TS) and Amount Set (AS) were not differentiated, results indicated that there was a slight tendency for subjects with the most work to work faster in the beginning and to show a greater work-rate decrement (Bills & Brown, 1929). When TS and AS were independently manipulated, these results were confirmed for the AS conditions only. TS did not affect either the initial rate or the rate decrement. Bills and Brown concluded that AS accounted for performance changes when TS and AS conditions are combined. These results suggest that TS and AS may differentially affect performance. Krueger (1937) however, found results that were at variance with Bills and Brown (1929) for both conditions. Using a similar addition task he found higher problem solving rates for both a smaller amount of problems and for shorter work periods, suggesting that TS may operate in the same way as AS.

Barmack (1939) sought to clarify further the findings of Bills and Brown (1929) and Krueger (1937), and to analyze interrelationships between the work curve, subjective fatigue, and length of the work period. Using TS conditions only, his results generally supported Krueger (1937) in that the rate of work for the one hour period was greater than for a comparable hour when subjects worked for longer periods. Barmack cautioned that while the results seemed to support Krueger's (1937) results they could not be interpreted as disconfirming Bills and Brown's (1929) findings. He suggested that since Bills and Brown used smaller units of work (minutes) subjects may work more quickly at the largest of those units, but that within larger time or amount units (hours) subjects may be "pacing" their efforts. Subjective fatigue and performance varied together in the Barmack (1939) study in all but the shortest work period. As people reported being more tired, performance rate also deteriorated. However, the subjective fatigue ratings for the longest period did not become unfavorable as rapidly as those of the one, two or three hour periods.

It is conceivable that when workers know they are about to complete a work session, they may report an increase in their subjective feelings of fatigue. Stave (1977) reported a similar finding in research with a complex psychomotor task. Using a helicopter simulator, he investigated the effects of different cockpit environments on long term pilot performance and subjective fatigue. Subjects "flew" for 3, 4, 6, or 8 hours. He found that as expected, subjective fatigue increased with time of "flight". In addition pilots seemed to adapt their feelings of fatigue to the time left to go in the mission. The highest fatigue ratings in most missions were essentially the same. In many individual cases, the fatigue ratings seemed to reach a plateau until the subjects realized the session was almost over. At that time the ratings would climb rapidly. However, performance on the instrument landing system tracking task improved, revealing a dissociation among the measures.

The Stave (1977) and Barmack (1939) results suggest that reports of subjective fatigue may be influenced by worker knowledge of end of session. Conditions in the Stave (1977) study were such that end-of-mission combined AS and TS. Both were known to the subjects who demonstrated a rapid increase in reported fatigue near the end of the mission.

TS and AS have been investigated in relation to performance and found to possibly have differential effects on initial work rates in mental tasks. The effects of knowledge of time, or of task limits as it affects fatigue feelings, have not been directly investigated. There have been some suggestions in observed results, however, that such knowledge may affect overall level of subjective fatigue and rate of change in feelings of fatigue near the end of a work session. The present study investigated the relationships between knowledge of end of session (accurate knowledge vs. inaccurate knowledge) and work set (amount set vs. time set) and subjective fatigue and performance. It was predicted that subjective fatigue would increase over time for all subjects and that it would increase at a greater rate for subjects in the accurate knowledge conditions. Subjective fatigue for amount set was expected to be greater than that for time set.

METHOD

Subjects

Thirty six subjects were drawn from undergraduate psychology student volunteers. Subjects were randomly assigned to four groups of nine each.

Materials

The addition task consisted of randomly generated pairs of six digit numbers arranged in both rows and columns with space provided on the sheet for answers. One page contained 77 problems.

The experimentally validated School of Aerospace Medicine Subjective Fatigue Checklist (Pearson & Byars, 1956) was used to measure subjective fatigue. The form is easily completed in less than one minute, and yields a fatigue score on a scale from 0 to 20.

Procedure

Subjects met at 1 p.m. in their assigned groups. They were instructed to work as quickly and efficiently as possible in adding pairs of digits. An example page of problems was provided and a five minute practice trial conducted to familiarize subjects with the task. They were instructed in standard procedures for completion of the subjective fatigue checklist and completed one checklist prior to the practice and trial blocks.

Four nine minute trial blocks were followed by one minute periods for completing the checklist. Subjects were signaled as to time limits by the experimenter. Data collection in each group was terminated after 40 minutes and a checklist obtained following the final trial block.

Subjects were randomly divided into one of four groups: Group I, Accurate Knowledge - Time Set, was informed that they were to work for 40 minutes but no set amount of problems was to be done. Group II, Inaccurate Knowledge - Time Set, received the same instructions as Group I except that they were told they were to be working for 1 1/2 hours. Group III, Accurate Knowledge - Amount Set, was given sufficient problems to occupy them for at least 40 minutes and were informed they would be working until the problems were completed. The number of problems was set so that no one was able to finish them in advance of the 40 minutes. Subjects in Group IV, Inaccurate Knowledge - Amount Set, were given sufficient problems to occupy them for 1 1/2 hours and were told they would be working until all problems were completed. All of the sessions ended after 40 minutes.

RESULTS

Subjective Fatigue Ratings

Changes in Subjective Fatigue Ratings relative to pretest scores were derived. The mean ratings of the groups' difference scores for the four trials are presented in Figure 1.

An ANOVA on the mean ratings of the groups' difference scores for the four trials revealed that the effect of Workset (Amount Set vs. Time Set) was marginally significant ($.05 < p < .10$). Time Set groups reported greater fatigue than Amount Set groups. A Scheffe post-hoc analysis revealed that the Accurate Knowledge - Amount Set group reported significantly less fatigue than each of the other groups. The Accurate Knowledge - Time Set group reported significantly greater fatigue than the Inaccurate Knowledge - Amount

Set group. In addition there was a significant effect of Trials ($p < .01$). All groups reported greater fatigue over time. As Figure 1 indicates, there was a fairly substantial increase in rated fatigue over time, corresponding to 15-20% of the range of the School of Aerospace Medicine Subjective Fatigue Checklist.

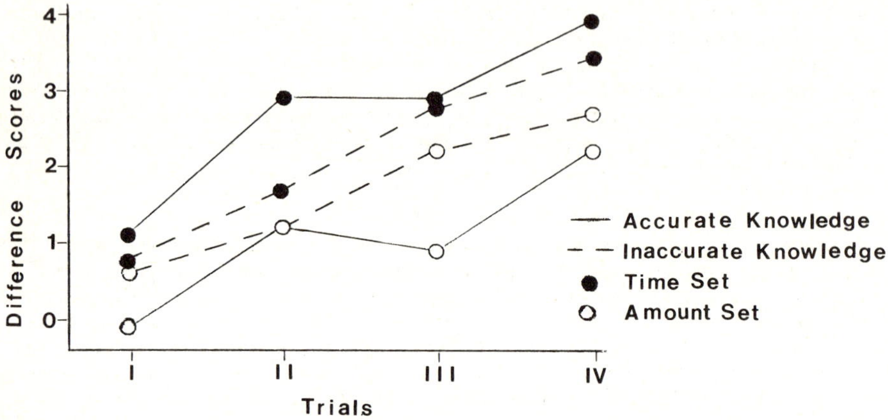

Fig. 1: Mean difference from pretest scores, subjective fatigue ratings.

Performance

The number of problems subjects attempted during a nine minute trial was recorded whether or not they were correct. To improve data reliability, the number of problems attempted was averaged over trials I and II and over trials III and IV. The mean values for the number of problems attempted are presented in Table 1.

Table 1. Mean values for number of problems attempted. Trials I & II and III & IV combined for analysis.

	Workset			
	Amount Set		Time Set	
Trials	I&II	III&IV	I&II	III&IV
Accurate Knowledge	52.61	54.38	60.94	60.38
Inaccurate Knowledge	57.05	57.22	55	58.5

Anova revealed a marginally significant effect of Trials ($.05 < p < .10$). For all but the Accurate Knowledge - Time Set group the number of problems attempted increased slightly or revealed little change. A three way interaction for Knowledge, Trials and Workset was also found. A Scheffe post-hoc analysis revealed that the Accurate Knowledge - Time Set group attempted significantly more problems than the Accurate Knowledge - Amount Set group.

There were no significant effects of the experimental variables on error rates.

DISCUSSION

The results support the validity of the School of Aerospace Medicine checklist for evaluating subjective fatigue in a prolonged task. As expected all groups reported more fatigue over time. In addition, the subjective ratings varied with performance for the fastest and slowest groups. There was also some dissociation among subjective fatigue ratings and performance rate. For subjective ratings, time on task and workset condition primarily influenced changes in reports of fatigue while knowledge of work session length had no significant effects.

The marginal effect of an increase in work-rate over trials can be interpreted in two ways. First the subjects may have evidenced some practice effect over time. Second, the one minute time allowed for completion of the subjective fatigue scale was observed to be more than sufficient time for most subjects, allowing for brief rest periods between trials. Rest periods as short as 1 minute are known to reduce effects of prolonged work on subjective fatigue and vigilance performance (Davies & Parasuraman, 1982; Davies et al., 1983).

The three way interaction for problems attempted is complex and difficult to interpret. The Anova and post-hoc analyses suggest that two effects were present. First the number of problems attempted increased for the Accurate Knowledge - Amount Set and Inaccurate Knowledge - Time Set groups. The reason for this is not clear. The second effect is fairly strong. When subjects had accurate knowledge of the 40 minute session length they attempted more problems in the Time Set than in the Amount Set condition but there were no differences when they had inaccurate knowledge. This is consistent with the view that subjects' expectations influenced their performance. All students reported to the experiment expecting to participate for approximately 1 1/2 hours. The Accurate Knowledge - Amount Set group may have treated the experiment as an untimed test on less material than they originally anticipated and set their work pace low, feeling they had as much time as needed and could work at a comfortable pace. The Accurate Knowledge - Time Set group may have treated the experiment as a timed test and worked at the highest rate. The inaccurate knowledge groups expected to be working for 90 minutes or to be doing more problems and instructions supported this expectation so that they may have paced their output to the full time.

Previous research suggested that knowledge of work session length as a cognitive variable could influence reports of fatigue where subjects in the longest sessions did not report increased fatigue at the same rate as those in the shorter sessions (Barmack, 1939; Stave, 1977). These results were not confirmed in the present study. The difference in findings suggests that task type (i.e. cognitive vs. complex psychomotor), task duration (minutes vs. hours), and extensiveness of practice may interact differently with cognitive factors of knowledge of work session length and workset.

One similarity with Stave's (1977) findings is that even as reported fatigue increased, performance did not deteriorate. Rates of work did not decline and improved slightly, in some cases. In addition, Barmack's (1939) findings of an increased work rate for shorter time set (Accurate Knowledge) was supported.

Future research with primarily mental tasks may benefit from the use of heavily practiced subjects who may be better able to assess and report feelings of fatigue based on information of task duration. Further research should also investigate these cognitive variables in relation to complex psychomotor tasks such as flying missions in flight simulators. This research might be further improved by using the dual-task mental workload assessment paradigm to measure the effects of these variables on performance and the allocation of attentional effort, as well as on the subjective experience of fatigue. Further investigations of the complex relationship between these cognitive variables and reports of subjective fatigue and performance may clarify the factors which may differentially influence subjective fatigue and performance.

REFERENCES

[1] Barmack, J., The length of the work period and the work curve, Journal of Experimental Psychology, 3 (1939) 109-115.

[2] Bills, A., and Brown, C., The quantitative set, Journal of Experimental Psychology, 12 (1929) 301-323.

[3] Davies, D. R., and Parasuraman, R., The Psychology of Vigilance (Academic Press, London, 1982).

[4] Davies, D. R., Shackelton, V. J., and Parasuraman, R., Monotony and boredom, in: Hockey, G. R. J. (ed.), Stress and Fatigue in Human Performance (Wiley, London, 1983).

[5] Krueger, W., The influence of amount limits and time limits upon the rate of work, Journal of Applied Psychology, 21 (1937) 113-118.

[6] Pearson, R., and Byars, Jr., G., (1956). The development and validation of a checklist for measuring subjective fatigue, Air University School of Aviation Medicine, USAF, Randolph AFB, Texas, (December, 1956).

[7] Stave, A., The effects of cockpit environment on long-term pilot performance, Human Factors, 19 (1977) 503-514.

[8] Thackray, R. I., The stress of boredom and monotony: a consideration of the evidence, Psychosomatic Medicine, 43 (1981) 165-176.

Trends in Ergonomics/Human Factors II
R.E. Eberts and C.G. Eberts (Editors)

TASK PERFORMANCES AS FUNCTIONS OF LEVEL OF PROCESSING, MORNINGNESS-EVENINGNESS, TIME OF DAY, AND GENDER

Georgia Kathryn Green
Ben B. Morgan, Jr.

Center for Applied Psychological Studies
Old Dominion University
Norfolk, Virginia
U. S. A.

This study was undertaken to examine the relations between human circadian activity patterns and information processing. Eighty subjects performed digit-sequence and analogies tasks in either morning or afternoon. A significant Gender x Circadian Type x Time-of-Day interaction was found for digit-sequence task performance. Time and gender effects were found for the analogies task. Results were discussed in the context of past literature.

INTRODUCTION

In the past few decades, increasing attention has been focused upon the psychological implications of human circadian rhythms (Finger, 1982). Especially prevalent are studies examining industrial work rhythms (Hildebrandt, 1980), adjustment to shift work in relation to circadian rhythm, sleep behavior with relation to the sleeper's circadian type, transmeridial jet travel, and continuous military operations (cf. Morgan & Coates, 1974). Researchers have been greatly concerned with the relationship between human circadian rhythms and the performance of different types of tasks (cf. Folkard, 1982). Of major interest have been questions concerning the extent to which different types of cognitive performances follow different circadian cycles and the extent to which these cycles are mediated by individual-difference variables such as morningness-eveningness, cognitive style, and gender of subjects. In this regard, Mackenberg, Broverman, Vogel, and Klaiber (1974) reported that topics requiring repetitive drill were more effectively learned in the morning hours, while subjects requiring restructuring and/or complex thought were more effectively learned in the afternoon. The current study was designed to further investigate the variables that influence the interaction between type of task and performance at different times of day.

Many studies have utilized measures of the immediate ordered recall of digit sequences presented auditorially in examining time-of-day effects in tasks requiring repetitive drill for acquisition. These studies (Baddeley, Hatter, Scott, & Snashall, 1970; Blake, 1967; Folkard, 1979; Gates, 1916a, 1916b) agree in demonstrating that performance improves from early (c. 8:00AM) to mid-morning (c. 10:30AM) but then declines for the remainder of the day until it is at its lowest in the evening (c. 9:00PM).

The immediate comprehension-based recall of prose information has been employed in studies focusing upon time-of-day effects in tasks requiring complex thought and restructuring for acquisition. These studies (Folkard & Monk, 1980; Folkard, Monk, Bradbury, & Rosenthall, 1977; Gates, 1916b; Laird, 1925) report an improvement in performance from mid-morning (c. 10:30AM) to afternoon (c. 1:30PM), despite the finding that performance decreases approximately 30% from early morning (c. 8:00AM) to evening (c. 8:00PM). A task which has received much documentation in the literature as requiring complex thought and restructuring for acquisition is the solution of analogy items (Sternberg, 1977); unfortunately, time-of-day effects in the performance of the analogy-solver have not been the subject of widespread study.

Horne and Ostberg (1976) developed an English-language self-assessment questionnaire to determine morningness-eveningness in human circadian rhythms. This 19-item instrument yields a single score which places the individual under one of five designations: (1) definitely morning type; (2) moderately morning type; (3) neither type; (4) moderately evening type; (5) definitely evening type. The Horne and Ostberg (1976) questionnaire was externally validated against the criterion of individual differences in circadian oral temperature variation.

Within the information processing approach to the study of human behavior, the concept of learning style has emerged as useful when seeking a comprehensive view of individual differences in learning. Schmeck (1983) defined learning style as "a predisposition on the part of some students to adopt a particular learning strategy regardless of the specific demands of the learning task" (p. 233). The Schmeck (1983) Inventory of Learning Processes (ILP) is a 72-item instrument which yields four scale scores (corresponding to four different learning styles) for each individual. The Deep Processing scale assesses students' usage of critical evaluation, comparison/contrast, and conceptual organization of studied information. Elaborative Processing assesses ability to translate information into personal experience, examples, and imagery. Fact Retention assesses ability to store specific items of detail in studied information. Methodical Study assesses usage of systematic techniques of study.

The purpose of the current study is to investigate more thoroughly the relations between human circadian activity patterns and information processing in terms of individual differences in task performances as functions of level of processing, morningness-eveningness, time of day, and gender of the individual learner. The gender variable is included due to findings indicating that females score significantly higher than males on the ILP Methodical Study scale (Watkins & Hattie, 1981) and that males are more likely than females to be classified as morning types by the Horne and Ostberg (1976) morningness-eveningness questionnaire (Green, 1983). It was generally expected that morning types would perform better than evening types in the morning condition and that evening types would perform better than morning types in the afternoon condition; that performance on the digit sequence task would be better in the morning than the afternoon and that performance on the analogies task would be better in the afternoon than the morning; that the relationship between gender, circadian type, and time of day in task performance would be interactive; and that task performance would be related to level of processing.

METHOD

Subjects

Eighty undergraduate university students served as voluntary and experimentally naive participants. Of the 80 subjects, 20 were male morning types, 20 were male evening types, 20 were female morning types, and 20 were female evening types. Morningness and eveningness were defined by use of the Horne and Ostberg (1976) English-language Self-Assessment Morningness-Eveningness Questionnaire.

Apparatus and Materials

The Horne and Ostberg (1976) Self-Assessment Morningness-Eveningness Questionnaire and the Schmeck (1983) Inventory of Learning Processes were administered to all subjects. Additionally, a tape recording of 22 lists of nine random digits per list was used to present the digit-sequence task; responses to this task were recorded on a written-response sheet consisting of 22 rows of nine boxes each. The analogies task employed a tape recording of 22 analogy items; subjects responded to this task by recording their answers on a written-response sheet consisting of 22 rows of four alphabetic characters each.

Procedure

The first part of the experiment consisted of the administration of the two questionnaires to a total of 344 subjects in large group sessions. The first questionnaire (Horne & Ostberg, 1976) addressed morningness-eveningness. The second questionnaire was the Schmeck (1983) Inventory of Learning Processes. Questionnaire order was counterbalanced across subjects. All subjects were informed that the study's purpose was to examine new questionnaire measures of college students' activities.

The second part of the experiment consisted of obtaining measures of task performance. On the basis of their questionnaire responses, subjects were classified into the four experimental groups (male morning type; male evening type; female morning type; female evening type), and 20 subjects were randomly selected from each group for further testing. This testing consisted of the performance of each subject of two tasks: (1) the immediate ordered recall of digit sequences presented auditorially; and (2) the immediate solution of analogies presented auditorially. Within each of the four groups, subjects were randomly assigned to either the morning (10:00AM EST) or the afternoon (5:00PM EST) condition such that half of the subjects in each experimental group were assigned to each condition. Within each time-of-day condition, the ordering of task performance was counterbalanced across subjects for each experimental group.

The immediate ordered recall of digit sequences presented auditorially was tested using a tape recording of 22 sets of nine random digits pronounced at the rate of one digit per second, with 15 s separating each set from the one following it and with the identifiable single "tick" of a metronome marking the beginning and end of each digit set. The separation between digit sequences was utilized by subjects to produce a written list of digits presented in the sequence that preceded the separation. Subjects recorded their responses on a sheet consisting of 22 rows of nine boxes each. The first two digit lists (of the 22) were designated practice lists.

The immediate solution of analogies presented auditorially was tested using a tape recording of 22 analogy items with four answer alternatives per item, and with 15 s separating each item from the item following it and the identifiable single "tick" of a metronome marking the beginning and end of each item. The separation between items was utilized by subjects to circle the letter of the chosen answer alternative on the answer sheet. The first two analogy items (of the 22) were designated practice items.

In a debriefing session, subjects were given a complete explantation of all aspects of the study. Subjects were asked to agree not to discuss this information with others.

RESULTS

In analyzing the data from the digit sequence task and the analogies task, a 2 x 2 x 2 (Gender x Circadian Type x Time-of-Day Condition) multivariate analysis of covariance with multiple covariates (the four Inventory of Learning Processes scale scores) was employed. This was followed by a separate 2 x 2 x 2 analysis of covariance with multiple covariates for each of the two tasks. The digit sequence task further required the computation of simple effects for a three-way interaction. The .05 alpha level was utilized throughout.

The results of the multivariate analysis of covariance with multiple covariates indicate significance for the three-way Gender x Circadian Type x Time-of-Day Condition interaction ($F(2, 67) = 4.23$, $p < .05$) and also for the second covariate, the Schmeck Methodical Study scale ($F(2, 67) = 5.16$, $p < .05$).

The results of the analysis of covariance with multiple covariates for the digit sequence task indicate significant three-way Gender x Circadian Type x Time-of-Day Condition interaction ($F(1, 68) = 6.60$, $p < .05$). The computation of simple effects based upon this significant three-way interaction yields significance for the Circadian Type x Time-of-Day Condition interaction for the female gender ($F(1, 68) = 7.65$, $p < .05$). In the morning condition, performance on this task by morning type females was superior to that of evening type females. Performance of the evening type females was better than that of the morning type females in the afternoon condition.

The results of the analysis of covariance with multiple covariates for the analogies task indicate significant effects due to time of day ($F(1, 68) = 4.39$, $p < .05$), gender ($F(1, 68) = 5.51$, $p < .05$), the covariates ($F(4, 68) = 2.68$, $p < .05$), and the Schmeck Methodical Study scale ($F(1, 68) = 6.94$, $p < .05$). Performance on the analogies task in the afternoon condition was superior to that of the morning. Analogies task performance by female subjects was superior to that of males. The significant negative covariance of the Schmeck Methodical Study scale with performance on the analogies task was such that high scores on this scale were associated with inferior performance, while low scores were associated with superior performance.

DISCUSSION

It was expected that morning types would perform better than evening types in the morning condition and that evening types would perform better than morning types in the afternoon condition. Circadian type and time-of-day condition did interact in the predicted manner for female subjects' performances on the digit sequence task. However, this relationship was not found for male subjects' performances on this task nor the performances of all subjects upon the analogies task. Circadian type apparently does interact with time-of-day condition but does not reliably enhance task performance, while both circadian type and time-of-day condition seem to exert different effects for the two genders. It is unclear why the digit sequence task and the analogies task do not both yield interactions of circadian type and time of day, but it seems possible that the task-specific information processing demands influence the phase of the circadian activity pattern differentially. The different effects of circadian type and time of day for the two genders may result in part from differences in the relative motivation of male and female subjects or from gender differences in performance expectancies. Further, motivational level and/or performance expectancy may in some way interact with circadian arousal pattern such that male subjects' task performances differ from those of females. Research on the female circadian arousal pattern is meager.

It was expected that performance on the digit sequence task would be better in the morning than the afternoon and that performance on the analogies task would be better in the afternoon than the morning. Performance on the analogies task was significantly better in the afternoon condition than in the morning condition, as predicted. However, performance on the digit sequence task exhibited no significant difference due to time of day. It may be suggested that the presence of a very slight nonsignificant performance superiority in the morning condition indicates nothing other than the possibility that the time-of-day effect reported by other researchers may be accounted for by a failure to consider the mediating actions of circadian type and/or gender.

A major hypothesis of this study was that the relationship between gender, circadian type, and time-of-day condition in task performance was interactive. This contention was supported by the results of the digit sequence task. Computation of simple effects based upon the significant three-way interaction for this task demonstrated that female subjects' performances exhibited a Circadian Type x Time-of-Day Condition interaction. The absence of such an interaction for male subjects may be explained tentatively in terms of gender-differentiating motivational and/or performance expectancy factors relating to the task, as previously discussed. Further, the analogies task may exert such different information processing demands from the digit sequence task that circadian activity patterns do not affect both tasks in the same manner, and a significant three-way interaction for the analogies task does not occur in the same time period.

The hypothesis that task performance was related to level of processing was supported by the results from the analogies task. The Schmeck Methodical Study scale covaried significantly with analogies task performances such that high scores on this scale were associated with inferior performance, while low scores were associated with superior performance. This result may seem surprising, but does logically follow from Schmeck and Grove (1979), who found that students with high college grade-point averages (GPAs) had significantly lower scores on Methodical Study than did students with low GPAs. This would seem to indicate that high scorers on this scale may not possess cognitive skills necessary for superior performance in tasks which require complex thought and restructuring, such as the analogies task.

Of additional note is the gender effect obtained for the analogies task. Female subjects exhibited markedly higher scores than did males. This finding is in accordance with the literature on the psychology of gender differences; females in general exhibit greater verbal ability than males (Maccoby & Jacklin, 1974).

The present study makes a major contribution in its consideration of gender, circadian type, and time of day as components of a complete view of task performance. There are several theoretical and practical implications. First, results found in studies using simple tasks such as the digit-sequence task to measure performance should not be equated with results found in studies measuring performance with tasks requiring complex thought and restructuring such as the analogies task. This distinction should be consistently noted both in research and in work-scheduling recommendations based upon such research. Second, this study highlights the desirability of including a measure of circadian type in experiments examining time-of-day effects on task performances so as to foster the eventual development of accuracy in task performance predictions for individuals from questionnaire data. Third, a lack of gender differences in circadian type and time-of-day performance effects should not be readily assumed, either in applying past research or designing future research. Finally, the interactive relationship between gender, circadian type, and time-of-day condition in task performance deserves special emphasis in terms of the generation of new research.

REFERENCES

[1] Baddeley, A.D., Hatter, J.E., Scott, D., & Snashall, A. (1970). Memory and time of day. Quarterly Journal of Experimental Psychology, 22, 605-609.

[2] Blake, M.J.F. (1967). Time of day effects on performance in a range of tasks. Psychonomic Science, 9, 349-350.

[3] Finger, F.W. (1982). Circadian rhythms: Implications for psychology. New Zealand Psychologist, 11, 1-12.

[4] Folkard, S. (1979). Changes in immediate memory strategy under induced muscle tension and with time of day. Quarterly Journal of Experimental Psychology, 31, 621-633.

[5] Folkard, S. (1982). Circadian rhythms and human memory. In F.M. Brown & R.C. Graeber (Eds.), Rhythmic aspects of behavior (pp. 241-272). Hillsdale, NJ: Erlbaum.

[6] Folkard, S., & Monk, T.H. (1980). Circadian rhythms in human memory. British Journal of Psychology, 71, 295-307.

[7] Folkard, S., Monk, T.H., Bradbury, R., & Rosenthall, J. (1977). Time of day effects in school children's immediate and delayed recall of meaningful material. British Journal of Psychology, 68, 45-50.

[8] Gates, A.I. (1916a). Diurnal variations in memory and association. University of California Publications in Psychology, 1, 323-344.

[9] Gates, A.I. (1916b). Variations in efficiency during the day. University of California Publications in Psychology, 1, 1-156.

[10] Green, G.K. (1983). Morningness-eveningness and personality factors: A preliminary investigation of adaptability. Unpublished manuscript, ODU, Norfolk, VA.

[11] Hildebrandt, G. (1980). Survey of current concepts relative to rhythms and shift work. In L.E. Scheving & F. Halberg (Eds.), Chronobiology (pp. 261-292). Alphen aan den Rijn, The Netherlands: Sijthoff & Noordhoff.

[12] Horne, J.A., & Ostberg, O. (1976). A self-assessment questionnaire to determine morningness-eveningness in human circadian rhythms. International Journal of Chronobiology, 4, 97-110.

[13] Laird, D.A. (1925). Relative performance of college students as conditioned by time of day and day of week. Journal of Experimental Psychology, 8, 50-63.

[14] Maccoby, E.E., & Jacklin, C.N. (1974). The psychology of sex differences. Stanford, CA: Stanford University.

[15] Mackenberg, E.J., Broverman, D.M., Vogel, W., & Klaiber, E.L. (1974). Morning-to-afternoon changes in cognitive performances and in the electroencephalogram. Journal of Educational Psychology, 66, 238-246.

[16] Morgan, B.B., Jr., & Coates, G.D. (Eds.). (1974). Sustained performance and recovery during continuous operations (ITR-74-2). Norfolk, VA: Performance Assessment Laboratory.

[17] Schmeck, R.R. (1983). Learning styles of college students. In R.F. Dillon & R.R. Schmeck (Eds.), Individual differences in cognition (vol. 1; pp. 233-279). New York: Academic.

[18] Schmeck, R.R., & Grove, E. (1979). Academic achievement and individual differences in learning processes. Applied Psychological Measurement, 3, 43-49.

[19] Sternberg, R.J. (1977). Intelligence, information processing, and analogical reasoning. Hillsdale, NJ: Erlbaum.

[20] Watkins, D., & Hattie, J. (1981). The learning processes of Australian university students. British Journal of Educational Psychology, 51, 384-393.

Trends in Ergonomics/Human Factors II
R.E. Eberts and C.G. Eberts (Editors)

A PRELIMINARY ASSESSMENT OF READING AS IT RELATES TO KNOWLEDGE WORKER PERFORMANCE

Richard L. Shell, O. Geoffrey Okogbaa, Nicholas A. Damachi
Department of Mechanical & Industrial Engineering
University of Cincinnati

As the proportion of knowledge workers increases, the mental work output and the work content, especially for those tasks that are cognitive in nature, will become more important in the productivity equation. This paper examines work output for reading tasks without scheduled rest breaks. The results of an experiment designed to determine the relationship between the actual time expended on the reading task and reading output is presented. In addition, the instrumentation used for data acquisition is described.

INTRODUCTION

A major impact of the increase in the sophistication and automation of work systems is the change in the nature of human work. This change has emphasized the importance of mental tasks and functions especially in the areas of information processing and decision making. Available evidence indicates not only massive technological growth but tremendous growth in the rate of learning and knowledge acquisition as well. The continued efficacy in the technology of production and services and the rate of growth of most industrialized economies suggest a knowledge worker dominated economy in the near future. According to recent projections, employment in the goods producing industries rose only 9 percent between 1965 and 1968 (US, BLS, 1981). In 1980, the total manufacturing work force numbered about 23 million which is only about 22 percent of the work force. During this period, however, significant gains in productivity without additional workers has occurred due to automated production and other technological breakthroughs. On the other hand, the number of knowledge workers has risen rapidly. Once a small proportion of the workforce, knowledge workers now make up about half the total. According to the U.S. Bureau of Labor Statistics, estimates for the total knowledge worker population, professional and technical workers will grow from 14.2 to 16.9 million in 1990 while managers and administrators will grow from 10.1 million to 12.2 million during the same period.

The implication of high levels of automation and other technological gains is that most jobs will involve a greater degree of mental activity. Consequently, as the proportion of knowledge workers increases, the work output and the work content for those tasks that are cognitive in nature (i.e. reading and text comprehension) will become more important in the productivity equation. The productivity of such a vital part of the work force will defy meaningful and effective measurement until a measure of their work output and actual work content are adequately defined.

One of the limiting factors in the measurement of knowledge worker output is mental fatigue. Except in a few well defined cases, no work activity can be defined as strictly physical or mental. While physical activities exact some demands upon the central nervous system, mental functions in some situations also result in some activity of the muscles. However, in those purely mental operations such as reading, mental arithmetic and the solution of verbal intelligence problems, the functional impairment of neuronic cells account for mental fatigue evident in work output decrement.

The current investigation focused on the cognitive aspect of mental processes as evident in reading tasks.

Reading

Humans have engaged in the act of reading since the beginning of civilized times. To the early peoples, reading was one of the most fascinating of arts. Today, in most cultures, the inability to read or write means to be outlawed in the commonwealth of learning. Until very recently, the test of true scholarship was the ability to speak or read in several languages. Writing in 1908, the late Professor Edmund Huey observed

> 'Reading is the means by which the world does a large part of its work. The printed page is a contrivance used hours daily by tens of millions of people. The slightest improvement either in the page or in the method of reading means a great service to the human race. It remains finally for us to consider an aspect of the psychology of reading which is of the greatest importance practically and pedagogically, the rate of reading and the factors which affect speed.'

Professor Huey's observations are as true today as they were then. The explosion in knowledge especially in the physical sciences means that people are reading more today than at any other time in history.

Reading, like any other physical or mental activity, exacts some demands upon the psychophysical being. In the case of reading, however, these demands are prominent in the eye, the mechanisms of speech and those cortical areas of the brain and nervous system associated with rapid functioning of attention, perception, feeling and imagery.

Reading Rate and Comprehension

Most reading specialists agree that the parameters that could affect reading rate and comprehension would include the time or duration of reading, the length of the text, difficulty level and interest appeal. There is, however, a noticeable absence of research on the relationship between reading rate/comprehension and time especially as regards workers whose job functions are sedentary and mental. Most of the available research contend that the effective reading output is constant over the reading or work duration. An examination of the time span for these studies shows that it is in the order of tens of minutes (120 seconds or 2 minutes maximum). For such a short time span, it is unlikely that any changes in work output will be observed. Additionally such intervals are highly unlikely in any meaningful reading situation and is atypical of the time spent by the general knowledge worker population on the reading aspect of their task assignments.

PROBLEM INVESTIGATION

In an initial study, a series of laboratory experiments were conducted to establish the nature of mental work output over a time span that is typical of that spent by knowledge workers in the course of their daily assignments. The tasks used in the experiment were representative of the bulk of mental activities, namely, abstract processes (Bennett, 1976; Beetchereva, 1981) and cognitive processes (Carver, 1977, 1982). Standardized test packages were modified and used to simulate the different mental processes. For abstract processes, arithmetic-logic tests were used while reading and comprehension tests were used for the cognitive processes. In these experiments work output was measured as well as physiological parameters such as heart rate and brain activities. Work output for the reading task was defined in terms of effective reading output and incorporates both reading rate and reading efficiency. For the arithmetic task, work output was defined as the ratio of the correct problems to the total number of problems attempted over the time period. Each given experimental session lasted five hours and was run under each of two types of conditions, namely rest and no rest. The results of the experiment showed that over considerable long periods (5 hours or more) reading output is nonlinear and that total work output shows considerable decrement.

In the reading part of the experiment, some of the alloted time was spent reading the text and the remaining part was spent answering questions about the text to demonstrate comprehension. A meaningful indicator of the time duration for the task should be the actual time spent reading the text.

To determine the actual duration of the reading task, a slightly different experiment was conducted. The length of the revised experiment was also five hours and was run without any rest breaks. The participants were junior level engineering students with an average age of 21 years. The subjects were considered homogenous as regards age and ability.

Experimental Design

The instrumentation for this experiment consisted of a 2-channel strip chart recorder and a metal pen and pen holder set. The strip chart was calibrated before the start of each experimental session. The recorder was connected to the pen and pen-holder through a switching mechanism. The switch closes or opens depending on whether the pen was in or out of the holder respectively. The setup was arranged in parallel so that the recorder continuously recorded even when the switch was open. The baseline was established with the pen in the holder. This also represented the time for actual reading task. When the pen is removed from the holder (to answer questions) the switch opens and consequently the baseline shifts to a different level. Upon replacing the pen in the holder, the baseline shifts back to the orginal position. From this both the actual time spent reading the text and the time required to answer questions were estimated by summing the appropriate elemental times (reference figure 1).

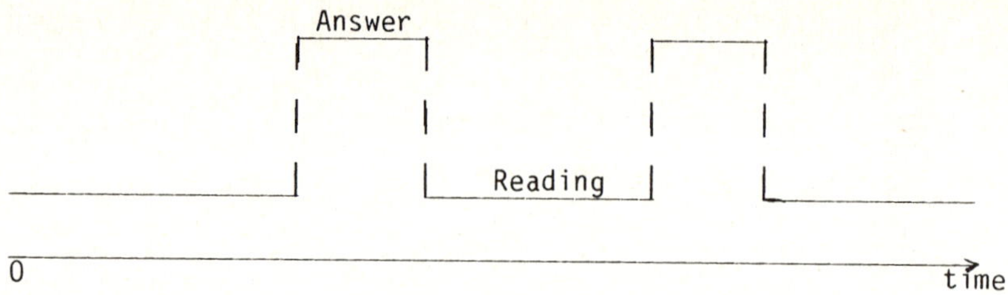

Figure 1
Typical output from strip chart recorder indicating alternative reading and answer times

RESULTS

The results show the raw reading rate decreased significantly after the first 50 minute period (reference table 1, figure 2). In addition, the work output or the effective reading rate (the product of raw reading rate and the efficiency) remained fairly constant where efficiency, E, is defined as

$$E = \frac{\text{No. of Correct Responses}}{\text{Total No. of Attempted Responses}}$$

The actual expended time on the reading task also increased. The percentage of actual reading time increased from an initial 52 to 63 percent at the end of the experiment (reference Table 2).

Cumulative Time (minutes)	Cumulative Reading Rate (words/min)	Effective Reading Rate (words/min)
50	95.8	43.6
105	86.9	42.1
155	85.5	42.5
210	84.1	43.1
260	85.8	41.5
300	86.7	42.7

Table 1
Work Output for Reading Task

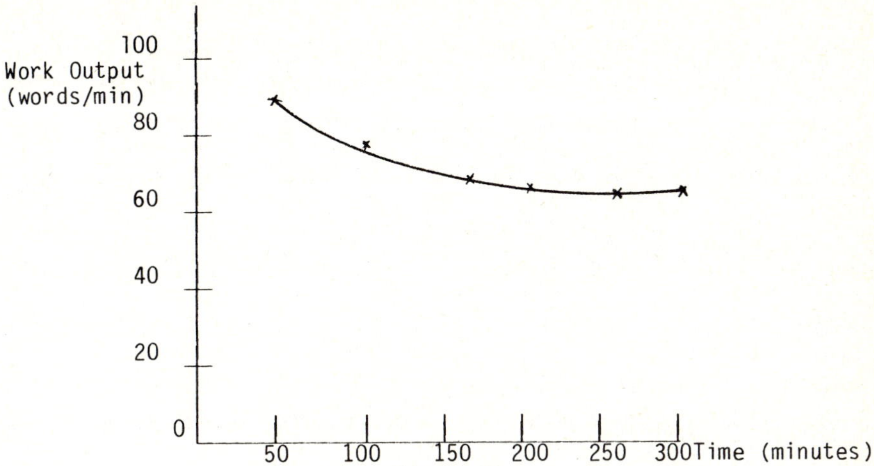

Figure 2
Effective Reading Output Over Time

Effective Reading Rate (words/min)	Actual Reading Cumulative Time (min)	Actual Reading Time As % of Total Reading Time
83.7	26	52
80.3	55	52
78.4	84	54
74.8	121	58
70.0	154	59
67.8	189	63

Table 2
Work Output for Actual Reading Time

CONCLUSION

The results demonstrated that the approach used in this investigation is a viable method to evaluating the relationship between reading output and time. In particular this study supported the findings of an earlier investigation which found that reading rate is nonlinear over time.

As was indicated this was a preliminary investigation and as such only a select few of the parameters of interest were examined. Any future experiment should investigate the effect of rest versus no rest task conditions and some physiological measures such as heart rate and brain activities.

REFERENCES

[1] Bennett, C.A.: Counteracting psychological fatigue effects by stimulus change. Proceedings of the 18th Human Factors Society Meeting, Huntsville, Alabama, 1976.

[2] Beechtereva, N.P.: Neurophysiological correlates of mental processes in man. Psychophysiology Today and Tomorrow (Beechtereva, N.P., Ed.), Pergamon Press, New York, 1981.

[3] Bureau of Labor Statistics (US BLS), United States Department of Labor, 1981.

[4] Carver, R.P.: Optimal Rate of Reading Prose. Reading Research Quarterly No. XVIII, Fall, 1982.

[5] Carver, R.P.: Towards a Theory of Reading and Rauding. Reading Research Quarterly, 1977.

[6] Farr, R.C.: Reading: What Can Be Measured?. International Reading Association, Research Fund, Newark, Delaware, 1969.

[7] Okogbaa, O.G.: An Empirical Model for Mental Work Output and Fatigue. Unpublished PhD Dissertation, University of Cincinnati, 1983.

[8] Rose, L.: The Reading Processes and Some Research Implications. Journal of Reading, Oct. 1969.

[9] Shell, R.L., Okogbaa, O.G.,: The Effect of Mental Fatigue on Knowledge Worker Performance. 1983 Fall Industrial Engineering Conference, Toronto, Ontario.

[10] Shell, R.L., Okogbaa, O.G., Mital, A.: Data Transformation Concepts and Empirical Model Development for Mental Work Output and Fatigue. Trends in Ergonomics/Human Factors I, A. Mital (Editor), Elsevier Science Publishers B.V. (North-Holland), 1984.

[11] Stauffer, R.G.: Directing the Reading and Thinking Process. Harper and Row Publishers, New York, NY, 1964.

[12] Sjoberg, H: Physical Fitness and Mental Performance During and After Mental Work. Ergonomics, 1980, Vol. 23, No. 10.

Trends in Ergonomics/Human Factors II
R.E. Eberts and C.G. Eberts (Editors)

THE NATURE OF SUBJECTIVE WORKLOAD

Yei-Yu Yeh and Christopher D. Wickens

Department of Psychology
University of Illinois, Champaign-Urbana

The structure and the reliability of subjective workload were examined in the present study. The hidden cognitive structure disclosed from a Multi-dimensional scaling (MDS) approach corresponds to the structure of processing resources (Wickens, 1984). The demand for perceptual/central resources was found to be a salient component of subjective workload from the results of both the MDS and regression analyses. Although post-trial and post-session ratings were fairly reliable, the latter ratings were more reliable than the former ratings.

INTRODUCTION

Subjective effort has been considered to be one crucial aspect of mental workload. Sheridan (1980) stated that subjective perceptions of cognitive effort constitute the essence of workload. Others suggested that "if an operator feels effortful and loaded, he is effortful and loaded" (Johannsen et al., 1979, p. 105). In addition to this face validity, subjective measures are considered nonintrusive, cheap, and convenient to be implemented in any environment. However, little effort has been made to understand the origins of subjective feelings of load (Moray, 1982) and the limitations of the techniques. Many important questions about subjective measures remain unanswered. The objective of the present study is to examine three aspects of subjective workload: its structure, its reliability, and its prediction on the basis of specific rating scales.

<u>The structure of perceived workload: A Multidimensional Scaling approach</u>. The structure of perceived mental workload has been investigated by a Multidimensional Scaling (MDS) approach (Derrick, 1981; Yeh & Wickens, 1984). In this MDS approach, tasks were judged by their similarity of difficulty in order to disclose the hidden cognitive structure. Results from these studies demonstrated that the multidimensional structure shows a good degree of correspondence to the dimensions of the multiple resource model of dual task performance. Of the three subjective dimensions observed by Derrick, two of these (input modality and resource competition) corresponded to aspects of the model while a third (time demand) did not. Both of the two dimensions revealed by Yeh and Wickens' study (processing codes and resource demand) corresponded to aspects of the model.

<u>The reliability of subjective ratings</u>. Yeh and Wickens (1984) showed that post-trial ratings were less reliable than post-session ratings even though both were fairly reliable. In the present study, this issue was

examined to replicate such a difference in the rating reliability because the implication of this issue is important. If operators are more consistent in rating when they have a chance to compare the demands of all the systems, then ratings should be collected in the context of all of the just-performed tasks. Ratings collected immediately after operators test a given system may be less reliable.

Predicting global workload from specific ratings. Using a multiple regression analysis, the important components of subjective workload can be revealed by examining the beta weights of the predictors that significantly account for the variance in the ratings of overall workload. Vidulich and Wickens (1984) found that only three specific scales (i.e., mental/sensory effort, response load, and stress level) contributed to explain the variance in the ratings of two global scales. Two of the scales are directly related to the resource dimensions described by the multiple resource model. In the present study, the regression analysis was adopted to provide converging evidence for understanding the structure of subjective workload.

GENERAL EXPERIMENTAL PARADIGM

Fifteen students of the University of Illinois were paid volunteers in the first experiment and fourteen students volunteered in the second experiment. All subjects were right-handed, native speakers of English, and had normal vision. Three tasks were employed across the two experiments, with various manipulations of difficulty in each task. These three tasks are as follows:

(1) Sternberg memory search task. Prior to each trial of the basic memory search task (BM), subjects viewed one alpha-numeric string (e.g., A26). The string was held in working memory for the next two minutes as a series of probes were presented, half of which were in the memory set. Subjects were asked to press the first key on the panel if a probe was in the memory set (i.e., a "go/no-go" response). The demand on perceptual, central, and response processing stages was increased via different manipulations. In the perceptual load condition (PM), a grid mask was superimposed upon each probe stimulus. In the central load condition (CM), memory set size was increased to three (e.g., A26, C59, R89). In the response load condition (RM), subjects had to make double responses for every positive or negative response, by pressing key 2 then key 4 for a positive response, or by pressing key 3 then key 5 for a negative response. Subjects had to release the first key completely before they pressed the second key and they had to press the second key within a window of 600 msec. Two other manipulations varied the qualitative processing demands. In the spatial memory task (SM), they held two sets of dot patterns in working memory and made a positive response when a probe was in the memory set. In the speech condition (SP), they remembered two sets of strings and said "yes" when a probe was in the memory set.

(2) Single-axis tracking task. In the easy condition (L), subjects were required to nullify an error cursor that was displaced horizontally by a random forcing function with a cutoff frequency of 0.32 Hz. The error cursor was controlled by the left-right movement of the joystick with the left hand. Only the first order system dynamics were utilized. The bandwidth was increased to 0.54 Hz in the difficult condition (H).

(3) Dual-axis tracking task. The demand of the basic task (BT) was to maintain an integrated cursor superimposed upon a stationary cross in the center of the display, by controlling a single joystick with the left hand. The cursor was driven in two dimensions by independent X and Y forcing functions with a bandwidth of 0.30 Hz. The dynamics of the tracking control were first order. In the "separated-display" condition (DT), two cursors signified the errors on each axis respectively and were tracked by the left joystick. In the "separated-control" condition (CT), subjects tracked the integrated cursor by moving two separate joysticks, one in the fore-aft direction and the other in the left-right direction.

The memory search task and the tracking task were chosen because of the differential demands they imposed upon various components of processing resources within the framework of the multiple resource model. A tracking task places heaviest demands upon manual responses and spatial codes. A memory search task places heaviest demands upon perceptual/central stages of processing and verbal codes (or spatial codes when the stimuli are dot patterns). In Experiment 1, subjects performed the single-axis tracking tasks, single memory search tasks (basic, perceptual, central, and response load), and various combinations of the two tasks. In Experiment 2, subjects performed the dual-axis tracking tasks, memory search tasks (basic, spatial, central, and speech), and dual-task combinations. The dual-task configurations will be designated by an abbreviation of two letters, the first letter indicates the type of tracking task (L - low bandwidth, H - high bandwidth, B - basic dual-axis, D - separated display, and C - separated control), and the second letter indicates the type of memory search task (B - basic verbal, P - perceptual, C - central, R - response, S - spatial, and P - speech).

In each experiment, subjects practiced all of the tasks extensively in the first two sessions over a period of two days. Prior to Session 3, subjects were given a list of all the rating scales (overall workload, perceptual effort, mental effort, response effort, time-pressure, stress, overall effort, performance, and excess capacity). The experimenter went through the list with the subjects and made sure that they understood the meaning of the scale and the way to do the rating. Subjects performed one task first (the basic verbal memory search task in Experiment 1 and the basic dual-axis tracking task in Experiment 2) and were asked to use the subjective experience of doing this task as the reference, which was arbitrarily assigned a rating of 100. Subjects then performed the other tasks and rated each one against the reference on the nine unidimensional rating scales. They were told to assign a rating of 200 to a just-performed task if the subjective experience was twice the experience of doing the reference task and to assign 50 if it was half. The unidimensional ratings were collected at the end of each trial immediately after subjects performed the task, and again at the end of the session.

In Session 4, subjects followed the same procedure and they also rated all the possible paired combinations of the tasks on a 9-point similarity scale by the task difficulty. Unidimensional ratings collected in this session were compared with the corresponding ratings recorded in Session 3 to estimate the reliability of the ratings. In Experiment 2, there were some changes in the scales used: (1) the perceptual effort and mental effort scales were combined into one scale, (2) the performance scale and the overall effort scale were not used in this experiment, and (3) a task complexity scale and a modified Cooper/Harper scale were added.

RESULTS

Rating reliability. For each subject, the rating reliability for each scale was obtained by computing the correlation between the ratings collected in Session 3 and those collected in Session 4 across all the conditions. For all the subjects, most post-session ratings were more reliable than the corresponding post-trial ratings even though both were fairly reliable. The mean post-session correlation value across all the subjects was 0.84 for both Experiments, whereas the mean post-trial correlation values were 0.60 and 0.70 for Experiments 1 and 2 respectively.

Predicting global workload from specific ratings. Most ratings were highly correlated with the ratings on the overall workload scale. A multiple regression analysis was conducted to determine which variables significantly predict the ratings on the overall workload scale. The results of the analysis are presented in Table 1. In Experiment 1, the combination of four predictors (perceptual effort, mental effort, time demand, and stress) accounted for most of the variance in the overall workload ratings (90%). The other predictors did not significantly improve the adjusted R square. Among the four predictors, mental effort had the highest regression weight and accounted for more variances than the other three predictors. In Experiment 2, 93% of the variance in the overall workload ratings were accounted for by the combination of perceptual/mental effort, response effort, stress, and task complexity. The perceptual/mental effort scale had the highest regression weight among the four predictors.

Table 1 - Predicting Overall workload from specific scales

Experiment 1 (Adjusted R square = 0.90)
Overall Workload = 0.095 + 0.263 PE + 0.319 ME + 0.154 TI + 0.292 ST
Experiment 2 (Adjusted R square = 0.93)
Overall Workload = 0.085 + 0.979 P/ME - 0.085 RE - 0.317 ST + 0.295 CM

The perceived structure of subjective difficulty. Similarity judgment data were analyzed by the SINDSCAL program. Figures 1 and 2 present the three subjective dimensions from the data of Experiment 1.

Figures 1 & 2 - The structure disclosed from the MDS data of Experiment 1

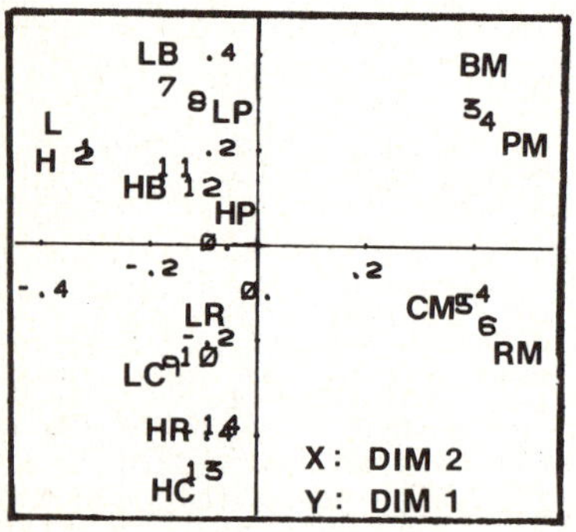

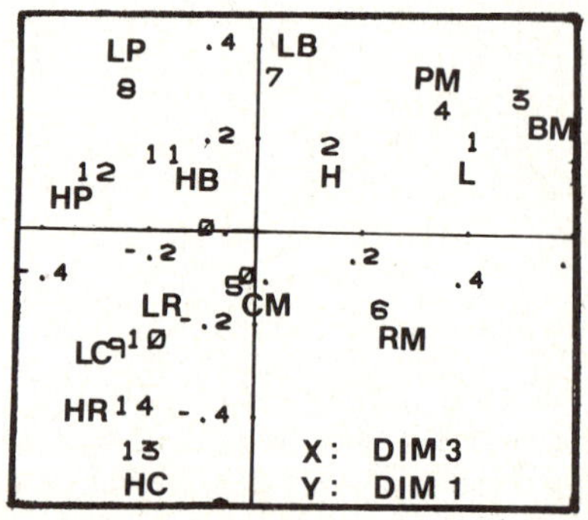

As Shown in Figure 1, Dimension 1 was related to the processing stage. Basic memory tasks and perceptual load memory tasks had positive weights on this dimension. Central load and response load memory tasks had negative weights. Dimension 2 was related to processing codes or to demands on working memory. All the single memory tasks, demanding both verbal codes and information held in working memory, had positive weights on Dimension 2. The two single tracking tasks, demanding spatial codes without holding information in working memory, had negative weights. Dual tasks, combining both the memory search task and the tracking task, were located between the two clusters. Figure 2 represents Dimensions 1 and 3 from the SINDSCAL solution of Experiment 1. Dual tasks that demanded more cognitive resources to coordinate (all except the condition in which a low-bandwidth tracking task was combined with a basic memory task - LB) had negative weights on Dimension 3. The central load memory task also had a negative weight on this dimension. The other tasks, demanding less cognitive resources, had positive weights on Dimension 3.

All the three dimensions disclosed from the SINDSCAL solution in the data of Experiment 2 were related to the demands on perceptual/central resources. The results are presented in Figures 3 and 4.

Figures 3 & 4 - The structure disclosed from the MDS data of Experiment 2

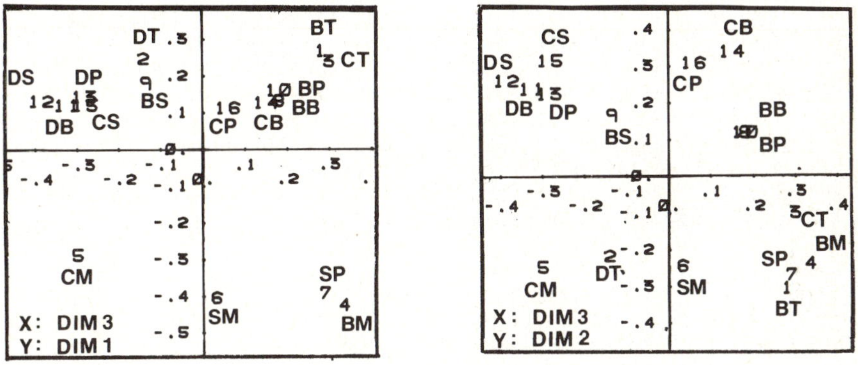

Dimension 1 in Figure 3 was tied to the demands for working memory. Independent of the processing codes and the response modes, all the single memory search tasks which required that information be held in working memory had negative weights. On the other hand, single tracking tasks which did not demand memory capacity were located on the top of the dimension. Dual tasks, combining the two types of tasks, were located between the two extremes. Dimension 2 (Figure 4) also appeared to be associated with the demands for working memory. All the single tasks had negative weights whereas dual-task conditions which required executive management to coordinate the time-sharing between the two tasks had positive weights on Dimension 2. On Dimension 3 (Figure 3), tasks that demanded more perceptual/central resources had negative weights. These tasks included central load memory task, tasks in which two cursors were displayed for the tracking task, and dual tasks in which a tracking task competed with a spatial memory task. The other tasks had positive weights on Dimension 3.

DISCUSSION

Using the multidimensional scaling approach, Derrick (1981) has first demonstrated that the structure of perceived task difficulty is related to the structure of processing resources as portrayed by the multiple resource model. The results of the present study supported this corresponding relationship. It appears that the origin of subjective workload comes from the resource demand imposed upon the human information processing system. Subjective workload is indeed multidimensional although its characteristics are not a direct analog of the multidimensional characteristics of performance. Furthermore, subjective workload seems to be dominated by the demand on working memory, independent of the processing codes and the response modes. Although Dimension 2 revealed from the data of Experiment 1 could be related to processing codes or to the demand on working memory, the results from Experiment 2 suggested that it was the demand on working memory that affected subjective workload. The spatial memory search task (SM), the basic verbal memory task with manual responses (BM), and the verbal task with speech responses (SP) all had similar weights on both Dimensions 1 and 2.

Due to the multidimensionality, ratings on a unidimensional scale may be the product of an averaging process and lose certain information about the differences in vector characteristics. Three examples are presented to illustrate this point. The first one is the difference between the central and response load memory search tasks in Experiment 1. Subjects indeed weighted the two tasks distinctively on the third dimension of cognitive demand. However, the two tasks were weighted closely on the first two dimensions. When subjects rated the overall workload of the two tasks, the difference in the cognitive demand may be reduced by the similarity of the resource demand on the other two dimensions. Thus, subjective workload may not correspond to the difference in performance and dissociation between performance and subjective measures may occur.

The second example is the difference between the display-separated and control-separated tracking tasks in Experiment 2. The two tasks were perceived differently on both Dimensions 2 and 3. Processing two cursors (DT) was perceived as demanding more perceptual/central resources on Dimension 3. On the other hand, using two control joysticks (CT) was perceived as closer to the other dual-task configurations on Dimension 2. Furthermore, these two tasks were perceived quite similar on Dimension 1 which was also related to the demand for working memory. Averaging the weights of the three dimensions would have made the tasks indistinguishable because the differences on Dimensions 2 and 3 would be cancelled out and the two were similar with regard to the demand on Dimension 1. Once again, the two tasks would not be distinguished by subjective ratings of workload regardless of the differences in performance.

Finally, the demand for the executive management to coordinate the time-sharing between the two tasks appears to be a dominant factor in assessing the workload of a dual-task configuration. Subjective workload was insensitive to the difference in the amount of resource competition on three dimensions: (1) Processing stages - the weights of the dual-task configurations were similar on both Dimensions 2 and 3 even though there was relatively more resource competition when a tracking task was performed concurrently with a response load memory task (LR or HR) than with a perceptual/central load memory task (LP/LC or HP/HC). (2) Processing codes - the weights were similar on both Dimensions 1 and 2 even though a tracking task was better time-shared with a verbal than with a spatial memory search task (BB vs. BS, DB vs. DS, and CB vs. CS). (3) Response modes - the weights were similar on both Dimensions 1 and 2 even though a tracking task was better time-shared with a memory search task that was responded with speech than the memory task that was responded manually (BP vs. BB, DP vs. DB, and CP vs. CB). The insensitivity of subjective workload to the difference in the amount of resource competition could produce the dissociation between subjective measures and performance in dual-task conditions (Wickens & Yeh, 1983).

In addition to being unable to preserve vector characteristics, subjective ratings of workload appear to be dominated by the demand for perceptual/central resources from the results of both the MDS and regression analyses. This finding is consistent with the model of verbal report data proposed by Ericsson and Simon (1980). According to this model, verbal report data such as introspective ratings of workload reflect information heeded in the working memory.

If it is plausible that information on some subjective dimensions is not fully preserved by the unidimensional ratings, what scales should then be used by workload practitioners? In theory, resource dimensions could be used to assess subjective workload since the structure of information processing is the common basis for both performance and subjective workload. However, whether a particular resource dimension will be revealed from the data depends upon the processing patterns imposed by the set ot tasks employed in a study. Since the exact structure of subjective workload in a set of tasks can only be disclosed *post hoc*, how may a workload practitioner choose the scales *a priori* to assess the load of different processing characteristics?

Apparently, a global scale such as the overall workload, capacity used, and modified Cooper/Harper scale may be used to tap the aggregate demands and the variance of all the components. However, the choice of specific scales should be based upon the nature of the important components of subjective workload. The correlations between subjective dimensions and unidimensional ratings as well as the results of the regression analysis suggest four scales: perceptual/mental effort, response effort, stress, and time-pressure. The first three scales were also found to be salient components ot task difficulty or overall workload in Vidulich and Wickens' study (1983). Time-pressure ratings were not recommended by Vidulich and Wickens (1983) because these ratings were insensitive to the effect of the manipulation that, a priori, should have exerted an influence. Therefore, the perceptual/mental effort, response effort, and stress scales are recommended to assess the demands on specific aspects.

Given that the demand on working memory is an important component of subjective workload, the immediate post-trial ratings may be subject to some phenomenal aspects that are task-specific and less stable. On the other hand, post-session ratings collected within the context of all of the just-performed tasks are more relative and the effect of task-specific features may be reduced. Consequently, post-trial ratings were in general less reliable than post-session ratings even though both were fairly reliable. For example, an alpha-numeric string is particularly easy to be remembered because it has a special meaning for a subject. When this subject is asked to rate the load immediately after he performs the task, his ratings will be very different from the ratings of the same condition with another memory set. When the subject has a chance to retrospect the experiences of doing each task at the end of the session, he may compare the differences in general processing characteristics and rank order the ratings for each condition. It is not argued that immediate ratings are hence less valid. It is only suggested that post session ratings, in the long run, may provide relatively stable and reliable ratings of the processing characteristics. Immediate ratings are more vulnerable to specific task features that fluctuate from time to time. If a workload practitioner is concerned with the relative workload from operating different systems, post-session ratings within the task context are recommended.

REFERENCES

[1] Derrick, W., The relationship between processing resource and subjective dimensions of operator workload, Proceedings of the 25th Annual Meeting of the Human Factors Society (1981) 532-536.
[2] Ericsson, K., and Simon, H. A., Verbal reports as data, Psychological Review, 87 (1980) 215-250.
[3] Johannsen, G., Moray, N., Pew, R., Rasmussen, J., Sanders, A., and Wickens, C. D., Final report of experimental psychology group, in: Moray N. (ed.), Mental workload: Its theory and measurement (Plenum Press, New York, 1979).
[4] Moray, N., Subjective mental workload, Human Factor, 24 (1982) 25-40.
[5] Sheridan, T. B., Mental workload - What is it? Why bother with it?, Human Factors Society Bulletin, 23.2 (1980) 1-2.
[6] Vidulich, M. D., and Wickens, C. D., Processing phenomena and the dissociation between subjective and objective workload measures (Tech. Rep. EPL-83-2/ONR-83-2), Engineering Psychology Laboratory, University of Illinois (1983).
[7] Wickens, C. D, Processing resources in attention, dual task performance, and workload assessment (Tech. Rep. EPL-81-3/ONR-81-3), Engineering Psychology Laboratory, University of Illinois (1981).
[8] Wickens, C. D., Processing resources in attention, in: Parasuraman, R., Beatty, J., and Davies, R. (eds.), Varieties of attention (Academic Press, New York, 1984).
[9] Wickens, C. D., and Yeh, Y.-Y., The dissociation between subjective workload and performance: A multiple resource approach, Proceedings of the Human Factors Society, 27th Annual Meeting (1983) 244-248.
[10] Yeh, Y.-Y., and Wickens, C. D., An investigation of the dissociation between subjective measures of mental workload and performance (Tech. Rep. EPL-84-1/ NASA-84-1), Engineering Psychology Laboratory, University of Illinois (1984).

ACKNOWLEDGMENTS

This research has been supported by NASA grant NAG 2-308, NASA AMES research center, Moffett Field, CA. The contract monitor is Sandra Hart.

Trends in Ergonomics/Human Factors II
R.E. Eberts and C.G. Eberts (Editors)

NUTRITION AS AN ERGONOMIC COMPONENT IN THE WORKPLACE

T. Vinciguerra*, S. S. Asfour**, and A. Genaidy**

*Industrial Safety Division
Frank B. Hall and Company
Coral Gables, Florida
U.S.A.

**Dept. of Industrial and Biomedical Eng.
University of Miami
Coral Gables, Florida
U.S.A.

The major objective of this research is to review, evaluate, and establish nutrition guidelines as an integral ergonomic component in industrial activities of various degrees of intensity. The ergonomic implications of nutrition are of vital importance in assessing strategies for enhancing maximum physiologic efficiency of a variety of occupational tasks which could lead to an overall increment in both human and industrial productivity.

INTRODUCTION

It has long been an accepted postulate of everyday life that in order to function at our best, it is most important to have a complete well-balanced diet, replete with 3 meals a day and encompassing the necessary intake of all food groups essential for maintenance of body metabolism, energy expenditure, and cell replenishment. The essential breakdown of foodstuffs to energy sources yields the 3 principals of carbohydrates, fats and proteins, all of which contain some combination of carbon, hydrogen and oxygen. According to Astrand and Rodahl (1977), at rest carbohydrates and fats contribute almost equally to the energy supply. Proteins have a limited role in energy supply, being utilized instead as cell and tissue building blocks. The difference in energy use is dependent on many factors, the most important of which are the type and severity of activity and relative availability of either carbohydrates or fats as an energy source.

The proper regulation of nutritional requirements for individual activities can be classified under the various interdisciplinary bases of ergonomics. The theories of ergonomics are intertwined with the theories of many other disciplines, with the primary objective to fit the demands of the workplace to the efficiency of the human "machine". The proper relationships should reap the desired benefits, such as a reduction of stress, greater efficiency and accuracy, and personal health and well-being. Thus it is that the quantity, quality, and timing of an individual's food intake, along with the basic requirements of one's basal metabolic rate and characteristic energy expenditure when engaged in a particular industrial task, can readily be seen to affect the workplace.

The present report will further explore the interrelationship between ergonomics and nutrition through a review of the research and the background of nutritional effects on physical activity, and an analysis of those industrial activities which could be of sufficient physical demand to necessitate a beneficial nutritional scheme.

NUTRIENTS

The most useable form of carbohydrates to an individual are in the form of glycogen, the storage component of glucose (the major carbohydrate) in the muscle cells. From the muscle, glycogen undergoes a series of reactions known as the Krebs critic acid and Coricycles, from which the direct "battery packs" known as Adensosine Triphosphate (ATP) are synthesized in quantity through the aerobic fragmentation of glycogen, or glycogenolysis. (Astrand and Rodahl (1977)). Glucose is also broken down anaerobically in the bloodstream, and will provide an ATP source for severe exercise lasting for 5 or 6 seconds at the commencement of activity or when all sources of energy in the bloodstream and muscle cells able to be mobilized and synthesized have been exhausted.

The most common form of fat storage in the individual takes the form of both triglycerides and free fatty acids (FFA) in the adipose tissues, which are primarily located in the peripheral skin folds. As an energy source, fats depend on oxygen as the hydrogen acceptor for free fatty acid oxydation. Without oxygen, carbohydrates are the only energy source, and as indicated, their use would be restricted to a short period of time.

The percentage participation of carbohydrates and fats in energy metabolism is usually determined by the use of a nonprotein respiratory quotient (R), which is the ratio of CO_2 volume produced/O_2 volume utilized. In general, the higher the R value the more carbohydrates are used as fuel and the greater the efficiency of the system (Astrand and Rodahl (1977)). An approximate R value for a system utilizing only fats as an energy source is 0.7; for a system utilizing only carbohydrates the value is 1.0. Of course, these values are absolute, and are very rarely attained in controlled studies.

FOODSTUFF INGESTION AND PERFORMANCE

It can be seen in the literature that carbohydrates are the major determinant of exercise to fatigue at levels ranging from 70% to 100% of individual VO_2 max on tasks completed until exhaustion, whereas for tasks under this demand level triglycerides would tend to maintain the level of performance, since at levels below 50% VO_2 max work activities can continue at infinitum until mental fatigue and boredom surface. It has been shown (Bonen et al. (1981)) that glucose feeding and carbohydrate loading for days before and even up until 2 to 3 hours before an activity, will result in proper depositing of ATP in muscle glycogen depots and a corresponding "tapping" of depots at the proper junction in the exercise, thus limiting presence of FFA in the bloodstream and increasing the R value (efficiency). However, glucose feeding 15 to 30 minutes before exercise at levels of 70-80% VO_2 max will result in a condition of hyperglycemia (depletion of blood glucose) and hyperinsulemia (too much insulin in the blood), since glucose tends to dominate as an energy source and is used up quickly in exercise (Pernow and Saltin (1971)). This produces a feeling of sluggishness during exercise that is not allayed unless glucose feeding is undertaken during exercise. After the initial "rush" of glucose from the bloodstream to the working muscles and the resultant hyperglycemia, plasma FFA levels are increased 200 to 300 percent, with the expected decrease in R value for the system. The reintroduction of glucose during the event allows the bloodstream to synthesize this substrate and deliver to the working muscles. At levels of 80% of VO_2 max, the respiratory quotient increases to .90-.95 and blood lactate levels are increased. The hyper-insulemia effect is avoided due to the even faster turnover of glucose than the amount ingested 30 minutes before exercise (Bonen et al.(1981)). This exercise during feeding also produces significant increases in endurance; a study by Coyle et al. (1983) resulted in enducance time increases of approximately 25% for subjects working a task at 70-79% VO_2 max to exhaustion, with glucose administration 20 minutes into exercise. The feeding prevented muscle glycogen depots from being utilized as rapidly.

A good summary of the fat-carbohydrate interaction is provided by Pruett (1970) who studied FFA mobilization and use for subjects at rest and at work to exhaustion at 20, 50, 70, and 90% of VO_2 max. Fifteen minutes after work stop a glucose solution was administered to ascertain the extent of reduction of FFA plasma levels.

The major conclusion of this study is that the direction of the change in plasma FFA level during exercise is dependent on the relative severity of the work in terms of VO_2 max, and not on work duration. In this research, the point at which plasma FFA levels stop increasing as the subject exercises to exhaustion lies between 70 and 85% of VO_2 max, the level previously indicated as triggering carbohydrate sources as energy.

WORKPLACE AND ENERGY REQUIREMENTS

It is no secret that the increasing automation of industry has resulted in a sizable decrease in manual workers. Whereas 100 years ago almost 80% of work was classified as moderately heavy to severe in terms of energy expediture, recent figures (Wirths (1976)) indicate a decrease to 30% of workers engaged in this type of work.

Occupations are rendered more taxing by variables such as temperature extremes, the age of the worker (VO_2 max decreases about 20% by age 60), lack of assistance from mechanical equipment for lifting and transport, and biomechanical postures necessary in performing an activity. Workers who must shovel or chop wood continuously in their work have almost a 50-60 percent increase in energy consumption over workers who maintain a sitting posture. (Grandjean (1980)). These variables together with large variations in the mental and physical status of the worker on a day-to-day basis are the crux of the ergonomic aspect of nutrition; that is, how to devise a scheme of food intake that may account for the stated variations in workers and task demands.

FOODSTUFFS AND DISTRIBUTION OF INTAKE

Table 1 indicates the relative energy supply provided by some of the vital foodstuffs commonly consumed. It is seen that the protein-rich foods such as vegetables and milk are those not commonly eaten to provide a feeling of satiety; rather, it is the meat, cheese, and pastas which are required to provide the overwhelming supply for energy needs.

It is obvious that in order to take in from 3500 to 4000 kcalper day, a worker would want to increase intake of proteins and fats to well above normal levels, as carbohydrates such as potatoes, pastas and sweets tend to be bulky and difficult to digest. It is also accepted logic that the 3 main meals should be fairly evenly spaced so that energy provisions are uniform across the entire day. The normal course of muscular efficiency as

TABLE 1 CALORIFIC REQUIREMENTS OF SOME IMPORTANT FOODSTUFFS (ADOPTED FROM GRANDJEAN (1980)).

Foodstuff	Quantity Eaten to Produce 100 Kcal of Energy
Green Vegetables	670 g
Full-Rich Milk	200 ml
Potatoes	150 g
Eggs	60 g
Meat	50 g
Cheese	45 g
Bread	42.5 g
Pastas	30-40 g
Sugar	25 g
Butter or Margarine	13.5 g

measured by the R value (CO_2 volume produced/O_2 volume utilized) follows a fairly definite path after each meal; for the first hour after the meal production begins sluggishly, then rises to a peak level gradually by the first hour. Thereafter, production decreases with a fall in the R value until the next meal, when the cycle commenced again. Studies show a decrease of 10-12% in both R value and production from the first to the fourth hour after a meal (Haggard and Greenberg (1935)). The solution to this consistent dropoff is in the provision of snacks to the workers; depending on the severity of the work task, a snack every 2 hours between meals has been shown to reduce the performance decrement and result in an overall increase in production over the course of the work day (Table 2).

TABLE 2 EFFECT OF FOOD ON OUTPUT IN A TENNIS-SHOE FACTORY (ADOPTED FROM HAGGARD AND GREENBERG (1935)).

Feeding	Number of Items Produced		
	During 1st Hour	During 4th Hour	Daily Total
Without Breakfast	144	156	1379
3 Main Meals	192	168	1455
3 Main Meals Plus 2 Snacks	193	186	1521

Table 3 indicates the calorie content of several common snacks. For sedentary workers a very light snack or no snack at all is suggested, since with a balanced diet and normal leisure activity the work environment will not unduly stress the R value beyond any expenditure utilized by mental processes. Coffee and tea are commonly considered due to their role in stimulating sympathetic nervous activity rather than as an energy source, but

TABLE 3 SUGGESTIONS FOR BETWEEN MEAL SNACKS, AND THEIR CALORIE CONTENT (ADOPTED FROM GRANDJEAN (1980)).

Type of Snack	Number of Kcal
1 Cup of Soup	10-15
1 Cup of Tea with Two Lumps of Sugar	35
1 Cup of Coffee with Milk and Two Lumps of Sugar	37
1 Cup of Yogurt or Milk	66
Bread (50 g)	120
Bread with Fruit	about 240
Bread with Cheese	about 300

snacks such as bread with fruit can be seen to be vital if provided in the morning and afternoon, providing manual workers with as much as 20% of their daily energy requirement.

Whereas a sedentary worker would be well-advised to limit meals to those foods which are natural and protein-enriched, such as vegetables, raw fruit, milk, and liver (Table 1), disturbances in the balance of energy intake and expenditure will lead to obesity in low energy consumers. The problem of obesity is exacerbated as time goes on, since the desire to engage in physical activity will decrease with corresponding increase in adipose tissue storage and often with age and various physical infirmities. In recent years there has been a rather intense push toward diets with a higher content of nutrients per energy unit, along with the institution of foods with polyunsaturated fats to reduce the amount of fats consumed which are difficult to emulsify. In addition, with an increase in energy output an individual can, without the risk of obesity, eat more and automatically get a greater supply of essential nutrients (Astrand and Rodahl (1977)).

SUMMARY

There are a number of ways by which active control of nutrition can optimize both the productive capacity of an individual and that individual's health and well-being.

1. Ensure that a person's energy uptake is consistent with the energy expenditure for that particular task. A recommended distribution of food intake for both sedentary and manual work whould be established, with leisure activity maintained at a constant level. This schedule would of course be flexible, depending on the individual metabolism and work schedule. However, it can be seen that the substance of meals and, more importantly, the provision of snacks is of vital importance to manual workers.

2. As indicated in Table 2, all meals including breakfast are necessary to maintain a consistent performance. However, it is also suggested that in order to mitigate the "lows" associated with the first hour after meals, individuals

should be encouraged to engage in some type of activity to help reduce the sluggishness of hyperinsulemia. Even tasks such as showering, dressing, and running errands after breakfast, or a short walk following lunch, will tend to allow the worker to begin work at a higher level of performance.

3. Should it be deemed impractical to limit an individual's energy (food) uptake, ways and means to increase energy expenditure should definitely be explored. Many large employers have instituted company fitness centers and mandatory exercise sessions in response to philosophies of "sound mind-sound body". A 2-kilometer walk per day would add 100 kilocalories to average daily expenditure, a small but often decisive difference in terms of the uptake/demand balance (Astrand and Rodahl (1977)).

4. Many work activities feature seasons or even individual days when task demands will be especially intense. Summer construction and mining work during peak periods will necessitate prolonged physical exertions, perhaps without a rest break. Just as an athlete participates in carbohydrate loading on the days preceding an event, so should a worker ingest ample quantities of carbohydrates several days before peak task demand, preferably in tandem with an increased amount of fluids.

5. A necessary ingredient of proper nutrition in the workplace, and indeed of optimum maintenance of body function, is the inclusion of a proper diet. According to Astrand and Rodahl (1977), fat content should range between 25 and 35 percent, with a significant amount of polyunsaturated fats; refined sugar in the diet should be kept as low as possible; and consumption of vegetables, fruit, berries, low-fat milk, fish, lean meat, and cereal products should be relatively high. This would ensure the necessary complement of protein, iron, calcium, and certain vitamins in the diet in addition to satisfying energy requirements.

REFERENCES

[1] Astrand, P.O. and Rodahl, K., Textbook of work physiology (McGraw-Hill, New York, 1977).

[2] Bonen, A., Malcolm, S.A., Kilgour, R. O., MacIntyre, K. P., and Belcastro, A. N., Glucose ingestion before enduring intense exercise, J. Appl. Physiol., 51 (1981) 766-771.

[3] Consolazio, C. F. and Johnson, H. L., Dietary carbohydrate and work capacity, The American Journal of Clinical Nutrition, 25 (1972) 85-90.

[4] Coyle, E. F., Hagberg, J. M., Hurley, B. F., Martin, W. H., Ehsani, A. A., and Holloszy, J. O., Carbohydrate feeding during prolonged strenuous exercise can delay fatigue, J. Appl. Physiol., 55 (1983) 230-235.

[5] Grandjean, E., Fitting the task to the man (Taylor and Francis Ltd., London, 1980).

[6] Haggard, H. W. and Greenberg, L. A., Diet and physical efficiency (Yale University Press, New Haven, 1935).

[7] Pruett, E. D. R., FFA mobilization during and after prolonged severe muscular work in men, J. Appl. Physiol., 29 (1971) 809-815.

[8] Wirths, W., Ist eine zwischenverpflegung wahrend der arbeitszeit ernahrungsphysiologish not wending? Ernahrungspad agogisches colloqium (Muhlenstelle, Bonn, (1976).

Trends in Ergonomics/Human Factors II
R.E. Eberts and C.G. Eberts (Editors)

METHODOLOGICAL CONSIDERATIONS IN THE ASSESSMENT OF VIGILANCE IN CHILDREN

Kathleen M. Krafft and Kim N. Dietrich
Department of Environmental Health
College of Medicine
University of Cincinnati

Selected reports incorporating the use of the vigilance paradigm in assessing sustained attention in children were reviewed. The efficacy of the vigilance paradigm for investigating the development of attention in normal children and clinical populations, and studying the effects of pharmacotherapy on attentional skills, was examined.

INTRODUCTION:

The role of attention in children's social and cognitive development has been a central issue in both psychological theory and research. The ability to attend for sustained periods of time has obvious significance for children. Developmental psychologists and other behavioral researchers have recently focused on the characteristics of sustained attention in both normal and exceptional children. As a result, various types of vigilance tasks have been established primarily by researchers who are unaware of the nuances of vigilance methodology or its theoretical framework. Likewise, researchers within the vigilance area are, for the most part, unaware of the magnitude and extent of vigilance studies in children. A major purpose of this paper is to attempt to remedy the latter, by providing a methodological review of selected studies of vigilance performance in children.

The results of studies are summarized below in three major groupings: 1) studies involved with a developmental assessment of sustained attention in children, 2) vigilance performance of special clinical groups, and 3) use of vigilance tasks in assessing pharmacotherapy. Table 1 presents a summary of the vigilance task parameters for most of these studies.

Developmental Studies of Normal Children

In the first study of this kind, Gale and Lynn (1972) used an auditory Continuous Performance Task (CPT) on children ages 7-13 1/2 years. While vigilance performance (misses) and age were linearly related, a *post hoc* analysis revealed that the greatest improvement occurred between 8 and 9 years. However intelligence and personality measures failed to correlate with vigilance performance. Levy (1980) used a visual CPT-X version to determine age norms for vigilance performance in children from 3 to 7 years. Levy found that 73% of the children under 3 1/2 years could not complete the CPT, while all children above 4 1/2 years were fully testable. Misses, false alarms (FAs), and reaction times (RT) significantly decreased with age. Simon (1982) measured performance of preschool children using a vigilance task in which a star on the fuselage of a plane changed from red

TABLE 1: Vigilance Parameters

Ref.	Task Type	Stimulus Type	Stimulus Duration (sec)	ISI (Sec)	CS/NS	Event Rate
11	Group CPT	Auditory	-	-	40/1560	40
22	CPT-X	Visual	2.0	1.5	20/60	17
35	Vigilance	Visual	0.5	-	-	-
16	Vigilance	Visual	0.5	75*	-	<1
18	Vigilance	Visual	0.5	.5	35/2430	30
21	CPT-AX	Visual	0.8	1.0	72/408	30
		Auditory	-	<1.0	90/510	60
14	Vigilance	Visual	0.5	5-120	16/-	1
23	CPT-X	Visual	2.0	1.5	20/-	-
41	CPT-AX	Visual	0.2	1.5	-	-
		Auditory	-	-	-	-
19	CPT-AX	Auditory	-	-	60/-	-
25	Vigilance	Auditory	40m sec	-	64/-	40
27	CPT-X	Auditory	200m sec	-	-	-
	CPT-AX	Auditory	200m sec	-	35/-	-
30	Vigilance	Visual	-	-	60/-	-
15	Vigilance	Visual	0.5	5-120	-	-
26	Vigilance	Auditory	40m sec	1.5	65/-	-
29	Vigilance	Visual	0.02	-	-	-
33	CPT-AX	Visual	0.10	-	32/-	-
36	CPT-AX	Visual	0.10	-	-	-
10	CPT-AX	Visual	100m sec	-	-	-
40	CPT-AX	Visual	0.2	1.5	-	-
		Auditory	-	-	-	-
2	Vigilance	Visual	0.2	2.0	60/900	30
38	Vigilanace	Visual	0.5	-	25/-	-

*Average ISI

to green (CS) at periodic intervals. Simon found fewer correct detections (CDs) for children below the mean age of the sample (5.6 years) than for those above the mean.

The developmental significance of sustained attention is explored in four studies which relate vigilance to academic performance. Kirchner and Knopf (1974) examined the relationship between vigilance and achievement test scores in a group of normal second graders. High achievers made significantly more CDs and fewer FAs than low achievers. In addition, the vigilance decrement was more pronounced among low achievers. Knopf and Mabel (1975) looked at the relationship between vigilance and reading ability in second graders. While vigilance performance was not associated with reading level overall, poorer readers in the lower socioeconomic class subsample appeared to exhibit a lower base rate of CDs. Simon (1982) used a visual task to determine the school readiness of preschoolers. The author found CDs significantly correlated with scores on the Metropolitan Readiness Test. Kupietz and Richardson (1978) studied the relationship between vigilance and classroom inattentiveness in children from 7 1/2 to 12 1/2 years of age. The authors found misses and FAs on the visual CPT to be positively correlated with off-task behavior in the classroom. Significant correlations were also found between FAs and "aggressivity" and "hyperactivity" factors on the Connors Teacher Behavior Rating Scale.

Although the development of sustained attention appears to be mediated by central nervous system maturation, other factors such as sex, social class, and intelligence have been considered as well. The higher representation of male children in clinical populations have led some investigators to hypothesize that, in general, girls attend better than boys due to genetic or social-rearing differences. However, the majority of vigilance studies on normal children have failed to find sex differences in sustained attention (Gale and Lynn, 1972; Harcherik et al., 1982; Kirchner and Knopf, 1974; Levy, 1980; Levy and Hobbes, 1979; Locke, 1970). Gale and Lynn found that girls consistently performed better than boys at earlier ages; but this difference tended to diminish after 9 years. Mohan (1982) also reports sex differences in vigilance, but the superior performance of girls in the oldest (12 year) age group was largely responsible for the significance of the sex factor. The inconsistency of these results indicate that assumptions concerning the superiority of girls in sustained attention requires re-examination.

Attempts to relate intelligence (IQ) to vigilance performance in normal children have generally failed to find a significant relationship (Gale and Lynn, 1972; Kupietz and Richardson, 1978). Herman et al. (1980) found a moderate correlation between IQ and CDs in a visual vigilance task. However, in this preschool sample, an IQ score at this age may reflect the child's ability to attend to a long task as well as intellectual ability. Therefore, the significance of IQ and vigilance performance may diminish with age.

Few developmental variables escape the influence of social class (Broman et al. 1975), and vigilance does not appear to be an exception. Knopf and Mabel (1975) compared middle and lower class second graders' performance on a visual vigilance task. Middle class subjects obtained a substantially higher percentage of CDs than their lower class counterparts. From a study by Levy and Hobbes (1979), it appears that middle class subjects not only perform better than their age mates, but also show an earlier development in this capacity. While most upper class children in their study could complete a visual CPT by four years, the majority of lower class children could not. Further, upper class children made significantly fewer misses through 7 years.

In summary, vigilance performance substantially improves with age and therefore is probably linked to central nervous system maturation. However, the development of sustained attention appears to be moderated by the child's environment, as the relationship between vigilance and socioeconomic status suggests. The association between vigilance and academic performance confirms the salience of sustained attention in the child's daily transactions. Developmental delay in this area is likely to lead to both academic and social difficulties. Finally, vigilance performance does not appear to be related to sex or IQ in normal children.

Vigilance in Special Clinical Groups

A large portion of the studies of vigilance in children attempt to demonstrate differences in vigilance performance between normal and exceptional children. These groups represent neurologically compromised children with characteristic diagnosed classifications, of largely unknown etiology. Children of schizophrenic parents, epileptic and mentally retarded subjects (Stores et al. 1978) have been studied and the data suggest they are at higher-risk for deficits in sustained attention. Children of schizophrenic parents, epileptic (Stores et al. 1978), and mentally retarded subjects have been studied, and the data suggest they are at higher-risk for deficits in sustained attention. This literature has been reviewed elsewhere in detail and will not be considered any further here (Garmezy, 1978; Warm and Berch, in press).

While these investigations contribute to the wide range of populations studied in terms of vigilance, the performance of hyperactive children has constituted the bulk of research in this area and will be reviewed in more detail here.

Hyperactivity is often characterized by overactivity, impulsivity, failure to sustain attention and various other behavioral problems. In a study by Porrino et al. (1983), motor activity (24 hr.) of hyperactive and normal boys in their natural environments was examined. A CPT task was used before activity monitoring to characterize, *a priori* differences in attentional skills between the two groups. Hyperactives made significantly more misses and FAs than controls. A discriminant analysis determined that a combination of overall school activity scores and misses on the CPT best discriminated between hyperactives and controls with approximately 88% accuracy. Levy and Hobbes (1981) also utilized a discriminant analysis involving measures from a CPT task to discriminate between hyperactive and control groups. This combination of variables correctly identified only 60% of hyperactives, demonstrating high specificity (few false positives) but low sensitivity by failing to identify 'true' hyperactives (false negatives).

While excessive motor activity is the more salient feature of the hyperkinetic disorder, attentional deficits represent another major class of symptoms extensively studied using a vigilance paradigm. Sykes, Douglas, and Morgenstern (1973) compared the performance of hyperactives and controls on visual and auditory versions of the CPT. The auditory CPT appeared to be more difficult (fewer CDs) than the visual CPT for both groups. Hyperactives made fewer CDs and more FAs than controls in both CPTs. The performance of all subjects on both tasks was related to age but not IQ measures.

Kupietz (1976) examined the rate of vigilance decrement in hyperactives and controls tested twice in a 4 day period. During the initial CPT, the hyperactive group made significantly more misses than controls, overall and obtained more misses in the second half of the session compared to the first. In the second CPT session, both groups made significantly more misses later in the session relative to the first portion.

A study by Loiselle et al. (1980) examined differences in cortical event related potentials (ERPs) normal and hyperactive boys. Tests of attention including an auditory vigilance task were conducted in two sessions. As in previous reports, hyperactives made fewer CDs, more FAs and longer RTs compared to controls. The ERPs of the hyperactives reflected their CPT performance and were thought to represent changes in neural processes associated with the differentiation and selection of incoming information. A similar tact was taken by Michael et al. (1981) in obtaining concurrent measurements of sustained attention and ERPs in normal and hyperactive children. Normals had superior vigilance performance under both CPT versions. Methylphenidate normalized the performance of hyperactives in the direction of the controls. Changes in the ERPs were associated with the deterioration of vigilance performance with time on task.

The performance of hyperactives was compared to hypoactives on the vigilance task where knowledge of results was provided for CDs (KR+) or FAs (KR-) in a study by Ozolins and Anderson (1980). Feedback conditions differentially affected performance of the two groups. Hyperactives made more CDs under the KR- condition compared to the no-feedback of KR+ conditions. Hyperactive subjects made more FAs than hypoactives under the KR+ condition. Kirchner (1976) also examined the effects of different types of feedback/distractors on vigilance performance. Controls made significantly more CDs compared to hyperactives. A vigilance decrement was evident for both groups. There were no differences between experimental conditions or groups with respect to FAs.

While it is clear that attentional deficits are not unique to hyperactive children, a pattern of vigilance performance characteristic of this group remains apparent. Compared to normal children, hyperactives made fewer CDs, more FAs and tend to show a faster rate of deterioration in vigilance performance with time on task.

Children labeled 'Learning Disabled" constitute another broad categorization of children with characteristics including low achievement scores, severe deficits in specific academic skills, short attention span and high distractibility. Comparisons of the vigilance performance between LD and normals yield results similar to those of hyperactives. That is, LD children made consistently more FAs and fewer CDs than controls (Anderson, Halcomb and Doyle, 1973; Doyle, Anderson, and Halcomb, 1976; Swanson, 1981; Danier et al. 1981). Differences in vigilance performance between remedial readers and controls were addressed in two studies (Lovich and Stamm 1983; Noland and Schuldt, 1971). The results were contradictory; Lovich and Stamm reported longer RTs for remedial readers, while Noland and

Schuldt reported that the control group made more CDs as well as a decrease in vigilance with time on task for both groups.

The lack of consistent findings in this area probably reflects the 'catch all' nature of the LD classification. As in previous studies, the selection of appropriate vigilance task parameters was not addressed. In some cases, the tasks may have been too easy and therefore not able to provide a distinction between LD and control subjects rather than failing to demonstrate attentional deficits. Likewise, consideration for the nature of their learning disability (reading skills) was not made in selecting the type of stimuli (letters) or modality (visual) for the vigilance task.

Effects of Drugs on Vigilance Performance

Vigilance tasks have also been used to evaluate the effects of drugs on the performance of hyperactive children. Amphetamine results in an improvement in vigilance performance of hyperactive children compared to placebo (Porrino et al. 1983; Rapoport et al. 1980; Sostek et al. 1980; Werry and Aman, 1975; Kupietz and Balka, 1976; Klorman et al. 1979) and improved vigilance in hyperactive and normals compared to no drug (Rapoport et al. 1978). Caffeine was associated with improved vigilance performance of normal boys, in a dose-related fashion (Elkins et al. 1981). Similar dose-related, but non-significant trends were reported for hyperactive children (Connors, 1979).

Methylphenidate-induced changes in vigilance performance for hyperactives have also been reported. Hyperactives made more CDs and fewer FAs under the drug condition compared to placebo (Sykes et al. 1971, 1972), and a disruption of the vigilance decrement under active drug conditions (Sykes et al. 1972).

Anderson et al. (1974) compared drug- to no-drug performance in hyperactive boys. The subjects were currently on medication (methylphenidate, amphetamine, phenobarbital or other drugs), so vigilance testing occurred once while on medication and once while off medication. Children in the younger age groups made more CDs on medication compared to off-medication. However, this was not true of the older children. The authors interpreted the absence of a drug effect of older children in terms of either a ceiling effect (task too simple) or that they were not responsive of the effects of the medication.

The final study included in this section, addresses a slightly different issue than those previously discussed; the effects of exposure to exogenous compounds in utero on later cognitive development. Streissguth et al. (in press) assessed vigilance performance in children with different levels of alcohol and nicotine exposure during gestation. There was a dose related increase in misses, FAs and RTs associated with maternal alcohol use during pregnancy. A similar relationship was found between vigilance performance and nicotine use.

Collectively these studies indicate that the vigilance task provides an objective and reliable measure with which to assess drug effects, dosage issues, and may also serve as a potentially powerful and sensitive assessment of the neurobehavioral effects of xenobiotics.

Discussion

The studies reviewed here suggest that the vigilance paradigm provides an objective and reliable measure of sustained attention in children. Developmental studies on normal children have demonstrated that vigilance performance substantially improves with age and therefore is probably linked to central nervous system maturation (Gale and Lynn, 1972; Levy, 1972; Levy and Hobbes, 1979; Mohan, 1982). However it appears that environmental influences (SES) also moderate the development of sustained attention (Knopf and Mabel, 1975). Vigilance performance has also been associated with academic performance, off-task classroom behaviors and behavioral ratings (Kirchner and Knopf, 1974; Kupietz and Richardson, 1978).

A large portion of the articles in this review have been concerned with the sensitivity of the vigilance task in identifying performance differences between special clinical groups and controls. A steeper decrement in vigilance performance over time for hyperactives has been demonstrated in some (Doyle et al, 1976; Kirchner, 1976) but not all cases (Kupietz, 1976). Other clinical groups which are characterized by attentional deficits (LD, epilepsy, retardation, schezophrenia) have exhibited inferior vigilance compared to controls, although due to the small sample sizes and procedural differences, characteristic patterns of vigilance have not been identified.

The major portion of this review was devoted to the use of vigilance type tasks in assessing attentional deficits in children. While these studies clearly demonstrate the potential usefulness of the vigilance paradigm as an assesment instrument. Caution must be taken in interpreting these findings for several reasons. First, the types of vigilance tasks vary widely between studies and preclude direct comparison. It can be seen from Table 1 that procedural details (type and duration of stimuli, ISI, number of critical to noncritical stimuli, and event rate) are often not reported. This makes it difficult not only for readers to compare results, but also for researchers to use similar tasks. Second, because of the lack of precision on describing vigilance parameters, researchers looking for an attentional measure may use a vigilance task which is not appropriate for their sample.

The psychophysics of vigilance has been extensively studied in adults. By varying specific psychophysical dimensions, investigators have associated superior vigilance efficiency with auditory stimuli, critical signals (CSs) which are of longer duration, more frequent and higher intensity, regular interstimulus intervals (ISI) and inter-critical-stimulus intervals, knowledge of results, greater cognitive involvement, and rest periods between trials (Davies and Parasuraman, 1982; Dember and Warm, 1979; Warm, Sprague, and Dember, 1981; Berch and Kanter, 1984). The psychophysics of vigilance in children has been less extensively studied. Although equivocal, the following parametric variations have been associated with more efficient vigilance performance in children; use of visual stimuli (Kupietz and Richardson, 1978; Sykes et al., 1973), Shorter ISIs (Locke, 1970), minimizing task difficulty Michael et al., 1981; Sykes et al., 1971), and knowledge of results (Ozolins and Anderson, 1980; Ware et al., 1962). Unfortunately, these findings are limited in that the parametric values associated with optimal performance change with age, diagnostic classification, or differentially affect performance of subgroups within a sample.

Third, methodological considerations make the interpretation of some studies difficult if not impossible. These short comings suggest that further investigations in the area of children's vigilance performance address task-specific issues, provide systematic replication, and continue to explore the validity of vigilance as a measure of attentional abilities which reflect academic performance.

ACKNOWLEDGEMENT:

The authors were supported by grants from the National Institute of Environmental Health Sciences (USPHS ES 01566-04, and 3T32 ES07073), and the Center for the study of the Human Environment (USPHS ES 00159) during the preparation of this manuscript.

The authors are grateful for the assistance of Mrs. Nancy Knapp who typed the manuscript.

References

1. ANDERSON, R.P., HALCOMB, C.G., & DOYLE, R.B. (1973). The measurement of attentional deficits. Exceptional Children, 39, 534-539.

2. ANDERSON, R.P., HALCOMB, G.C., GORDON, W.G., & OZOLINS, D.A. (1974). Measurement of attention distractibility in LD children. Academic Therapy, 9, 261-266.

3. BERCH, D.B. & KANTER, D.R. (1984). Individual differences. In J.S. Warm (Ed.) Sustained attention in human performance, Chichester, England: John Wiley and Sons.

4. BROMAN, S.H., NICHOLS, P.L., & KENNEDY, W.A. (1975). Preschool IQ: Prenatal and early developmental correlates. Hillsdale, N.J.: Erlbaum.

5. CONNERS, C.K. (1979). The acute effects of caffeine on evoked response, vigilance, and activity level in hyperkinetic children. Journal of Abnormal Child Psychology, 7(2), 145-151.

6. DAINER, K.B., KLORMAN, R., SALZMAN, L.F., HESS, D.W., DAVIDSON, P.W., & MICHAEL, R.L. (1981). Learning-disordered children's evoked potentials during sustained attention. Journal of Abnormal Child Psychology, 9(1), 79-94.

7. DAVIES, D.R. & PARASURAMAN, R. (1982). The Psychology of Vigilance, New York: Academic Press.

8 DEMBER, W.N., & WARM, J.S. (1979). Psychology of perception (2nd ed.). New York: Holt, Rinehart & Winston.

9. DOYLE, R.B., ANDERSON, R.P., & HALCOMB, C.G. (1976). Attention deficits and the effects of visual distraction. Journal of Learning Disabilities, 9(1), 59-65.

10. ELKINS, R.N., RAPOPORT, J.L., ZAHN, T.P., BUSHBAUM, M.W., WEINGARTNER, H., KOPIN, I.J., LANGER, D., & JOHNSON, C. (1981). Acute effects of caffeine in normal prepubertal boys. American Journal of Psychiatry, 138(2), 178-183.

11. GALE, A., & LYNN, R. (1972). A developmental study of attention. British Journal of Educational Psychology, 42, 260-266.

12. GARMEZY, N. (1978). Attentional processes in adult schizophrenia and in children at risk. Journal of Psychiatric Research, 14, 3-34.

13. HARCHERIK, D.F., CARBONARI, C.M., & COHEN, D.J. (1982). Attentional and perceptual measures: Developmental changes. Schizophrenia Bulletin, 8, 349-355.

14. HERMAN, C.S., KIRCHNER, G.L., STREISSGUTH, A.P., & LITTLE, R.E. (1980). Vigilance paradigm for preschool children used to relate vigilance behavior to IQ and prenatal exposure to alcohol. Perceptual and Motor Skills, 50, 863-867.

15. KIRCHNER, G.L. (1976). Differences in the vigilance performance of highly active and normal second-grade males under four experimental conditions. Journal of Educational Psychology, 68(6), 696-701.

16. KIRCHNER, G.L., & KNOPF, I.J. (1974). Differences in the vigilance performance of second-grade children as related to sex and achievement. Child Development, 45, 490-495.

17. KLORMAN, R., SALZMAN, L.F., PASS. H.L., BORGSTEDT, A.S., & DAINER, K.B. (1979). Effects of methylphenidate on hyperactive children's evoked responses during passive and active attention. Psychophysiology, 16(1), 23-29.

18. KNOPF, I.J., & MABEL, R.M. (1975). Vigilance performance in second graders as a function of interstimulus intervals, socio-economic levels, and reading. Merrill-Palmer Quarterly, 21(3), 195-203.

19. KUPIETZ, S.S. (1976). Attentiveness in behaviorally deviant and non-deviant children: I Auditory vigilance performance. Perceptual and Motor Skills, 43, 1095-1101.

20. KUPIETZ, S.S., & BALKA, E.B. (1976). Alternations in the vigilance performance of children receiving amitriptyline and methylphenidate pharmacotherapy. Psychopharmacology, 50, 29-33.

21. KUPIETZ, S.S., & RICHARDSON, E. (1978). Children's vigilance performance and inattentiveness in the classroom. Journal of Child Psychiatry, 19, 145-154.

22. LEVY, F. (1980). The development of sustained attention (vigilance) and inhibition in children: Some normative data. Journal of Child Psychology and Psychiatry, 21, 77-84.

23. LEVY, F., & HOBBES, G. (1981). The diagnosis of attention deficit disorder (hyperkinesis) in children. Journal of the American Academy of Child Psychiatry, 20 376-384.

24. LOCKE, J.L. (1970). Acoustic vigilance behavior in four-year-old children. Psychonomic Science 2(2), 111-112.

25. LOISELLE, D.L., STAMM, J.S., MAITINSKY, S., & WHIPPLE, S.L. (1980). Evoked potential and behavioral signs of attentive dysfunctions in hyperactive boys. Psychophysiology 17(2), 193-201.

26. LOVRICH, D., & STAMM, J.S. (1983). Event-related potential and behavioral correlates of attention in reading retardation. Journal of Clinical Neuropsychology 5(1), 13-37.

27. MICHAEL, R.L., KLORMAN, R., SALZMAN, L.F., BORGSTEDT, A.D., & DAINER, K.B. (1981). Normalizing effects of methylphenidate on hyperactive children's vigilance performance and evoked potentials. Psychophysiology, 18(6), 665-677.

28. MOHAN, J. (1982). An experimental developmental study of performance reminiscence and vigilance. Studia Psychologica, 24(1), 71-76.

29. NOLAND, E.C., & SCHULDT, W.J. (1971). Sustained attention and reading retardation. The Journal of Experimental Education, 40(2), 73-76.

30. OZOLINS, D.A., & ANDERSON, R.P. (1980). Effects of feedback on the vigilance task performance of hyperactive and hypoactive children. Perceptual and Motor Skills, 50, 415-424.

31. PORRINO, L.J., RAPOPORT, J.L., BEHAR, D., SCEERY, W., ISMOND, D.R., & BUNNEY, W.E. (1983a). A naturalistic assessment of the motor activity of hyperactive boys. Comparison with normal controls. Archives of General Psychiatry, 40, 681-687.

32. PORRINO, L.J., RAPOPORT, J.L., BEHAR, D., ISMOND, D.R., & BUNNEY, W.E. (1983b). A naturalistic assessment of the motor activity of hyperactive boys. II. Stimulant drug effects. Archives of General Psychiatry, 40, 688-693.

33. RAPOPORT, J.L., BUCHSBAUM, M.S., & WEINGARTNER, H. (1980). Dextroamphetamine: Cognitive and behavioral effects in normal and hyperactive boys and normal adult males. Psychopharmacology Bulletin, 16, 21-23.

34. RAPOPORT, J.L., BUCHSBAUM, M.S., ZAHN, T.P., WEINGARTNER, H., LUDLOW, C., & MIKKELSON, E.J. (1978). Dextroamphetamine: Cognitive and behavioral effects in normal prepubertal boys. Science, 199(3), 560-563.

35. SIMON, M.J. (1982). Use of a vigilance task to determine schoolreadiness of preschool children. Perceptual and Motor Skills, 54, 1020.

36. SOSTEK, A.J., BUCHSBAUM, M.S., & RAPOPORT, J.L. (1980). Effects of amphetamine on vigilance performance in normal and hyperactive children. Journal of Abnormal Child Psychology, 8(4), 491-500.

37. STORES, G., HART, J., & PIRAN, N. (1978). Inattentiveness in school children with epilepsy. Epilepsia, 19, 169-175.

38. STREISSGUTH, A.P., MARTIN, D.C., BARR, H.M., SANDMAN, B.M., KIRCHNER, G.L., & DARBY. B.L. (in press). Intrauterine alcohol and nicotine exposure: Attention and reaction time in four-year-old children. Developmental Psychology.

39. SWANSON, L. (1981). Vigilance deficit in learning disabled children: A signal detection analysis. Journal of Child Psychology and Psychiatry, 22(4), 393-399.

Trends in Ergonomics/Human Factors II
R.E. Eberts and C.G. Eberts (Editors)

TASK COMPLEXITY, SUBJECT SEX, AND NOISE MEANINGFULNESS AS DETERMINANTS OF COGNITIVE VIGILANCE PERFORMANCE

Thomas K. Noonan, Daniel W. Ash, and Michel Loeb
University of Louisville

and

Joel S. Warm
University of Cincinnati

Subjects performed vigilance tasks of four levels of complexity during exposure to "Quiet" (40 dBA ambient noise) or to 70 dBA levels of White Noise or Speech. Generally RT increased over time and decreased during noise, though effects depended on subject sex. Hits tended to decline over time, but there were complex effects involving subject sex and type of noise. FA's increased with complexity, more so in noise, especially during Speech and in women. Implications and relationships to previous results are discussed.

Warm and his colleagues at the University of Cincinnati have presented or published several recent studies utilizing a "cognitive" vigilance task, in which the signals to which subjects respond represent a primarily conceptual change rather than the more commonly employed physical changes characterizing the "sensory" vigilance tasks more often employed. (See Warm, Sprague, & Dember, 1981; Warm, Dember, and Lanzetta, 1984.) They report that while with simple tasks of this type they obtain the decrement over time characterizing the sensory tasks, with more complex tasks they obtain a vigilance increment--an improvement in performance over time. The nature of this improvement has not been constant, however: With an LED display they obtained an increase in detections over time (Warm, et al., 1981) and with a CRT display they obtained no such change in detections but did find a progressive decrease in reaction time (Warm, et al., 1984). At Louisville we have failed to find either kind of increment, though with more complex tasks we often have found no decrement. (See Ash, Noonan, Loeb, & Warm, 1983; Noonan, Ash, Loeb, & Warm, 1984.) The difference in reaction time (RT) results may reflect a lower level of precision in measurement of RT; we shall address that in a later paper. Some interesting effects of noise on cognitive vigilance have also been reported at the Cincinnati laboratory. (See Lysaght, 1982; Lysaght, Warm, Dember, & Loeb, 1984.) Generally low levels of noise have been found to have either no effects on simple (mostly sensory) vigilance tasks or beneficial ones (presumably due to arousal), even though, on complex tasks involving scanning and performed in more intense noise several investigators, beginning with Broadbent (1953) have reported detrimental effects. (See Loeb & Alluisi, 1984, for a review of relevant literature.) Lysaght, et al. reported in a recent (1984) paper that the cognitive increment occurred only in women and not in men; further, broadband noise had no effect on performance of either men or women, but a complex noise containing unintelligible speech and machine noise reversed the increment to a decrement for women working at 70 dBA and produced a steady but lower level of performance for women working at 90 dBA. Performance of men was not affected. On the other hand in a recent experiment at Louisville, Noonan, Ash, Loeb, and Warm (1984) reported that a general tendency toward a decrement in simple and cognitive vigilance tasks was abolished by the presence of 70 dBA broadband noise.

In the experiment to be described we investigated the effects of several kinds of noise (varying in level and meaningfulness) on cognitive tasks of several levels of complexity.

PROCEDURE

Four different vigilance tasks, which differed in level and complexity, were employed in this experiment. Three different noise conditions were employed with each. A different group of subjects, with an equal number of males and females, was assigned to each combination of conditions.

In each task a ten-minute training session was followed by a 50-minute actual vigilance session. In each the event rate was 20/minute and 5 signals were presented in each successive 10-minute period. Event durations were 1.5 seconds. Intersignal interval was random with the restriction that no two successive events could be signals. The Level 1 task was essentially the same as the "Simple" condition employed by Warm, et al. (1981, 1984): 2 numbers were presented at the bottom of a CRT every three seconds, and a signal, to which subjects were to respond, was defined as the instance in which the numbers were the same or differed by no more than 1. The Level 2 condition was similar to Warm, et al.'s "Complex" condition--signal events not only met the requirements defined for Level 1 but also had to sum to a total within a target range defined by a pair of numbers separated by a dash at the top of the CRT. The numbers always differed by 10 (e.g., 4-14, 7-17). The target range changed at intervals ranging from 30 to 60 to 90 seconds, randomly. (This differs from Warm, et al.'s condition in that their target range always was 4-14.) In the Level 3 condition the target ranges were not necessarily 4-14, though the bounding numbers always differed by 10 (e.g., 2-12 or 6-16). There were two pairs of numbers at the bottom of the CRT and two target ranges, which changed at intervals ranging from 30 to 60 to 90 seconds, randomly. A signal could occur at either the left or right pair of numbers, though not both simultaneously. Again the numbers at the bottom had to differ by no more than 1 and sum to a total within the target range. In the Level 4 condition there were 4 pairs of numbers at the bottom and 4 target ranges; otherwise the condition resembled Level 3.

The three noise conditions were Quiet (40 dBA inside an IAC double-walled soundproof chamber), Random Noise (70 dBA, presented through speakers), and Speech (70 dBA, produced by feeding a tape recording of G. B. Shaw's Don Juan in Hell through an amplifier and speakers).

Subjects were students in Introductory Psychology classes at the University of Louisville, participating for credit. Eight men and eight women were assigned to each combination of conditions. A very few were rejected after the training session since they had trouble following the instructions.

RESULTS

As in almost all vigilance experiments the scores were Reaction Time (RT),

Detections or Hits (H), and False Alarms (FA). For RT there were significant main effects of Complexity ($F = 126.1$, $df = 3/168$, $p < .001$), Sex ($F = 7.06$, $df = 1/168$, $p < .01$), and Noise ($F = 3.51$, $df = 2/168$, $p < .05$). Mean RT values for Levels 1, 2, 3, and 4 were 1.03, 1.67, 1.77, and 1.84 seconds respectively. (While most of the Complexity effect probably is attributable to the difference between Level 1, the Simple condition, and the other Levels, it should be noted that when the data were analyzed without Level 1 values there still was a complexity effect significant beyond the .01 level.) Mean RT's for Quiet, White Noise, and Speech were 1.63, 1.58, and 1.53 seconds respectively. The mean RT for men was 1.54 seconds; for women, 1.62 seconds. There were also a significant Sex X Noise interaction ($F = 4.30$, $df = 2/168$, $p = .015$), a significant linear trend across periods ($F = 5.40$, $df = 1/168$, $p = .02$), and a linear interaction of Noise X Periods ($F = 6.03$, $df = 1/168$, $p = .015$). (See Table 1.) For men noise, especially Speech, reduced RT; for women Speech reduced RT to a mean value slightly below that of men and White Noise had little effect on them. Generally mean RT tended to increase across periods in Quiet and Speech and to decline during White Noise, though the overall trend across periods was an increase.

Table 1. Reaction Time Effects

(Values are mean RT's in seconds. Q, S, and W refer to Quiet, Speech, and White Noise respectively; M and F to Male and Female. P1 refers to Period 1, P2 to Period 2, etc. The numbers on the last line of the table are overall means for periods.)

a. Sex X Noise Interaction

Q	W	S	Q	W	S
1.59	1.48	1.54	1.67	1.68	1.51

b. Noise X Period Interaction

P1			P2			P3			P4			P5		
Q	W	S	Q	W	S	Q	W	S	Q	W	S	Q	W	S
1.59	1.66	1.47	1.66	1.60	1.49	1.57	1.57	1.55	1.63	1.63	1.54	1.69	1.57	1.50
	1.54			1.58			1.57			1.60			1.60	

For Hits there was a highly significant variance component attributable to Complexity ($F = 106.4$, $df = 3/168$, $p < .001$). In addition there was a significant linear trend across periods ($F = 4.63$, $df = 1/168$, $p < .05$), and a quadratic interaction of Periods with Complexity, Sex, and Noise ($p < .05$). Generally Hits declined as Complexity increased (Figure 1), and there was a tendency for Hits to decrease in later periods. There appear to have been some tendencies for the sexes to exhibit different trends as a function of exposure to different noises at different levels of Complexity. For example, White Noise apparently depressed performance relative to Quiet at the most complex level, especially for women, but Speech, if anything, apparently increased Hits for women and produced a less pronounced tendency for Hits to decline at this level than was apparent for men. (See Figure 1.)

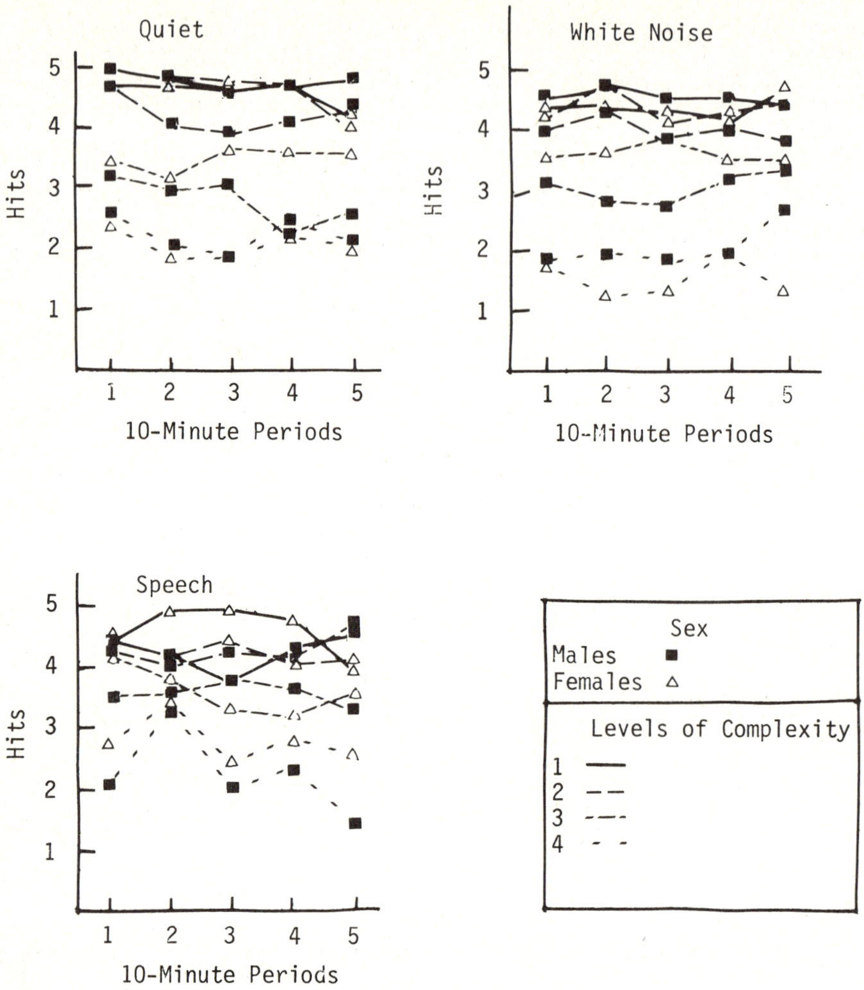

Fig. 1. Mean hits as a function of complexity, sex, type of noise, and time on task.

The most interesting results were obtained for False Alarms. There was significant variance attributable to Complexity ($F = 6.20$, $df = 3/168$, $p < .001$), and to Noise ($F = 3.22$, $df = 2/168$, $p < .05$). There were significant linear and quadratic trends across periods ($p = .005$ and $.007$, respectively), as well as a linear interaction of periods with subject Sex and Noise ($p = .017$) and a significant quadratic interaction of Periods with Complexity ($p = .03$). Figure 2 plots means for FA's. It will be noted that in the Level 1 (Simple) condition that they were negligible. In the Level 2 (Single Range) condition they increased appreciably, especially for men in White Noise and women in White Noise and Speech. In Level 3 (Dual Range) the FA's in Quiet appreciably exceeded those in the comparable condition in Levels 1 and 2. White Noise at Level 3 did not increase the decrement in men; if anything it decreased it. While the FA rate for women in Quiet was low, it increased considerably in both White Noise and Speech, but more so in White Noise. At Level 4 in Quiet men maintained a lower FA rate than at Level 3, and White Noise had little effect on that, but Speech at this high level of complexity produced a very high FA rate for men. Women at this level, on the other hand, exhibited a high FA rate in Quiet, and it increased somewhat in White Noise and still more during exposure to Speech.

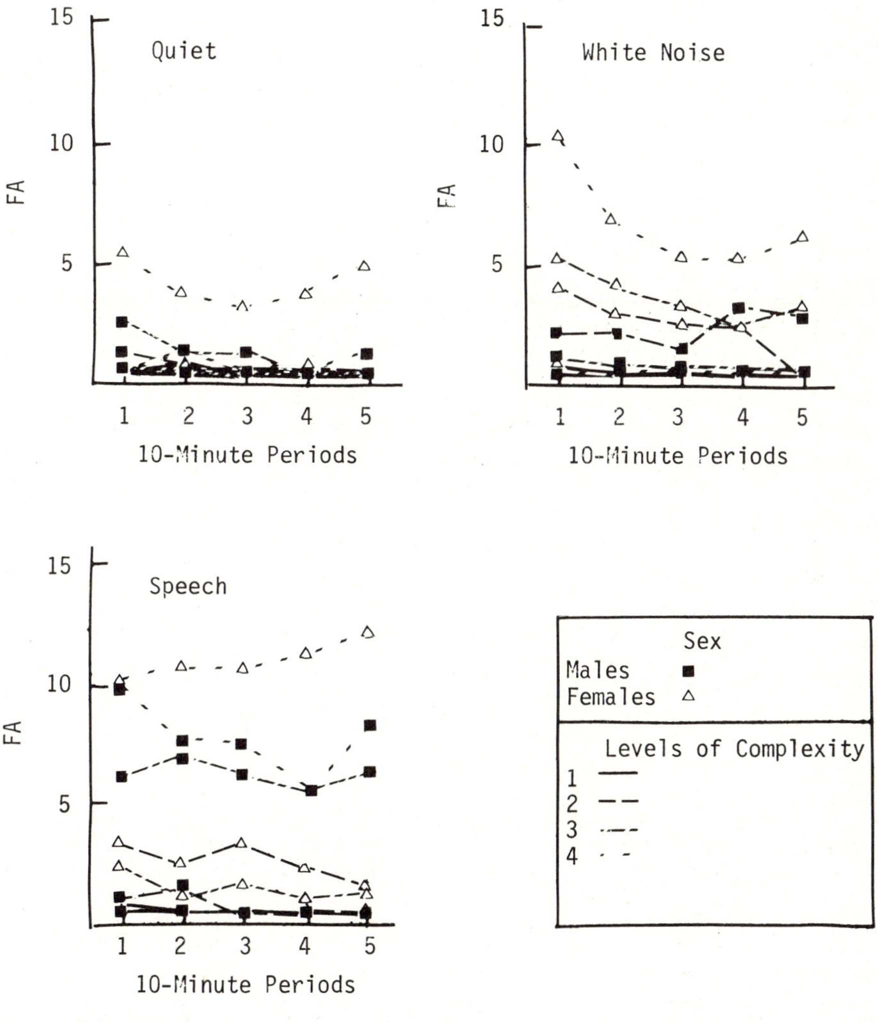

Fig. 2. Mean FAs as a function of complexity, sex, type of noise, and time on task.

In all conditions (except in Level 1, where they were negligible) FA's decreased with time on task, a general finding for vigilance tasks.

It appeared that at low levels of complexity that White Noise and Speech have somewhat comparable effects, generally impairing performance, but sometimes (in the case for White Noise for men at Level 2) having no effect or a facilitating one. At very high levels Speech appears to be more detrimental than White Noise.

DISCUSSION

In this experiment as with most others at Louisville, we failed to confirm the "cognitive increment." As we indicated earlier the increment may not occur for Hits with CRT displays and the failure to find it for RT may reflect the lower level of precision used in this experiment. In another experiment which we shall report later in which precision was improved, there is some evidence of such an increment, though not exactly under the same conditions.

The noise effects are similar in some ways to some earlier results reported by Buckner and McGrath (1963). In that experiment interesting irrelevant stimulation increased performance at low event rates but impaired it at high rates, presumably because subjects could not cope with the increased load when attending to the irrelevant stimulation. In this experiment the effects of speech became particularly detrimental at the higher levels of complexity. The RT data and the FA data obtained in Quiet at higher complexity levels suggests that possibly the task is more complex at given levels for women than for men. It will be noted that women in Quiet at Level 3 exhibited a lower FA rate, perhaps reflecting a greater tendency to concentrate on the task and avoid FA's, but at Level 4 they may not have been able to do this, and the rate went up. Men in Quiet at Level 3 did not exhibit this tendency, but at Level 4 in Quiet their FA rate similarly rose.

The data are also somewhat in accord with those of Lysaght, et al. (1984), who also found differential trends for men and women in noise, probably reflecting differences in effective task load. In this instance, however, the effects were more apparent in the FA data than in the Hit data.

Finally, the effects of noise were very complex and not very predictable. This also is very much in accord with previous findings.

REFERENCES

[1] Ash, D. W., Noonan, T. K., Loeb, M., and Warm, J. S., Is processing mode or degree of challenge the critical variable in vigilance? Paper presented at the Annual Meeting of the Southern Society for Philosophy and Psychology, Atlanta (1955).

[2] Broadbent, D. E., Noise, paced performance, and vigilance tasks, Brit. Jrnl. of Psychol. 45 (1953) 295-303.

[3] Loeb, M. and Alluisi, E. A., Theories of vigilance, in: Warm, J. S. (ed.), Sustained Attention in Human Performance (Wiley, Chichester, England, and New York, 1984).

[4] Lysaght, R. J., The effects of noise on sustained attention and behavioral persistence, Ph.D. Thesis, Dept. of Psychol., University of Cincinnati (1982).

[5] Lysaght, R. J., Warm, J. S., Dember, W. N., and Loeb, M., Effects of noise and information-processing demand on vigilance performance in men and women, in Mital, A., (ed.), Trends in Ergonomics/Human Factors (North-Holland, Amsterdam, 1984).

[6] McGrath, J. J., Irrelevant stimulation and vigilance performance, in Buckner, D. N. and McGrath, J. J., Vigilance: A Symposium (Wiley, New York, 1963).

[7] Noonan, T. K., Ash, D., Loeb, M., and Warm, J. S., Task complexity, noise and cognitive vigilance performance, in Mital, A. (ed.), Trends in Ergonomics/Human Factors (North-Holland, Amsterdam, 1984).

[8] Warm, J. S., Dember, W. N., and Lanzetta, T. M., Cognitive demand and vigilance performance, in Mital, A., Proceedings of the 1st Mid-Central Ergonomics/Human Factors Conference (University of Cincinnati Press, Cincinnati, Ohio, 1984).

[9] Warm, J. S., Sprague, R. L., and Dember, W. N., The congitive increment function in sustained attention, Paper presented at the Annual Meeting of the Psychonomic Society, Philadelphia (1981).

Trends in Ergonomics/Human Factors II
R.E. Eberts and C.G. Eberts (Editors)

PERCEPTION OF CHALLENGE AND DIFFICULTY IN A COGNITIVE VIGILANCE TASK

Thomas K. Noonan, Waldemar Karwowski, Michel Loeb

University of Louisville
Louisville, Kentucky 40292
U. S. A.

The purpose of the present study was to determine if subjects' ratings of difficulty and challenge reflect the complexity of a cognitive vigilance task. Also of interest was the possibility that a challenging task might sustain or improve performance over time. Both difficulty and challenge ratings increased at increasing levels of task complexity. However, no evidence was found that a challenging task serves to alter performance over time.

INTRODUCTION

Recently, research conducted by Warm and his colleagues have found performance on a "demanding" cognitive vigilance task to either improve or remain stable over time while performance on a simple task resulted in a decrement over time (Dember, Warm, Bowers, & Lanzetta, 1984; Warm, Howe, Fishbein, Dember, & Sprague, 1984). Dember, et al. (1984) have suggested that their findings may be due to intrinsic motivation. They propose that a subject's interest is aroused by complex, challenging tasks which serve to improve or sustain performance over time. Deci (1971, 1972) earlier suggested that intrinsic motivation is fostered by challenging tasks. Although Warm and his colleagues have found results supportive of Deci's contention, they have not directly measured the interest or challenge experienced by subjects.

This paper concerns an effort to measure the challenge as well as the difficulty experienced by subjects participating in a cognitive vigilance experiment. Given these earlier findings (Dember et al., 1984; Warm, et al., 1984) as well as Deci's concepts on which Warm's interpretations are based, it is expected that the more complex conditions will be rated as more challenging and more difficult. Further if challenging tasks enhance motivation thereby improving or sustaining performance over time, then both the difficulty and challenge ratings should correlate with such a measure.

METHOD

Subjects

A total of 192 undergraduates (96 males and 96 females) participated in this experiment to fulfill a requirement in an undergraduate psychology class.

Apparatus

The experiment was run using an Apple II + microcomputer which controlled the timing and presentation of stimuli as well as the recording of responses. Subjects viewed the stimuli on a NEC monitor and responded to stimuli on the screen by pressing a hand-held button.

Procedure

Upon arrival at the experimental setting, the subject was familiarized with the sound chamber. Each subject was given instructions appropriate for the condition to which he/she had been assigned. Following this the experiment began. The experiment consisted of a ten-minute practice session followed by a fifty-minute vigilance task. Subjects were exposed to one of four levels of complexity of a cognitive vigilance task in either quiet (50 dbA), white noise (70 dbA) or human speech (a recording of G. B. Shaw's "Don Juan in Hell").

Tasks

The first level of complexity (1) was one in which two numerals (each ranging from one to nine) appeared simultaneously on the screen. A signal was defined as any presentation during which the two numbers were either equal or one digit apart (e.g., 3 & 4, 4 & 4, etc.). The second level of complexity (2) was one in which a pair of numbers appeared at the top of the screen above the bottom pair of numbers. This top pair represented an eleven-digit range. In this condition, a signal was defined as any presentation in which the sum of the two bottom numbers fell within the range as well as differing from each other by no more than one digit.

The next most complex condition (3) was one in which two pairs of numbers appeared at the top of the screen. Directly below each of these pairs was another pair of numerals. Each of the top pairs represented separate ranges. A signal was defined as a presentation in which the two numbers in either bottom pair differed by no more than 1 and the sum of which fell within the above range. On no occasion did both bottom pairs satisfy the criteria. Finally, the most complex condition (4) was one in which four pairs of numbers, each representing a separate range, appeared at the top of the screen. Below each range was a pair of numbers. The criteria for a signal was the same as in the two-range condition. As in the two-range condition, no more than one pair could fulfill the criteria on any one stimuli presentation.

In all four conditions, the stimuli at the bottom of the screen changed every 3 seconds and remained on the screen for only 1.5 seconds. The ranges in the three more complex conditions changed every minute on the average (i.e., every 30, 60, or 90 seconds) and remained visible until the new ranges appeared. The multi-range conditions were programmed so that the ranges changed simultaneously. In all conditions, a total of 200 stimuli occurred every 10 minutes. Five signals occurred at random intervals within each ten-minute period. In the multiple-range conditions, the location of the "target pair" varied randomly. The subject was told to respond by pressing a hand-held button whenever a signal occurred.

Following the vigil, the subject was asked to indicate on analog scales the perceived difficulty as well as the perceived challenge of the task. The scales were divided into nine equal intervals indicated by dashes ten millimeters apart. By simply marking through the scales, a difficulty and challenge rating was ascertained. In addition to the rating scale data, performance data was also gathered. The number of correct detections (Hits), median response time of the correct detections (RT) as well as responses to non-signals (FAs) were recorded for each ten minute period of the task.

RESULTS

The mean challenge and difficulty ratings are presented in Tables I and II.

TABLE 1
DIFFICULTY RATINGS

	LEVEL OF COMPLEXITY	1	2	3	4
SEX					
MALE		3.73	3.77	4.10	5.15
FEMALE		2.83	3.04	3.98	4.08

TABLE 2
CHALLENGE RATINGS

	LEVEL OF COMPLEXITY	1	2	3	4
SEX					
MALE		3.44	4.19	4.62	5.44
FEMALE		3.02	3.46	4.44	5.42

A multivariate analysis of these data indicated significant differences due to complexity as well as sex ($F(6, 332) = 4.52$, $p < .001$ & $F(2, 167) = 3.14$, $p. < 05$ respectively). Univariate analyses of the data revealed a significant difference in both the challenge and difficulty ratings attributable to task complexity ($F(3, 168) = 7.64$, $p<.001$ & $F(3, 168) = 4.82$, $p<.005$ respectively). Both difficulty and challenge ratings increased significantly with complexity. Further, difficulty ratings varied significantly due to sex of the subject ($F(1, 168) = 6.32$, $p<.05$).

Males tended to rate the task as more difficult. The mean difficulty rating for males was 4.19 while the mean difficulty rating for females was 3.48. One interaction (complexity x sex x noise) neared significance ($F(6, 168) = 2.03$, $p\approx.07$). Performance data suggests that males responded quicker and made fewer false alarms than females (see Noonan, Ash, Loeb, & Warm, 1985). No sex differences occurred for correct detections of signals.

A correlation was computed between difficulty and challenge ratings. The Pearson product-moment correlation coefficient was moderately high and significant ($r = .48$, $df = 190$, $p<.001$) suggesting that the terms may have had somewhat similar connotations for the subjects. A measure of the performance trend over time was computed by subtracting the number of hits in the first post-training ten-minute period from the number of hits in the last ten-minute period. A positive score indicates an improvement in performance over time while a negative score is indicative of a performance decrement.

Neither the difficulty nor challenge ratings correlated with this measure. Further both the difficulty and challenge ratings were negatively correlated with an index of overall performance. This index was simply the average number of hits during the fifty minute vigil. The negative correlations of the difficulty and challenge ratings with performance were significant ($r = -.224$, $df = 190$, $p<.05$ & $r = -.27$, $df = 190$, $p<.05$ respectively). Thus high difficulty and challenge ratings were associated with relatively poorer performance.

DISCUSSION

The present study supports Deci's contention (1971, 1972) that complex tasks are intrinsically motivating. Both challenge and difficulty ratings increased with complexity. However, the ratings were not correlated with temporal performance trends. The performance trends of subjects who rated the task as challenging failed to differ from those who did not. Thus, although subjects may have felt challenged by the task, their performance over time did not differ from those who rated the task as less challenging.

When a measure of overall performance (average number of hits) was computed, this correlated negatively with both challenge and difficulty ratings. Thus subjects who performed well tended to rate the task as less challenging and less difficult. Those who rated the task as both challenging and difficult performed poorly. These results are not in line with Dember et al. (1984) proposal that challenging tasks (those providing intrinsic motivation) serve to sustain and possibly to improve performance. It may be that the increasing difficulty of the more complex conditions "cancelled out" any motivating effects of the tasks. However, given the overall level of performance at the various levels of complexity, the lack of differential trends suggests that challenge fails to alter performance over time.

The significant correlation between challenge and difficulty ratings suggests that the terms may have held similar connotations for the subjects despite the fact that the terms were distinctly defined for the subjects. It is possible that the subjects rated the task as difficult or challenging in terms of their own personal definitions. Though subjects may have found the task simple, being asked to continue unabated for fifty minutes may have affected the ratings. If this is the case, the ratings may reflect tediousness rather than difficulty or challenge.

The findings that males tended to rate the task as more difficult than females was unexpected. The performance data failed to explain this effect. Females did not perform better than the males (see Noonan et al. 1985). Given the results of this and earlier studies (Lysaght, Warm, Dember & Loeb, 1984) further research into sex differences in sustained attention tasks seems warranted.

In summary, subjects did rate tasks of increasing complexity to be more challenging as well as more difficult. This tends to support Deci's (1971, 1972) notion that complex tasks are intrinsically motivating. This study failed to support the findings of Dember et al. (1984) that intrinsically motivating tasks serve to either sustain or improve performance over time. In fact, challenge and difficulty ratings were not correlated with performance trends and were negatively correlated with overall performance.

REFERENCES

[1] Deci, E. L., Effects of externally mediated rewards on intrinsic motivation, Journal of Personality and Social Psychology 18 (1971) 105-115.

[2] Deci, E. L., Intrinsic Motivation (Plenum, New York, 1972).

[3] Dember, W. N., Warm, J. S. Bowers, J. C., and Lanzetta, T., Intrinsic motivation and the vigilance decrement, in: Mital, A. (Ed.), Trends in Ergonomics/Human Factors I (North-Holland, Amsterdam, 1984).

[4] Lysaght, R. J. Warm, J. S., Dember, W. N., and Loeb, M., Effects of noise and information-processing demand in men and women, in: Mital, A. (Ed.), Trends in Ergonomics/Human Factors I (North-Holland, Amsterdam, 1984).

[5] Noonan, T. K., Ash, D., Loeb, M., and Warm, J. S., Task complexity, sex and noise meaningfulness as determinants of cognitive vigilance performance, in: Eberts, R. (Ed.), Trends in Ergonomics/Human Factors II (North-Holland, Amsterdam, in press).

[6] Warm, J. S., Howe, S. R. Fishbein, H. D. Dember, W. N., and Sprague, R. L., Cognitive demand and the vigilance decrement, in: Mital, A. (Ed.), Trends in Ergonomics/Human Factors I (North-Holland, Amsterdam, 1984).

Trends in Ergonomics/Human Factors II
R.E. Eberts and C.G. Eberts (Editors)

EFFECTS OF KNOWLEDGE OF RESULTS ON PERFORMANCE IN SUCCESSIVE AND SIMULTANEOUS VIGILANCE TASKS: A SIGNAL DETECTION ANALYSIS

Mary Lynne Dittmar, Joel S. Warm and William N. Dember
Department of Psychology
University of Cincinnati, Cincinnati, Ohio 45221
USA

This study assessed the effects of knowledge of results (KR) on sensing and decision making in simultaneous- and successive-type vigilance tasks. The vigilance decrement appeared in both tasks, and in each it reflected a decline in perceptual sensitivity (A') and an increase in caution in responding (B") over time. KR modified the temporal characteristics of both performance measures. However, the effects were complex and depended upon whether KR was given in terms of correct detections, false alarms or misses. The effects of KR were independent of the type of task employed.

Knowledge of results (KR), or feedback, is an important factor in the control of vigilance performance. Considerable evidence indicates that KR enhances the speed and accuracy of signal detections and tends to attenuate the temporal decline in performance that occurs in most monitoring tasks (Davies & Parasuraman, 1982; Warm & Jerison, 1984). Although the beneficial effects of KR in vigilance seem to be clearly established, the precise manner in which KR achieves its effects is poorly understood. Chinn and Alluisi (1964) have suggested that the effects of KR might be specific to the information given and that KR to correct detections, missed signals and false alarms might have differential effects upon performance efficiency. Moreover, the manner in which any of these forms of KR leads to improved performance is unclear. One possibility is that KR induces subjects to attend more carefully to the task and thereby enhances their ability to discriminate signals from noise. Such an effect would have direct perceptual consequences. Alternatively, the effects of KR may reside in subjects' decision-making strategies through changes in their willingness to emit a detection response.

These alternatives could be evaluated by applying a signal detection analysis to vigilance performance in the presence of different forms of KR. The theory of signal detection offers a technique for the independent assessment of the observer as a sensor and as a decision-maker; it has been used extensively in vigilance research (Davies & Parasuraman, 1982; Warm & Jerison, 1984), but to date only two attempts have been made to apply such an analysis of the effects of KR. Mackworth (1964) reported that KR to missed signals increased perceptual sensitivity (d') and also led to the adoption of a more lenient response criterion (β). Williges & North (1972), using KR to correct detections or false alarms, showed that both forms of KR enhanced perceptual sensitivity but had no effect upon the response criterion. These studies differed in the kinds of KR information given as well as in the types of discrimination required. What is needed is a comparison of the three forms of KR in a common vigilance task. Such a comparison constituted one aspect of the present investigation.

A second aspect of this study was an attempt to extend the implications of a taxonomy for vigilance tasks developed by Davies and Parasuraman (1982). That taxonomic scheme emphasizes the types of judgements subjects must make in detecting critical signals. Within the scheme, vigilance tasks are categorized as involving either successive or simultaneous discriminations. Successive discrimination tasks are essentially absolute judgement tasks in which the subject must determine if a stimulus configuration differs from a remembered standard. In simultaneous discrimination tasks the information needed to make a discrimination is present in the stimuli themselves.

Davies and Parasuraman have marshaled a good deal of evidence to support the meaningfulness of these categories. They have shown that the effects of several important vigilance parameters, including time on watch, the sensory modality of signals and individual differences, need to be considered in terms of the successive/simultaneous dichotomy. It would be useful to determine if the successive/simultaneous dichotomy has meaning for other vigilance parameters, in particular, KR, which plays an important role in the quality of sustained attention. Moreover, an examination of the effects of KR along these might help resolve the ambiguity in the mode of action of KR outlined earlier. The study by Mackworth (1964), which found evidence for changes in the response criterion in the presence of KR, employed a simultaneous discrimination task in which critical signals for detection were brief interruptions in the continuous radial movement of a target line. By contrast, the Williges and North (1972) study, which reported that KR had no effect upon the response criterion, used a successive discrimination task in which critical signals were increments in the luminance of a repetitively flashing light. The present study provided a direct comparison of the effects of different forms of KR (information about correct detections, misses or false alarms) in the contexts of successive or simultaneous discrimination vigilance tasks.

METHOD. One hundred and sixty students, 85 women and 75 men, served as subjects. All were enrolled in psychology classes at the University of Cincinnati and participated in the experiment in order to fulfill a course requirement.

Two types of vigilance tasks (simultaneous and successive) were combined factorially with four KR conditions (KR to correct detections, misses, and false alarms, and a no-KR control). Twenty subjects were assigned at random to each of the eight combinations of task type and KR.

All subjects participated in a 60-minute vigil. In the successive discrimination conditions, they monitored the repetitive presentation of a single 32 mm x 2 mm green line which appeared against a dark green background in the center of a 20 x 25 cm CRT screen. The line was oriented along a vertical vector, and was flashed on the screen for a duration of 150 msec. at a rate of one exposure every two seconds (30 events/min.). Critical signals for detection were 3 mm increments in the height of the line.

In the simultaneous discrimination conditions, subjects were asked to monitor the repetitive presentation of a pair of 32 mm x 2 mm green lines which was centered on the CRT screen. Again, the lines were oriented vertically and were exposed for a duration of 150 msec. once every 2 sec. They were separated laterally by a distance of 24 mm. Critical signals for detection were occasional 2 mm increments in the height of one of the lines. The spatial position of the higher line (left or right member of the pair) was varied at random for each subject in the simultaneous conditions throughout the watch. The luminance of the stimuli was maintained at 1.21 cd/m2 throughout the experiment.

Pilot work assured that the tasks were equated for difficulty at the outset of the vigil. Under alerted conditions, perceptual sensitivity was quite high, as indexed by A', a non-parametric measure which incorporates correct detections and false alarms. In both the simultaneous and the successive tasks, the average A' value was 0.96. Knowledge of results was provided by a 1,000 Hz acoustic indicator presented at 68dBA SPL, as measured at the subject's head. Depending on experimental condition, the tone sounded immediately after (a) a correct detection - "Hit KR", or (b) a false alarm - "False Alarm KR", or (c) a miss - "Miss KR" or (d) after each response made by the subject - "Response Acknowledgement". In the latter condition, the tone controlled for the presence of accessory auditory stimulation which occurred through the delivery of feedback, but carried no evaluative information.

The vigil was divided into six continuous ten-minute periods of watch. Critical signals for detection occurred on the average of one per minute under all experimental conditions, with inter-signal intervals (ISIs) of 24, 48, 72, 96, and 120 seconds. These intervals were varied at random for each subject within each ten-minute period with the constraint that critical signals were separated three times by ISIs of 24 seconds and ISIs of 72 seconds, twice by an ISI of 48 seconds, once by an ISI of 96 and once by an ISI of 120 seconds. Subjects were tested individually in an Industrial Acoustics sound chamber. They were seated before a table which held a CRT monitor positioned at eye-level and responded to critical signals by depressing a push-button switch held in the preferred hand.

Stimulus presentations, the occurrence of critical signals and the presentation of the appropriate form of KR or response acknowledgement were controlled by an Apple IIe microprocessor. The computer also recorded the subjects' responses and calculated the percentage of correct detections and false alarms for each 10 min. period during the vigil. Responses occurring within 1.85 seconds after the onset of a critical signal were recorded automatically as correct detections; all other responses were recorded as false alarms.

RESULTS. Non-parametric indices of perceptual sensitivity (A') and response criterion (B") were computed from the percentages of hits and false alarms for each subject in all experimental conditions. Distribution-free measures were employed in response to recent arguments that the assumptions governing the use of their parametric analogs, d' and β, (normality and equal variance in the noise and signal plus noise distributions) are not likely to be met by data gathered in vigilance experiments (Warm & Jerison, 1984).

Perceptual Sensitivity. An analysis of variance of the A' scores indicated that there was a small but statistically significant difference in perceptual sensitivity favoring the successive task, $F(1,152)=7.44$, $p<.01$. The mean A' score for the successive task was 0.89 while the comparable value for the simultaneous task was 0.86. In addition, significant main effects were found for feedback, $F(3,152)=5.59$, $p<.005$, and for periods of watch, $F(5,760)=20.75$, $p<.001$, and there was a significant Feedback x Periods interaction, $F(15,760)=4.75$, $p<.001$. None of the remaining interactions in this analysis was significant.

The interaction between feedback and time on task is shown in Figure 1. Sensitivity scores are plotted as a function of periods of watch with type of KR as the parameter. It is evident in the figure that perceptual sensitivity deteriorated with time on watch in the Control and Miss KR conditions. In contrast, the A' scores for the Hit KR and False Alarm KR groups tended to remain stable throughout the 60-minute vigil. Tests of the simple effects of periods within each feedback condition confirm these impressions. The A' scores for the Control and Miss KR conditions declined significantly

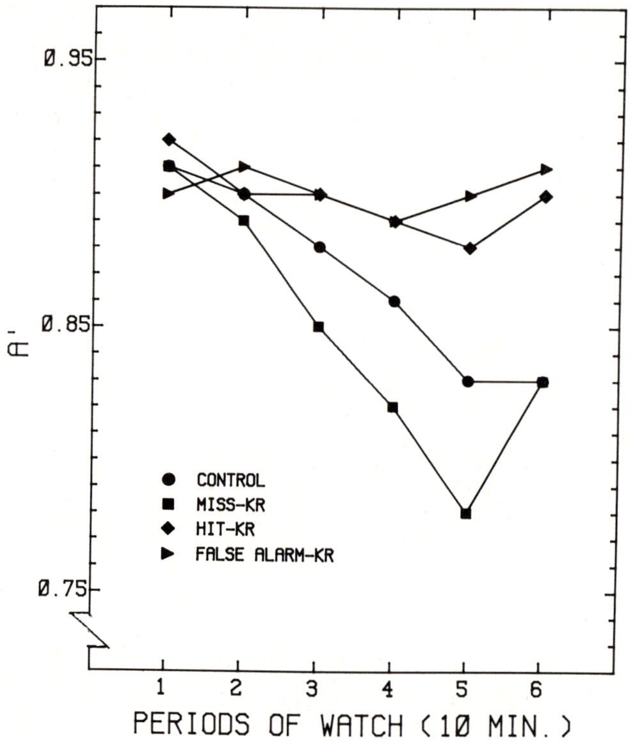

Fig. 1. A' scores for several feedback conditions as a function of time on watch.

over time, $\underline{F}(5,760) > 10$, $\underline{p} < .001$ in each case, while those for the Hit KR and False Alarm KR conditions did not vary significantly over time, $\underline{F}(5,760) = 0.63$ and 1.34, respectively, $\underline{p} > .05$ in both cases.

<u>Response Bias</u>. An analysis of variance of the B" scores revealed significant main effects for feedback, $\underline{F}(3,152) = 2.97$, $\underline{p} < .05$, and periods of watch, $\underline{F}(5,760) = 20.93$, $\underline{p} < .001$. As in the previous analysis, a significant interaction between these factors was also indicated, $\underline{F}(15,760) = 2.72$, $\underline{p} < .005$. No other source of variance in this analysis was significant.

The Feedback x Periods interaction is shown in Figure 2. B" scores across time are shown for each of the feedback conditions. It is evident in the figure that the subjects in all but the Hit KR group tended to become considerably more conservative in responding as the vigil progressed. Those in the Hit KR condition increased their response criterion only slightly, if at all. Tests of the simple effects of periods within each feedback condition indicated a significant increment in B" across time for the Control and False Alarm KR groups, $\underline{F}(5,760) > 8$, $\underline{p} < .001$ in each case, as well as for the Miss KR group, $\underline{F}(5,760)=3.29$, $\underline{p} < .01$. No effect for periods was noted in the B" scores of the Hit KR group, $\underline{F}(5,760)=1.37$, $\underline{p} > .05$.

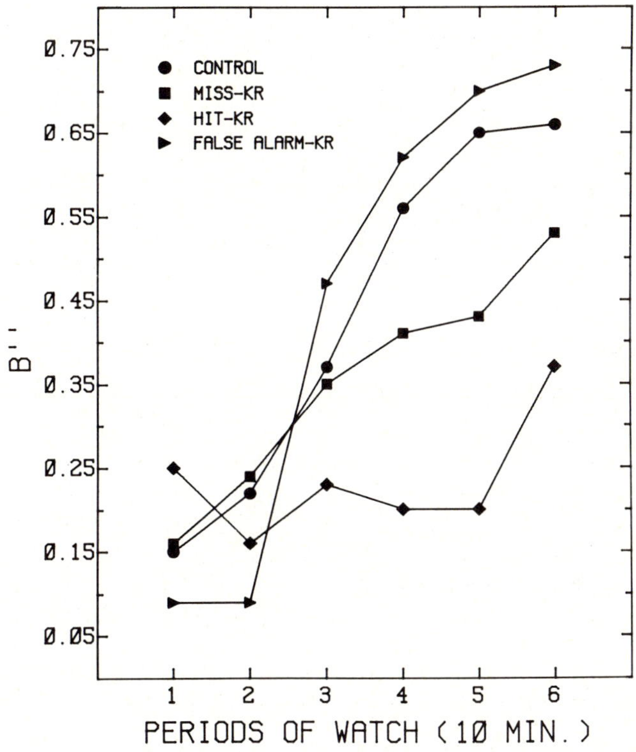

Fig. 2. B" values for several feedback conditions as a function of time on watch.

Although the B" scores for the Control, False Alarm KR and Miss KR groups increased significantly over time, further examination of Figure 2 suggests that the rate of gain in conservatism may have differed in these three groups. To assess this possibility, the least squares procedure was used to determine the best fitting linear function for the data of the Control, Miss KR and False Alarm KR groups. The slopes of these functions were 0.12, 0.07 and 0.15, respectively. Tests of significance indicated that the slope for the Control group was significantly steeper than that for the Miss KR group, $\underline{t}(8)=4.55$, $\underline{p} < .01$, while the slope of the function for the False Alarm KR group did not differ significantly from that of the Control group, $\underline{t}(8)=0.03$, $\underline{p} > .05$. Thus, the rate of gain in conservatism was more gradual in the Miss KR as compared to the control group while the rate of gain in conservatism among the False Alarm KR subjects was similar to that of the controls.

DISCUSSION. The present study explored the effects of knowledge of results on sensing and decision making in a vigilance task. It was prompted by the fact that two prior attempts to pursue this line of investigation have led to differing outcomes. Mackworth (1964; 1970) reported that KR produces changes in both perceptual sensitivity and response bias while Williges and North (1972) found effects of KR only on perceptual sensitivity. Broadly speaking, the present results confirm Mackworth's. However, they are not straightforward; complex interactions occur when different signal detection indices are combined with different types of KR.

The performance of the control group in the present study showed a decline in perceptual sensitivity (a drop in A') and increased caution in responding (a rise in B") as time on watch progressed. These temporal changes, characteristic of vigilance tasks (cf. Davies & Parasuraman, 1982; Warm & Jerison, 1984), were modified by KR but not in the same way by each KR format. Information about correct detections stabilized both signal detection indices over the course of the watch. Feedback about false alarms also stabilized perceptual sensitivity but had little impact upon temporally-linked elevations in the response criterion. Information about misses had no effect upon perceptual sensitivity, but it did retard the rate at which the observers became more cautious in responding over time. Thus, while the present findings generally support Mackworth's (1964;1970) conclusion that sensing and decision-making aspects of vigilant behavior can be modified by KR, they also indicate, as Chinn and Alluisi (1964) have suggested, that the effects of KR are dependent upon the specific information given to the observer.

An explanation of the present results with respect to KR begins with three assumptions: (1) KR in a vigilance task can have cue properties as well as motivational value (cf. Warm & Jerison, 1984), (2) observers accommodate themselves to the specific type of KR given (Salmoni, Schmidt & Walter, 1984) and (3) temporal variations in vigilance performance are multiply determined (cf. Loeb & Alluisi, 1984). Given these assumptions, and borrowing heavily from Chinn and Alluisi (1964), the explanation might proceed as follows.

At the outset of a watchkeeping session it is typical for observers to be well able to distinguish signal and non-signal events. As the vigil wears on, however, state changes occur, such as declines in cortical arousal (cf. Loeb & Alluisi, 1984) or increases in blurred or misdirected observing (Warm & Jerison, 1984). These changes add "noise" to the perceptual system and lead to a reduction in signal discriminability. In addition, as time on watch progresses, observers may come to realize that their initial expectations regarding the frequency of signals were too high. As a result, they increase the level of the response criterion (Craig, 1978). When Hit KR is provided, observers receive direct and positively-reinforcing information about the correctness of a response and also indirect information about incorrectness. As Chinn and Alluisi point out, in a Hit KR condition if one is not told that a detection response was correct, one can infer that it was incorrect. As a result, Hit KR may have continuously reoriented the observers to the character of the signals to be detected and in this way offset the deleterious effects of time-linked state-dependent perceptual changes. Moreover, since there was no direct penalty for errors of commission, observers in the Hit KR conditions may have maintained their initial level of responding in order to maximize the probability of receiving direct information that they were doing well.

When KR is provided in a false alarm format, observers are given direct, negatively reinforcing information about the incorrectness of a response as well as indirect information about correctness. Not being told that a response was a false alarm implies that it was correct. Hence, False Alarm KR may be viewed as the reciprocal of Hit KR. Consequently, this form of feedback may also have been able to reiterate the character of the signals to be detected for the observer and thus counter the degrading effects of a build-up of state-dependent perceptual changes. However, since False Alarm KR also provided negative feedback about errors of commission, it might be expected to inhibit responding and lead to a more conservative response bias.

Unlike Hit and False Alarm KR, feedback in the Miss format does not provide the observer with complete information about the correctness or incorrectness of a response. It merely indicates that a response should have occurred, but does not necessarily specify what the overlooked signal was like. Thus, Miss KR may not have provided specific enough information to be perceptually useful, but still through its negatively reinforcing property may have induced observers to avoid the "response of not responding" and, hence, retarded the rate at which response conservatism increased over time.

In addition to exploring the general effects of feedback on sensing and decision making functions in vigilance, this study also sought to identify the way in which KR influences performance in the simultaneous and successive tasks described by Davies and Parasuraman (1982). The absence of any significant Task x KR interactions in this study implies that while the effects of KR were information-specific, they were not task-specific.

Although this investigation was unable to link the effects of KR to task differences, it does have important implications for the vigilance taxonomy offered by Davies and Parasuraman (1982). Recall that a major feature of the taxonomy is the claim that the vigilance decrement is perceptual in origin only when successive discriminations are combined with a fast event rate. In all other combinations of event rate and task type, the decrement is rooted in an elevation of the response criterion. The present study also addresses this issue since the event rate employed (30 events/min.) falls into what Davies and Parasuraman would consider the "fast" category. Therefore, temporally-linked declines in A' would be expected to be limited to the successive task. Clearly, this did not occur. While A' declined significantly over time, the absence of any interactions between time and task indicates that it did so in a similar manner when both simultaneous and successive discriminations were involved.

Due to its importance for the validity of Davies and Parasuraman's taxonomic analysis, the finding of an A' decrement in the simultaneous task warrants careful consideration. One element in the experimental situation which may have led to a perceptual decrement in the simultaneous task is the presence of parallax. When discriminations of spatial extent are involved, parallax changes, stemming from head movements, can lower the visibility of signals in a simultaneous-type vigilance task (Warm, Wait & Loeb, 1976). Spatial discriminations were involved in the current investigation, and in post-experimental interviews, more than half of the observers reported that head movements impaired their ability to distinguish between critical signals and neutral events. A close examination of the experiments used by Davies and Parasuraman (1982) to build their taxonomy revealed that parallax was not likely to be a factor in any of these investigations. For the most part, they required acoustic discriminations or the detection of changes in the illumination, saturation or diameter of visual stimuli. In addition, the stimuli in these experiments were typically much longer in duration than the brief 150 msec. flashes used in this investigation. Consequently, it is possible that the combination of parallax-induced distortions in signal quality coupled with very brief exposure times burdened the perceptual system sufficiently in this study to produce a perceptual decrement in both the simultaneous and the successive conditions. An account along these lines is consistent with a recent study by Nuechterlein, Parasuraman and Jiang (1983) which also reported a perceptual decrement in a simultaneous-type vigilance task when the perceptual system was taxed by highly degraded stimuli. Taken together, the present study and the one by Nuechterlein, et al. suggest that the Davies and Parasuraman taxonomy does not completely describe the conditions under which the vigilance decrement is rooted in perceptual and non-perceptual factors. Some simultaneous discrimination tasks may be more vulnerable than others to perceptual as well as decision-making changes over time. The problem of defining precisely this subset of simultaneous tasks remains for the future.[1]

References

[1] Chinn, R. McC., and Alluisi, E. A., Effect of three kinds of knowledge-of-results information on three measures of vigilance performance. Perceptual and Motor Skills 18 (1964) 901-912.

[2] Craig, A., Is the vigilance decrement simply a response adjustment towards probability matching? Human Factors 20 (1978) 441-446.

[3] Davies, D. R., and Parasuraman, R., The Psychology of Vigilance. (London, Academic Press, 1982).

[4] Loeb, M., and Alluisi, E. A., Theories of vigilance, in: Warm, J. S. (ed.), Sustained Attention in Human Performance (Chichester, Wiley, Ltd., 1984).

[5] Mackworth, J. F., Vigilance and Attention, (Baltimore, Penguin, 1970).

[6] Mackworth, J. F., The effect of true and false knowledge of results on the detectability of signals in a vigilance task, Canadian Journal of Psychology 18 (1964) 106-117.

[7] Nuechterlein, K. H., Parasuraman, R., and Jiang, Q., Visual sustained attention: Image degradation produces rapid sensitivity decrement over time, Science 220 (1983) 327-329.

[8] Salmoni, A. W., Schmidt, R. A. and Walter, C. B., Knowledge of results and motor learning: A review and critical reappraisal. Psychological Bulletin 95 (1984) 355-386.

[9] Warm, J. S. and Jerison, H. J., The psychophysics of vigilance, in: Warm, J. S. (ed.), Sustained Attention in Human Performance (Chichester, Wiley, Ltd., 1984).

[10] Warm, J. S., Wait, R. G., and Loeb, M., Head restraint enhances visual monitoring performance, Perception & Psychophysics, 20 (1976) 299-304.

[11] Williges, R. C., and North, R. A., Knowledge of results and decision making performance in visual monitoring, Organizational Behavior and Human Performance 8 (1972) 44-57.

[1]No attempt has been made to account for the finding that perceptual sensitivity was significantly higher in the successive than the simultaneous task for two reasons: (1) the magnitude of this effect was quite small and (2) it was not replicated in a recently completed study using essentially the same stimuli. The second experiment did, however, replicate the finding of an A' decrement in the simultaneous task.

Trends in Ergonomics/Human Factors II
R.E. Eberts and C.G. Eberts (Editors)

HEMISPHERIC ASYMMETRY AND SUSTAINED ATTENTION

Mark W. Scerbo, Barbara E. McDermott, James A. Griffin
Barbara C. Houston and Joel S. Warm
Department of Psychology
University of Cincinnati, Cincinnati, Ohio 45221
USA

Subjects monitored bursts of white noise for brief (5 msec) interruptions during a 48-min vigil. Stimuli were presented either (1) binaurally, (2) monaurally to the left or to the right ear, or (3) to one or the other ear in a randomly alternating fashion. Intersensory alternation of stimulation enhanced the temporal stability of vigilance performance. Neither auditory gap detection nor the vigilance decrement were found to be as clearly lateralized as previously believed.

A considerable amount of evidence is available to indicate that there are general asymmetries in brain function. Broadly speaking, the left cortical hemisphere dominates in the processing of verbal material and does so in an analytic manner, while the right hemisphere dominates in processing visual-spatial information and handles this material in a holistic manner (Allen, 1983; Bradshaw & Nettleton, 1983).

Recently, cerebral laterality has been implicated as a factor in the performance of vigilance or sustained attention tasks in which observers must maintain their focus of concentration and remain alert to stimuli for prolonged periods of time. Reivich, Gur and Alavi (1983) have indicated that metabolic activity is greater in the left as compared to the right parietal lobe when subjects are engaged in a vigilance task. These neurophysiological findings, which point to greater right hemispheric involvement in the capacity to maintain alertness, are reinforced by performance measures in commissurotomized and normal individuals. Diamond (1979a; 1979b) has reported that split-brain patients consistently perform more efficiently with right than with left hemisphere signals in tests requiring the detection of auditory, tactual and visual stimuli. Significant hemispheric differences in the vigilance performance of normals have been described by Warm and his co-workers (Warm, et al., 1976; 1980). In order to identify such differences, these investigators made use of the fact that the auditory system is constructed so that each ear projects predominantly to the contralateral hemisphere (Bradshaw & Nettleton, 1983; Casseday & Diamond, 1977). Thus, they inferred asymmetries in hemispheric function from differences in the responses to signals presented to the two ears. Using this approach, Warm and his colleagues (1976) found the right hemisphere to be more sensitive than the left to variations in the temporal uncertainty of signals. They also found that the vigilance decrement--a progressive deterioration in performance efficiency over time that characterizes vigilance tasks (cf. Davies & Parasuraman, 1982; Warm, 1984)--was limited to left ear (right hemisphere) signals. Performance with right ear (left hemisphere) signals remained stable over time.

Warm, et al. (1980) have offered a cooperative interaction model (cf. Allen, 1983) to account for laterality effects in vigilance. They suggest that both hemispheres are involved in sustained attention but that the right hemisphere predominates and is the source of the vigilance decrement. While it fits the available data, this model may be premature. All of the studies just described used non-verbal stimuli which may typically be processed by the right hemisphere. A different picture of hemispheric specialization for vigilance may emerge if "left-brain" stimuli were used. It is conceivable that the hemisphere which typically handles the information processing required by a task also shows the greatest deterioration in performance over time. Contrary to the suggestion of Warm et al. (1980), the root of the vigilance decrement may lie in the hemisphere that is specialized for processing task-specific information. To test this possibility, it is necessary to select a task, preferably a non-verbal one, for which the left hemisphere normally dominates and adapt it for use in a vigilance situation. If the decrement function remains associated predominantly with right hemispheric stimulation under these conditions, the notion of a right hemisphere "vigilance center" would be strengthened. By contrast, if the decrement now occurs predominantly with left hemispheric stimulation, it would appear that the decrement is not limited to the right hemisphere.

Vroon, Timmers and Tempelaars (1977) have developed a non-verbal task involving auditory gap detection which seems to reflect a strong right ear (left hemisphere) advantage. They asked subjects to detect brief interruptions of 1 to 4 msec in a three-second burst of white noise and found higher levels of detection for the right as compared to the left ear at all gap durations. According to a review by Hammond (1982), the study by Vroon, et al. provides some of the strongest evidence for hemispheric asymmetries in temporal resolution. Consequently, we used the gap detection task within a vigilance framework in this investigation to explore the possibility that cerebral laterality in the decrement function is task specific.

Procedure. Forty students from the University of Cincinnati served as subjects. They participated in order to fulfill a course requirement. The subjects were free of any known hearing impairments and were selected for lateral dominance in order to maximize the possibility of finding hemispheric differences. All subjects were right-dominant as indexed by a minimum laterality quotient of +.67 on the Edinburgh handedness inventory (Oldfield, 1971) and by consistent "right" preferences in tests of eye and foot dominance.

Twenty individuals were assigned at random to each of two groups. One group served under conditions that were representative of auditory vigilance experiments; they received binaural stimulation throughout the vigil. A second group of subjects were presented with stimuli (critical signals and nonsignals) in either the left or the right ears on a randomly alternating basis throughout the experiment. The random alternation of stimuli to each ear permitted an assessment of hemispheric asymmetries in performance in a manner identical to that employed by Vroon, et al. (1977).

All subjects participated in a continuous 48-minute vigil. Throughout the session, 2.0 sec bursts of white noise were presented once every 3.0 seconds. These constituted nonsignal events, to which no overt response was required. The critical signal for detection was a burst that contained a small gap or a brief interruption. The duration of the gap was 5 msec.

When present, the gap appeared 1 sec into the burst. Acoustic stimulation was presented through a set of headphones at an intensity of 62 dBA SPL. Appropriate shunting circuitry and an electronic switch permitted either binaural stimulation or stimulation of the left or right ears without switching transients.

The session was divided into three 16-min periods. In both the binaural and random alternation groups, eight critical signals occurred within each period. The intervals between critical signals--intersignal intervals (ISI's)--ranged from 30 to 210 sec with a mean of 120 sec. The ISI's were varied at random for each subject within each period of watch. In the random alternation group, half of the critical signals appeared in each ear during each 16 minutes of watch. Subjects responded to critical signals by pressing a button located directly in front of them.

An Apple II Plus computer was used to control stimulus presentations and to record the subjects' responses. Responses occurring within 2.0 sec after the appearance of a critical signal were considered correct detections, all other responses were considered errors of commission (false alarms). Each subject was tested individually in an Industrial Acoustics sound chamber. Practice sessions prior to the start of the vigil insured that all subjects reached a level of proficiency of at least 75% correct detections before they began the main portion of the experiment. There were no systematic differences in performance during practice among any of the experimental conditions.

Results and Discussion. Percentages of correct detections and false alarms were determined for each subject within each period of watch in all conditions. A preliminary examination of the data revealed that false alarms were rare. The average rates of false alarms in the binaural and random alternation groups were only 4.77% and 4.29%, respectively, and many subjects in both groups made no false alarms at all. Consequently, only detection scores will be considered further.

Mean percentages of correct detections are shown in Table 1. It is evident in the table that overall detection rates were similar in the binaural and random alternation groups. On the other hand, the groups seem to differ in the temporal course of performance. While the scores in the binaural group started out at a somewhat higher level than those in the alternation group, they tended to decline more steeply over time. The drop in detections from the first to the third period in the binaural group (14 points) was 65% greater than that in the alternation group (8.5 points). Table 1 also reveals that within the alternation condition, performance was essentially identical for left-and right-ear signals.

The data of the binaural and random alternation groups were converted to arcsins and subjected to separate analyses of variance in order to assess the effects of time on watch in the binaural group and the effects of this factor and that of the auditory channels in the alternation group. The analyses of variance revealed that detections suffered a significant temporal decline when stimulation was binaural, $F(2,38)=4.60$, $p < .05$, but not when stimulation was alternated between ears, $F(2,38)=1.17$, $p > .05$. In addition, no significant overall differences were noted between auditory channels in the alternation condition and the Ears X Periods interaction also lacked significance, $p > .05$.

Table 1

Mean Percentages of Correct Detections in all Experimental Conditions

Condition	Periods (16 - min)			
	1	2	3	M
Binaural	86	76	72	78
Random Alternation				
Left Ear	81	79	74	78
Right Ear	82	77	72	77
M(RA)	81.5	78	73	77.5

This study was designed to determine if the vigilance decrement would be more prominant in the left as compared to the right hemisphere when a supposedly left-dominant task, auditory gap detection, was employed. Unfortunately, no conclusion can be drawn about this issue in the absence of any ear (hemispheric) asymmetries.

On the basis of the report by Vroon, et al. (1977) we would have anticipated at least an overall right ear (left hemisphere) advantage in the early portions of this experiment. However, cerebral laterality is a dynamic effect which often depends upon specific task characteristics (cf. Allen, 1983; Bradshaw & Nettleton, 1983). Consequently, our inability to replicate the findings of Vroon and his co-workers may be based on the presence of different task demands in the two experiments. Although we tried to follow their methodology as closely as possible, there were two substantial modifications: (1) our task was experimenter-paced while self-pacing was used in the earlier report (2) the signal probability in our study (.026) was much lower than in the case of Vroon, et al. (.67). Experimenter-pacing and low signal-probability are characteristic features of vigilance tasks (Warm, 1984), and therefore, were necessary procedural changes. It is conceivable that such changes altered the manner in which information about stimulus interruption was processed in the brain. To the extent that this is the case, the auditory gap detection task may not be as clearly lateralized as Vroon, et al. (1977) and Hammond (1982) may have been led to believe.

The absence of a significant vigilance decrement in the random alternation condition is surprising. The decrement occurs so often in vigilance experiments that its absence is of interest in its own right. It is important to realize that the decrement is not immutable; it can be arrested and even reversed under special circumstances (Warm & Jerison, 1984). The random alternation condition used in this experiment may be such a circumstance. As noted earlier, it is customary for auditory vigilance studies to present stimuli binaurally. In the random alternation procedure that we used, stimulation was monaural at a given time and was shunted at random between ears. Since a monaural vigilance decrement has been described by Warm, et al. (1980), it seems likely that the shunting component of the random alternation condition may have led to a reduction in the vigilance decrement. This issue is addressed in a second experiment.

Experiment 2

This experiment was designed to test the possibility that interaural shunting was responsible for the absence of a significant vigilance decrement in the random alternation condition of Experiment 1. Toward this end, stimulation was delivered to the left or to the right ears as in the initial study, but under consistently monaural conditions. Thus, unilateral stimulation was retained while shunting was eliminated. If interaural shunting was responsible for the lack of a significant decrement, its absence should permit the reappearance of that effect. In addition, Experiment 2 provided a second opportunity to examine hemispheric asymmetries in the detection of stimulus interruptions in an auditory vigilance setting.

Procedure. Forty students from the University of Cincinnati participated as part of a course requirement. All of the students were free of any known hearing impairments and met the same criteria for right-dominance as used in Experiment 1. Twenty students were assigned at random to either a left or a right ear group. In both groups, eight critical signals were presented within each 16 minutes of watch.

Results and Discussion. The percentages of correct detections and false alarms were determined for each subject within each 16 minutes of watch. As in the first study, only detection scores are reported because false alarms were rare.

Mean percentages of correct detections for the left and right ear groups are plotted in Figure 1 as a function of time on watch. The figure shows a tendency for an overall left ear advantage in detections and a trend for performance to decline over time in both the left and right ear groups. An analysis of variance based upon an arcsin transformation of the data of Figure 1 indicated that the decrement was significant, $F(2,76)=7.49$, $p<.01$, but that the overall difference between groups was not, $F(1,38)<1$. In addition, the Groups X Periods interaction also lacked significance $F,(2,76)<1$.

The results confirm the notion that interaural shunting was responsible for the absence of a significant vigilance decrement in the random alternation condition of Experiment 1. With monaural stimulation maintained but with shunting eliminated, the decrement reappeared in this study. The finding that random alternation of stimulation between ears can retard the decrement is reminiscent of an earlier report by Gruber (1964) who found a similar result when alternating stimulation between the auditory and visual channels. Gruber stressed the importance of intersensory alternation in maintaining alertness. Our results suggest that alertness can also be maintained by intrasensory alternation or alternation between channels in

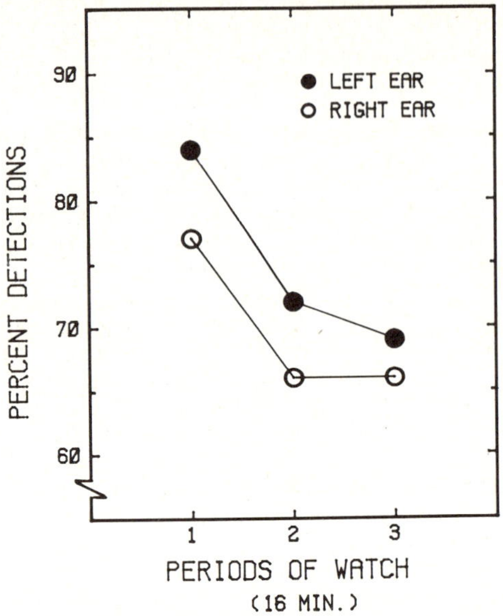

Figure 1. Percentage of correct detections as a function of periods of watch for left and right ear signals.

the same modality. Findings such as these might be explained by arguing that channel alternation (either inter or intrasensory) produces an arousing or dishabituating effect (cf. Parasuraman, 1984) which prevents the accumulation of inhibitory processes that may underlie the decrement. Along these lines, it is worth noting that Warm and his associates (Warm, et al., 1980) also used an alternation procedure to achieve unilateral stimulation of the left or right ears. However, in their work, alternation was not random and the input channel was predictable. Evidently, channel uncertainty may be a necessary element in maintaining alertness through interaural alternation.

Once again, no evidence was obtained for ear (hemispheric) asymmetries in the overall level of performance or in the vigilance decrement. This result confirms our earlier suspicion that auditory gap detection may not be as strongly lateralized as Vroon et al. (1977) have indicated. Similarly, the finding of equal vigilance decrements in the two ears implies that the decrement may not be as dependent upon right hemispheric functioning as Warm et al. (1980) have suggested. Our results, it should be noted, are complimented by the fact that other investigators have also been unable to uncover lateralization effects in vigilance (Andreassi, Rebert & Larsen, 1983; Ellenberg & Sperry, 1979).

The reasons why some studies find laterality effects while others do not are unclear. The experiments which have probed the functional organization of the cerebral hemispheres in regard to vigilance have used a wide variety of tasks and conditions. Thus, it is difficult to compare them directly.

Nevertheless, a lead to one factor that might be involved in these equivocal findings may be gained by considering a potentially important task characteristic which differentiates our study from that of Warm, et al. (1980).

Davies and Parasuraman (1982) have distinguished between successive and simultaneous vigilance tasks. The former are those in which subjects separate signals from non-signals by determining if a stimulus differs from a remembered standard. Simultaneous tasks are those in which all the information needed to discriminate signals from non-signals is present in the stimuli themselves. According to Davies & Parasuraman, successive tasks are more capacity demanding than simultaneous tasks, and the decrement is based on different mechanisms in these tasks. The experiment by Warm et al. (1980) made use of a successive-type task (subjects were required to determine when an acoustic pulse was longer in duration than usual), while the task used in this investigation falls within the simultaneous category (either a noise burst was interrupted or it was not). Evidence is available from other situations indicating that laterality effects are more robust when a comparison to memory rather than a simple differential discrimination is required (Bradshaw & Netleton 1983). Thus, it is possible that a cooperative interaction model, such as that offered by Warm, et al. (1980) may characterize the functional organization of the hemispheres when successive-type vigilance tasks are involved while the less demanding information processing requirements of simultaneous tasks are handled equally by each hemisphere. This possibility warrants further investigation.

References

[1] Allen, M., Models of hemispheric specialization, Psychological Bulletin 93 (1983) 73-104.

[2] Andreassi, J. L., Rebert, C. S. and Larsen, F. F., Hemispheric asymmetries in a signal detection task, Perceptual and Motor Skills 57 (1983) 923-929.

[3] Bradshaw, J. L. and Nettleton, N. C., Human Cerebral Asymmetry (Prentice-Hall, Englewood Cliffs, 1983).

[4] Casseday, J. H. and Diamond, I. T. Symmetrical lateralization of function in the auditory system of the cat: Effects of unilateral ablation of cortex, Annals of the New York Academy of Sciences 299 (1977) 255-263.

[5] Davies, D. R. and Parasuraman, R., The Psychology of Vigilance (Academic Press, London, 1982).

[6] Diamond, S. J., Performance by split-brain humans on lateralized vigilance tasks, Cortex 15 (1979a) 43-50.

[7] Diamond, S. J., Tactual and auditory vigilance in split-brain man, Journal of Neurology, Neurosurgery and Psychiatry 42 (1979b) 70-74.

[8] Ellenberg, L. and Sperry, R. W., Capacity for holding sustained attention following commissurotomy, Cortex 15 (1979) 421-438.

[9] Gruber, A., Sensory alternation and performance in a vigilance task, Human Factors 6 (1964) 3-12.

[10] Hammond, G. R. Hemispheric differences in temporal resolution, Brain and Cognition 1 (1982) 95-118.

[11] Oldfield, R., The assessment and analysis of handedness: The Edinburgh inventory, Neuropsychologia 9 (1971) 97-113.

[12] Parasuraman, R. The psychobiology of sustained attention, in: Warm, J. S. (ed.), Sustained Attention in Human Performance (Wiley, Chichester, 1984).
[13] Reivich, M., Gur, R. and Alavi, A., Positron emission tomographic studies of sensory stimuli, cognitive processes and anxiety, Human Neurobiology 2 (1983) 25-33.
[14] Vroon, P. A., Timmers, H. and Tempelaars, S., On the hemispheric representation of time, in: Dornic, S. (ed.), Attention and Performance VI (Erlbaum, Hillsdale, 1977).
[15] Warm, J. S., An introduction to vigilance, in: Warm, J. S. (ed.), Sustained Attention in Human Performance (Wiley, Chichester, 1984).
[16] Warm, J. S. and Jerison, H. J., The psychophysics of vigilance, in: Warm, J. S. (ed.), Sustained Attention in Human Performance (Wiley, Chichester, 1984).
[17] Warm, J. S., Schumsky, D. A. and Hawley, D. K., Ear asymmetry and the temporal uncertainty of signals in sustained attention, Bulletin of the Psychonomic Society 7 (1976) 413-416.
[18] Warm, J. S., Richter, D. O., Sprague, R. L., Porter, P. and Schumsky, D. A. Listening with a dual brain: Hemispheric asymmetry in sustained attention, Bulletin of the Psychonomic Society 15 (1980) 299-232.

IV

TRACKING, TRAINING AND MOTOR LEARNING

Trends in Ergonomics/Human Factors II
R.E. Eberts and C.G. Eberts (Editors)

RULE-BASED AND SPATIAL MODELS OF TRACKING

Ray E. Eberts and Susan E. Burlak

Purdue University
School of Industrial Engineering
West Lafayette, IN

The methods and problems of assessing the internal model of an operator in a tracking task are reviewed. From past research, two representations of the tracking task have been hypothesized: a rule-based and a spatial-based representation. Results from a previous set of experiments are reviewed which indicate that the representation of the internal model depends on the information given during training. Methods to model these data are proposed; modeling the data in the manner described can get around many of the problems usually associated with modeling techniques in this kind of study.

INTRODUCTION

The ability to track an object underlies many of the skills which we have developed; driving a car, watching the ball in a tennis match, and controlling the faucets on a shower are all common examples of tracking behavior. Many of the complex tasks which operators must perform in industry or military settings must be tracked; piloting an aircraft, landing a space shuttle, and controlling a nuclear power plant are examples of these complex skills. Consequently, tracking performance has been studied extensively with the goals of trying to predict system performance, trying to automate the tracking task, and trying to design equipment so that it is easier to control. Of particular interest in these areas is the determination of the methods that the operator uses to represent the tracking task; this representation is often called the internal model (e.g., Jagacinski, et al., 1978; Veldhuyzen & Stassen, 1976). In this article, we survey the methods that have been used to determine the representation of the internal model and, based upon previous experimental work, outline an approach which has been undertaken to differentiate between alternative representations.

THE INTERNAL MODEL

Two kinds of tracking tasks are typically studied. In pursuit tracking, the operator is required, by moving a joystick, to keep a cursor on a moving target. The cursor is controlled by the input from the joystick along with a random disturbance input. In the other kind of tracking, compensatory tracking, the operator is required to place the cursor on a stationary target. The distance between the cursor and the stationary target is the same as the distance between the cursor and the moving target in the pursuit display. As a consequence, the system input and command output are inseparable in a compensatory display. Tracking in these kinds of tasks can be quite difficult for novices but, with enough practice, the task can be mastered. Expertise that is gained with practice is usually equated with the development of an operator's internal model of the task. The internal model can be defined as the conceptual representation formed by the operator, through interaction with the system and with system documentation, about how the system works. This model is then used to control behavior and make decisions about the system dynamics or possible causes for malfunctions that can occur. In some cases, we say that the operator has internalized the system dynamics (Eberts, 1983).

Three methods have been used to assess the operator's internal model: recall protocols, modeling, and transfer of training. Each of these methods will be discussed.

One of the simplest and most popular methods used to assess the internal model is to ask the operators how they do the task; the operators' responses are called protocols. Crossman and Cooke (1974), Bainbridge (1974), and Kragt and Landeweerd (1974) all used verbal protocols from expert operators to try to determine how they performed their process control tasks. Another example of the use of verbal protocols was that of Hollan et al. (1980). Before designing a simulator for the Navy (STEAMER) to train operators in the control of a steam plant, they asked expert operators how they internalized the steam plant control task. For all of these examples, the reporting of the verbal protocols was very informal and not rigorously analyzed.

A variation of the protocol method is to ask the subjects to draw the way they think something works. McCloskey (McCloskey, 1983; McCloskey, Caramazza, & Green, 1980) used this kind of graphic protocol to determine the ways that subjects internally represent movement characteristics. As an example, he gave subjects a picture of a ball rolling towards a cliff and asked the subjects to draw the motion that the ball would take once it rolled off the cliff.

The protocol method has many problems associated with it. Some of the problems are: 1) the reports can be biased by how the experimenter interprets the answers and by the choice of questions that are asked; 2) the subject may not be able to verbalize the task because it has become automatized for the subject; 3) the nature of the task may be such that it cannot be easily verbalized; and 4) verbalization of the task may increase the workload of the task or change the task so that it is done differently than normal. These problems can be minimized if the experimenter is aware of the problems and if the right kinds of questions are asked at the right time (see Ericsson & Simon, 1980, for a discussion). The protocol method has been used extensively in other areas to transfer knowledge of problem solving to automated expert systems (Newell & Simon, 1972).

The second internal model assessment technique, modeling, has three variations. In the first variation, human behavior and the internal model are described in the same kind of engineering equations that are used to describe the system in general. As an example, Veldhuyzen and Stassen (1976) analyzed total system performance in terms of differential equations. The operator's internal model was equated with any nonrandom behavior exhibited by the operator. In the second variation, the internal model is modeled by parameter estimation. As an example of this variation, Jagacinski et al. (1978) hypothesized that the operators were using three psychologically plausible parameters to control a system and estimated values for these parameters to show how they changed with practice and how they compared to optimal performance. In the third variation, human performance is compared to an ideal mathematical or computer solution. This method has been used by Rouse to analyze the human's internal model for stochastic estimation tasks (Rouse, 1977) and in cognitive prediction tasks (Rouse, 1973, 1976). The internal model is assessed by analyzing how human performance deviates from the optimal performance which is described by the mathematical solutions to the problem.

Two main problems exist with the modeling approach. First, an assumption is made in some models that the human can be described in the same engineering terms as the physical system. Several researchers (e.g., Jagacinski et al., 1978) have pointed out that these kinds of parameters have little psychological validity. Often, the human is described in the terms of the latest engineering and mathematical models; Gaines (1969) points out that many examples of premature mathematization of human behavior are discarded when new branches of physics and mathematics turn up. The second problem with modeling, somewhat related to the first, is that with enough free parameters almost any performance can be modeled and can account for a large percentage of the variance. Again, the parameters used must have some psychological validity and comparison of models, with different numbers of parameters, is difficult.

The third internal model assessment technique is transfer of training. Using this method, subjects are first trained on the task and then transferred to one or more other tasks. An example of the transfer method to assess the internal model of the operator is the Kessel and Wickens (1982) experiment. In this experiment, one group of subjects was trained on a system by monitoring it and another group was trained by controlling it. In a transfer task, the accuracy of the internal model was evaluated by having the subjects indicate when a change occurred in the output of the system. The group trained by controlling the system had a more accurate internal model than the other group. If the tasks are carefully chosen, then important aspects of the internal models can be evaluated. As an example, Eberts (1983) trained subjects for six hours on a tracking task and then transferred them to several transfer tests. One of the tests had the subjects try to choose the correct joystick position, with the track frozen on the screen, so that they would move the system to a certain state. Misconceptions in the subjects' internal model of the relationship between the joystick position and the expected movement of the system could be determined.

The transfer of training method has a few problems associated with it. First, the transfer tests must be carefully chosen so that they evaluate what they should be evaluating. Usually, the tests are not chosen randomly but are chosen from the predictions of a particular theoretical perspective. If no theoretical perspective is used in the design of the tests, then choosing the appropriate test may be difficult. Second, the transfer of training method can be time and resource consumptive to run. It requires a training period for several subjects and then requires a variety of transfer tests for assessment.

REPRESENTATION OF THE INTERNAL MODEL FOR TRACKING

Characterizations of the operators' internal model of a tracking task have been postulated and can be divided into three main types. The first type is what has been called the optimal control model (Kleinman, Baron, & Levison, 1971). In this model, the operator's internal model is postulated in the same engineering terms which are used to describe the physical system. The optimal control model does have high descriptive value, it can account for a high percentage of the data, but as Jagacinski et al. (1978) point out, the parameters used have little psychological validity. The second type of characterization is called the bang-bang method of tracking. In this method, the operator will place the joystick at one extreme, wait a short time until the vehicle moves past the target, place the joystick at the other extreme, wait until the vehicle changes directions and moves past the target, and then repeat the sequence. This is a very effective strategy in uncertain situations with uncertain control dynamics, and it can be used if the operator has a very poor or limited internal model of the system dynamics. It is also used if the operator is given only two possibilities, such as two buttons instead of a continuous joystick, for control inputs (e.g., Gill et al., 1982). The third type of characterization of the internal model hypothesizes that the subjects form a perceptual representation of the system dynamics. For this type, an assumption is made that the operator tries to determine the trajectory of the vehicle and this prediction is then used to determine when control inputs should be made (Jagacinski, et al. 1978). These latter two types of representations of the internal model will be further considered in this article.

These latter two representations will be referred to as rule-based and spatial-based. For a rule-based representation, the actions can be stored in IF-THEN rules. To apply the rule, the subject searches through a hypothetical list of rules in memory until the conditional matches the environmental situation and then carries out the action specified (the THEN part of the rule). Newell and Simon (1972) outline how this representation is used in human problem solving; systems of rules can be programmed, so-called expert systems, so that the computer can solve problems (see Eberts et al., 1984). For a spatial-based representation, information is stored in images or templates. The image can be applied to a problem and a solution can be found (see Anderson, 1978, and Eberts, 1984, for a discussion). The bang-bang tracking can be seen as the implementation of a simple rule system. Although the rule can change with the kind of system used, the rules for bang-bang tracking could state: if the track is moving to the right then place the joystick to the extreme position at the left; if the track is moving to the left then place the joystick to the extreme position at the right. Thus, two very simple rules could be used to control tracking behavior. Similarly, the prediction representation assessed by Jagacinski et al. (1978) is a spatial-based model in which the subjects image a prediction of the future path of the system and then use this to decide if and when changes need to be made through a control input. Recently, Eberts (1983) has found evidence that different groups of subjects will develop these two kinds of internal model representations for the same system depending on the kind of training given. A discussion of this experimentation also illustrates two of the three assessment techniques mentioned previously and shows how the third assessment technique can be used without the problems normally associated with it.

In the Eberts (1983) experiments, three groups of subjects were given six hours of training in tracking a second order system. One group was provided with augmentation that allowed them to see a parabola predictor of the future path, another group was provided with a point predictor, and the third group was provided with no predictor. After training, the three groups performed in a series of transfer tests which were designed to evaluate the internal models developed. It was found that the parabola predictor group developed a spatially based internal model of the task through the manipulation of an image of the parabola. The other two groups developed non-spatial rule-based internal models. Informal verbal reports, or protocols, were taken and the transfer tests were evaluated. On the one transfer test when subjects were required to choose a control input to move to a certain place, the parabola subjects claimed that they imaged a parabola that would fit between the current and future positions while the subjects in the other two groups claimed that they tried several different strategies, invoking several different rules, until the right rules were found. The results of the transfer test corroborated these verbal reports. The two different representation schemes found in this experimentation, rule-based and spatial-based, are similar to two of the three representations of the internal model outlined previously and the representation developed is dependent on the kind of training given.

MODELING OF THE DATA

We are currently in the process of applying the third internal model assessment technique, modeling, to the data as further corroboration of the differences in the internal models developed. A production rule system has been used in the past to model performance in which rules are used to represent the task (e.g., Newell & Simon, 1972). We are in the process of applying a similar production rule system to this data. In the model, the decisions made by the subject will be simulated in the LISP (LISt Processing) language. Production rules are being formulated from the protocols and the results from the transfer tests; these production rules will be the inference engine for the LISP program. LISP has convenient structures so that the program can read through a list of conditionals, evaluate which condition is true, and then carry out the action (a new joystick position) associated with the condition. In the simulation, the LISP program will communicate with the FORTRAN program which was used to run the track much as the subject interacts with the display. In particular, the joystick position chosen from the LISP rules will be read by the FORTRAN program, the FORTRAN program will update the position of the track accordingly, the LISP program will take this new track position as part of the conditional, and will then use this information to choose a new joystick position. This sequence will be repeated until the trial is completed.

The other model which will be used is a spatial-based model. For this model, it is assumed that the subject looks at the curve of the track on the screen, images the parabola that would fit that curve, and then chooses the joystick input which would produce that parabola. Subject protocols and the kinds of training given support this model of the task. To model this behavior, we are programming the computer so that it can find the parabola that best fits the track curve and, then, in a look-up table, the appropriate joystick input can be applied.

Combining the three assessment techniques will allow us to get around some of the problems with the methods which were reviewed earlier. In particular, the protocol method has allowed us to query the subjects about the methods which they use to represent the tracking task. The transfer tests have allowed us to experimentally evaluate these possible parameters to make sure that the informal verbal reports were accurate. Finally, the modeling technique will use the parameters identified by the previous methods to determine how well they can fit the data. We would expect that the production rule system would fit the data of the point predictor and nonpredictor groups but not the parabola predictor group. Similarly, we would expect that the spatial model would fit the data of the parabola predictor group but not of the other two groups. Instead of showing that one model fits all the data better, as has been done in the past, we will attempt to show that one model describes the data for one of the groups and the other model describes the data for the other two groups. Showing this "interaction" will alleviate the problem of what constitutes a good fit; with the "interaction," a good fit for a model will be shown to be relative to the particular training given.

The modeling is still in its initial stages. This article has attempted to outline some of the methods which are used to assess the internal model and to point out some of the potential problems with these methods. A particular technique which makes use of all three methods has been proposed and is in the process of being carried out.

REFERENCES

1. Anderson, J. A., Arguments concerning representations for mental imagery, Psych. Review 86 (1978) 395-406.
2. Bainbridge, L., Analysis of verbal protocols from a process control task, in: Edwards, E. and Lees, F.P. (eds.), The Human Operator in Process Control (Halsted Press, New York, 1974).
3. Crossman, E. R. F. W., and Cooke, E. J., Manual control of slow-response systems, in: Edwards, E. and Lees, F.P. (eds.), The Human Operator in Process Control (Halsted Press, New York, 1974).
4. Eberts, R. E., Internalizing the system dynamics for a second order system, Ph.D. Dissertation, Univ. of Illinois, Champaign, IL.
5. Eberts, R. E., Augmented displays for problem solving, in: G. Salvendy (ed.), Human-Computer Interaction (Elsevier, Amsterdam, 1984).

6. Eberts, R. E., Nof, S. Y., Zimolong, B., and Salvendy, G., Dynamic process control: Cognitive requirements and expert systems, in: Salvendy, G. (ed.), Human-Computer Interaction (Elsevier, Amsterdam, 1984).

7. Ericsson, K. A., and Simon, H. A., Verbal reports as data, Psych. Review, 87 (1980) 215-251.

8. Gaines, B. R., Linear and nonlinear models of the human controller, Int. J. of Man-Machine Studies, 1 (1969) 333-360.

9. Gill, R., Wickens, C. D., Donchin, E., and Reid, R. E., Pseudo-quickening: A new display technique for the control of higher order systems, in: Edwards R.E. (ed.), Proc. of the Human Factors Society (Human Factors Society, Santa Monica, CA, 1982).

10. Hollan, J., Stevens, A., and Williams, N., STEAMER: An advanced computer-assisted instruction system for propulsion engineering, paper presented at the Summer Simulation Conference, Seattle, 1980.

11. Kessel, C. J., and Wickens, C. D., The transfer of failure-detection skills between monitoring and controlling dynamic systems, Human Factors, 24 (1982) 49-60.

12. Kleinman, D. L., Baron, S., and Levison, W. H., A control theoretic approach to manned-vehicle systems analysis, IEEE Trans. on Automatic Control, AC-10 (1971) 824-832.

13. Kragt, H., and Landeweerd, J. A., Mental skills in process control, in: Edwards, E. and Lees, F.P. (eds.), The Human Operator in Process Control (Halsted Press, New York, 1974).

14. Jagacinski, R. J., and Miller, R. A., Describing the human operator's internal model of a dynamic system, Human Factors, 20 (1978) 425-433.

15. McCloskey, M., Naive theories of motion, in: Gentner, D. and Stevens, A.L. (eds.), Mental Models (Erlbaum, Hillsdale, NJ, 1983).

16. McCloskey, M., Caramazza, A., and Green, B., Curvilinear motion in the absence of external forces: Naive beliefs about the motion of objects, Science, 210 (1980) 1139-1141.

17. Newell, and Simon, H. A., Human Problem Solving (Prentice-Hall, Englewood Cliffs, NJ, 1972).

18. Rouse, W. B., A model of the human in a cognitive prediction task, IEEE Trans. on Systems, Man, and Cybernetics, SMC-6 (1973) 473-477.

19. Rouse, W. B., A model of the human as a suboptimal smoother, IEEE Trans. on Systems, Man, and Cybernetics, SMC-6 (1976) 337-343.

20. Rouse, W. B., A theory of human decisionmaking in stochastic estimation tasks, IEEE Trans. on Systems, Man, and Cybernetics, SMC-7 (1977) 274-283.

21. Veldhuyzen, W., and Stassen, H. G., The internal model: What does it mean? In: Sheridan, T.B. and Johannsen, G. (eds.), Monitoring Behavior and Supervisory Control (Plenum Press, New York, 1976).

Trends in Ergonomics/Human Factors II
R.E. Eberts and C.G. Eberts (Editors)
Elsevier Science Publishers B.V. (North-Holland), 1985

Operator Tracking Performance with Time-Compressed and Time-Integrated Moving Target Indicator (MTI) Radar Imagery[1]

John R. Bloomfield and Raleigh K. Little
Honeywell Systems and Research Center (MN55-1000)
2600 Ridgway Parkway
Minneapolis, Minnesota 55413

This experiment explored the effects of intermittent radar coverage on the ability of operators to track moving targets. Frames of simulated moving target indicator (MTI) radar imagery were presented, either in time compression or in time integration, to groups of six operators. Interruptions in radar coverage of 15 and 30 seconds produced little effect on tracking performance. However, longer interruptions produced performance decrements: moderate with 60 seconds, pronounced with 120 seconds. Contrary to expectation, the use of time integration did not produce better tracking performance than time compression.

INTRODUCTION

This is the third study in a series of investigations of the situation in which a moving target indicator (MTI) radar is switched back-and-forth between two areas, in order to provide coverage of both of them. In the previous two studies (Bloomfield and Little, 1984; Little and Bloomfield, 1984), frames of simulated moving target indicator (MTI) radar imagery were presented in time compression. This mode of presentation of radar imagery was first demonstrated by White, in 1956. With it, a series of previously obtained frames are presented to an operator in the same order in which they were acquired, but at a faster rate, so that a target appears on the display as a steadily moving dot.

In the first study, Bloomfield and Little (1984) investigated the effects of switching away from one area, and of varying the length of the interruptions. They found that the interruptions in coverage did produce decrements in the tracking performance of operators. With small, ten-vehicle target units, the decrements occurred with interruptions of 45 seconds or longer. With larger target units, containing 25 or 50 vehicles, the decrements occurred with 120-second interruptions.

In the second study, Little and Bloomfield (1984) presented two, or more, consecutively collected frames of MTI radar imagery before each interruption in coverage occurred. With this procedure, the decrements associated with the interruptions in coverage were not as large. For the larger, 25- and 50-vehicle target units, it was necessary to present four consecutively collected frames, in order to produce a similar effect.

The experiment reported here, in this the third study in the series, was conducted in order to determine whether improvements, like those obtained in the second study, could be achieved in a different way. Instead of increasing the number of frames before each interruption, the mode of presentation was changed from time compression to time integration. With time integration, when each frame of imagery is played back, successive frames do not replace one another. Instead, they are added together. As a result, a target appears on the operator's display as a rapidly lengthening line of dots. Scanlan (1975) used the term "laydown" to describe the time-integration mode of presentation.

METHOD

Imagery. The MTI radar imagery used in this study was developed and presented on a simulator designed by Lorence, Kittleson, Ottoson, Baum, and Graffunder (1980). Basically this simulator consists of the hardware and software necessary for simulating a radar ground station, and of a set of computer models capable of simulating radar returns and of converting those returns into imagery.

The frames of imagery used in this study, and the previous two in the series, were taken from a set of 27, specially-produced, master sequences of simulated MTI radar frames. Each master sequence consisted of 96 frames of imagery, assumed to have been collected at 15- second intervals, throughout a 24-minute period of time.

We wanted our experimental imagery to be representative of the kind of imagery that would be collected by an MTI radar in a realistic operational setting. At the same time, we needed to control the characteristics of the target and some nearby units. These goals were not completely compatible. However, by basing our new imagery on the "Stand-Off Target Acquisition System Movement Scenario" developed by Stephens, Dolby, Plocher and Little (1980), we were able to satisfy each of them to a large extent. Three target areas were chosen from the Movement Scenario area of operations. Within each area, a highly-travelled route was identified. We determined which segments of those routes would be masked from the radar by intervening higher terrain, so we could select target paths that varied in the extent to which they were masked.

A tracking task was used in this study. During each trial, the operator was asked to track a designated target as it moved along its particular path. This was not easy, both because of the masking, and because of the presence of other simulated units moving in the vicinity. On each trial, there were four of these other, distractor units. They were of the same size, containing 10, 25, or 50 vehicles, were in the same single-column formations, and moved at the same speed, 10, 20, or 30 kph, as a target unit. In addition, the paths travelled by these units were selected in such a way that they would either intersect, or come close to, the path of the target unit at a point in the trial where it was passing through a masked area. Deliberately, these units were made difficult to distinguish from the target. The operator had to rely on the coherence of motion, and direction of that motion, in deciding which of the units emerging from a masked area were the distracting units, and which the designated target.

Operators. The operators used in this study were taken from a pool of 21 veterans (10 ex-Army, 6 ex-Air Force, 3 ex-Navy, and 2 ex-Marine), who closely match the population of U.S. Army radar operators. Two groups of six operators were used in the current experiment. All twelve had taken part in at least one of the previous experiments in this series.

Design. Two presentation modes - time compression and time integration - were used in this experiment. For both modes of presentation, there were 108 unique trials. The trials were produced by combining each of four interruption conditions, with each of three target areas, three target unit sizes, and three target unit velocities.

The time-compression condition was essentially a replication of the first experiment conducted by Bloomfield and Little (1984). The 108 trials with this presentation mode were given to each of six operators. In a second experimental session, the same six operators were given 108 trials in the time-integration mode. In addition, six different operators were presented with the 108 time-integration trials.

The four interruption conditions used with both presentation modes, were obtained by omitting various frames from the 27 master sequences of simulated MTI radar imagery. These interruption conditions are shown in Table 1.

INTERRUPTION CONDITION	RATE OF UPDATE (SECONDS)	NUMBER OF FRAMES OMITTED	FRAME NUMBER																
			1	2	3	4	5	6	7	8	9	10	11	12	13	14	15	16	17
A	15	0	X	X	X	X	X	X	X	X	X	X	X	X	X	X	X	X	X
B	30	1	X		X		X		X		X		X		X		X		X
C	60	3	X				X				X				X				X
D	120	7	X								X								X

TABLE 1. Interruptions of Zero, One, Three, or Seven Frames.

The Table shows the update rate for each condition, the number of consecutive frames omitted, and the particular frames retained of the first 17 in the 96-frame master sequence.

It should be noted that the interruptions were interruptions in coverage. The omitted frames were simply left out of the presentation sequence. The retained frames of imagery were presented at a rate of five per second. Thus, the more frames were omitted, the faster the sequence was replayed, and the faster the target and the distracting units appeared to move.

Procedure. At the start of each experimental session, the operator was seated in front of a video display terminal, while the experimenter read the instructions for the experiment and answered any questions the operator might have. At the beginning of each trial, the initial frame of imagery was displayed and a cursor was centered on the lead vehicle of the target. The imagery was presented in either time compression or time integration, at a rate of five frames per second. At the end of the trial, the last frame in the sequence remained on the screen, and the operator indicated where he or she thought the target unit had stopped by moving the cursor. Trials could not be repeated. The operator was told whether the responses were correct or incorrect.

RESULTS AND DISCUSSION

In both the time compression and time integration conditions, for each operator, each of the 108 trials was scored as "correct" or "incorrect." By collapsing over the three target areas and the three target unit velocities, a block of nine trials was formed for each of the 12 combinations of interruption condition and target unit size. Thus, the number correct for each block varied between zero and nine. In order to compare presentation modes, the results were averaged further across target unit sizes, and the resultant means are shown on Figure 1. The time integration data are almost identical for both groups of operators, showing a clear decrement in performance with interruptions of 120 seconds. The data obtained from the first group when they received the imagery in time compression shows a similar pattern. A comparison of the two modes of presentation shows, clearly, that the employment of the time integration mode did not result in the hoped-for prevention of decrements in tracking performance, when there were interruptions of 120 seconds: in fact, if anything, there was a smaller decrement when time compression was used.

For both modes of presentation, the effects of target unit size were similiar to those observed by Bloomfield and Little (1984) in their first experiment. When the duration of the interruption changed from 15 to 30 seconds, there was little effect on tracking performance. With a 60 second interruption, performance deteriorated for the smaller ten-vehicle target units, and, when the interruptions were as long as 120 seconds, there was a decrement for all targets irrespective of size.

The three studies in this series have shown that, as far as tracking performance is concerned, there is no particular advantage in either updating MTI radar imagery more frequently than every thirty seconds (Bloomfield and Little, 1984, and the current study), or in presenting the frames of imagery in time integration rather than time compression (current study). Also, if four consecutive frames of imagery can be acquired before the radar is switched to cover another area, interruptions of up to 120 seconds can be sustained without a statistically significant decrement in performance, even for the smaller ten-vehicle target units (Little and Bloomfield, 1984).

---

[1] This work was carried out for the U.S. Army Electronics Research and Development Command under contract No. DAAK20-80-C-0254.

REFERENCES

(1) Bloomfield, J.R. and Little, R.K., Tracking with intermittent radar coverage I: Interruptions after each collected frame of imagery, in: Lee, G.E. and Ulrich, T.E. (eds), Proceedings of the Ninth Symposium: Psychology in the Department of Defense (U.S. Air Force Academy Report No. USAFA TR84-2, Colorado Springs, 1984) 445-449.

(2) Little, R.K. and Bloomfield, J.R., Tracking with intermittent radar coverage II: Interruptions after two or more consecutively-collected frames of imagery, in: Lee, G.E. and Ulrich, T.E. (eds), Proceedings of the Ninth Symposium: Psychology in the Department of Defense (U.S. Air Force Academy Report No. USAFA TR84-2, Colorado Springs, 1984) 450-454.

(3) Lorence, L.R., Kittleson, C.G., Ottoson, K.O., Baum, D.R., and Graffunder, K., SOTAS ground station simulator design plan, Honeywell Systems and Research Center Technical Report. Prepared for U.S. Army Electronics Research and Development Command under Contract No. DAAK20-80-C-0254, 1980.

(4) Scanlan, L.A., Visual time compression: spatial and temporal cues, Human Factors 17 (1985) 337-345.

(5) Stephens, J.A., Dolby, M.B., Plocher, T.A., and Little, R.K., Functional and operational characteristics of the stand-off target acquisition system (SOTAS): Simulation update/SOTAS movement scenario, Honeywell Systems and Research Center Technical Report prepared for U.S. Army Electronics Research and Development Command under Contract No. DAAK20-80-C-0254, 1980.

(6) White, C.T., Time compression of radar and sonar displays, Naval Electronics Laboratory Report No. 61, San Diego, 1956.

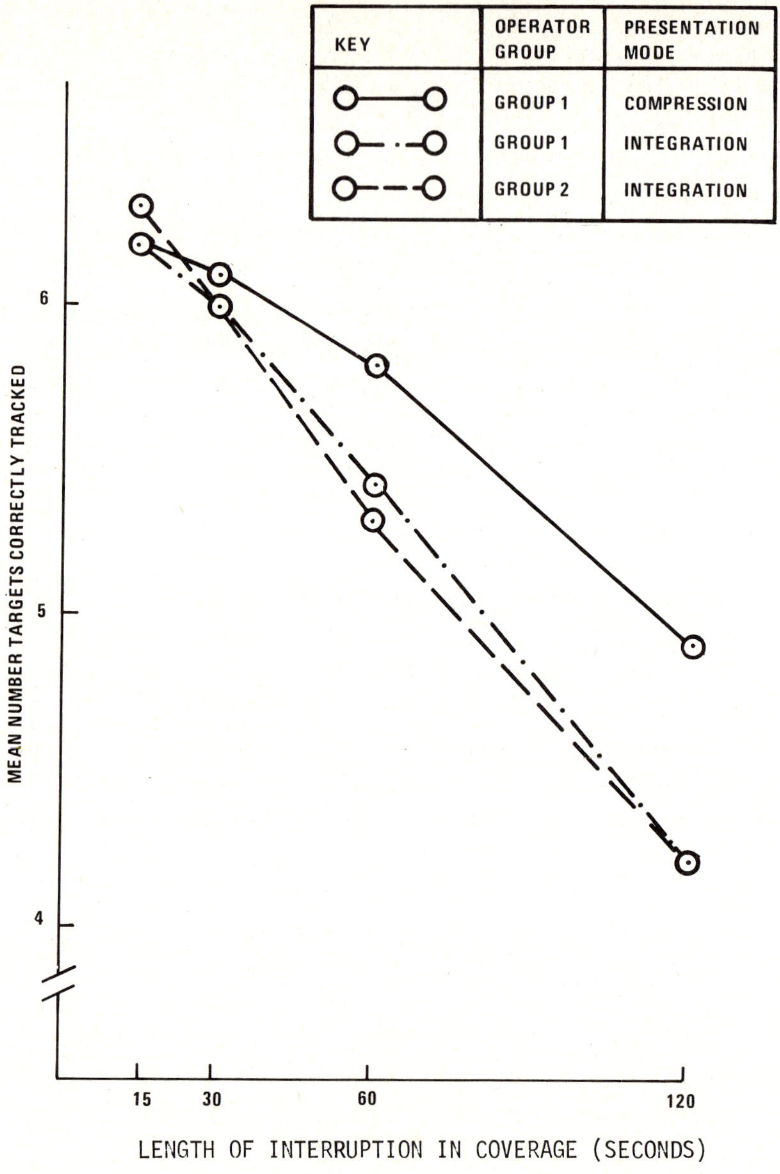

FIGURE 1

Tracking performance as a function of duration of interruption in coverage: data averaged across six operators in each group.

Trends in Ergonomics/Human Factors II
R.E. Eberts and C.G. Eberts (Editors)

DESIGN OF A SOFTWARE PACKAGE FOR A PSYCHOMOTOR TRACKING TASK USING A MICROCOMPUTER

Helmut T. Zwahlen and Chandrashekaran Viswanathan
Department of Industrial and Systems Engineering
Ohio University
Athens, Ohio 45701
USA

The design of a user friendly software package for a preview tracking task on the screen of a regular Apple IIe computer with two disk drives is described. To make the graphics animation feasible, the program was coded in 6502 assembler language. The top view of a two lane road with tangent and curve sections represents the reference input. The highway is scrolled from top to bottom continuously in the vertical direction. The operator is required by means of a paddle to continuously control the lateral position of a circle or of a silhouette of the top view of the car.

The experimenter can select from a menu the road width, the relative frequency of curves, the preview distance, the speed, the gain and the lag of the control input, the object shape (circle or top silhouette of car), and the sampling rate. For each sampling point, the position of the left edge of the road, the front left corner of the object, and the control input from the paddle is recorded. 8000 samples consisting of three values each can be stored in the RAM and subsequently transferred onto a flexible disk for subsequent statistical analyses. The software package is easy to use, is an effective tool to demonstrate various aspects of human tracking performance, and is well suited to conduct tracking experiments in human factors engineering classes.

INTRODUCTION

Human tracking, or the continuous control of some system such as driving an automobile represents a common everyday human activity and is therefore discussed in most human factors engineering textbooks. McCormick and Sanders [1] and Sheridan and Ferrell [2] discuss a number of aspects of human tracking operations to provide the reader with some background and knowledge regarding the concepts involved in tracking, the complexity involved in the analysis of tracking operations, and the many variables associated with tracking performance. The primary requirement of a tracking operation is to execute correct control movements at the correct instances in time. Tracking can be paced by the person doing it, such as in driving an automobile at a selected speed, or it can be externally paced where the person has no control over the rate at which a tracking operation has to be performed.

In a pursuit system both the instantaneous reference input and the instantaneous controlled process output are displayed to the person separately and independently on a display. In a compensatory system only a single input, the error, or the difference between the actual system response and the reference input, or ideal response, is displayed to the person on a display. Many tracking operations provide the operator with a preview which makes available to the operator a true display of the reference input from the present time until some future time (time prior to when the person will have to make a control response to a corresponding segment of reference input). As Sheridan and Ferrell [2] point out, preview control such as driving an automobile is more representative of an everyday tracking operation than pursuit or compensatory control which involves only two points or only one point on a display.

In the past it was usually not feasible for instructors of human factors engineering courses to incorporate realistic student laboratory projects dealing with tracking operations. This was mainly due to the fact that the necessary experimental apparatus to demonstrate and to investigate tracking operations in the classroom and laboratory was too expensive, difficult to operate by students, and just not readily available. The workbook by Sanders and McCormick [3] is a case in point. It contains a number of student projects, most of which require little or no apparatus to carry out the project and none of the projects deal with human tracking. The major objective of this project was to design a user friendly software package for a preview tracking task for a microcomputer which could be used in a human factors engineering class as an effective tool to demonstrate various aspects of human tracking performance and which could also be used to conduct human tracking experiments.

REQUIREMENTS

A regular Apple IIe computer (64K RAM) with a regular screen, two disk drives, and a paddle was to be used. A preview tracking task was to be used. A top view of a two lane road with tangent sections and curves, scrolled continuously from top to bottom in the vertical direction of the screen, was to be used as the reference input. Figure 1 illustrates the preview tracking task displayed on the screen. The object used for tracking was to be either a circle or the silhouette of a car viewed from the top. There was to be a

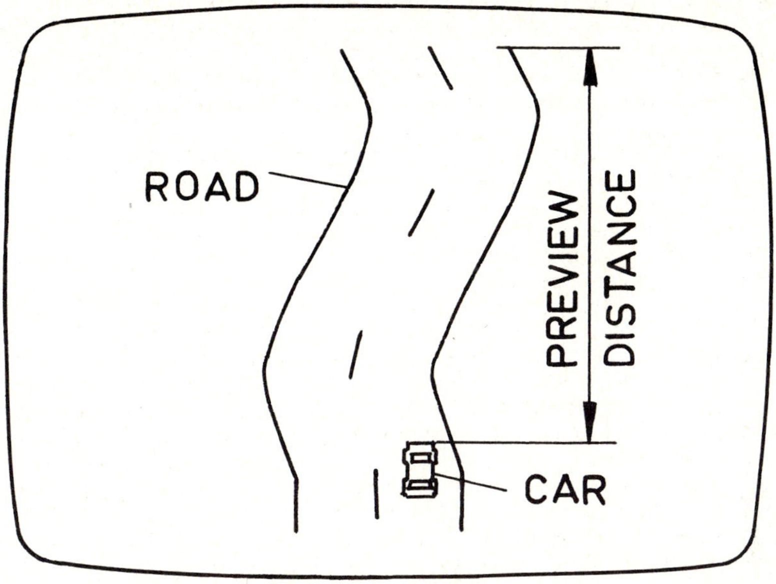

Figure 1 Preview Tracking Task Displayed on the Screen

broken center line in order to produce a visual speed effect. The curves had to be fairly smooth. The animation and update rate of the displayed task had to be fast and smooth enough in order to give the appearance that the tracking operation took place in real time. This could only be done by coding all tracking software in 6502 assembler language. The software was to be user friendly and no programming knowledge on the part of the experimenter was to be required. At the beginning of a tracking experiment, the experimenter was to be presented a menu to select the shape of the object, the road width, the preview distance, the gain for the control input from the paddle, the lag or time delay for the control input from the paddle, the relative frequency of curves along the road, the sampling rate, a repeatable or non-repeatable sequence of randomly generated left and right curves, and the speed. The computer was to use a specific auditory sound to indicate that the object was off the roadway. Further, the left road edge of the road, the left front corner of the object, and the input value from the paddle was to be measured for each sampling interval and stored in the RAM. After an experimental run was conducted transfer of the recorded data from the RAM to a flexible disk was to be done so that the data could be statistically analyzed at a later time.

PROGRAMMING APPROACH

A thorough understanding of the Apple IIe computer system

is essential in order to create graphics and animation on the screen. The memory map for high resolution graphics contains the locations $2000 to $3FFF and $5000 to $6FFF. Since we have an extremely limited amount of memory for the program and to store the data, only the locations $2000 to $3FFF are used for graphics. Whenever the graphics mode is activated, the video image produced by the apple is based on the values found in these memory locations. The high bit of each byte determines which color group to display for that byte. The remaining 7 bits are used for plotting the pixels. Hence, each byte controls 7 pixels on the screen. These bits are displayed in what seems to be a backward fashion. The lowest bit of each byte plots the left most pixel of the set, the highest plotting bit plots the right most pixel. The hex values, 01, 02, 04, 08, 10, 20, 40, in sequence turn on each pixel of the byte in order from left to right. To put up any specific pattern of seven bits, one merely changes the pattern into ones and zeroes and then converts this binary value to hex. The Apple IIe reference manual [4] was very helpful and essential in the development of the tracking task program. Figure 2 presents a general overview of the functions the tracking task program has to perform. The program consists of 6 modules which interact with each other and which are briefly described below.

The first module of the program generates specific pixels that need to be lighted up on the top most line of the screen. This module generates the left curves, the right curves, the tangent sections, and the broken center line. Initially a random number in the range 0 to 255 is generated using the Apple supplied random number generator. The module checks whether the random number generated is greater than a certain integer value which is determined by the relative frequency of curves specified by the experimenter. If the random number generated is greater than this value , then a tangent section is generated. If the random number is lower than this value, then curves are generated. The curve could be either a right or a left curve and this again is dependent upon the value of another random number between 0 and 255. The length of each left curve, right curve, and tangent section is determined by the initially generated random number which specifies the number of memory iterations or lines displayed. Both the left curves and the right curves use the same graphics algorithm. For a right curve, the horizontal coordinates of the left edge of the road are incremented whereas for a left curve the horizontal coordinates of the left edge of the road are decremented. The horizontal coordinates for the right edge of the road are always a given distance shifted to the right from the horizontal coordinates for the left edge of the road. At every iteration the module checks whether a curve is within 37 pixels of the horizontal boundaries of the screen (256 pixels of the 280 horizontal pixels used). Whenever a curve gets that close to the left or to the right horizontal boundaries of the screen, a new random number is generated and checked against the integer value based upon the relative frequency of curves. If it is larger a straight section

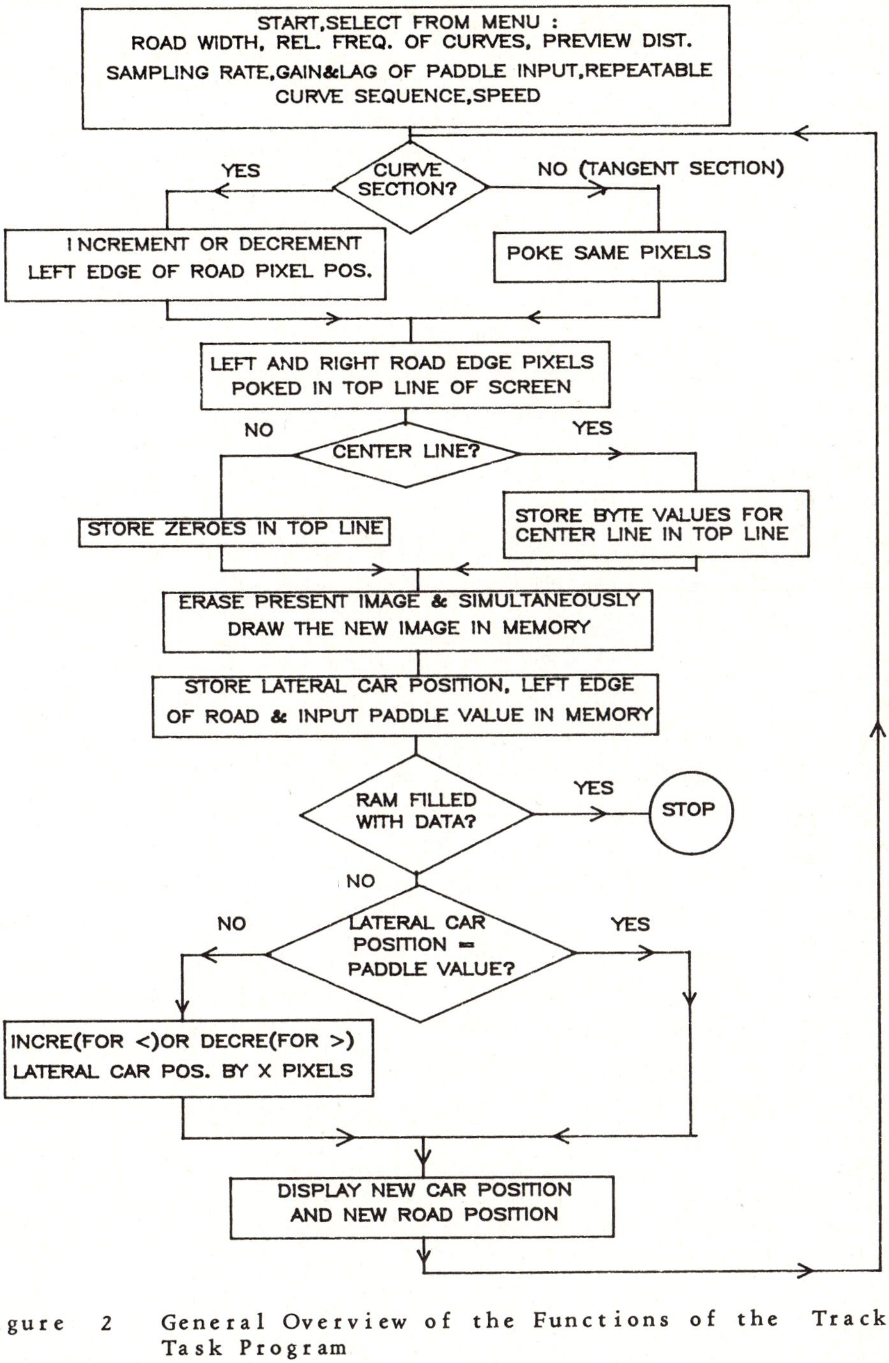

Figure 2 General Overview of the Functions of the Tracking Task Program

follows, if it is smaller a curve in the opposite direction towards the center of the screen follows. The present horizontal coordinate value for the left edge of the road is subjected to the modulus 7 routine in order to determine which of the 40 horizontal bytes needs to be poked. The value that needs to be poked in this byte is found from a look up table, with the bit values generated by the modulus 7 routine as the index.

The second module of the program does the scrolling. The speed of the car depends upon how fast the road is moving down the screen, which in turn depends on how many lines are scrolled at one time. Scrolling is a specialized form of byte manipulation. To make the screen scroll down, the bytes in the next to last line are moved to the last line, the bytes from the third from the last line are moved to the next to the last line and so on. New data is created to fill the vacated top line. Since scrolling the whole screen (192 lines) was very time consuming, only the bytes which were displayed in the previous iteration are scrolled. To save additional time only each alternate line on the screen is displayed. This means that in the vertical direction of the screen only every second pixel of the left road edge, the right road edge, and the broken center line is lighted up. For every iteration, three pointers, one for the left road edge byte, one for the left road edge bit, and one for the center line byte, are modified so that the new values are stored one on top of another in the memory. This kind of algorithm which effects scrolling without actually scrolling is done 254 times. Every 254 times a complete memory move takes place where the contents of the present 192 memory sequence are transfered to the initial 192 memory locations.

An Apple paddle is interfaced to the game I/O controller of the Apple computer. The paddle can take on values between 0 and 255 and the center of the paddle position is 127. The third module of the program accesses the game I/O controller of the apple and determines the value the paddle has. Based on the paddle value, the position of the car is incremented or decremented until the car's horizontal coordinate value coincides with the paddle value. The gain and the lag that is to be incorporated in the paddle input value is a part of this paddle routine. The program checks whether the lag is to be incorporated and if so, a variable for the lag is decremented for each iteration until it is zero. Only after the lag variable value has been decremented to zero, the paddle value is read. Until this time the car stays in the same position on the road in the display. The experimenter can choose lag values from 0 to about 2.5 seconds. Approximately 120 iterations are necessary for the program to create a lag of 2.5 seconds (based on a display update rate of 47 Hz). For a given gain value the paddle value is compared with the current position of the car or the circle. If the position of the car or the circle is different from the paddle value, then the car's or the circle's position is decremented or incremented by a value stored in the memory location for a given gain value. The gain value can be between 1 and 7, which is identical to the number of pixels the car is going to be shifted horizontally, either left or right, for each iteration. If the experimenter selects a gain value of 5, then the lateral position of the car or the circle is displaced during each iteration by five pixels. The program makes sure that the car or the circle does not go past the paddle value by checking the car's or circle's position at every iteration. If the car or circle position value does go beyond the paddle value, then the car or circle takes the value of the paddle.

The fourth module of the program draws the shape of the car, or the shape of the circle, on the screen at the specified coordinate location based upon the input value from the paddle. The byte values of the car, or the circle, that are to be displayed are determined and stored sequentially in the RAM as shape tables. However, since we must access these shapes byte by byte, whenever the shape is to be moved, the shapes would have to jump by seven pixels. Lubar [5], [6], recommends the use of pre-shifted shape tables starting from each of the bits in a byte so that the shape can be moved on a pixel by pixel basis which is much to be preferred. The seven displaced car and circle shapes were created using graph paper and actually drawing the shapes seven times for each shifted position and then calculating the corresponding byte values. The seven shifted versions for each shape are placed in a table in the RAM. Depending upon the position of the left front edge of the object, a correctly shifted shape is accessed from the pre-shifted shape tables and stored in the graphics memory. The overall dimensions of the car are 23 pixels long and 14 pixels wide, the circle has a diameter of 14 pixels. In order to display a stable image, erasing of the previously displayed car or circle and displaying the new car or circle is done line by line in the graphics memory. Erasing of the previously displayed car or circle is done by simply storing zeroes in those memory locations where the car or circle was previously displayed. For displaying the car or circle in the new position, bytes from the pre-shifted shape tables are accessed and poked onto the graphics memory where the car or circle is to be displayed.

The fifth module of the program checks whether the car or circle is off the road. The car's or the circle's left front corner is compared with the left and right edge of the road. If the car or circle is inside the road, then the module produces an auditory sound which is created by accessing the speaker once for every iteration. If the car or circle is off the road, then the speaker is accessed rapidly 112 times per iteration to produce a distinctly different auditory sound to inform the subject that the car or the circle is off the road.

The sixth module of the program collects the current input value from the paddle, the current left edge of the road, and the current left front corner of the car or circle and stores these values in the RAM. In order to use all the available memory, the bank switching memory (between RAM and ROM) which is available on the Apple IIe is used to store the data. There are 24K bytes of memory available to store the data. 8K bytes of this memory are allocated for the paddle input value, 8K of this memory are allocated for the left road edge and 8K of this memory are allocated for the left front corner of the car or circle. The data for each sampling interval is stored in a contiguous fashion until the whole RAM is filled. The experimenter has the option to stop the experiment at any time by pressing the escape key. If the experimenter does allow the whole RAM to get filled, then the module stops the program and terminates the experimental run and asks the experimenter whether the data collected in the RAM is to be stored on a flexible disk as sequential files or whether it is not to be stored. After an experimental run is completed and the data is stored as sequential files on a flexible disk one can access the flexible disk and perform a statistical analysis of the data. Such a statistical analyses may include the calculations of lateral car or circle position histograms and summary statistics, control input reversals, autocorrelations, cross-correlations, spectral distributions for input and output values, and the proportion of the distance where the car or the circle was off the road.

MENU AND OPTIONS

The experimenter is first shown a display on the screen which asks him whether he wants a car (Press C) or a circle (Press O). After this decision has been made, the experimenter is presented with a new display on the screen representing the main menu. By pressing the numbers 1 through 8 the experimenter can select and specify the following options from the main menu:

(1) Select the Road Width
(2) Select the Preview Distance
(3) Select the Gain for the Paddle
(4) Select the Lag for the Paddle
(5) Select the Relative Frequency of Curves
(6) Select the Sampling Rate
(7) Select a Repeatable or Non-Repeatable Sequence of Curves and Select the Speed
(8) Quit the Experiment

For each number (between 1 and 7) pressed the experimenter is presented with a new display or sub-menu which lists the options available. For example, one can select eight different road widths, select any integer value for a preview distance (car or circle positioned at a different distance from the top of the screen) between 2 and about 130 feet (based upon a 40 feet cycle length for the broken center line, consisting of a 15 feet line and a 25 feet space and where the 15 feet are represented by 15 lines or 7 lighted pixels), select a gain integer value from 1 to 7, where 1 is the smallest gain (1 bit per paddle digit) and 7 is the largest gain (7 bits per paddle digit), select a lag or delay integer value between 1 and 100%, where 1 represents almost no lag and 100% represents a lag of about 2.5 seconds, select the relative frequency of curves along the road from 1 to 100%

where 1 represents a road which is practically straight, 100% represents a road which is made up of a sequence of continuous curves and an integer value of 50% which would represent a road made up of an equal number of curved and tangent sections. The experimenter may also select one of six sampling rates which can either be every iteration or update of the display (1 collects data for about 3 mins.), every second iteration (2), every 4th (3), every 8th (4), every 16th (5), and every 32nd (6, collect data for about 90 mins.), he may also select either a repeatable (Press 0, and input seed for random number generator) or a non-repeatable sequence of curves (Press R), and finally the experimenter may select one of ten speeds (number 0 for lowest and 9 for the highest). After each of the first 6 options has been specified the experimenter is always presented with the main menu on the screen. After specifying the speed in option 7 and by pressing the return key the tracking experiment is ready to be run by the subject. Pressing the return key will bring the road in motion. A special option permits the experimenter or the subject to alter the selected speed during the experiment to any of the other programmed speeds by pressing the appropriate number key from 0 to 9.

CONCLUSIONS

The hardware memory and computation speed limitations of the Apple IIe computer made the design of this software package rather difficult. This user friendly software package can be used as an effective tool to demonstrate various aspects of human tracking performance and to conduct human tracking experiments in human factors engineering classes.

REFERENCES

[1] McCormick, E.J. and Sanders, M.S., Human Factors in Engineering and Design (McGraw-Hill, New York, Fifth Edition, 1982).

[2] Sheridan, T.B. and Ferrell, W.R., Man - Machine Systems: Information, Control and Decision Models of Human Performance (MIT Press, Cambridge, 1974).

[3] Sanders, M.S. and McCormick, E.J., Workbook for Human Factors in Engineering and Design (Kendall/Hunt, Dubuque, Iowa, 1976).

[4] Apple II : Reference Manual for IIe only, Apple Computer, Inc., Cupertino, California, 95014 (available from the U.S. Government Printing Office, Washington, DC, 20402, stock number 004-000-00345-4).

[5]
[6] Lubar, D., The Graph Paper; Part Seven: Scrolls and Other Creatures and Part Eight: A Bit of a Shift, Creative Computing, vol. 9, no. 1, Jan. 1983, 290/300, and vol. 9, no. 2, Febr. 1983, 242/250.

Trends in Ergonomics/Human Factors II
R.E. Eberts and C.G. Eberts (Editors)

THE EFFECT OF STRATEGY ON THE RESOURCE DEMANDS OF SECOND ORDER MANUAL CONTROL

Christopher D. Wickens and Barry Goettl

Institute of Aviation, Aviation Research Laboratory
University of Illinois-Willard Airport
Savoy, IL U.S.A.

Subjects performed a second order manual control task using either a "double-impulse" or a continuous control strategy. Concurrently, they performed a Sternberg memory search task using auditory and visual material of a verbal or spatial nature. The pattern of task interference suggests that the double-impulse strategy demanded perceptual-cognitive resources to update an internal model of system dynamics. The continuous strategy demanded response-related resources used to modulate the continuous control.

INTRODUCTION

The human operator is not a passive filter and transmitter of information, but rather is an active participant who can and will select strategies of processing that are appropriate to external circumstances and to imposed performance criteria. On the one hand, this flexibility and adaptability is precisely what makes the human a valuable and perhaps irreplaceable element in complex systems. On the other hand, this same flexibility presents a problem for the system designer who would like to predict the workload demands imposed by a system on the human's limited cognitive resources. In the first place, different strategies of performance can influence the quantitative demand for resources (Welford (1978); Sperandio (1978); Rasmussen (1981)). Some strategies may be quite precise and resource-consumptive; others may be crude but resource-efficient. It is also possible that different strategies may influence the qualitative demand for resources. That is, the nature of the resources used within the multiple resources available may be different under different strategies.

The present research uses the multiple resources model developed in our laboratory as a framework for examining strategy-induced resource changes. This particular model, shown in Figure 1, has been proposed to account for variance in dual task performance (Wickens (1984a)). According to data summarized by the model, the degree of interference between two tasks will increase to the extent that the tasks in question demand common levels along the three dichotomous dimensions shown in the figure: processing codes (verbal vs spatial), processing modalities (auditory vs visual), and processing stages (perceptual-cognitive vs response). Most of the data that have been used both to formulate and test the model have been derived from experiments that manipulate task characteristics (e.g., whether a task uses spatial vs verbal material). However, our interest in the present research is to assess whether similar differences in time-sharing performance result from the strategies deployed to perform a task. As one example, performance of a working memory task may emphasize either imaginal or phonetic characteristics of a word. Based upon data obtained by Klee and Eysenck (1974), these different rehearsal strategies should disrupt concurrent performance of visual and auditory secondary tasks, respectively.

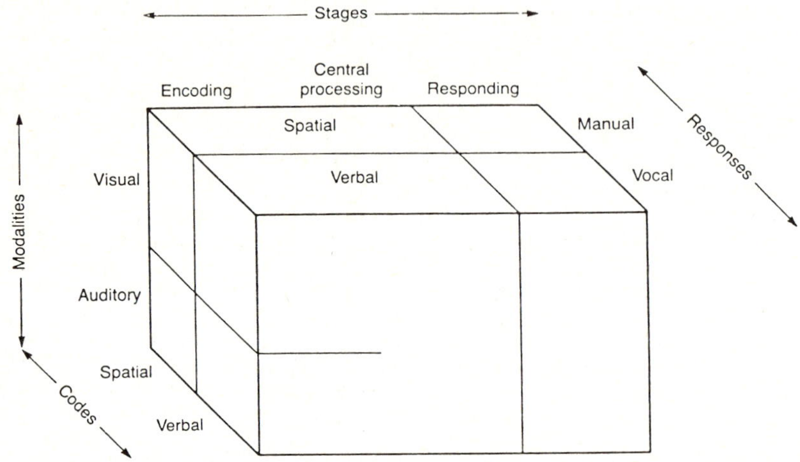

Figure 1
The structure of the multiple resources model

The present study examines the impact of using two different strategies of second order manual control on the resource demands of tracking, with particular emphasis placed on the stage of processing dimensions of Figure 1. Second order, or acceleration control systems are sluggish and unstable and impose unusually high demands upon the operator's limited resources (Wickens, (1984b); Wickens, Gill, Kramer, Ross, and Donchin (1980)). Analysis has revealed two alternative strategies that may be employed to cope with these demands (Wickens (1984b); McRuer et al. (1968)). Using a continuous control strategy, the operator responds with a control position directly proportional to the first derivative of the tracking error, thus requiring the direct perception of error velocity to exert continuous manual control. This strategy is known as "generating lead" (McRuer and Jex (1967)). A time-optimal double-impulse strategy in contrast requires the operator to apply maximum stick deflection (acceleration) in a direction to reduce an existing error, followed by maximum velocity in the opposite direction (deceleration) for an equal duration (Young and Meiry (1965); Hess (1979)). This strategy is time-optimal because, if the times of the two corrections are appropriately chosen, the strategy will succeed in reducing the error in the minimum possible time. However, in real system control, this strategy will be inefficient in fuel consumption, and to the passenger in a vehicle it will produce high levels of lateral acceleration.

The objective of the present research is to determine the processing stages whose resources are demanded by these two strategies of control. Analysis of the two reveals that this choice cannot be made a priori. For example, the continuous strategy requires direct perception of velocity and its changes, a perceptual ability that humans do not readily perform (Fuchs (1962); Wickens (1984b)). Yet the same strategy also requires continuous analog responding with a need to modulate the velocity and position of the control. In contrast, the double-impulse strategy places periodic intensive demands on the response system, akin to those in a reaction time task. Yet at the same time, the timing of these responses must be precisely controlled by an internal model of the system dynamics, a characteristic associated with working memory. These memory demands in turn are normally associated with perceptual-cognitive resources (Wickens et al. (1980)).

In the present study, subjects were trained to perform each of the two strategies to equivalent levels of tracking proficiency. Then both strategies, along with a first order control tracking condition, were varied. Since the Sternberg task is assumed to place heaviest demands on perceptual-cognitive or "early" processing resources (Wickens, Sandry, and Vidulich (1983)), it was anticipated that this task would interfere most with whichever strategy also placed its demands on "early" processing. In addition, the modality and code of the Sternberg task material were varied using visual or auditory stimuli of a verbal or spatial nature. The rationale behind this manipulation was as follows: If a given strategy places heavy demands on early resources, then these scarce resources should be at a premium in dual task performance. Hence, changes in a concurrent task structure that improve the efficiency of sharing these resources will produce enhanced benefits in time-sharing efficiency. These changes will be those that use a different modality (auditory) and a different code (verbal) from the visual-spatial resources required in tracking. In short, we predict an interaction between Sternberg task modality (or code), and processing strategy, to the extent that the two strategies demand different stage-defined resources.

METHOD

Eight righthanded subjects were paid $3.35/hr to participate in the experiment. The total experiment consisted of two 1 1/1 hour practice sessions and four 1 hour data collection sessions. Subjects performed a single axis compensatory tracking task, attempting to nullify an error that was perturbed by a .32 Hz forcing function. Control was exercised with a hand-held spring-loaded joystick.

Four versions of the Sternberg task were employed, each representing one of the four modality x code combinations. In the visual and auditory verbal conditions, subjects were given four letters to hold in working memory prior to each trial. During the subsequent 3 minute trial, subjects were presented a series of probe stimuli at an interstimulus interval that ranged between 3 and 5 seconds, and were required to indicate with a button press whether a given probe stimulus was or was not a member of the memory set. In the two spatial conditions, subjects responded to a dot or a tone in a 3 x 4 matrix. In the visual version of the task, a 4 row x 3 column matrix (3 cm x 2 cm) appeared on the screen. The task was to memorize the position of 2 dots presented in the matrix and indicate whether subsequent stimuli were members of the memory set. The auditory matrix was composed of 4 tones (250, 400, 650, 1000 Hz) which could occur at three positions relative to the head (left ear, right ear, and mid-plane) generating 12 possible stimuli. Subjects were required to memorize the location (i.e., pitch and position) of 2 tones (i.e., set size of 2), and indicate whether subsequent probe tones were from the memory set. For dual task configurations, when the Sternberg task was visually presented, it appeared immediately (1-1.5 degrees visual angle) below the tracking display.

During the two practice sessions, subjects received extensive training on the various single and dual task combinations. In addition, they were instructed as to the nature of the two second order control strategies. The advantages and disadvantages of each were described. Subjects were told that when using the double-impulse strategy, they were to emphasize quick error correction with large rapid stick deflections. For the continuous strategy, subjects were instructed to keep error minimal but give the cursor a 'smooth ride.'

During the four experimental sessions, all four Sternberg task types were completely crossed with the three tracking trial types (first order, double-impulse, and continuous). Both single and dual task performance was assessed during each session. In addition, task emphasis was manipulated. On sessions 1 and 3, subjects were instructed to allocate 30% of their resources to one task and 70% to the other. During sessions 2 and 4 this emphasis was reversed.

RESULTS

Prior to examining the interference effects, it was first necessary to ensure that the two strategies of performance were actually trainable and employed by the subjects. This conclusion was strongly supported by the analysis of the mean absolute control velocity, which was over twice as great during dual task conditions when the double-impulse strategy was employed, as when the continuous strategy was used ($F(1,7) = 37.28$, $p < .01$). Furthermore, under single task conditions, both strategies yielded precisely the same level of tracking error. Thus, it may be stated with confidence that equivalent levels of performance were obtained with different styles of behavior.

To examine the effects of the two different strategies on resource demand, RMS error and the four Sternberg RT measures measured in the first order tracking condition were subtracted from each of the corresponding 8 second order conditions (2 strategies x 4 Sternberg types). This subtraction was done at each priority level. Thus, each decrement measure may be interpreted as reflecting the impact that adopting a particular strategy imposed on performance of the task in question.

These decrement scores are presented in Figures 2a (tracking) and 2b (reaction time). The left half of each figure shows the performance data when the perceptual strategy was adopted while the right half shows the same effects for the response strategy. Within each half of the graph, the leftmost points indicate performance when tracking was time-shared with the verbal Sternberg tasks while the rightmost points are for the spatial tasks. The open circles and dashed lines indicate visual formats, and filled circles with solid lines denote auditory formats. Finally, the sloping lines running through each point indicate the effects of the task emphasis manipulations. The point on the right of each line is that observed during Sternberg emphasis conditions (lower for RT, higher for tracking error). The points in the middle of each sloping line represent

(a)

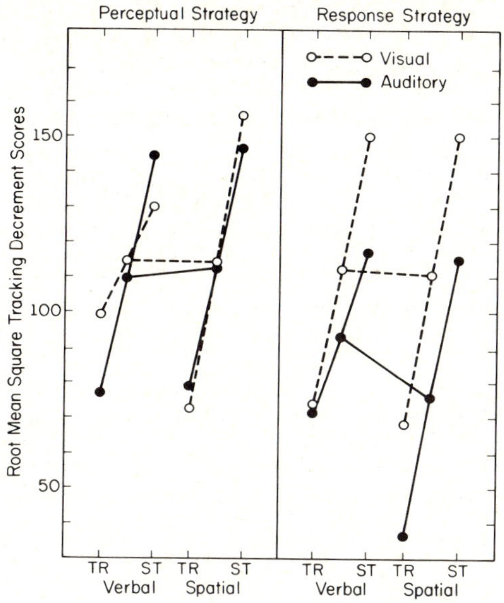

(b)

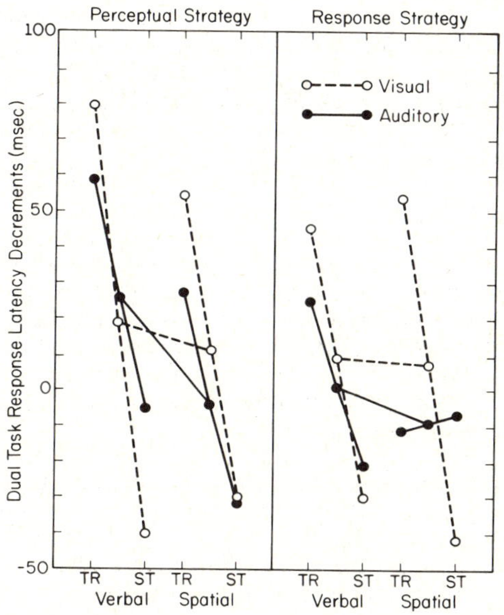

Figure 2
The effects of second order control strategy, Sternberg task code and modality, and task emphasis on the performance decrements from first order control. Figure 2a: Tracking error; Figure 2b: Reaction time.

the mean performance of the two bias conditions. The effects portrayed in these figures will be described in the context of 5 way (code x modality x strategy x bias x practice) ANOVAs that were performed on each of these dependent variables.

There was a large reliable effect of emphasis on the decrement scores ($F(1,7) = 29.11$, $p < .01$; RT, $F(1,7) = 7.65$, $p = .03$). These results indicate that the emphasis effect was enhanced when resources were more scarce under the second order control conditions. For the Sternberg task, this bias effect was greater with the visual than the auditory modality ($F(1,7) = 13.85$, $p < .01$). Second, it is evident that there is no main effect of strategy on either variable (both ps > .10). Third, it appears for both dependent variables that there is an interaction between modality and strategy (defining the position of the four "Hs" in each figure), such that the performance is improved by using separate modalities (auditory Sternberg), when the double-impulse strategy is employed, but not when the continuous control strategy is used. This effect was reliable for the tracking decrement ($F(1,7) = 6.48$, $p = .04$), but not for the Sternberg task data. Finally, the main effects of processing code approached the statistical level of reliability for the Sternberg task ($F(1,7) = 5.26$, $p = .06$), indicating smaller decrements with the spatial than with the verbal task. This effect however did not interact with strategy.

While there were also some effects of practice on tracking error and Sternberg RT, as well as reliable effects of some of the variables on Sternberg task error rate, none of these effects involved the strategy manipulations. Therefore, it can be assumed that this manipulation was not influenced by practice, nor did it produce a speed-accuracy tradeoff. More details of these analyses and a discussion of some individual differences in performance can be found in Goettl and Wickens (1984).

DISCUSSION

The present results were only partially successful in revealing the effects of strategy within the multiple resource space. Based upon the logic presented in the introduction, the double-impulse strategy appears to demand resources that are perceptual/cognitive in nature. These demands result from the need to maintain an internal model of the second order system dynamics in working memory, employed as a basis for coordinating the timing of the discrete responses. This need places perceptual-cognitive resources at a premium and therefore provides an advantage for that condition which utilizes those resources most efficiently (i.e., that with the auditory Sternberg task display). This finding is consistent with conclusions drawn by Wickens et al. (1980) using event related brain potentials to assess the cognitive demands of double-impulse control.

This interpretation however is less than entirely satisfactory because the effect of the double-impulse strategy was to produce a benefit for auditory processing rather than to eliminate a cost for visual processing. That is to say, there was no overall cost associated with the supposed resource competitive perceptual-cognitive stratey. Instead, there was a non-significant benefit to employing the double-impulse strategy. To accommodate this second finding, one must assume that, in addition to a scarcity of cognitive resources, there may also be an overall benefit to the double-impulse strategy in terms of its resource demands at some other location in the processing system. Thus, the data shown in Figure 2 can be seen to represent the additive effects of modality-specific resource competition for the double-impulse strategy and competition from some other non-perceptual/cognitive source resulting from the continuous control strategy. It is likely that this latter source represents increased competition for response-related resources that are demanded by the need to exercise continuous analog manual control, a more exacting response demand than that imposed by simply moving the control to the physical stop, as done in the double-impulse strategy (Kerr (1973)). This increased demand for response-related resources will increase competition with the discrete manual responses of the Sternberg task, and hence lower the overall level of performance. While we did not vary the response modality in the Sternberg task, using the same logic applied here, the present hypothesis would predict that a change in response modality should have a greater impact on performance when the continuous control strategy is in effect. These effects were observed by Wickens, Sandry, and Vidulich (1983).

The present results then provide one demonstration of how strategy choice can influence the qualitative nature of resource demands. Their implications to workload assessment technology and task configuration derive from realizing that different workload measures may be appropriate, and different concurrent task benefits may result, depending upon which strategy is deployed. As an example, the present results suggest that auditory displays to interface side tasks may be useful only when double-impulse strategies are in force. If employment of these strategies is situation-dependent, and therefore not constant for a particular system, then measurements under one circumstance, may not be appropriately applied to others.

ACKNOWLEDGMENTS

This work was sponsored by the Office of Naval Research, Engineering Psychology Program, under contract #N000-14-78-C-0658. Gerald Malecki was the technical monitor.

REFERENCES

[1] Fuchs, A., The progression regression hypothesis in perceptual-motor skill learning, Jrnl. of Exptl. Psych. 63 (1962) 177-192.

[2] Goettl, B. and Wickens, C.D., The effect of strategy in second order manual control on resource competition with a Sternberg memory search task, University of Illinois Engineering Psychology Laboratory Technical Report, Dept. of Psych. (1984, in press) (EPL-84-3/ONR-84-2).

[3] Hess, R.A., A rationale for human operator pulsive control behavior, Jrnl. of Guidance & Control 2 (1979) 221-227.

[4] Klee, H. and Eysenck, M.W., Comprehension of abstract and concrete sentences, Jrnl. of Verbal Learning & Verbal Behavior 12 (1973) 522-529.

[5] Kerr, B., Processing demands during mental operations, Memory & Cognition 1 (1973) 401-412.

[6] McRuer, D. et al., New approaches to human-pilot vehicle dynamic analysis, Air Force Flight Dynamics Laboratory (AFDL-TR-67-105), Wright-Patterson AFB, Dayton, OH (1968).

[7] McRuer, D.T. and Jex, H.R., A review of quasi-linear pilot models, IEEE Trans. on Human Factors in Electronics 8 (1967) 231.

[8] Rasmussen, J., Models of mental strategies in process plant diagnosis, in: Rasmussen, J. & Rouse, W.B. (eds.), Human Detection and Diagnosis of System Failures (Plenum Press, New York, 1981).

[9] Sperandio, J.C., The regulation of working methods as a function of workload among air traffic controllers, Ergonomics 21 (1978) 193-202.

[10] Welford, A.T., Mental workload as a function of demand, capacity, strategy, and skill, Ergonomics 21 (1978) 151-167.

[11] Wickens, C.D., Processing resources in attention, in: Parasuraman, R. & Davies, R. (eds.), Varieties of Attention (Academic Press, New York, 1984a).

[12] Wickens, C.D., Engineering Psychology and Human Performance (Charles Merrill, Columbus, OH, 1984b).

[13] Wickens, C.D., Gill, R., Kramer, A., Ross, W., and Donchin, E. The processing demands of higher order manual control, in: Beszy, A. & Lyman, J. (eds.), Proceedings, 17th Annual NASA Conference on Manual Control, Los Angeles, CA, June (1981).

[14] Young, L.R. and Miery, J.L., Bang-bang aspects of manual control in higher order systems, IEEE Trans. on Automatic Control 10 (1965) 336-341.

Trends in Ergonomics/Human Factors II
R.E. Eberts and C.G. Eberts (Editors)

OPTIMIZING USER TRAINING
ALTERNATIVES AND COMPARISON

Richard F. Kobylak

Human Factors Engineering
AT&T Communications
Cincinnati, Ohio
U.S.A.

Much time, effort and money is spent training personnel in industry. User training is critical in the successful use of new EDP systems, robotics, assembly design and establishing manual procedures. The user cannot be expected to operate efficiently without proper training. The delivery and type of training is critical to convey information to the user. There is a gap between basic research in theory of training and case study research. The present paper describes results from ten years of experience with AT&T Communications in determining and delivering user training and may help to bridge the gap.

INTRODUCTION

There are four major methods of providing training to a user. They are: Videotape, Teletraining, Computer-Based Instruction (CBI) and Instructor-Led. All four types may be intermixed in a training session. A comparison of these methods will be presented in this paper with guidelines and cost considerations for determining the proper mixture in various training situations. There are six major organizations that are responsible for delivering user training: Corporate Training Centers, Contracted Trainers, Independent Company Trainers, Human Factors Engineering, Key Instructors and Local Instructors. This paper will present guidelines to determine under what alternatives they should be used. It will weigh the benefits of each.

TRAINING METHODS

Videotape

This method can be used to convey concepts to large groups in many locations or as a lead-in for more detailed training using one or all three of the other methods. A videotape must cover the major points in an allotted time. However, videotapes by themselves have no provision to determine if all the points were understood by the user.

A videotape must be produced for eye contact by the student. Color combinations on graphics and diagrams must be easy on the eye. The speaker must have a pleasant but firm voice. In many cases a professional speaker is used in making the tape. The professional while using a script must come across as an expert.

Preparing the script, artwork, rehearsal and editing time must be included in the cost of preparing a videotape along with studio and camera persons' time. A ten minute tape will take an average of three to four weeks to prepare and complete.

Teletraining

This method is used to convey information to students at various locations. It eliminates the expenses of travel time by the students and instructor to a centralized or user location. A fully equipped teleconference room is the best setting for this type of training. This means a teleconference facility must be constructed at major user locations. The facility is in a soundproof room with video monitors, array microphones, electronic blackboard and a telephone control panel. Telephone bridging is used to hook up the microphones and electronic blackboard for two-way conversation and displays. Student size depends on the size of the room. The instructor will have problems in determining if the major points are getting across. Therefore, question and answer periods must be part of the training plan.

Viewgraphs and slides may be used during the training session. In this case the material must be distributed in advance of the training session. Any student material must be sent in advance to allow time for duplication.

A teletraining session to be effective should be one to one-and-a-half hours in length. If a session is scheduled to exceed this length, a break period is recommended. Additional two-way dialogue is also recommended. This will allow the students time to refresh their minds.

Training which requires interaction with an EDP system is a problem using teletraining. Once again room size limits the number of VDT's that can be made available. The instructor being in a remote location cannot be sure if all students are on the same display. This limits the effectiveness of using teletraining for EDP hands-on training. Therefore, it is recommended that a person be at each location to monitor the training. That person should have been trained on the subject matter. He/she can make sure that students are not having any problems.

Multipoint video teletraining is now coming on the scene. Video equipment, facilities and two-way audio are required for this media. Justification depends on the number of locations to be involved in the training sessions. Rental cost of video equipment and facilities to each location must be considered. Facilities may be satellite, microwave or land lines. Production cost must also be considered. This includes use of electronic graphics, a production crew, set up, etc.

Computer-Based Instruction (CBI)

CBI provides a self-paced individualized sequence means for a person to train themselves using their VDT. The delivery system should be user-friendly. Students should require minimal training in how to use the CBI system. The system should lead the student through the course with minimal help from the person responsible for administering the course.

Periodic as well as pre- and post-tests are part of the training to insure that the major points are conveyed to the user. The on-line test results are calculated for review by the person monitoring the training. The system should also be used to provide data on how long the student took on each section.

CBI should provide conditional, defined branching. This branching capability will allow authors to send a student to different levels of tutorial depending on the student's answers to a question or a student-initiated request. For example, if a student misses more than an author-defined limit on test questions, the system can send them into a more detailed description of the material to provide a better understanding. The system should provide the ability to hold the place of an individual student so that they can leave the system and come back at a later time to continue where they left off.

A one-hour CBI course will require approximately two to three weeks to prepare. This course will primarily consist of text with limited use of pictures, animation and graphics. More sophisticated use of pictures, animation and graphics will increase the time to four or five weeks. Industrial standards use the ratio of 100 to 125:1. However, this ratio deals primarily with use of text. It is estimated that 80% of the text preparation time is required to cover 20% of the exceptions that may cause the student major problems.

The course author will require at least one week of training on using the system to enter the training material. Studying the course material will very depending on the complexity of the methods and procedures to be conveyed on CBI. At least two to three months should be allotted for studying the material and holding meetings for clarifications.

A new CBI technique is coming on the scene using video discs, video disc systems and PC's. It is estimated that production cost using electronic graphics and full video capability averages $2,000 per one minute of instruction. A video disc master costs approximately $3,000 with an average of $60 per disc thereafter. Further study using this method will be conducted during 1985.

Instructor-Led

Instructor-Led training was practically the only source of user training ten years ago. This required the instructor to be in a classroom setting with students. The proper instructor/student class size had to be considered to convey the data to the students.

This is still the most popular method of training. The instructor may be from a headquarters location, part of the local staff or an outside concern. The instructor will require a training room equipped for proper delivery of the training. Normally this includes an easel marker board, viewgraph, TV monitor/player, 35 mm projector and VDT's for EDP-type training.

Student workbooks and course material must be provided. Preparation time for this material must be considered for timely course delivery. This time varies from 4 to 12 weeks depending on the complexity of the training.

Cost considerations must include travel time and accommodations for the instructor and/or students. Normally if the instructor is traveling, he/she will probably train at several locations at a specified time. This is done in order to make training timely for all users.

DELIVERY

Instructor-Led user training can be provided by six major organizations:

- Corporate Training Center
- Contracted Training
- Independent Company Trainers
- Human Factors Engineering
- Key Instructors
- Local Instructors

Corporate Training Center (CTC)

Primarily the CTC trains on courses they develop. The manager of the center consults with the various users on their training requirements. A course development team is established for each course requirement. They then proceed to collect course data, write, test and pilot the course before it is offered.

The training facilities are modified to meet the particular needs of the course. Student workbooks and material are prepared in advance of the training. In this case the students normally travel to the CTC. There are instances when a course will be suitcased to a location provided the necessary facilities are available.

Contracted Training

Outside concerns are contracted to provide training for a corporation when a large user population must be trained in a short time interval. This organization must first be trained on the course material before they can do any instructing. In many cases the course will be modularized to enable each person to become an expert on particular modules rather than the entire course. These persons may assist in rewriting the training modules when old procedures have to be enhanced based on policy, methods and procedures or EDP changes.

Several instructors will be sent to a user location to train on their particular modules. Their travel time and accommodation cost is part of the contract fee paid to their company. Long term contracts may result in the organization assisting in preparing and teaching new courses.

Independent anization

There are many cases when an independent company may develop a course or hold a seminar to provide a specific business need. Persons within a company dealing with that need are encouraged to attend the session. This session is usually held at a hotel in a large city. A fee is charged by the company to cover its cost.

Circulars are sent out by the company advertising the session. These sessions are a great opportunity to meet people from other companies and colleges. There are occasions when a large user population within an organization has a similar need. In this case the independent company will be contracted to provide the seminar for that particular group of people.

Human Factors Engineering (HFE)

HFE is part of the project team dealing with manual and EDP enhancements. Based on the team concept, HFE is involved in front-end analysis, input and output design, edit messages, testing and training.

In system development HFE will prepare the complete training package for the user. Depending on the complexity of the training, HFE may travel to each user location to provide the training. The other alternative is to train Key Instructors (see below).

For manual methods, HFE will assist in training development. The training will normally be modularized based on the expertise of the individual team members.

Key Instructors

Key Instructor training is used when information has to be conveyed back to a large user group in a short period of time. This information normally consists of an EDP or manual methods change that is important but not a major revision.

These instructors are brought into a central location and are trained on the new procedure normally in two days or less. They will be provided the material to train their people normally in less than one day.

Pre- and Post-tests are given to the students. These tests are analyzed by the course developer and the Key Instructors. They are used to determine how well the student comprehended the material and if any follow-up is required.

Local Instructors

These instructors are on each user site as part of the staff. They are responsible for training on local methods and manual and EDP methods for newly hired persons. These people normally will be the Key Instructors covered above.

GUIDELINES

The following guidelines may be helpful in determining the media and who should be responsible.

1. Videotapes should be used to convey concepts. It should be followed by one of the other training methods.

2. Teletraining should be used to convey training on a particular subject for users at different locations at the same time. Teletraining room setup costs must be considered.

3. CBI preparation cost is high but their method is excellent for self-paced courses as long as sufficient time is allotted the student to use this media.

4. Instructor-Led training is still the best method of training to insure all points are conveyed and understood by the user.

5. Corporate Training Centers, Contracted Training and HFE must be staffed to develop and provide the training courses for which they are responsible in a timely manner. Understaffed organizations will result in poor user training because short cuts will have to be taken.

6. Key Instructors should provide training on manual and EDP enhancements but not on major changes to manual methods or new EDP systems. A key instructor cannot be expected to learn in a short time period what other people have been learning and modifying for months. They cannot be expected to remember all they will be taught in the allotted period and go back and properly train their people. They should be made part of the team in order to jointly develop the training if the other organizations are understaffed or training must be completed in a short time period.

7. Local (Key) Instructors will become the on-site subject matter experts. Therefore, their managers must insure they attended all training provided that concerns their organization.

8. There will be times when an affiliated company or a subsidiary will be contracted to develop or enhance an EDP system. If the contract does not include user training, an organization must be given the responsibility to work with that company to develop and deliver the training.

CONCLUSION

The key to a well-trained person is that they are provided the necessary training to do their job in a timely manner. A training path must be developed for each employee, especially the newly hired. All persons working on user training must make sure the correct media and delivery is used to train the user to become proficient in the shortest possible time. Only in this way will training be cost-effective.

Trends in Ergonomics/Human Factors II
R.E. Eberts and C.G. Eberts (Editors)

CONTEXT AND DEPENDENCY EFFECTS IN SKILL ACQUISITION

Gavan Lintern

Aviation Research Laboratory
University of Illinois at Urbana-Champaign
Savoy, Illinois 61874
U.S.A.

Two contrasting effects in motor learning are described. One, known as contextual interference, shows a pattern of poor performance during training and good performance during transfer. The other, referred to as contextual facilitation, shows a pattern of good performance during both training and transfer. These apparently conflicting trends can be rationalized by an appeal to principles of perceptual learning. The potential impact of these manipulations on the acquisition of motor skills is examined. There is strong support for the use of contextually facilitating manipulations with synthetic training devices. Dependency effects pose a threat, but they can be avoided. As contextual facilitation is generally possible only in synthetic training devices, such devices offer training benefits in addition to their cost advantage.

INTRODUCTION

Synthetic training devices provide one means of improving cost effectiveness in large training programs. While the issue of fidelity has received considerable attention, the effectiveness of training might be enhanced considerably by consideration of training principles, many of which cannot be implemented except in synthetic devices. Some training principles might compensate for reduced fidelity, while others could suggest deliberate departures from it. Thus, design and use of training devices have become important issues, and the traditional approach of maximizing fidelity is no longer adequate. However, there are few guidelines for design and use of training devices. This paper examines two patterns of data from skill learning studies for the insights they might offer.

Transfer refers to improved performance as a result of prior training. Although transfer issues dominate the following discussion, data from motor retention studies are used to bolster the arguments. There is an important similarity between transfer and retention experiments. Both have subjects learning a new skill in an early phase, and both test proficiency in a later stage. Retention experiments that use multiple training strategies with independent groups, and that test subsequent performance on a common set of conditions, can be viewed as transfer studies if the retention interval approximates the training-transfer interval used in transfer studies. Several such studies are discussed, and for consistency within this paper, their acquisition and retention phases are referred to as training and transfer phases.

CONTEXTUAL INTERFERENCE

Battig (1966) noted from verbal learning research that manipulations that depress performance in training could produce better transfer. He predicted a similar pattern for motor learning. Shea and Morgan (1979) confirmed that prediction and referred to the effect as contextual interference. Note that the interference occurs in training and the pattern in transfer is one of facilitation.

Task variety is one manipulation that has produced contextual interference. Although training performance is generally poorer, transfer to a different version of the same task has been shown to be better after varied training with substitution tasks (Dashiell, 1924), memory tasks (Duncan, 1958), paired-associate learning tasks (Morrisett and Hovland, 1959), and category search tasks (Schneider and Fisk, 1983).

A task variation effect of this type is one of the foundations of schema theory as proposed by Schmidt (1975). Variation in training was considered as a means of strengthening the schemata that guide behavior so that transfer to novel task variations should be better. Shapiro and Schmidt (1982) have reviewed the literature in relation to this prediction, and have found strong support from experiments with children, and some support from experiments with adults.

Lee and Magill (1983) offer a slightly different view that explains contextual interference in terms of greater cognitive effort in training. They verified the earlier result of Shea and Morgan (1979) in which contextual interference was produced by randomizing task variations in training. For this sequence, no task variation was ever repeated in succession. In contrast, contextual interference was not observed when task variations were repeated in succession within blocks of trials. A barrier collision task (subjects used a hand to knock over small barriers in specified sequences) was employed by both pairs of investigators.

In addition, Lee and Magill (1983) tested a serial order (no task variations repeated in succession) and it produced trends similar to those from the randomized order. They proposed that contextual interference resulted from frequent switches between tasks during training. They argued that both random and serial sequences forced subjects to regenerate an action plan before each trial while a blocked series of events would most frequently permit subjects to rely on the action plan generated for the previous trial. The regeneration of action plans is a type of cognitive effort, or active participation, that could depress performance during acquisition but enhance it during transfer.

CONTEXTUAL FACILITATION

Evident in the reviews of augmented-feedback research (Lintern and Roscoe, 1980), and part-task training research (Wightman and Lintern, 1983) is a pattern of improved performance both in training and in transfer. Contextual facilitation seems to be an appropriate term because the pattern it designates contrasts sharply with that of contextual interference.

Furthermore, successful augmented-feedback and part-task manipulations appear to reduce contextual ambiguity during training. For example, Lintern's (1980) augmented cues provided compelling visual guidance in a situation in which students normally find it difficult to interpret the natural visual information that is used for guidance in the approach to touchdown. Wightman and Lintern's (1984) backward-chaining manipulation permitted students to learn the fine control movements that are essential in the final stages of a carrier-landing approach, without the confusion arising from errors that accumulate earlier in the approach through insensitivity of the glideslope guidance system at that distance. In these studies the experimental manipulation resulted in improved performance both in the training and the transfer phases.

FACILITATION VERSUS INTERFERENCE

Conflicting notions of contextual facilitation and contextual interference have emerged from the research literature and both appear to have considerable empirical support. Either one, if shown to be relevant to the acquisition of operational skills, could have a substantial impact on the design and use of synthetic training devices. A plausible case can be made to rationalize the apparent conflict in terms of processes that underlie perceptual learning (Lavery, 1962; Lintern, 1985). Manipulations that lead to both contextual interference (e.g., Lavery, 1962; Lee and Magill, 1983) and to contextual facilitation (e.g., Lintern, 1980; Wightman and Lintern, 1984) are consistent with perceptual learning principles outlined by Gibson (1969). On the other hand, superficially similar manipulations that do not produce these effects (e.g., Lee and Magill, 1983; Lavery, 1962; Lintern, 1980) exclude some critical elements thought to promote perceptual learning. These observations imply that perceptual modification dominates the learning processes associated with acquisition of psychomotor skills.

Nevertheless, the available data are not compelling in their convergence on a perceptual learning explanation. Type of experimental task, and amount of variation in the transfer test, might impact the trends of context effects. For example, contextual interference has been observed only with discrete psychomotor tasks, while contextual facilitation is more commonly found with continuous tracking tasks. Also, contextual interference has been found in experiments in which the content of the transfer test has been varied, while support for contextual facilitation has been found in experiments in which the content of the transfer test has been constrained to one consistent set of conditions. In that sense, the transfer tests of contextual facilitation have not reflected the changing demands and conditions normally found in operational tasks.

An experiment by Lintern, Thomley, Nelson, and Roscoe (1984) had some features that make it particularly relevant to this discussion. Augmented feedback and task variation were manipulated as training variables. In previous research the first had produced a contextual facilitation effect, and the second a contextual interference effect. A complex and operationally relevant task (30-degree bombing approach from an 8000-foot altitude circular pattern) was taught in a flight simulator. The transfer test included task variations similar to those used as training manipulations and that represented some of the variations found in operational flight.

Task variation in training was accomplished by scheduling subjects to fly with two different representations of the outside visual world. One was a well-detailed landscape scene of flat terrain with fields, towns, and roads. The other, an improverished representation of the real world, was a white grid on a green background. The varied training group flew 80 training trials in eight-trial sequences of three trials with one scene, three with another, and then one with each. The number of switches between task variations approximated the number used by Lee and Magill (1983) in their randomized and serial conditions.

There was a strong contextual facilitation effect of augmented feedback for pilots with no previous bombing experience. Those trained with augmented feedback outperformed those trained without it, both in the training phase and later in the transfer phase in which augmented cues were no longer available (Figure 1). Note, however, that pilots with some weapons experience (approximately 60 prior bombing runs in an aircraft) did not show differential transfer effects.

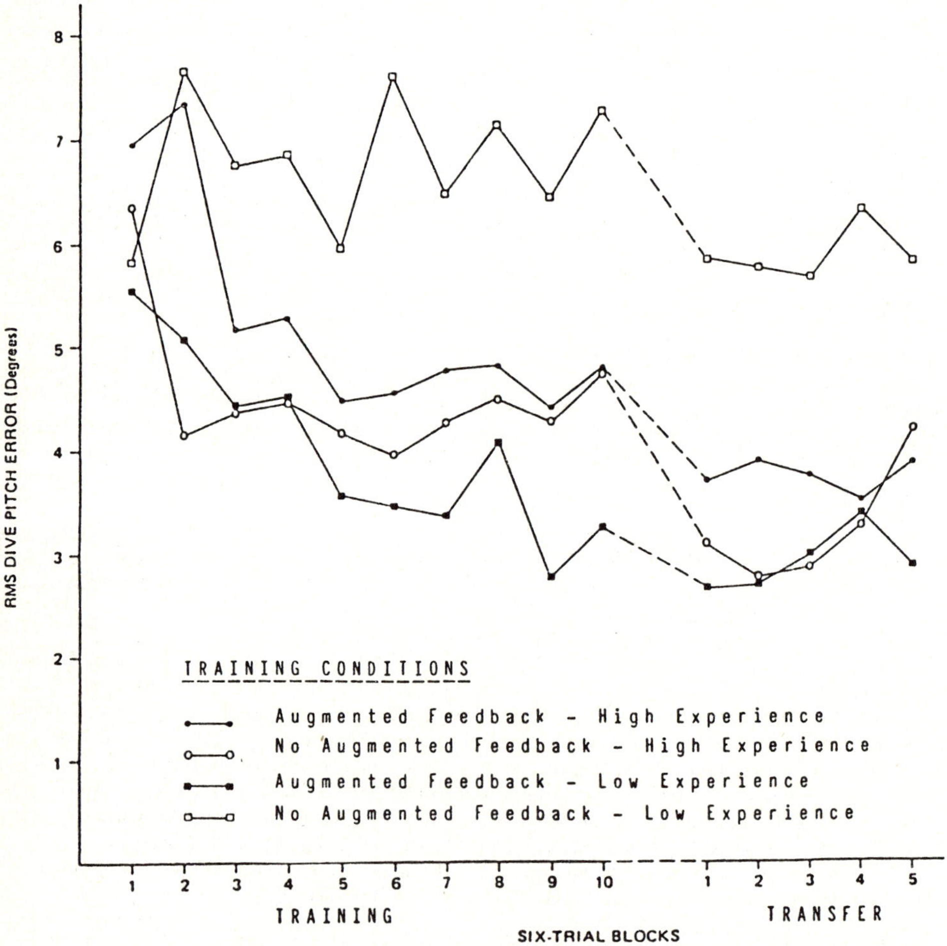

Figure 1. Effects of augmented feedback and pilot experience on the acquisition of bombing skills (Lintern, Thomley, Nelson, and Roscoe, 1984).

There was no contextual interference effect as a result of varied training, but there was a differential transfer effect due to type of scene used in training. Two consistently-trained groups had been tested. One had been trained with the high-detail landscape, and the other with the grid pattern. Those trained with the high-detail landscape performed better in training and in transfer than those trained with the grid pattern or with both scenes (Figure 2). The result of this manipulation of scenes in training was also suggestive of a contextual facilitation effect.

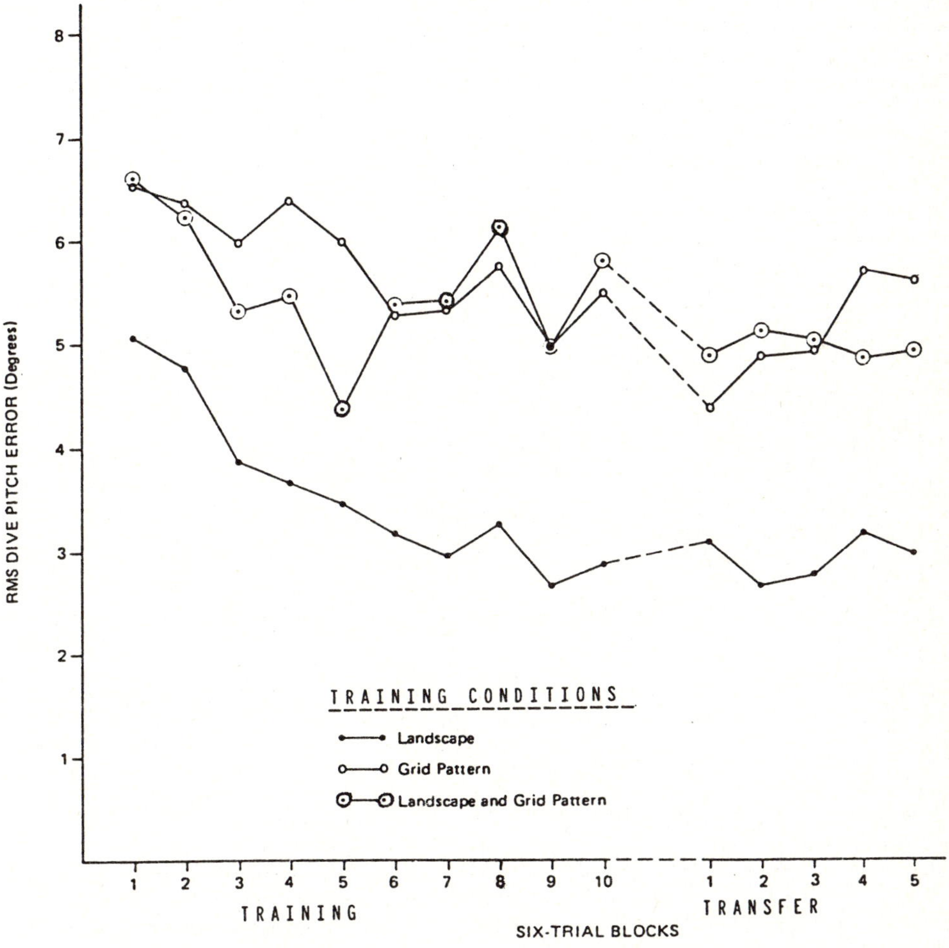

Figure 2. Effects of variety in training and of training scene type on the acquisition of bombing skills (Lintern, Thomley, Nelson, and Roscoe, 1984).

In summary, there was a general facilitating effect from providing better visual information in training, an observation consistent with a perceptual learning approach to skill acquisition (Lintern, 1985). The earlier speculation that contextual facilitation would occur only if the transfer test was constrained to one consistent set of conditions was not supported. However, there is no resolution of the role that task complexity plays in the production of contextual effects.

DEPENDENCY

Although some question remains about the value of contextual interference for teaching operational tasks, contextual facilitation does appear to be useful. Nevertheless, discussions of facilitation in acquisition often raise the possibility that unwanted dependencies will develop (Lee and Magill, 1983; Lintern, 1980). Schmidt (1982, p550) has used a "crutch" analogy to illustrate the danger. Overreliance on the crutch can disrupt later performance without it. Contextual facilitation similarly raises a threat of unwanted dependencies. Where it is continued for too long, or is so strong that it obscures the criterion task, dependencies that completely negate any potential benefit of the manipulation may develop.

Sharp performance decrements are often observed in transfer from an easy version of a criterion task (Lavery, 1962; Lee and Magill, 1983; Micheli, 1966). However, other evidence showing improved transfer after facilitation in learning suggests that dependency effects can be avoided. Williams and Briggs (1962) tested an augmented feedback schedule in which supplementary cues were provided only when the subject was performing poorly. Thus, the subject could not rely on assistance from the supplementary cues during periods of good performance. Transfer to a nonaugmented condition showed no signs of disruption. In the Lintern et al. (1984) study, in which an off-target schedule was also used to present the supplementary cues, there was an interaction between augmented feedback and pilot experience (Figure 1). Inexperienced pilots profited from the augmented feedback but the more experienced pilots did not. It appears that the supplementary cues speeded early learning, but that the effect diminished as learning progressed.

These data suggest that the crutch analogy should be extended. A crutch is first used for early (and often essential) support, and becomes a problem only if it is used past the time of need. Similarly, facilitation in acquisition will aid retention if it is scheduled at the appropriate time and to the correct degree, and will cause problems only if it is used for too long or if it overwhelms other essential elements of the task.

SUMMARY

Two contrasting effects in motor learning were described. One, known as contextual interference, shows a pattern of poor performance during training and good performance during transfer. The other, referred to as contextual facilitation, shows a pattern of good performance during both training and transfer.

These apparently conflicting trends can be rationalized by an appeal to principles of perceptual learning. However, they could also be an artifact of methodological differences; specifically, the complexity of the experimental task and the amount of variety in the transfer task. Data considered here indicate that transfer-test variety is not an important variable but the role of task complexity was not resolved.

Although there is some doubt that contextual interference could be used to improve the effectiveness of synthetic training devices, there is good support for the development of contextually facilitating manipulations. Dependency effects pose a threat, but they can be avoided. As contextual facilitation is generally possible only in synthetic training devices, such devices offer benefits in addition to their cost advantage. The challenge to the human factors community is to determine how these devices can be used to maximize training effectiveness.

ACKNOWLEDGEMENT

This work was performed under Contract N-81-C-0105 with the Naval Training Equipment Center, Orlando, Florida. Dr. Dennis C. Wightman was the Contracting Officer's Technical Representative. Dr. Walter Schneider and Dr. Amir Mane of the University of Illinois reviewed an earlier version of this paper.

(1) Battig, W.F., Facilitation and interference, in Bilodeau, E.A. (Ed.) Acquisition of skill (New York: Academic Press, 1966).

(2) Dashiell, J.F., An experimental isolation of higher level habits, Journal of Experimental Psychology, 1 (1924) 391-397.

(3) Duncan, C.P., Transfer after training with single versus multiple tasks, Journal of Experimental Psychology, 55 (1958) 65-72.

(4) Gibson, E.J., Principles of perceptual learning and development (New York, Appleton-Century-Crofts, 1969).

(5) Lavery, J.J., Retention of simple motor skills as a function of type of knowledge of results, Canadian Journal of Psychology, 16 (1962) 300-311.

(6) Lee, T.D., and Magill, R.A., The locus of contextual interference in motor-skill acquisition, Journal of Experimental Psychology; Learning, Memory, and Cognition, 9 (1983) 730-746.

(7) Lintern, G., Transfer of landing skill after training with supplementary visual cues, Human Factors, 22 (1980) 81-88.

(8) Lintern, G., A perceptual learning model of skill transfer for manual control (Orlando, FL: Naval Training Equipment Center, NAVTRAEQUIPCEN 81-C-0105-12, 1985).

(9) Lintern, G., and Roscoe, S.N., Visual cue augmentation in contact flight simulation, in Roscoe, S.N., Aviation psychology (Ames, IA: Iowa State Univ. Press, 1980).

(10) Lintern, G., Thomley, K.E., Nelson, B.E., and Roscoe, S.N., Content, variety, and augmentation of simulated visual scenes for teaching air-to-ground attack (Orlando, FL: Naval Training Equipment Center, NAVTRAEQUIPCEN 81-C-0105-3, 1984).

(11) Micheli, G.S., Augmenting feedback and transfer of training (Orlando, FL: Naval Training Device Center, NAVTRADEVCEN IH 41 1966).

(12) Morrisett, I.J., and Hovland, C.I., A Comparison of three varieties of training in human problem solving, Journal of Experimental Psychology, 58 (1959) 52-55.

(13) Schmidt, R.A., A schema theory of discrete motor skill learning, Psychological Review, 82 (1975) 225-260.

(14) Schmidt, R.A., Motor control and learning: a behavorial emphasis (Champaign, IL: Human Kinetics Press, 1982).

(15) Schneider, W., and Fisk, A.F., Attention theory and mechanisms for skilled performance, in Magill, R.A. (Ed.) Memory and control of motor behavior (North Holland, Amsterdam, 1983).

(16) Shapiro, D.C., and Schmidt, R.A., The schema theory: Recent evidence and developmental implications, in Kelso, J.A.S., and Clark, J.E. (Eds.) The development of movement control and coordination (New York: John Wiley and Sons, 1982).

(17) Shea, J.F., and Morgan, R.L., Contextual interference effects on the acquisition, retention, and transfer of a motor skill, Journal of Experimental Psychology: Human learning and memory, 5 (1979) 179-187.

(18) Wightman, D.C., and Lintern, G., Part-task training of tracking in manual control (Orlando, FL: Naval Training Equipment Center, NAVTRAEQUIPCEN 81-C-0105-2, 1983).

(19) Wightman, D.C., and Lintern, G., Part-task training strategies in simulated carrier landing final approach training, in Proceedings of the 28th Annual Meeting of the Human Factors Society (Santa Monica, CA: Human Factors Society, 1984).

(20) Williams, A.C., and Briggs, G.E., On-target versus off-target information and the acquisition of tracking skill, Journal of Experimental Psychology, 64 (1962) 519-525.

Trends in Ergonomics/Human Factors II
R.E. Eberts and C.G. Eberts (Editors)

EFFECTS OF PRIOR MUSCULAR EXERTION ON PERFORMANCE OF A LINEAR-POSITIONING TASK

James D. St. James

Department of Psychology
University of Illinois at Urbana-Champaign
Champaign, Illinois
U.S.A.

Two experiments are reported that examine the effects of prior muscular exertion on performance of a linear-positioning task. After each block of three trials, subjects either rested for 30 seconds or spent that time exercising the arm used for the positioning task. A decrement in accuracy occurred with prior exercise, but reaction time to start a movement was unaffected. The pattern of results suggests that the decrement in accuracy is due to changes in feedback control of the movement, and is not due to changes in the ballistic, programmed components of the movement.

Experience alone has demonstrated to us all the effects on motor performance of prior exertion, and a large body of research evidence confirms the observation. Decrements in massed practice in comparison to distributed practice, presumably attributable to the exertion during massed practice, have been shown in such tasks as inverted-alphabet printing (Kimble and Horenstein (1948)), the rotary pursuit test (Adams and Reynolds (1954)), and the Bachman ladder (Stelmach (1969)). Recovery from the performance decrement following rest (reminiscence) occurs with all of these tasks, and it has been interpreted as recovery from the effects of prior exertion. The experiments reported here attempt to separate possible effects of exertion on the ballistic (programmed, non-feedback controlled) and feedback-controlled components of simple, graded linear movements.

The effect of prior exertion or exercise may well be attributable to the processes often labelled as <u>fatigue</u> (e.g., Godwin and Schmidt (1971)). While such an assertion seems reasonable, the terms "prior exertion" and "exercise" will be used here, since no test of the level of fatigue (such as a measure of strength decrement) was made in these experiments.

Experiment 1 relies on the use of reaction time (RT) as a measure of the ballistic component of the movement used. An increase in RT with prior exertion would suggest that the effect of that exercise on performance might be to slow the programming and initiation of the movement. If, on the other hand, fatigue affects the feedback-regulated component of movement (where information about the limb's position is used to guide the limb to the intended location), then greater errors of positioning should occur following exercise, when visual guidance of the movement is not allowed. In an experiment relevant to these points, Lance and Chaffin (1971) had subjects point as quickly as possible to one of a row of lights when it came on. Exertion (holding a 20-pound weight suspended from the wrist until "voluntary termination") produced no change in the time to initiate the movement, but doubled the time to make final corrective movements to stabilize the hand pointing at the target. They concluded that "muscle exertion changes the neuromuscular function required to maintain a specific position."

In the studies that follow, visual guidance of the movement was precluded, forcing subjects to rely largely on proprioception for feedback about the target position. If the effects of exercise are related to proprioception, then the effects of that exercise on performance should be larger than when vision is allowed, as in Lance and Chaffin's (1971) study, and should be reflected in increased positional error, rather than in increased time to stabilize the limb at the target.

The purpose of Experiment 1 was to determine whether prior exertion affects the programmed and/or the feedback-regulated parts of a graded positioning movement.

EXPERIMENT 1

METHOD

Subjects. Subjects were 75 undergraduate males at the University of Illinois, who participated as part of a requirement for an introductory course in psychology. All were right-handed, by self-report.

Apparatus and procedure. A linear positioning device was used that required the subject to move a slide down a straight track from right to left. The slide was mounted inside a large wooden box with an opening through which the subject reached to the slide. A drape over the subject's arms effectively prevented the use of visual cues about limb position. The total distance the slide could move was 35.5 cm. The slide moved on ball bearings with little resistance.

The subject gripped a handle on the slide so that he could push down a button on top of the slide with his index finger. This button released the slide, allowing it to move. At the end of the movement, the subject released the button, locking the slide in place. The score was then recorded and the knowledge of results (KR) given, after which the experimenter returned the slide to the starting point. The subject returned his hand to a cushion near the slide after completing each movement.

On each trial, "Hand" signalled the subject to grasp the handle. As soon as that was done, the experimenter said "Ready" (the command to push down the button to release the slide) and pressed a switch to start a timer. One, two, or three seconds later (in random order) a buzzer sounded. Subjects were instructed to "begin the movement as quickly as possible after the buzzer sounds, then move at your own pace to where you think the target is." Subjects were told to make a single, smooth movement, avoiding small oscillating corrections at the end.

Verbal KR was given immediately after each trial; for example "Reaction time was 22, and accuracy was minus 5." Reaction time (RT) KR was in 1/100ths of a second, recorded on a stop clock from the time the buzzer began to the beginning of movement. A displacement of the slide of 1 mm stopped the clock. The unit of KR for accuracy was 3.8 mm, determined by the apparatus, which locked in place at that interval. A KR report on positioning error of "minus 5" indicated an undershoot of five units, while a score of "plus 10" indicated an overshoot of 10 units. The criterion score (error = 0) was 25.2 cm.

Subjects received 75 trials of the positioning task, with a 10-second interval between trials in a block. Trials were in blocks of three, with 30 seconds between blocks. Subjects were randomly assigned to one of three conditions, defined by activity during the break between blocks. During these breaks, subjects either sat quietly (No Exercise), or did one of two levels of exercise (Low or High). Exercise was done on a device that consisted of a handle that moved down a straight track through a distance of 35.7 cm. A cable attached to the handle ran over a pulley and permitted the attachment of weights. The device was situated to the subject's right, so that a slight turn in the chair allowed him to reach it easily. Subjects moved the handle back and forth in time with a bell that rang every 1.5 seconds during exercise.

Before the trials began, each subject's grip strength was measured using a steel-bow hand-grip dynamometer. The weight used for the exercise was adjusted accordingly. Table 1 gives the weight used and work done during each 30 seconds of exercise.

Table 1
Levels of Exercise

		Grip Strength (kg)		
		198	198-275	275
Low Exercise	Weight (kg)	0.50	1.00	1.50
	Work (kgm)	7.14	14.28	21.42
High Exercise	Weight (kg)	1.00	2.00	3.00
	Work (kgm)	14.28	28.56	42.84

Note: Weights moved during exercise and work done (20 moves of 71.4 cm x weight) for different levels of grip strength. Work includes both positive and negative work.

RESULTS

Since improvement in performance due to learning occurs across time, and since the effect of the exercise is also likely to accumulate across time, the best test of the effect of the exercise on performance of the task is at the end of the trials, where it is presumed that learning is largely complete, and that the effect of the exercise has set in. For this reason, each dependent measure was calculated as the mean of the scores on the last 10 blocks (30 trials).

Analysis across the 25 blocks, treated as 25 levels in a repeated-measures design, were performed, and confirm that improvement occurred across trials as expected in a task with each movement followed by KR. These analyses are not discussed, since they offer no direct test of the hypotheses.

Mean scores on the dependent variables for each group are reported in Table 2. These means are collapsed across the last 30 trials.

Table 2
Mean Scores, Last 30 Trials, Experiment 1

	No Exercise	Low Exercise	High Exercise
RT	190	208	192
AE	12.73	15.43	16.45
VE	10.94	14.10	14.78
CE	-1.29	2.85	1.03

Note: Mean scores for RT (in msec) and for accuracy (AE, VE, and CE in mm). N = 25 for each group.

Reaction time. RT is the time from the buzzer, signalling the subject to move, until the completion of 1 mm of movement. The three groups did not differ in mean RT, $F(2,72) = 1.02$, $p > .05$.

Absolute error (AE). AE is the mean of the unsigned deviation of movements about the 25.2 cm criterion. The instructions to the subjects to "keep your score as close to zero as possible" emphasized AE, which is the primary dependent variable in the analysis of accuracy scores. There was a significant difference in AE across the levels of exercise $F(2,72) = 3.79$, $p < .05$, with means reflecting an increase in error of positioning with increased workload.

Constant error (CE). CE is the mean of the algebraically signed error scores, and reflects bias in responding. The KR, as an agent of learning, should reduce CE to low values for all groups. Though CE was generally small, the error differed across levels of exercise $F(2,72) = 3.57$, $p < .05$, with the Low Exercise group showing a positive CE (overshoot), and the No Exercise and High Exercise groups showing negative CEs (undershoot).

Variable error (VE). VE is the standard deviation of the error scores, and reflects variability in responding. Group means differed significantly, $F(2,72) = 4.81$, $p < .05$, with the higher levels of exercise showing greater error.

DISCUSSION

The finding of a graded increase in AE for positioning accuracy with increased exercise, with no changes in RT, suggests that at levels of exercise insufficient to prevent the movement being made, there are effects on the feedback-regulated part of the movement, but not the ballistic, programmed part. This former effect could occur either through degrading the feedback from the proprioceptors or through degrading the utility to adjust the motor commands to compensate for the effects of the exercise. The current experiment does not separate these mechanisms.

The finding of no change in RT with exercise confirms the study of Lance and Chaffin (1971). Other evidence on RT comes from studies of fractionated reaction time, where no changes in simple RT were found under high levels of fatiguing exercise (Kroll (1973); Hayes (1975)).

EXPERIMENT 2

The interpretation of increased AE with prior exertion in Experiment 1 is that it suggests degradation of the feedback control of movement. As a check on this, Experiment 2 combines exercise with increased resistance to movement (spring tension), which is a variable known to affect proprioception. Adams, Gopher, and Lintern (1977) found improvement in performance on a linear positioning task with added spring resistance, which they interpreted as improved proprioception.

An interaction of heightened proprioception with fatigue would support the interpretation of Experiment 1. Accordingly, this experiment requires subjects to learn the linear positioning task with or without exercise interpolated between blocks of trials, and with or without a spring attached to the positioning slide.

In this experiment, no measure of RT was used because the effects of exercise on it that I found confirm findings in the literature and seems secure. Additionally, omitting the RT instruction permits a check on the possible relations between speed set on movement initiation and final positioning accuracy.

METHOD

Subjects. Subjects were 88 undergraduate males at the University of Illinois, who participated as part of a requirement for an introductory psychology course. All were right-handed, by self-report.

Apparatus and procedures. The positioning apparatus was the same as for Experiment 1, except that for some conditions, resistance was added to the slide by attaching a spring. This provided a resistance of 805 gm at the criterion distance of 25.2 cm.

Subjects were instructed to "move at your own pace to where you think the target is." Each trial was started by the commands "Hand" (instructing the subject to move his hand from a cushion to the handle of the slide and press down the button on top of the slide, releasing it), then "Ready." About one second after the "Ready" command, white noise was presented at 100 dB from a Grason-Stadler noise generator (Model 901B), through Grason-Stadler headphones (Model TDH39-3002), which were worn by the subject throughout the experiment. The 100 dB noise remained on until the subject completed the movement, effectively blocking auditory feedback from the slide.

Subjects performed 75 trials of the positioning task with a 10-second inter-trial interval. Trials were in blocks of three, with 30 seconds between blocks.

The experiment employed a 2 x 2 design, with resistance versus no resistance as one factor, and exercise versus no exercise as the other. During the 20 seconds between blocks of positioning trials, the exercise groups worked in the same manner as in Experiment 1. The device and levels of work were the same as for the High Exercise group of Experiment 1 (see Table 1). The no-exercise groups rested between blocks of trials.

RESULTS

As in Experiment 1, the analyses use AE as the principal dependent measure. Analyses of CE and VE are also reported. Group means of all dependent measures are reported in Table 3.

Table 3
Mean Scores, Last 30 Trials, Experiment 2

		Movement Condition	
		Control	Resistance
	AE	13.15	11.09
No Exercise	VE	11.59	10.53
	CE	-0.49	-0.15
	AE	13.53	15.12
Exercise	VE	12.96	13.45
	CE	-1.14	-1.90

Note: Mean scores for accuracy (AE, VE, CE in mm). N = 22 for each group.

All reported analyses are for mean error over the last 30 trials. As in Experiment 1, analyses of block means as the 25 levels of a repeated-measures factor confirm the improvement in accuracy across blocks of trials. Further discussion of these analyses is omitted, since they do not directly test the effects of interest.

Absolute error. The main effect of resistance was not significant, $F < 1$. The main effect of exercise was significant, $F(1,84) = 6.35$, $p < .05$, as was the interaction of exercise and resistance, $F(1,84) = 4.51$, $p < .05$. Examination of the means reported in Table 3 shows that accuracy was degraded for the exercise groups with the interaction reflecting that this effect was limited to the groups who did the positioning task against spring resistance.

Constant error. No significant differences were found between group means for CE. For the main effect of exercise, $F(1,84) = 1.10$, $p > .05$. For the main effect of resistance and the interaction of resistance and exercise, $F < 1$.

Variable error. The main effect of exercise was significant, $F(1,84) = 6.94$, $p < .05$, reflecting an increase in error with prior exertion. Neither the main effect of resistance nor the interaction of resistance with exercise was significant, $F < 1$ in both cases.

DISCUSSION

The results of Experiment 2 confirm the effect of prior exertion found in Experiment 1 and make it clear that the effect is not dependent upon the instruction for speeded initiation of the movement. The minimal effect of prior exertion on the no-resistance group is surprising, and I can offer no good conjecture as to why the effect of exercise was largely restricted to the resistance groups.

GENERAL DISCUSSION

Taken together, the experiments suggest that, at levels of exercise that do not prevent movement execution, there is still a deficit in positioning accuracy. Had RT increased with exercise, the mechanism could be either an increase in programming time (perhaps due to the time required to compensate for muscular changes by recruitment of more motor units) or a slowing of force generation by the muscle. At the levels of exercise used, no change in RT occurred, suggesting that the initial ballistic component of movement, under programmed control, is not changed. The decrement in accuracy seems likely, therefore, to be due to changes in the feedback-controlled component of the movement. Changes in the quality of the sensory return from the muscles (proprioception) may be implicated as a source of the decrement, with prior exertion changing the "feel" of the movement.

It should be noted that the use of KR on all trials of the experiments is likely to have minimized the effects of the exercise, since subjects could at least partially compensate on the next trial for error due to any source, including prior exertion. If perception of limb position is distorted by exercise, verbal KR could not completely overcome the deficit. When visual control was allowed, Lance and Chaffin (1971) found an increase in stabilizing time with prior exertion in their pointing task, indicating a deficit in control that could be compensated for by visual feedback (which is, of course, unaffected by the exercise), but only at the cost of increased time to complete the movement.

The ubiquity of fatigue-like phenomena such as the effects of prior exertion demonstrated here, as well as the common finding of a deficit in movement control with massed practice in a wide variety of motor skills, suggest that motor control theorists ignore these phenomena at their peril. Our lack of good understanding of these effects of prior exertion on motor performance leaves a large gap in our understanding of that performance. The experiments reported here suggest that a change in feedback control is a strong candidate as a mechanism mediating these effects.

REFERENCES

[1] Adams, J.A. and Reynolds, B., Effect of shifts in distribution of practice conditions following interpolated rest, Jrnl. of Exptl. Psych. 47 (1954) 32-36.

[2] Godwin, M.A. and Schmidt, R.A., Muscular fatigue and learning a discrete motor skill, Res. Quart. 42 (1971) 374-382.

[3] Hayes, K.C., Effects of fatiguing isometric exercise upon achilles tendon reflex and plantar flexion reaction time components in man. Eur. Jrnl. of Apl. Physiology 34 (1975) 69-79.

[4] Kimble, G.A. and Horenstein, B.R., Reminiscence in motor learning as a function of length of interpolated rest. Jrnl. of Exptl. Psych. 38 (1948) 239-244.

[5] Kroll, W., Effects of local muscular fatigue due to isotonic and isometric exercise upon fractionated reaction time components, Jrnl. of Motor Behav. 5 (1973) 81-93.

[6] Lance, B.M. and Chaffin, D.B., The effect of prior muscle exertions on simple movements, Human Factors 13 (1971) 355-361.

[7] Newell, K.M., Hoshizaki, L.E.F., Carlton, M.J., and Halbert, J.A., Movement time and velocity as determinants of movement timing accuracy, Jrnl. of Motor Behav. 11 (1979) 49-58.

[8] Stelmach, G.E., Efficiency of motor learning as a function of intertrial rest, Res. Quart. 40 (1969) 198-202.

Trends in Ergonomics/Human Factors II
R.E. Eberts and C.G. Eberts (Editors)

ANALYSIS OF SIMPLE THROWING

Timothy P. King
and
Ronald L. Huston

Department of Mechanical and Industrial Engineering
University of Cincinnati
Cincinnati, Ohio 45221-0072

Procedures for studying simple throwing motion are presented. Kane's equations are used to develop the governing dynamical equations. Constraint equations are developed for the desired throwing motion. The "orthogonal complement" of the constraint equations is then used to reduce the dynamical equations to a consistent set of governing equations. Numerical solutions for the case of simple dart throwing is presented.

INTRODUCTION

Recently there has been increased interest in sports mechanics with an emphasis upon both improved performance and reduced injury. Clearly, one of the most important sporting maneuvers is throwing. In this paper, we outline procedures intended to provide new insight into throwing dynamics. Specifically, procedures for determining joint forces, inertia forces, and the kinematics of simple throwing are discussed.

The procedures are based upon recent advances in multibody dynamics. The arm is modelled as a hand, forearm, and upper arm. As such, it has seven degrees of freedom. In general six degrees of freedom are required to throw an object. Hence, there is a redundant degree of freedom. This redundancy has, in the past, precluded comprehensive and quantitative studies of throwing.

The governing dynamical equations are derived using Kane's equations (1968, 1980) as exposited by Huston, et. al. (1974, 1978, 1979). Constraint equations are developed from the motion of the thrown object. An orthogonal complement to the matrix of constraint equations is then used to reduce the dynamical equations to a consistent system. The "orthogonal complement" array (see Hemami and Weimer, 1981) is developed from a "zero eigenvalues" algorithm (see Walton and Steeves, 1969) as exposited by Kamman and Huston (1984). Resulting equations are then solved numerically for two cases of dart throwing.

PRELIMINARY CONSIDERATIONS: ARM MODELLING

We model the arm as a series of three bodies consisting of the upper arm, the forearm, and the hand as depicted in Figure 1. We model the shoulder as a spherical joint, the elbow as a pin or hinge joint and the wrist as a

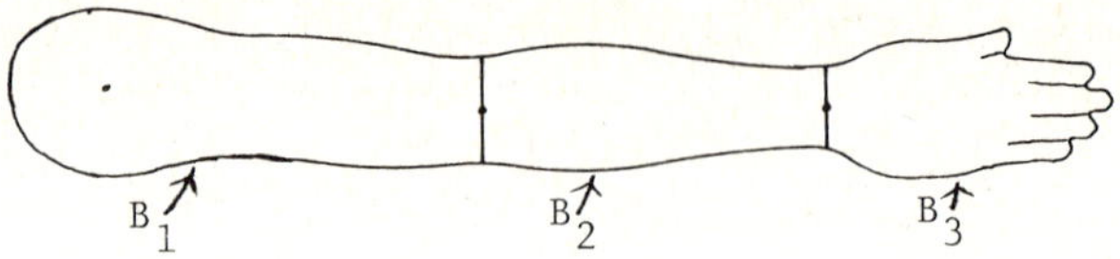

Figure 1. Arm Model

spherical joint, incorporating the axial rotation of the forearm. Hence, the system has 3+1+3, or 7, degrees of freedom.

For analysis purposes it is convenient to describe the relative orientation of the bodies with Euler parameters as discussed by Huston, et. al. (1978). Use of Euler parameters avoids singularities encountered with orientation angles. Finally, components of the relative angular velocity vectors are used as the fundamental kinematic quantities.

GOVERNING EQUATIONS

From the addition theorem for angular velocities, the angular velocities of the bodies may be written in the form:

$$\underset{\sim}{\omega}_1 = \hat{\underset{\sim}{\omega}}_1, \quad \underset{\sim}{\omega}_2 = \hat{\underset{\sim}{\omega}}_1 + \hat{\underset{\sim}{\omega}}_2, \quad \underset{\sim}{\omega}_3 = \underset{\sim}{\omega}_1 + \underset{\sim}{\omega}_2 + \underset{\sim}{\omega}_3 \tag{1}$$

where $\hat{\underset{\sim}{\omega}}_j$ represents the angular velocity of B_j relative to its adjacent lower numbered body B_i. Let $\hat{\underset{\sim}{\omega}}_j$ be expressed in terms unit vectors $\underset{\sim}{n}_{im}$ (m=1,2,3) fixed in B_i, as:

$$\hat{\underset{\sim}{\omega}}_j = \hat{\omega}_{j1}\, \underset{\sim}{n}_{i1} + \hat{\omega}_{j2}\, \underset{\sim}{n}_{i2} + \hat{\omega}_{j3}\, \underset{\sim}{n}_{i3} \tag{2}$$

The $\hat{\omega}_{jm}$ (j=1,2,3; m=1,2,3) are taken as the generalized coordinate derivatives (see Kane, 1968) and are designated by y_ℓ (ℓ=1,...,7). By using transformation matrices and by substituting from Equations (2) into (1), the angular velocities of the arm bodies relative to the shoulder may be written in the form (see Huston, 1978):

$$\underset{\sim}{\omega}_k = \omega_{k\ell m}\, y_\ell\, \underset{\sim}{n}_{om} \quad (k=1,2,3;\ \ell=1,\dots,7;\ m=1,2,3) \tag{3}$$

where the $\underset{\sim}{n}_{om}$ (m=1,2,3) are unit vectors fixed in the shoulder. (Regarding notation, there is a sum for repeated indices over the range of the index.)

Similarly, the velocities of the mass centers of the bodies may be expressed as:

$$\underset{\sim}{v}_k = v_{k\ell m}\, y_\ell\, \underset{\sim}{n}_{om} \quad (k=1,2,3;\ \ell=1,\dots,7;\ m=1,2,3) \tag{4}$$

Finally, if P is a point on the hand, say at the finger tips, the velocity of P relative to the shoulder may be expressed as:

$$\underset{\sim}{v}_P = v_{P\ell m}\, y_\ell\, \underset{\sim}{n}_{om} \tag{5}$$

Using Kane's equations (1978, 1980) and the procedures outlined by Huston, et. al. (1978), the dynamical equations may be written as

$$a_{\ell q}\dot{y}_p = f_\ell \quad (\ell,q=1,\ldots,7) \tag{6}$$

where $a_{\ell p}$ and f_ℓ are elements of the "generalized mass" and "generalized forcing function" arrays. They are functions of the $\omega_{k\ell m}$, the $v_{k\ell m}$, and the physical and geometrical properties of the system (see Huston et. al., 1974, 1978, 1979). In matrix form Equations (6) may be written as:

$$A\dot{y} = f \tag{7}$$

When the throwing motion is known, $\underset{\sim}{v}_P$ as given by Equation (5), and $\underset{\sim}{\omega}_3$, as given by Equation (3) are known. Hence, their components may be expressed as:

$$v_{P\ell m}y_\ell = g_m(f) \quad \text{and} \quad \omega_{3\ell m}y_\ell = g_{m+3}(f) \quad m=1,2,3 \tag{8}$$

where g_n $(n=1,\ldots,6)$ are known functions of time. In matrix form, Equations (8) may be written as

$$By = g \tag{9}$$

SOLUTION METHODS

Equations (7) subject to Equations (9) constitute a set of 13 equations for the 7 y_ℓ. The question thus arises: How should Equations (7) and (9) be combined to form a consistent set of equations? Kamman and Huston (1984) have shown that the dynamics equations can be reduced by premultiplying Equations (7) by the "orthogonal complement" of the matrix B. Specifically, a matrix C is sought such that

$$BC = 0 \tag{10}$$

The governing equations are then

$$C^T A\dot{y} = f \quad \text{and} \quad B\dot{y} = \dot{g} - \dot{B}y \tag{11}$$

Note that B is a 6X7 array and that C is a 7X1 array. Note further that since BC is zero, C defines the direction in the 7-dimensional space for which the system can move. Finally, the orthogonal complement array C may be identified as the eigenvector associated with the zero eigenvalue of the matrix B^TB as discussed by Walton and Steeves (1969).

J. Kamman (1983) has developed a computer program called DYNOCOMBS with the objective of numerically generating and solving the governing equations of constrained multibody systems. DYNOCOMBS is thus readily applicable for generating and solving Equations (11). Results of such application for simple dart throwing are presented in the following section.

APPLICATION: SIMPLE DART THROWING

To illustrate the foregoing ideas, simple dart throwing was simulated. The physical and geometrical properties of the throwing arm, used in the simulation, are listed in Table I.

Body	Mass	Principal Inertia	Length	Mass Center Location
B_1:Upper Arm	2.1095kg	(0.00236, 0.0143, 0.0143) kgm^2	0.297m	0.095m
B_2:Forearm	1.1658kg	(0.000710, 0.00518, 0.00518) kgm^2	0.284m	0.122m
B_3:Hand	0.4941kg	(0.000356, 0.000778, 0.000778) kgm^2	0.0962m	.0481m

Table I. Physical and Inertia Properties of the Throwing Arm

Two motions were examined: Case 1: The dart was thrown at 30° inclination to the horizontal, and Case 2: The dart was thrown at 10° inclination to the horizontal. In both cases the target was at the hand release level at a horizontal distance of 2.381m.

Figures 2., 3., and 4, show results for the shoulder, elbow, and wrist angles for the two cases.

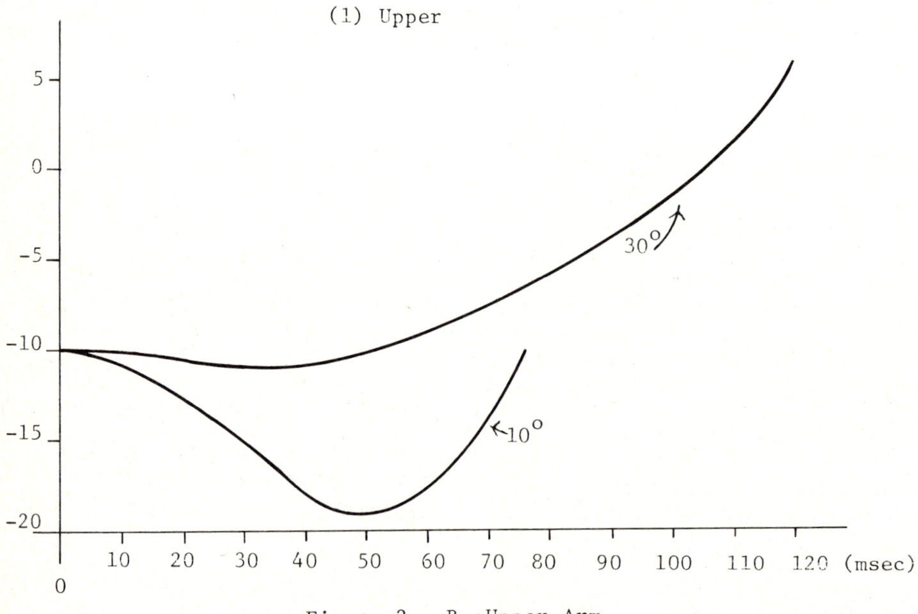

Figure 2. B_1:Upper Arm

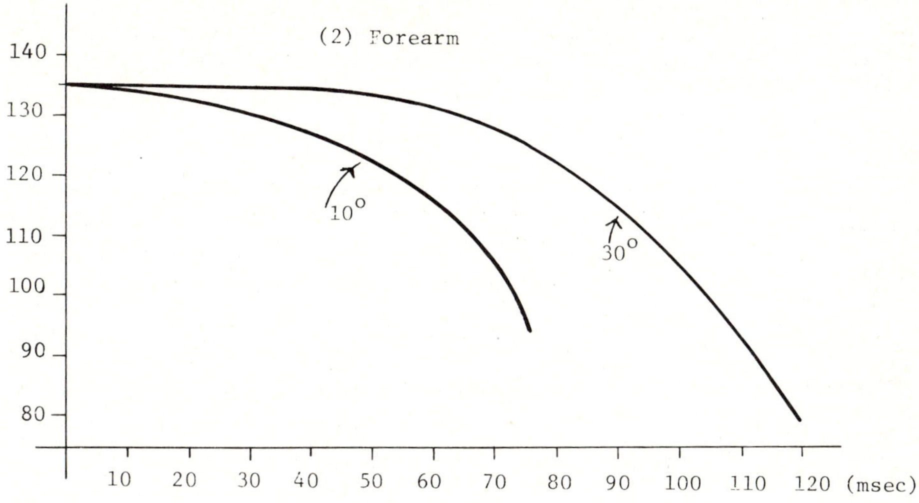

Figure 3. B_2:Forearm

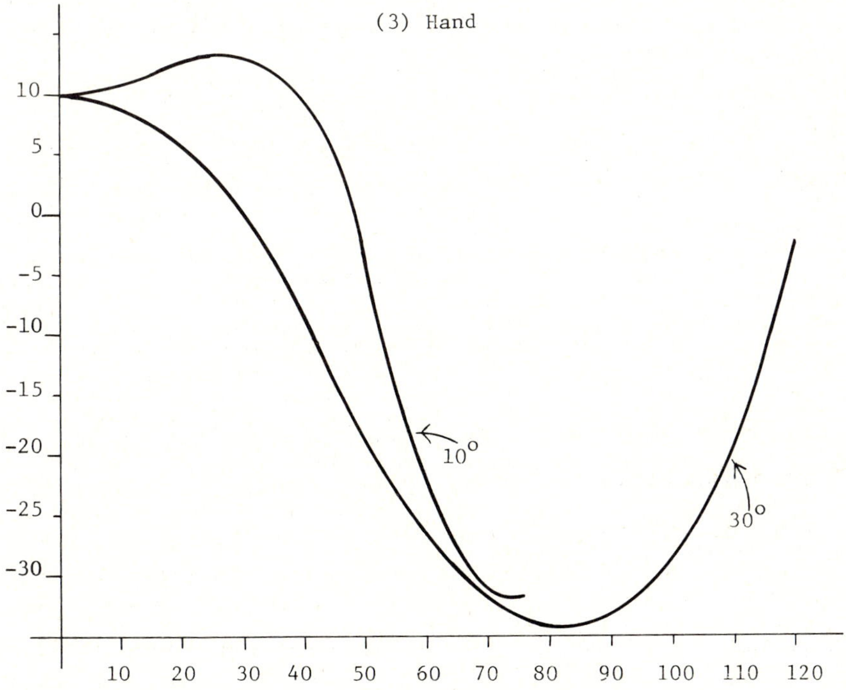

Figure 4. B_3:Hand

DISCUSSION

The analysis of throwing motion has four phases: 1) First, the movement of the hand for the desired throwing motion is determined. 2) Next, the hand motion is used to develop the constraint equations. 3) The governing equations are then solved to determine the joint motions. 4) Finally, knowing the joint motion, the joint moments required to produce the motion are determined. If joint forces are also desired, they may be obtained through analysis of successive free-body diagrams of the arm segments.

This same procedure is applicable for more complex throwing motions where movement of the torso and even the legs are important. Indeed, the equations herein are written in a form so that they may be easily generalized. Similar procedures are applicable in studying kicking. Such analyses are planned. Results are expected to be reported in future papers.

REFERENCES

[1] Huston, R. L., and Passerello, C. E., On the dynamics of chain systems, ASME Paper 74-WA/Aut 11, 1974.

[2] Huston, R. L., Passerello, C. E., and Harlow, M. W., Dynamics of multi-rigid-body systems, ASME Journal of Applied Mechanics, 1978, 45, 889-894.

[3] Huston, R. L., and Passerello, C. E., On multi-rigid-body system dynamics, Computers and Structures, 1979, 10, 439-446.

[4] Kane, T. R., and Levinson, D. A., Formulation of equations of motion for complex spacecraft, Journal of Guidance and Control, 1980, 3, 99-112.

[5] Kane, T. R., Dynamics, Holt, 1968.

[6] Hemami, H., and Weimer, F. C., Modelling of nonholomonic dynamic systems with applications, ASME Journal of Applied Mechanics, 1981, 48, 177-182.

[7] Walton, W. C., Jr., and Steeves, E. C., A new matrix theorem and its application for establishing independent coordinates for complex dynamical systems with constraints, NASA Technical Report TR R-326, 1969.

[8] Kamman, J. W., and Huston, R. L., Constrained multibody system dynamics - an automated approach, Computers and Structures, 1984, 18, 6, 999-1003.

[9] Kamman, J. W., On the Dynamics of Constrained Multibody Systems, Ph.D. Dissertation, University of Cincinnati.

ACKNOWLEDGEMENT

Partial support for this research has been provided by the National Science Foundation under Grant MEA-8313439. Support by Mr. Dimitrios Mavriplis in preparation of the graphs is also acknowledged.

Trends in Ergonomics/Human Factors II
R.E. Eberts and C.G. Eberts (Editors)
© Elsevier Science Publishers B.V. (North-Holland), 1985

S-R COMPATIBILITY IN A CONTINUOUS MOTOR TASK

Richard D. Miller II

Psychology Department
Miami University
Oxford, Ohio

Performance on a target acquisition task was measured under varying conditions of stimulus-response compatibility (SRC), internal model, and knowledge of the stimulus-response relationship. Different levels of SRC were obtained by varying the angle between cursor movements on a CRT screen and corresponding joystick movements. The major result was that the task was most difficult when cursor displacement was 90 degrees. The nature of the interaction among the experimental variables indicates that SRC in the task examined was complexly determined and not the result of a singular automatic process.

INTRODUCTION

As today's worker spends more and more time interacting with computer systems, there is a growing importance of and need for information about this interaction. The user is being called upon to employ a larger variety of input devices (e.g., mouse, lightpen, joystick). Each device entails a concomitant change in the perceptual-motor requirements of the user, including changes from discrete to continuous movements.

There are potential problems inherent in such perceptual-motor tasks. When using a pointing device to move a cursor on the screen, a user must monitor movements in two separate spatial planes, one for the pointing device and one for the cursor. Moreover, a human's processing load is increased when the spatial orientation of the plane containing the pointing device is different from that of the cursor, as when a rightward movement of the pointing device results in an upward motion of the cursor. This creates an incompatible stimulus-response situation.

The problem of incompatibility has been addressed in the literature, first by Fitts and Seeger (1953), under the rubric of S-R compatibility (SRC). Morin and Grant (1955) devised a mathematical model for measuring SRC and found that reaction time decreased as the absolute value of the correlation between 8 stimulus lights and their 8 associated response keys increased. If reaction time is plotted as a function of the correlation, their data produce an inverted U-shaped function. Although there has been extensive research on SRC, most of it deals with reaction times (RTs) in discrete, push-button tasks, and neglects the study of continuous tasks.

Studies dealing with adaptation to discordant stimuli fill some of the void in SRC research. An incompatible, continuous motor task is at least somewhat analogous to prism-induced sensorimotor discordance (Howard, 1981). In both paradigms, the relationship between kinesthetic and visual signals can be viewed as atypical, requiring some recalibration of the visual-motor system. The form taken by the recalibration depends upon the nature and role of different sensory modalities. Taub and Goldberg (1973) found that recalibration in monkeys wearing prism spectacles was facilitated by removal of joint proprioception. Two prism studies on humans (Canon, 1970; Kelso et al., 1975) suggest that locus of attention is an important factor in determining the site of adaptation. In both studies, the attended modality dominated the unattended one during the recalibration process. In addition, recalibration was facilitated when subjects attended to visual, rather than auditory (Canon) or kinesthetic (Kelso et al.) stimuli. It appears that, in at least some incompatible situations, vision is dominant. Attention has been found to be closely linked to visual dominance in non-adaptation studies as well (e.g., Posner et al., 1976). Consistent with the idea that recalibration occurs in the nondominant modality, adaptation is said to occur in the system's weakest link, that is, where there is most uncertainty (Howard, 1981).

How can one facilitate adaptation to a task involving an incompatibility between stimuli and responses? One approach is to employ a specific internal model, i.e., to try to adopt a particular strategy to control a system (Eberts, 1982). If the general goal of an internal model is to improve performance, then one can think of an internal model as the cognitive operating mode which an individual adopts in an attempt to reach optimal system performance. An appropriate internal model should provide a shortcut to optimal performance, as opposed to just a gradual tuning of the perceptual-motor system through practice. In the case of an incompatible S-R task, an internal model might reduce the informational conflict between the visual and kinesthetic modalities by directing attention away from kinesthetic input and toward visual input, thereby permitting an individual to take advantage of visual dominance.

The present experiment addresses the issues raised above, namely S-R compatibility in a continuous motor task and the effects of an attention-based internal model on task performance. A target acquisition task was chosen in which the subject moved a cursor with a joystick from the center of a CRT screen to a randomly located target. Time to target was the dependent measure. This task was selected because it was very similar to the zero-order positioning tasks involved in the operation of modern computer systems which employ pointing devices.

SRC was varied within subjects by displacing cursor movement a certain number of degrees (0, 45, 90, 135, and 180) counterclockwise from movement of the pointing device. For example, in the 90-degree condition, a movement of the pointing device to the east resulted in cursor movement to the north. This type of incompatibility is similar to the real-world situation in which the user shifts his bodily orientation relative to a display screen. Based on Morin and Grant's results, the compatibility curve should resemble an inverted U-shaped function, with time to target peaking at 90 degrees. The 90-degree condition in the present study is analogous to the zero-correlation condition (which was the peak of the inverted U-shaped curve) in the Morin and Grant study.

The internal model employed by subjects was a between-subjects variable. They either attended only to the visual modality and ignored information from the kinesthetic modality, or they attended to both modalities. Consistent with the above discussion of the role of an internal model, subjects in the unimodal condition were expected to reach the target more quickly by attending only to visual signals. Related to the internal model was another variable introduced in the experiment, information about the S-R relationship. The extent to which information about SRC influenced performance might depend on the internal model adopted by a subject. Bimodal subjects would seem to be more susceptible to the information because an understanding of the relationship between movement of the joystick and movement of the cursor is more central to their internal model.

METHOD

A 6809-based microcomputer was used to generate the target screens and record movement times. An RCA XL100 color television set was used as the display screen. The target and cursor were green on a black background. The joystick was used as a zero-order control.

A 2 X 2 X 5 mixed factorial design was employed, with the latter five-level factor of SRC as the only repeated-measures variable. The sequence of SRC conditions for each subject was randomly determined. Two levels of internal model type were employed: (a) unimodal (visual) attention and (b) bimodal (visual and kinesthetic) attention. Also, two levels of quantity of information were used (a) S-R relationship known and (b) S-R relationship unknown. Subjects were randomly assigned to the four conditions defined by combinations of internal model and quantity of information factors.

Each experimental session began with a brief explanation of the experiment's purpose--to investigate human performance with pointing devices. A zero-transformation trial was demonstrated and then given to the subject as a practice trial. The sequence of events for each trial was as follows: (a) A message on the text screen instructed the subject to "center pointing device". This was done in order to insure that the cursor would be in the center starting position when the graphics screen appeared. (b) When the joystick was positioned near the X-Y origin a low-frequency tone sounded and the graphics screen appeared with the cursor in the center and the target located in a random position on an imaginary rectangle surrounding the starting point. The distance that the joystick moved in order for the cursor to reach the target remained constant, irrespective of the position of the target. (c) When the graphics screen appeared, subjects immediately pressed the button on the joystick. This sounded a middle-frequency tone and started a timer, signalling the subject to begin pursuit of the target. (d) When the cursor was placed inside the target, a high-frequency tone sounded, stopping the timer, ending the trial, and returning the subject to the text screen. (e) The subject pressed the button on the joystick when he was ready for the next trial, and the sequence started over. Each subject participated in twelve such trials per condition.

The internal model variable was manipulated by instructing each subject to employ a "strategy" for the experimental task. Unimodal subjects were instructed to attend only to the visual modality and to ignore stimulation from the kinesthetic modality. Bimodal subjects, on the other hand, were instructed to attend to both the visual and kinesthetic modalities.

Subjects who received information prior to each SRC condition were told the angular displacement of the cursor and were given examples of cursor movements for given movements of the joystick (these examples were not in the form of practice trials, however). Unimodal subjects were told to keep the S-R relationship in the "backs of their minds" while still following the unimodal strategy. Subjects who did not receive information about the S-R relationship were told simply that in four of the five conditions the cursor will move in a direction different from the joystick.

Subjects (8 males and 8 females) were recruited from introductory psychology courses as part of their course requirement. All subjects were naive as to the specific purposes of the experiment.

RESULTS

Means for all conditions of the experiment are given in Table 1. Overall means for each level of SRC are shown in the bottom row of the table. The main effect of S-R compatibility was statistically significant ($F=24.2$, $p<.0001$, $df=4,48$). Analytical comparisons were performed to reveal the source of the main effect of compatibility. Results showed that pooled mean target acquisition times for the 90- and 135-degree conditions were significantly different from those of the 0-, 45- and 180-degree conditions ($F=16.56$, $p<.05$, $df=1,48$). Also, combined times for the 45-, 90- and 135-degree conditions were significantly different from combined times for the 0- and 180-degree conditions ($F=12,93$, $p<.05$, $df=1,48$). These results lend support to the prediction that target acquisition times will increase as compatibility approaches 90 degrees. The compatibility curve does not seem to be symmetrical, however, because times for the 90- and 135-degree conditions were significantly higher than times for the 45-degree condition ($F=5.93$, $p<.05$, $df=1,48$).

The other main effects of internal model ($F=1.68$, $p<.22$, $df=1,12$) and knowledge of S-R relationship (F 1.0, $p<.36$, $df=1,12$) were not significant. Also, there were no significant first-order interactions, but there was a significant second-order interaction between SRC, internal model, and S-R knowledge ($F=3.0$, $p<.03$, $df=4,48$). Simple interactions were examined to locate the source of the second-order interaction between compatibility, internal model, and S-R knowledge. Results showed that a significant interaction between internal model and S-R knowledge at the 135-degree level of compatibility ($F=4.65$, $p<.05$, $df=1,12$) accounted for the second-order interaction. This simple interaction was due to the extremenly poor performance of the bimodal, no-information subjects in the 135-degree compatibility condition.

Table 1 Mean target acquisition times

Internal model	SR knowledge	Target acquisition times				
Bimodal	yes	3.41	9.46	30.47	13.04	6.06
	no	5.75	13.88	27.32	34.94	5.86
Unimodal	yes	3.02	11.04	19.75	19.02	5.37
	no	3.71	7.50	24.98	14.64	6.80
Overall		3.97	10.47	25.62	20.41	6.02
Degree of cursor displacement		0	45	90	135	180

DISCUSSION

As mentioned previously, it was expected that the graph relating movement time to compatibility, or cursor displacement, would resemble the inverted U-shaped curve seen in Morin and Grant's (1955) experiment. The fact that the overall compatibility curve for the present experiment has an inverted U-shape and peaks at 90 degrees supports the results of Morin and Grant. The compatibility curve, however, is not symmetrical around 90 degrees. This may be because the 135-degree condition involved some type of more difficult double cognitive transformation (i.e., first reversing the cursor direction, then deflecting movement by 45 degrees) than the 45-degree transformation, which could simply be an allowance by the subject for 45 degrees of tolerable error in making movements. An example of this kind of allowance is the behavior of a driver whose steering wheel is slightly misadjusted.

The second-order interaction between compatibility, internal model, and knowledge of S-R relationship suggests that the compatibility curves for unimodal and bimodal subjects were different, but that they did not become significantly different until their internal models were combined with information about the S-R mapping. In other words, bimodal subjects, whose strategy was closely tied to information about the relationship between joystick and cursor movement, were hindered by the lack of information whereas unimodal subjects employed a strategy that did not rely on S-R information. In particular, there was a large difference in movement times between bimodal subjects with and without information at the 135-degree level of compatibility (34.9 sec. vs. 13 sec.). Since the 135-degree condition was perhaps the most difficult incompatibility to discover, the bimodal subjects, whose strategy implied discovery, needed the relationship to be pointed out to them in order to perform the task adequately. The implication of these results for applied settings is that when faced with an incompatible situation, the operator may be able to improve performance by using an attentional bias, depending upon what information about the task is available.

The failure of the first-order interactions to reach significance may be related to a less than optimal effectiveness of the internal model manipulation. In future studies, subject variability can be reduced by screening subjects for experience in pointing tasks and by making internal model instructions more explicit. These measures could result in significant first-order interactions, for the present data show a trend in that direction. Future research on SRC will include manipulation of such variables as control-display ratio, type of pointing device, and bodily orientation.

ACKNOWLEDGEMENTS

The author acknowledges with thanks the assistance of Professors Allan Pantle and Leonard Mark in providing help, support, and advice throughout both the study and the writing of this paper.

REFERENCES

(1) Canon, L. K., Intermodality inconsistency of input and directed attention as determinants of the nature of adaptation, Journal of Experimental Psychology, 84 (1970), 141-147.

(2) Eberts, R. E., The development of accurate internal models for high order systems, Proceedings of the Human Factors Society, 1982.

(3) Fitts, P. M. and Seeger, C. M., S-R compatibility: spatial characteristics of stimulus and response codes, Journal of Experimental Psychology, 26 (1953), 199-210.

(4) Howard, I. P., Human Visual Orientation, (Wiley, New York, 1981).

(5) Kelso, J. A., Cook, E., Olson, M. E. and Epstein, W., Allocation of attention and the locus of adaptation to displaced vision, Journal of Experimental Psychology: Human Perception and Performance, 3 (1975), 237-245.

(6) Morin, R. E. and Grant, D. A., Learning and performance on a key-pressing task as function of the degree of spatial stimulus-response correspondence, Journal of Experimental Psychology, 49 (1955), 39-46.

(7) Posner, M. I., Nissen, M. J. and Klein, R. M., Visual dominance: an information-processing account of its origins and significance, Psychological Review, 83 (1976), 157-171.

(8) Taub, E., and Goldberg, I. A., Prism adaptation: control of inter-manual transfer by distribution of practice, Science, 180 (1973), 755-756.

V

COMPUTER-HUMAN INTERACTION

Trends in Ergonomics/Human Factors II
R.E. Eberts and C.G. Eberts (Editors)

A TASK ORIENTED HELP SYSTEM: HUMAN FACTORS ISSUES AND ACTIVITIES

Jack A. Alford, Jr., Janet E. Thomas, and Bill Witham

IBM Endicott Development Laboratory
Endicott, New York 13760
U.S.A.

This paper describes a Task Help Menu system developed for IBM's VM/SP operating system, and the development activities directed towards ensuring its ease of use. The Task Menus improve access to command-based Help information for new VM/SP users who may not know the specific command names needed to perform their tasks. Human factors and usability specialists first worked with Developers on the design and organization of the menu set. Then, during human factors testing, subjects used the Task Menus to locate various command Help texts. Subjects were able to find the appropriate command Help texts without major difficulty, indicating that Task Menus can be a useful way for new users to access Help text.

INTRODUCTION

Users of large multi-user computer systems often need information about how to use their system, but may be unwilling or unable to consult reference manuals. Most computer systems provide some form of online information to address this need. Online information can consist of complete copies of user manuals which can be printed on system printers, or tutorials which instruct the user in a step-by-step manner. Alternatively, the information can be structured so as to provide concise answers to a user's specific information requests. This kind of online information is commonly referred to as system Help.

In this paper, we describe a new user interface to the Help system for an IBM product, and present an example of how it is used. We discuss some of the human factors issues related to this Help interface, as well as the activities of human factors and usability specialists during product design and development. We hope this discussion will serve both as an overview of a useful technique for guiding new users to online Help, and as a glimpse into the daily activities of human factors specialists in a large product development laboratory.

CURRENT VM/SP HELP SYSTEM. IBM's Virtual Machine/System Product (VM/SP) is an interactive operating system for medium and large mainframe computers. It contains an extensive set of Help files. VM/SP's Help system is command-based: A request for help results in information for one of several hundred operating system commands. The request is made in one of three ways:

1. Type the word HELP, followed by the correctly-spelled command name. The Help system then displays a portion of a reference manual in which that command is described in detail.

2. Type the word HELP alone. The user then receives a menu that is simply a list of components of VM/SP.

3. Press a terminal function key. The user then receives either the component panel described above, or a menu panel that displays an alphabetized list of commands related to a given component.

To make good use of the VM/SP Help system, then, the user must have enough system knowledge to either request the correct command, or to select the component that holds the correct command. However, new VM/SP users, whether they are novice or experienced computer users, may not be able to use such a system fully, simply because they are not able to access the information within it. That is, these users may have a clear idea of the tasks they want to do, but they are not likely to know the names of the commands which support those tasks.

THE TASK MENUS. The Task Menus[1] were designed to be used by new VM/SP users who need Help information, but who do not know specific commands. They are hierarchically organized, with an initial menu describing very general tasks, such as editing a file and communicating with other users. Making a selection from the initial menu leads the user to a lower-level menu describing more detailed tasks, such as ending an editing session or sending a file to another user. Ultimately, the user is led to the detailed reference information about the specific command.

USING THE TASK MENUS. The following scenario, which was one of ten presented to subjects during human factors testing, illustrates how the Task Menus are used:

> A co-worker told you "I'm sending you a file that you'll need. Look for it in your reader." Please find the HELP file that tells you how to get the file from your reader and put it into your permanent storage.

(The "reader" is a temporary disk storage facility for spool files.) To use the Task Menus, the user merely enters the command "HELP." The following menu is then displayed[2]:

```
==> HELP MENU <=====>        HELP  INFORMATION        <=====>line ===> 1 of 14

To select HELP information on one of the following topics, place the
cursor under any letter of the topic you want. Then press PF 1 or the
ENTER key. Enter 'HELP HELP' to view the format of the CMS HELP command.
Note: Message HELP files are viewed by keying in the message number in
the following manner: 'HELP MSG message number'.

Note: The SQLDS item is operational only if you have the SQL/Data
System (5748-XXJ) Program Product installed on your system.

*CMS       *Debug     *Exec2     *SQLDS     :Tasks
*CP        *Exec      *REXX      *Xedit     :HELP
* * * End of File * * *

1= Help    2= Top    3= Quit   4= Return        5= Clocate      6= ?
7= Backward 8= Forward 9= PFkey 10= Backward 1/2 11= Forward 1/2 12= Cursor

====>
```

The new user who chooses TASKS from this panel sees:

```
==> TASK TASK <=====> HELP  INFORMATION <=====>line ===> 1 of 16
Each VM task below leads to a more detailed
list of tasks for you to choose from.

Move the cursor to the task that you want.
Then press the ENTER key or the PF 1 key.


Create or change (edit) files
Manage files (copy, rename, erase, etc.)
Communicate with other users
Develop and debug programs and EXECs
Ask questions about the system
Customize your system
Print files
Use disks
Use tapes
* * * End of File * * *

1= Help      2= Top     3= Quit   4= Return          5= Clocate       6= ?
7= Backward 8= Forward 9= PFkey 10= Backward 1/2 11= Forward 1/2 12= Cursor

====>
```

From here, the correct Help file can be reached in several different ways. If, for instance, a user conceptualized this task as involving communications between users, then he or she might choose "Communicating with other users." The result of that choice is another menu, with a choice labeled "Work with your reader files." The word "reader" could cue the user that this is an appropriate selection.

Another user might consider this to be a file management task. This user might choose "Manage files," and would be presented with a menu with the choice, "Work with files sent by other users." The result of this choice is the reference text for the RDRLIST command. (This command displays a detailed list of all files currently in the user's reader, and allows the user to transfer a file to the user's personal disk area by pressing a single key.) We observed both strategies in user testing, demonstrating the usefulness of providing multiple paths to the same information, rather than maintaining a strict hierarchy.

DEVELOPMENT ACTIVITIES

Responsibility for developing the Task Menus was shared among several groups. Planners and designers identified user requirements for improvement to the Help system, based on requests from user groups and other sources of information about users. Program developers were responsible for modifying the Help system code to accommodate the Task Menus and other enhancements. Design and wording of the menus was the responsibility of information developers. Development activities were reviewed periodically by a Usability Committee, consisting of a usability coordinator, an information usability specialist, a human factors engineer (the authors), and other members of the development team.

The authors' involvement in menu development began during early design reviews, at which designers explained the specific user needs and requests being addressed, and the software functions to be provided. At this time, there were few constraints on the content of the menus, other than general agreement that they should be task-oriented. A series of Usability Committee meetings were held, in which the first task was to determine who the primary users of the Task Menus are likely to be. It was concluded that new computer users, and experienced computer users who are new to VM/SP, would benefit most from a task-oriented means of accessing Help information. It was also agreed that the overall hierarchy of menus would be based on the design of an existing product, the Interactive Productivity Facility, Release 5 (IPF). IPF is a full-screen, task-oriented interface to VM/SP which had done well in extensive human factors testing some months earlier. This overall organizational scheme was further refined by considering the organization and terminology of the CMS Primer, a highly successful beginner's guide to basic VM/SP use. In addition, data on frequency of command use, gathered in an internal user survey, were used to modify the organization of menus and to help determine the ordering of items in menu lists.

An information developer was then assigned to write the text of the individual menu panels. During this phase of development, the authors served as a review team, evaluating menu drafts and consulting with the developer on such issues as verbal and visual design, and stylistic consistency among menus. After a lengthy review-rewrite cycle, it was agreed that the Task Menus were ready for human factors testing.

HUMAN FACTORS TESTING

Method

Eighteen subjects, college students and part-time IBM secretarial employees, participated in the test. Most had little or no computer experience, other than brief exposure through a programming class, some home computer use, or video games. All subjects were informed of the nature of the study and their rights as human subjects, after which they read and signed an informed consent agreement.

Data collection began after they received background information about computers in general and the Help system in particular, and had practiced using the keyboard. Subjects read a brief description of a task which a new VM/SP user might want to perform. They were instructed to use the Task Menus to locate the Help reference information that would explain how to use VM/SP commands to complete the task. Each task began with the initial (component) Help menu displayed on the terminal. To find the correct Help text, subjects had to make the selection on that menu that would lead to the first Task Menu, and then select from subsequent Task Menus until the correct Help text for a command was displayed. Subjects were not required to perform any of the tasks described on the instruction sheets, or even to read the Help text, but only to locate the correct text panel. Subjects performed as many Help tasks as they could within the time available, up to a maximum of 10. Task instructions were presented in random orders to balance the effects of learning on performance.

Since testing began before software development was complete, a software simulation tool was used to create a realistic simulated Help system. The simulator contained facilities that allowed the authors to create and change menus and logic flows for the simulation quickly and without programming support. During data collection, the simulator also maintained a detailed log of each subject's interactions. Videotaped records of the entire experimental session were also made.

Results

Our general finding was that subjects were able to locate the correct Help reference material with little difficulty. Nine out of ten tasks were completed in less than a median time of 4 minutes. Completion times ranged from 39 seconds for the "Reader Files" task described above, to 9 minutes, 15 seconds for a task involving changing command defaults. Subjects quickly learned to use the menu system, as evidenced by a decrease in median completion time from the first (randomly presented) task to the second of from nearly 7 minutes to less than 3 minutes, with further decreases down to approximately 1 minute per task towards the end of the session.

In addition to completion time, the number of menus and other panels selected and viewed by subjects while attempting to complete a task was used as a measure of ease of use. After calculating the minimum number of panels needed to reach the correct Help file for each task (assuming error-free performance), we then subtracted this number from the mean number of panels actually displayed by our subjects. In this way, we derived a measure of the extent to which subjects made incorrect menu selections or were confused by the Task Menus. This "extra panels" measure ranged from less than one (0.9) unnecessary panel to the "Receive Files" task, up to 13.1 extra panels for the particularly troublesome "Defaults" task. The ordering of tasks on this measure corresponded quite closely to their ordering by completion time, further indicating that some tasks were described clearly by the menus, while other menus needed more work.

A further measure of the task menus' ease of use was a subjective questionnaire, completed by each subject at the end of the session. These subjective responses were generally positive, indicating that the menus were perceived as a useful source of information about VM/SP. However, responses to several items indicated a need to clarify the wording of some menus and to provide more instruction about how to use them. Subjects' written comments reinforced the need for these revisions.

CONCLUSIONS AND POST-TESTING ACTIVITIES

The most clear-cut finding arising from more detailed analysis of the data was that subjects had their greatest difficulties with the initial Help component menu. This menu was not a part of the new Task Menu set being tested, but once it became apparent that it was responsible for an excessive share of task times and errors, one of the authors embarked on an extensive revision which ultimately led to the design shown next.

```
==> HELP TASK <=====> HELP  INFORMATION <=====>line ===> 1 of 18
(c) Copyrighted IBM 1984 (from IBM Form SC19-6209)

To use VM Help, move the cursor to any topic below,
then press the ENTER key or the PF1 key.


TASKS       - Helps if you don't know VM commands.
              Good choice for beginners.
MESSAGE     - Explains how to get help for VM messages.
HELP        - Explains the HELP command.
COMMANDS    - Lists all VM commands that you can use.
CMS         - Shows only CMS commands.
CP          - Shows only CP commands.
XEDIT       - Lists System Product Editor items.
REXX        - Helps you use the REXX language.
EXEC 2      - Shows EXEC 2 language items.
DEBUG       - Helps you debug programs.
SQLDS       - Shows SQL/Data System items.
1= Help     2= Top     3= Quit    4= Return       5= Clocate      6= ?
7= Backward 8= Forward 9= PFkey 10= Backward 1/2 11= Forward 1/2 12= Cursor

====>
```

Most of the other changes that we recommended involved replacing jargon (e.g., "defaults") with text that would be more meaningful to inexperienced users, or clarifying and disambiguating the wording of menu items. We also suggested several changes to the menu hierarchy which resulted in a better match between subjects' conceptualizations of tasks and the organization of the menus. These changes added several new menus to the set.

During this period, the authors worked with the information developer directly, reviewing menus and making revisions as a group. It was also necessary to meet with program developers in order to implement the necessary changes to the main Help panel. Whenever possible, test results were used to justify proposed modifications.

OTHER ISSUES

There are several interesting issues which our study did not address. The first has to do with the amount of Help information which is appropriate for users of varying experience levels. As observed earlier, the Help information provided by VM/SP is quite detailed and technical. Consequently, sophisticated VM/SP users find it very useful. On the other hand, this level of detail, as well as the sheer volume of text, may tend to obscure the basic information that the novice user wants. Although the Task Menus help the novice user to access needed information, the information ultimately received may be more than the novice can assimilate. This possibility was not investigated in our study, since subjects were not required to read the reference text because of limited time.

Another question which we did not study involves the effects on task performance of traversing several Task Menus to reach Help information. It is quite possible that the distraction imposed by the need to navigate through several levels of menus can have negative effects on task performance once the user leaves the Help system and resumes the task. If the user could remain in the task context while accessing and using Help information, there might be less danger of losing track of why Help was being sought in the first place. Again, our study provides no information on this question, since users did not actually perform any tasks, and there was no comparison with any other Help system design.

FOOTNOTES

1 The Task Menus are part of Release 4 of VM/SP, which was announced in August of 1984 and will be available to customers in 1985.

2 This was the original menu design, which was changed as a result of user testing. The revised menu appears in the Conclusions and Post-Testing Activities section.

Trends in Ergonomics/Human Factors II
R.E. Eberts and C.G. Eberts (Editors)

COMPUTATIONAL ROLES FOR HUMANS IN INTERACTIVE DECISION SYSTEMS

Ronald L. Rardin, Ray E. Eberts and Mark A. Paz

School of Industrial Engineering
Purdue University
West Lafayette, Indiana 47907
U.S.A.

It is increasingly common to find interactive, man-computer systems for dealing with large, combinatorial decision problems in engineering and management that are too complex for solution by computer alone. The human has several roles in such systems, including some that are purely computational. Rational design of such roles requires an understanding of computational tasks (if any) at which humans can innately outperform computers. In this paper we evolve a framework for posing experimental hypotheses that could yield insight on such tasks, and then discuss some exploratory experimental results.

INTRODUCTION

An enormous variety of important design, planning and control problems in engineering and management can *in principle* be represented and solved as mathematical optimization problems. However, to do so often poses both an impossibly intricate data assembly task to develop a model suitable for the problem and a hopelessly difficult computational task to solve the derived model. Thus, in recent years such problems have often been addressed via interactive hybrids of computerized analysis and human judgment that we term *interactive decision systems.*

The human's role in an interactive decision system is to overcome the data and computational limits that make exact mathematical resolution impractical. For example, the human may serve as a *partial model substitute* by implicitly introducing historical perspectives, data uncertainties, and tradeoffs among goals into the interactive process of reaching a decision.

Naturally, as much as possible of the needed computation is assigned to the computer part of the hybrid. But even a partial model may pose too great a computational task for any available or anticipated computer. Thus, interactive decision systems often also employ the human as a *partial algorithm substitute,* i.e. they leave for the human some concisely formed, but computationally taxing subtasks. For example, the human may periodically be asked to conceive an escape from a "computational corner" in which a computer algorithm has become trapped or to recognize a "pattern" beyond the repertoire of the computer code.

Are such systems designs rational? To know, we require an answer to the very fundamental question "Are there strictly computational tasks to which humans are innately better adapted than computers?" In this paper we evolve a framework for new research on this deep question and discuss some exploratory experimental results.

PSYCHOLOGICAL RATIONALE

Whether computers can replicate all human intellectual skills is, of course, a well-known and highly controversial issue in artificial intelligence/cognitive science (see for example [2]). We hypothesize that if there are special human computational skills, they may very well be centered in the perceptual and pattern-matching capabilities that have proved so difficult to accomplish with computers.

At least two lines of past psychological research have provided rationales for believing there may be something qualitatively different in the way (at least some) humans accomplish such tasks. One such body of research relates to training of perceptual and pattern-recognition skills. Schneider and Shiffrin [6,7] have observed that experts and novices process information in qualitatively different ways they term *controlled* and *automatic*. Controlled processing is characterized by slow, generally serial, effortful, capacity-limited, human-controlled modes that must be used to deal with novel or inconsistent information. Automatic processing is a fast, parallel, and fairly effortless. It is not limited by short-term memory capacity, is not under direct human control, and performs well-developed skilled behaviors. With training, research has shown that tasks which are controlled in novice subjects can be made automatic in experts.

Another line of research indicating special human skills is experimentation showing that selected individuals are able to process complex stimuli much quicker and more accurately than others. Cooper [1] has classified these two groups as *Type I and Type II processors.* In her studies, she had the subjects perceive random-shaped geometric figures and respond whether two figures were the same or different. She found that for Type I processors, the "same" and "different" responses were quite rapid and did not vary with the complexity of the figure. Type I processors apparently perceived the figures in some "holistic" fashion by matching their perception to a template. Type II processors, on the other hand, had slower responses than those of Type I's, and response varied with the complexity of stimuli. Cooper concluded the Type II's compare figures analytically, feature by feature, until a match is found.

COMPUTATIONAL COMPLEXITY THEORY

The theory of *computational complexity* has evolved over the past decade a fairly complete answer to the related question "To what computational tasks are digital computers well adapted?". This theory (see for example the tutorial in [3,4]) defines a *problem* as a collection of *instances* varying in size and numerical input values. For example the classic *knapsack problem* arises in deciding what to include in a soldier's knapsack. Formally, it is posed

> Given a weight limit W, a value requirement V, and m candidate items with weights $w_1,w_2,\ldots,w_m$ and values $v_1,v_2,\ldots,v_m$, does any subset of the items have total weight at most W and total value at least V?

Each choice of constants V, W, etc. produces a different instance of the knapsack problem.

Observe that instances may have different *size,* i.e. different frequencies and magnitudes of input data items. Since it is reasonable to expect that larger size instances will require more solution time than small ones, complexity theory classifies problems according to the way computation grows with problem size. For example, we say the problem requires *order* $O(n^3)$ if computation grows with the cube of problem size. Computation is measured by the number of elementary steps that must be take by some abstract (but inclusive) model of a digital computer such as a *Turing machine.*

Some problems can the shown to be *undecidable* for digital computers, i.e. not solvable in any finite time. Fortunately, the problems encountered in management and engineering models are not in that most difficult category. They can almost always be solved (in principle) by some form of complete enumeration. For example, an $O(2^m)$ procedure for the knapsack problem is to try all possible subsets.

Such an *exponentially* growing amount of computation is finite, but it becomes prohibitive for even moderately large values of m. Thus, computational complexity theory has come to regard a problem as *tractable* by digital computers only if the required computation is at most of *polynomial* (e.g $O(n^2)$, $O(n^{5.5})$, etc.) order.

The formal theory is incomplete, but the above knapsack problem is regarded as *intractable* by this widely accepted criterion. However, we can obtain a polynomially solvable relative of the knapsack problem by relaxing the implied "all or nothing" character of its solutions. The *continuous knapsack problem* is stated

> Given a weight limit W, a value requirement V, and m candidate items with weights $w_1,w_2,...,w_m$ and values $v_1,v_2,...,v_m$, does any combination of fractions of the items have total weight at most W and total value at least V?

When fractions of items are allowed, the problem is easily solved by choosing items in decreasing order of v_i/w_i until the weight limit is exactly reached (perhaps using only a fraction of the last item chosen). Either this solution produces the required total value, or there is no way to do so.

Turn now to the negatives of problems like the two knapsack forms we have defined. For polynomially solvable cases the available algorithm completely resolves every instance, regardless of whether the answer is "yes" or "no". But formally hard or intractable problems lack (so far as is known) this symmetry of positive and negative instances. Although a complete enumeration will resolve either "yes" or "no" cases, "yes" instances can be settled much more rapidly with a lucky guess. In the knapsack problem, for example, when there is a (non-fractional) solution, we can report "yes" as soon as we stumble on to it. To conclude "no", however, we must essentially check all possibilities.

RESEARCH APPROACH

We propose to study whether humans are subject to the same computational limits as digital computers by comparing observed human performance to that which would be predicted by complexity theory for computers assigned the same tasks. If performance and prediction agree, we have strong evidence for the lack of unique human computational skills; disagreement would suggest such skills do exist.

Our approach is not an easy one because of the "asymptotic" nature of computational orders. At small values of problem size an exponential algorithm can seem much more efficient that a polynomially bounded one. Furthermore, empirically distinguishing between polynomial and exponential time growth would necessitate large and very delicate experiments.

We believe, however, that the "yes"-"no" symmetry issue outlined above may be easier to research. If humans perform like digital computers then they should find positive and negative trials equally difficult when the problems are derived from a polynomially solvable class. On the other hand, when instances are taken from a computer-intractable class, positive trials should require considerably less time to solve than negative ones because the former are resolved as soon as the subject discovers a solution, while the latter require (for computers) some form of exhaustive enumeration.

SOME EXPLORATORY RESULTS

We are only now organizing for serious experimentation on this symmetry question in human computational performance. However, some experimental results originally addressed to another question provide insight on symmetry.

In a series of experiments detailed in Paz[5], subjects were presented with a color graphics display of a planar graph and asked to decide the question

> Does the given graph contain a subgraph homeomorphic to K_4, the complete graph on 4 vertices?

A given graph has this K_4-*homeomorph* property if any four of its vertices (or nodes) are mutually connected by paths that have no intermediate vertices in common. Although certainly a laboratory abstraction, this task is representative of the sort of "find a combinatorial pattern in a given figure" ones that are often dedicated to the human in an interactive decision system.

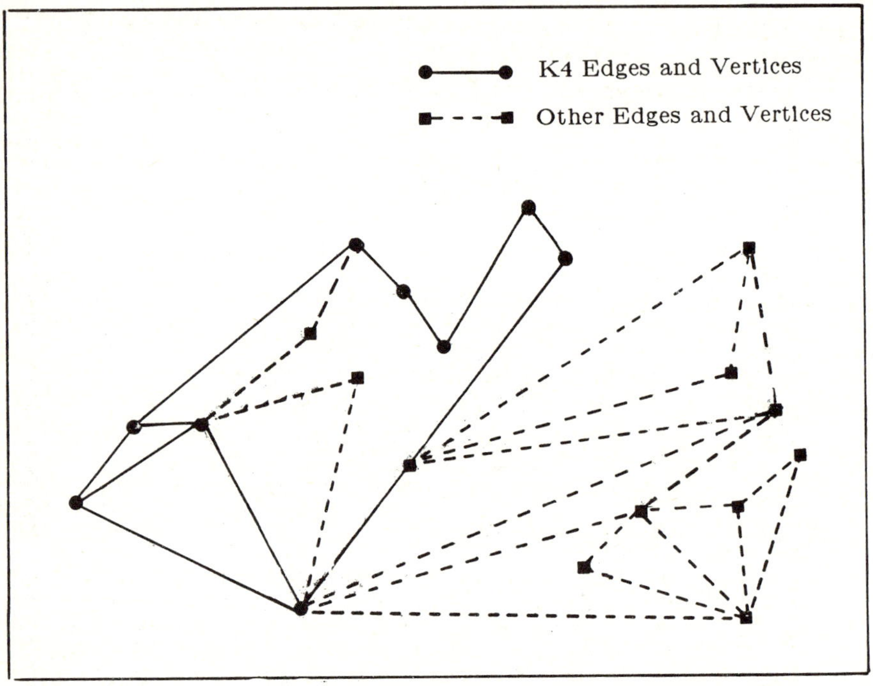

Figure 1. A Graph Containing a K_4-Homeomorph

Figure 1 illustrates for the type of instances actually presented to subjects. The 4 solidly colored vertices are mutually connected by the disjoint paths shown with solid lines. (Of course, the stimulus graph actually presented to subjects had all lines and vertices colored alike.) Thus, the correct answer for the instance in Figure 1 is "yes".

The experiment has several dimensions, but one phase employed 8 subjects, all of whom were graduate or undergraduate students in Industrial Engineering. After a brief introduction to the apparatus and the notion of a K_4-homeomorph, subjects were presented (in several sessions) with 18 blocks of 20 trials each. During the beginning and ending blocks of the sequence subjects also performed a *dual task* designed to require short term memory. Stimulus graphs for the trials were randomly generated and randomly selected in a way assuring a probability of 0.5 that the trial was positive (correct answer "yes").

Both *detection time* and *accuracy* were tabulated automatically by the computer program conducting the trials. After each block subjects were given feedback on their performance and encouraged to work as rapidly as possible while maintaining consistently high accuracy.

Figure 2 plots mean detection times for all 8 subjects by block. Separate means are shown for positive and negative trials.

Whether a given graph contains a K_4-homeomorph can be decided by a known computer algorithm in O(m) time, where m is the number of vertices in the given graph. Thus, the task of this experiment is one tractable to computers, and computers can resolve positive and negative trials with equal ease.

Are the data of Figure 2 consistent with humans having the same computational limits as computers? If so, we would expect times for positive and negative trails to require approximately equal detection times.

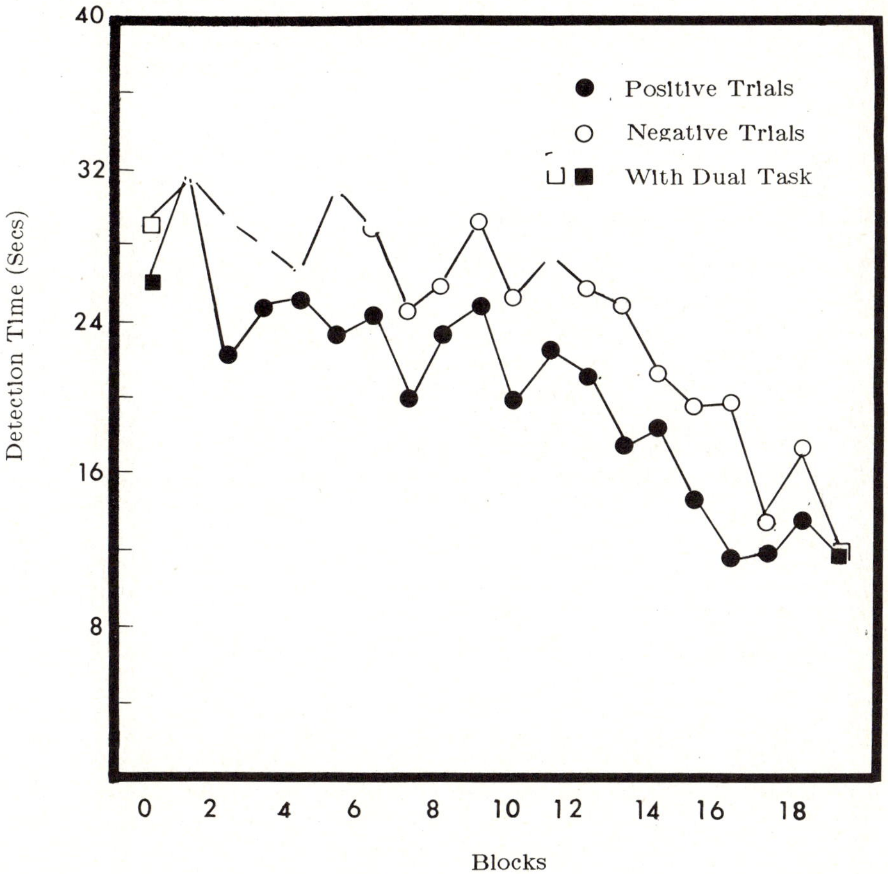

Figure 2. Sample Mean Detection Times for Positive and Negative Trials by Block

The proper conclusion to draw from these data is not entirely clear. Positive and negative trial times do follow parallel learning trends, and the two lines are never very far apart. However, the disparity is (with one early exception) uniformly in the direction of positive trials being faster.

It is particularly interesting that the distinction between positive and negative times virtually disappears in the last block, when subjects were doing both the K_4 task and a competing dual task. Perhaps the pressure on cognitive resources introduced by the dual task forced subjects to rely entirely on whatever polynomial-like mental algorithm they had evolved.

In the light of the overall pattern and this experience with the dual task, we believe these exploratory results at least suggest that humans exhibit some of the same behavior as computers on computational decision tasks computers find tractable. Research now underway is seeking to clarify these findings on a different task and to address the more interesting case of how humans perform on tasks formally intractable to computers.

REFERENCES

1. Cooper, L. A., Individual differences in visual comparison processes, Perception & Psychophysics, 12 (1976) 433-444.
2. Dreyfus, H. L., What computers can't do (Harper Colophon, New York, 1979).
3. Parker, R. G. and Rardin, R. L., An overview of complexity theory in discrete optimization: Part I. Concepts, AIIE Transactions, 14 (1982) 3-10.
4. Parker, R. G. and Rardin, R. L., An overview of complexity theory in discrete optimization: Part II. Results and implications, AIIE Transactions, 14 (1982) 83-89.
5. Paz, M. A., The acquisition of K_4 subgraph recognition skill, M. S. thesis, School of Industrial Engineering, Purdue University (August 1984).
6. Schneider, W. and Shiffrin, R. M., Controlled and automatic human information procession: I. Detection, search, and attention, Psych. Review, 84 (1977) 1-66.
7. Shiffrin, R. M. and Schneider, W., Controlled and automatic human information processing: II. Perceptual learning, automatic attending, and a general theory, Psych. Review, 84 (1977), 127-190.

Trends in Ergonomics/Human Factors II
R.E. Eberts and C.G. Eberts (Editors)

TOWARD AN UNDERSTANDING OF DOCUMENTATION USAGE: REFERENCING AND PROBLEM SOLVING STRATEGIES

Mark W. Scerbo
University of Cincinnati

Arthur D. Fisk and Joseph J. Limanowski
AT&T Communications

The present paper is an initial report on research aimed at determining how technical reference documentation is used to resolve simulated on-the-job problems. A reference document consisting of most reference sources expected to be contained in an efficient computerized referencing system was used in the experiment. The strategies and solution paths of documentation users were examined by videotaping them. The data are described in terms of time users spend in each class of documentation, type of question to be answered, first moves and false start moves. It is argued that if subjects make false starts such that they drift from the relevant problem space, the "heuristic" value of the subjects' strategy is eliminated. The data suggest that "computerized books", to be most efficient, should guide the user into the appropriate solution space.

INTRODUCTION

It is safe to say that interest has spread throughout the human factors community concerning how humans retrieve information from reference documents. This increased interest is due, to a large degree, to interest in on-line (computerized) documentation. Unfortunately, research on documentation retrieval strategies often employs only one class of documentation for a given experiment and often does not closely examine subjects' referencing strategies. The computer user, with an efficient on-line referencing system, should have ready access to indices, menus/contents, glossaries, standard textual reference material, logic/decision trees, etc., when resolving problems.

We addressed the question of how humans retrieve (or search for) information from a solution-oriented reference document by using a source document containing the above mentioned referencing tools (i.e., menus/contents, glossaries, both fact and procedural textual reference material, and logic/decision trees). This paper is an initial report of an on-going investigation that is aimed at identifying how to best structure various reference documentation sources; determining when a particular reference document is most efficient (and why); isolating major patterns of errors users make while resolving problems with the documentation; and developing an understanding of the strategic use of reference documentation.

As more and more reference material is made available on computer systems, it is critical that this material is not simply computer-based imitations of traditional paper books. The most powerful "computerized books" will be the ones that effectively integrate the sophistication of the computer with the information seeking skills of the humans. To answer how these "computerized books" should be designed, we must understand the referencing strategies that humans use when resolving problems. (These "problems" could be answering questions concerning how to engineer a telephone network; where the fifth president of the U.S. was born; etc.) We need to map out when people are successful in finding information (and why) and, as important, why they fail at solving a problem when using reference documentation.

When attempting to solve a problem, a person will be successful to the degree of his/her grasp of the problem space. That is, the person may fail to solve a problem if he or she has: 1) an inferior knowledge of the problem elements; 2) an incorrect understanding of the initial state of the problem or the final goal; and 3) lack of procedural knowledge concerning how to reach the final solution. We will argue, and indeed our research strategy assumes, that book referencing skills (for technical problems) are, in essence, problem solving skills and should follow the same principles.

The primary goal of the research described in this paper was to map out in a broad descriptive sense how people use documentation sources to solve problems and to uncover referencing strategies used to resolve problems. We were particularly interested in the errors made by our subjects. However, our focus was not to investigate how the subjects' referencing skills developed. We attempted to separate systematic errors from careless errors (see Van Lehn, 1983). It is our belief that careless errors will result from an inappropriate application of a heuristic while systematic errors should result from the consistent use of an inappropriate algorithm. Clearly, the strategy that the documentation user is likely to come to use depends on his/her experience in similar "problem domains" and on the structure of the task environment (see Simon, 1979, for a discussion). The task environment should provide the context to drive the user's strategy and substrategy choices. Investigation of systematic errors should be particularly important since it allows an understanding of the user's missing knowledge, incomplete understanding of a particular "system", an inappropriate "mental model", etc.

When a person solves a problem, he or she typically makes numerous choices of the solution path to follow. Newell and Simon (1972) point out that the task environment tends to reduce the space of acceptable (or legal) solution paths. We will seek to explain much of our subjects' error performance in the referencing task by showing that most failures to solve our particular referencing problems are due to repeated "false starts" that drive the person out of the appropriate problem space. The remainder of the chapter contains: 1) a description of the task and the material; 2) a presentation of the method used to gather and score the data; and 3) a discussion of our results.

METHOD

Subjects. Four AT&T Communications employees (two males) were used as subjects. The subjects were given a two-hour training session with the documentation one day prior to the experimental session. All of the subjects were unfamiliar with the documentation and the problems prior to the experiment.

Materials. Five distinct types of documentation were used as stimuli in this study. All of the documentation types were bound together into a single manual and separated from each other by colored and labeled tabs. The first section was an Overview explaining the organization of the manual. The Overview was followed by a table of Contents. The next section was comprised of individual references which provided the user with certain "how to" information. The Reference section was followed by the Workflow section. Each workflow was simply a logical decision tree that supplied the user with "what to do" information. The final section contained a glossary of terms and acronyms found throughout the manual. The manual contained approximately 400 pages.

A total of 20 questions were constructed from the manual. The questions addressed each of the documentation types and were prepared on individual cards. Some of the questions (Constrained) required the subjects to use a specific documentation type while others (Unconstrained) allowed the subjects the freedom to use the documentation type of their choice.

Figure 1 shows an overhead plan of how the studio was arranged for the study. Two cameras were used to videotape the sessions. The primary camera, positioned directly in front of the subject, recorded a head-to-waist view. A secondary camera was used to record page numbers of the documentation. A cameraman ran the second camera to follow the movement of the pages. The two signals generated from each camera were then mixed so that the page information appeared in the upper left-hand corner of the subject and the interviewer. After the sessions had been completed, a video-clock was superimposed on each of the tapes. The clock was accurate to one tenth of a second.

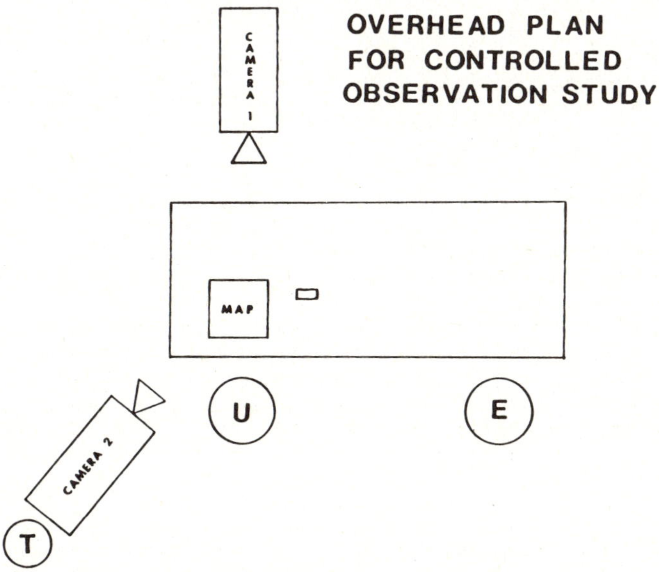

Figure 1

Procedure. The subjects were all tested individually. Each subject sat at a table with the documentation in front of them. The manual was affixed to the table to ensure that the page numbers were recorded. The subjects were told that they would be given a number of questions and to answer them to the best of their ability. They were instructed to use the documentation to answer all of the questions even if they believed they knew the correct answer. In addition, the subjects were asked to verbalize their thought processes or to "think out loud" in an effort to help identify their decision strategies.

The interviewer sat next to the subjects and handed them each of the questions individually. The session commenced with the appearance of the first question and terminated when the subject had answered the last question. The subjects were allowed a maximum of six minutes to answer each question so that the session would not exceed one hour. (Pilot work indicated that six minutes was the maximum amount of time a subject needed to answer a question.) The subjects indicated that they were finished answering by closing the cover of the manual.

Tape Coding. A video-editor was used to help code the tapes. The editor made it possible to view the tapes at variable speeds either forwards or backwards. This provided the capability of tracking gross eye movements and obtaining precise readings from the clock for a particular frame.

The data from the tapes were then coded according to changes in eye position. Four primary categories of eye fixations emerged; Interviewer, Question, Tabs and Manual. Values for the duration of each instance of eye position were determined by the video-clock. Thus, the Interviewer and Question categories refer to portions of time where the subject's eyes were positioned on the interviewer and question, respectively. The Tabs category refers to times where the subject either literally looked at the tabs in the manual or paged through the manual too quickly to obtain the page numbers from the manual. The Manual category refers to times where the subject's eyes were positioned on the manual. This category was further broken down into times where the eyes were positioned on individual pages within each of the documentation types. Finally, all subjects' comments were transcribed and recorded within boundaries of the appropriate fixations.

RESULTS

The session can be broken down into five major areas according to where the subjects' eyes were positioned; the Question, Interviewer, Tabs, Manual, and a miscellaneous category. The major portion (63%) of the session was spent looking at the documentation. Nineteen percent of the time was spent looking at the question and 12% looking at the tabs. Time spent looking at the interviewer only accounted for 2% of the session. The remainder of the session was comprised of times where: the subject looked away; it was not possible to determine where the subject verified the response they had just given.

The time spent looking at the manual can be subdivided according to documentation type. Three sections accounted for 98% of the total time with the largest block (37%) reflecting time spent looking at the Contents section. The times spent looking at the Workflow and Reference sections were approximately the same, 29% and 30%, respectively. The remainder is comprised of time spent primarily looking at the Glossary and Overview.

In examining the answers to the questions, no subject answered all of the questions correctly. (Since some questions had multiple parts, percentages are based upon 25 questions and not 20.) The number of questions answered correctly ranged from a high of 80% to a low of 52% with an average of 63%. There were six questions that every subject answered correctly and two questions that no subject answered correctly. On the average, the subjects voluntarily gave up on about three questions per session.

The particular type of question did not seem to make much difference. The subjects were able to correctly answer 69% of the unconstrained questions and 59% of the constrained questions. The success of answering a question was tied to the first move the subjects made after reading a question. The data indicate that subjects who turned to the Contents section first answered 73% of the questions correctly. However, the success rate dropped to 54% if they turned to any other section first. The 20% advantage for first moves to the Contents section was consistent for both constrained and unconstrained questions.

In addition to the first move data, false starts were examined. There are two actions a subject could take that would be classified as a false start. First, any time a subject consulted with the table of Contents and proceeded to an inappropriate section of the manual, it was categorized as a false start because an incorrect solution path was transversed. Thus, all instances where subjects gave wrong answers, by definition, were classified as false starts. Also, upon initial reading of the question, if a subject went to any section of the manual other than the Contents or Glossary, it was defined as a false start (see optional path discussion below). The only exception to this rule occurred when a subject went immediately to the correct section; however, this happened only five times during the course of the study. The subjects made a total of 79 false starts with an average of about one per question. There were no differences in the number of false starts made on either constrained or unconstrained questions.

<u>Solution Paths</u>. There are a number of ways a subject could obtain the answer to a question using the documentation. One way would be to carefully inspect every page of the manual. This strategy would guarantee an answer to the question (given that the question was thoroughly understood) and would be equivalent to transversing all solution paths until the appropriate answer was located.

A more practical strategy would be to adopt some heuristic. This procedure has the effect of greatly reducing the number of solution paths. While this strategy may not initially yield the correct answer, successive iterations decrease the possibilities until the solution is eventually determined.

Figure 2 is an example of an optimal path or heuristic for answering the question, "How does one determine if a new NI is acceptable?". (Optimal path is defined as the solution requiring the least number of steps.) The subject first reads the question. If the subject is unfamiliar with any terms, the next step is to locate a listing within the table of

Contents. Once the appropriate listing is found, the subject turns to the corresponding pages in the Reference section to locate the answer.

Figure 3 illustrates the actual solution path adopted by the user #1. In this instance, the subject read the question and turned immediately to the table of Contents. The subject then turned to reference #19 which was incorrect. Next, the subject returned to the Contents section and selected Reference #13. Unable to locate the correct answer, the subject began examining successive references to no avail. The subject once more consulted the table of Contents and finally arrived at the solution.

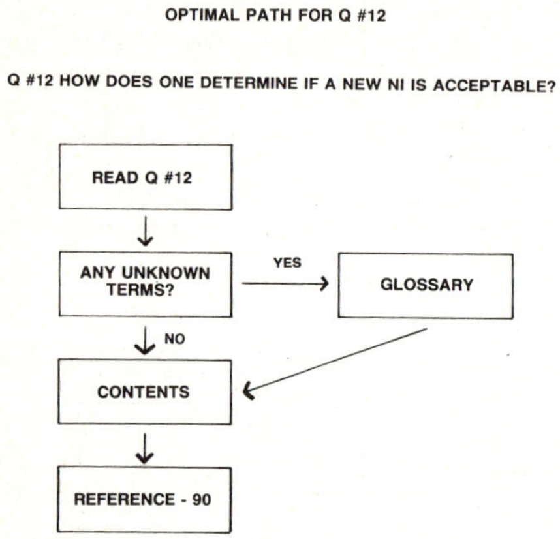

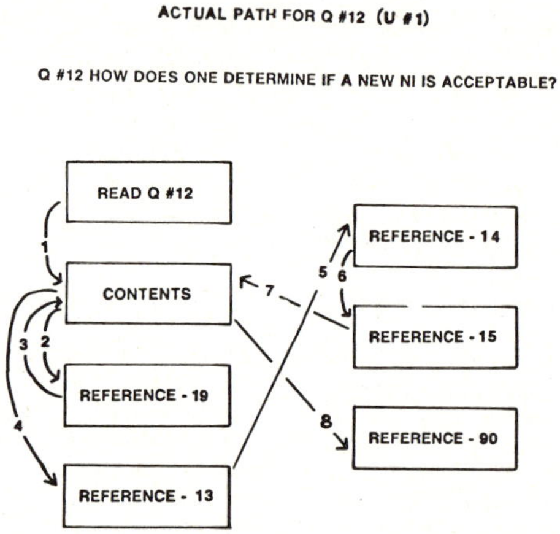

Figure 2 (left); Figure 3 (right)

This example was typical of most users and highlights some key points. First, the user attempted two different strategies. Steps five and six represent the algorithmic approach of examining successive pages to locate the answer. Steps two, four, and eight are iterations of the heuristic, although steps two and four would be considered false starts since they led to incorrect solutions. The heuristic was successful on the first iteration a total of 21 times throughout the study.

The previous example depicted one user's strategy for answering question #12. That question was answered correctly by all of the subjects. Figure 4 illustrates the optimal solution path for question #16, a question that no subject answered correctly.

The actual solution path (of subject #2) is shown in Figure 5. The dominant strategy appears to be examining pages within the Workflow section. The subject only consults the table of Contents on the fifth step and follows eight more incorrect solution paths before finally giving up. This pattern is representative of the other subjects.

DISCUSSION

We have set as our goal the development of an understanding of how a person uses various types of reference documentation to resolve technical problems. This paper is a report on our initial efforts to meet that goal. The present study was an initial attempt to investigate document retrieval strategies. The results are important for a number of reasons. First, the subjects who participated in this study had no prior experience with the documentation. However, they were able to use the present combination of documentation sources to solve problems encountered in real world situations. Thus, with a minimal amount of training, sufficient problem solving strategies were adopted and applied successfully.

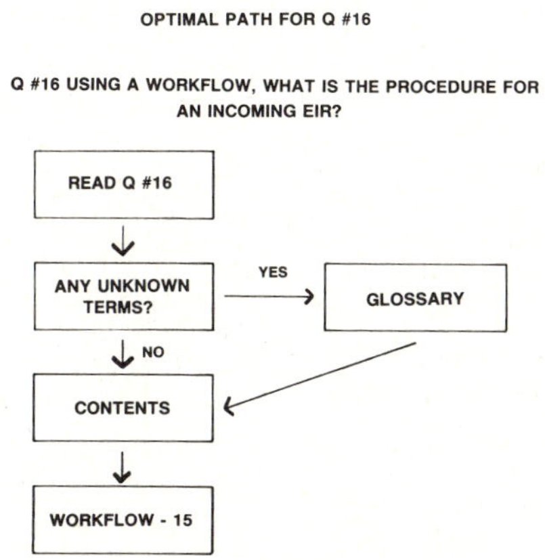

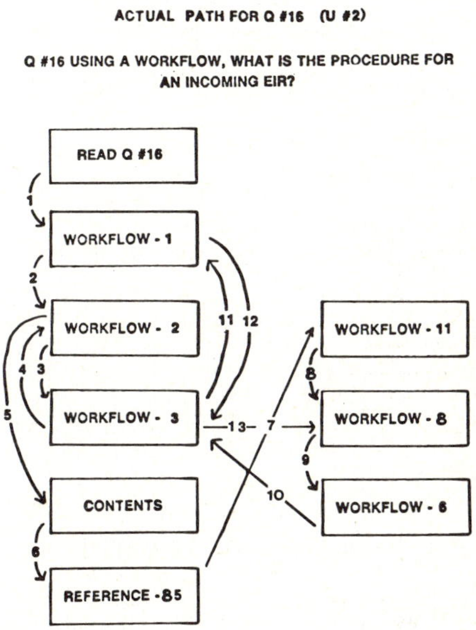

Figure 4 (left); Figure 5 (right)

Secondly, the study illustrated some of the similarities between problem solving and document retrieval strategies. In particular, when an appropriate heuristic was adopted, solutions were found in a minimal number of steps. However, a poor understanding of the problem or use of an inappropriate algorithm resulted in many times more the number of steps needed to reach a solution, if indeed a solution was reached at all. The false start data clearly suggests that "computerized books", to be most effective, should guide the user to the appropriate problem space. (That is, an expert guidance system should be considered if it is critical that a user find the correct answer.)

While these results have added to our understanding of documentation usage, it is clear that more work is needed. We are currently investigating other aspects of retrieval strategies. We have begun to examine strategies and searching patterns of expert documentation users. In addition, we are exploring issues in training; particularly how supplying novices with general heuristics for documentation usage affects their performance, strategic patterns, and very long-term memory of the heuristic training.

REFERENCES

(1) Newell, A., and Simon, H. A., Human Problem Solving, Englewood Cliffs, New Jersey, Prentice-Hall, 1972.

(2) Simon, H. A., Models of Thought, New Haven, Connecticut, Yale University Press, 1979.

(3) Van Lehn, K., Falicity conditions for human skill acquisition: Validating an AI-based theory, (Tech Rep CIS-21), Palo Alto, California, Xerox Palo Alto Research Centers, 1983.

Trends in Ergonomics/Human Factors II
R.E. Eberts and C.G. Eberts (Editors)

ELICITING EXPERT BEHAVIOR FOR SOFTWARE: A COGNITIVE PSYCHOLOGY PERSPECTIVE

Woodrow Barfield and Richard Koubek
School of Industrial Engineering
Purdue University
West Lafayette, Indiana 47907 USA

and

Sheue-Ling Hwang
Department of Industrial Engineering
National Tsing-Hua University
Hsin-Chu, TAIWAN, R.O.C.

The task of knowledge acquisition for the development of an expert system requires the extraction of expert knowledge and also clear definitions of expert and non-expert behavior. This study attempts to identify performance differences among programmers of varying experience levels on programming comprehension/recall tasks and in the mental representation of a programmer's knowledge base. Results indicate that superior expert performance may be due to chunking of programming code into meaningful clusters whereas non-experts store and recall lines of code individually. Implications for knowledge engineering are discussed.

INTRODUCTION

The field of knowledge engineering is concerned with the entire process of building expert knowledge based systems. The knowledge engineer attempts to develop an expert system based on the extraction of knowledge from an expert human. This knowledge is usually in the form of factual information, procedural rules, meta-rules, heuristics and relations among all of these for a specific knowledge domain (Hamill, 1984; Hopkin, 1984).

Currently, extraction of knowledge from an expert human is one of the bottlenecks in developing an expert system (Kidd, 1983; Boose, 1984). Associated with this is the problem of first differentiating between expert and non-expert behavior. This problem is due, in part, to the lack of guidelines and methodologies for extracting expert knowledge and the gaps which remain in our knowledge about human problem solving. This suggests the need for a more quantifiable method of extracting expert knowledge and for distinguishing between expert and non-expert behavior in specific knowledge domains.

Since knowledge based systems seek to implement the problem solving skills of human experts, studies on the cognitive strategies of experts in complex problem domains are particularly relevant. Current research has shown that expert behavior is not based on elaborate algorithms but on the mental storage and use of large incremental catalogues of pattern-based rules (Michie, 1980). Therefore, considerable effort has been expended in attempts to understand the nature and organization of human memory for problem solving and knowledge acquisition (Anderson, 1983; Wilensky, 1983). For example, McKeithen et. al. (1981) has previously shown that experts and novices differ in their ability to process large amounts of meaningful information. Chase and Simon (1973) and Mayer (1979) attributed these differences to an expert's ability to organize and categorize information more efficiently than novices (see also Shneiderman, 1976, 1977).

A particular knowledge domain where experts have been shown to differ from non-experts is in the field of software development. Expert computer programmers can recall far more information than novices can when performing program recall tasks (McKeithen, et al, 1981). Shneiderman (1977) suggests that this difference is based on the manner in which subjects perceive and store program information. While novices remember individual pieces of information, experts organize the information into meaningful chunks. We define a chunk as a meaningful group of items which are recalled together, such as functional units of programming code (see also Davis, 1984).

Adelson (1981), studying expert-novice software differences among programmers found that experts clustered the statements according to the algorithm the statements were composed of, whereas novices used a syntax-based organization. Thus, while novices were attending to the syntactic (surface) structure of the statement, the experts considered the statements at deeper levels of its algorithmic function (Curtis, et. al., 1984). Furthermore, Shneiderman (1976) found that experienced programmers could recall more lines of programming code than inexperienced programmers when the program was in executable order. However, when the program consisted of random lines of code, the two groups performed at similar levels.

Thus, insight may be gained into the cognitive processes of programming by examining the differences in recall for experts versus non-experts during the memorization and recall of a program. In addition, knowledge of how experts comprehend and organize functional units of programming code during recall may aid in the development of an expert system for software. Therefore, a memorization task was used in the following study to assess the chunking patterns of experts and non-experts in a software knowledge-base domain.

METHOD

Experimental Design

A Posttest Only Different Treatment Design was used. Programmer experience, at four levels (naive, novice, intermediate, expert), and program organization at three levels (executable order, random chunks, random lines) served as independent variables in the experiment. All factors were between-subject variables.

Subjects

A total of 221 students from Purdue University participated as volunteers in the study. The students were classified according to their previous experience in BASIC programming based on the following scheme (Shneiderman, 1976). Subjects with no previous programming course were classified as Naive (N=42) while those with one course in BASIC were placed into the Novice group (N=80). Subjects with at least one course in BASIC and two or three additional computer science courses were classified as Intermediate level programmers (N=73) while those programmers with at least one course in BASIC and either graduate student in Computer Science or faculty member status in the Computer Science Department were considered Experts (N=26).

Materials

Subjects completed a background form in which they identified their present educational level, the number of BASIC courses taken, and any additional computer science courses completed.

The 25 line BASIC program used as stimulus material was written to facilitate chunking for experienced programmers. The chunks presented in Table 1 have the following functions. Lines 10-30 initialize the arrays and variables. Lines 40-70 and lines 80-150 use for-next loops to input and sort data. Lines 160-170 compute the median and lines 180-200 use a for-next loop to print the values. Lines 210-230 compute the mean and print the mean and the median. Finally, lines 240-250 end the BASIC program.

The program was organized as test stimuli in the following three ways: (1) executable order, (2) random chunks, in which the chunks of the program were completely randomized (see Table 1 for an example of the program chunks), and (3) random lines, in which the lines of code were completely randomized. Original indentation of each line was left intact for each condition.

Procedure

All subjects were tested in groups which consisted of different combinations of experience levels. Subjects first completed the background form and were then given one treatment condition of the 25 line BASIC program. They were allowed five minutes to memorize the program and six minutes to write down their recall of what they had memorized.

RESULTS

In general, more lines and chunks are recalled as level of programming experience increases for all levels of program organization. In addition, line recall is superior when the program is presented in executable order than when the chunks or lines of the program are randomized. Line recall for the random chunk condition is also superior to the random line condition. In general, chunk recall is superior when the chunks are presented in executable order rather than in random order for the intermediate and expert BASIC programmers.

Figure 1 shows the number of lines recalled for each experience group across each level of program organization. The arcsine transformation was used for the number of lines recalled in order to meet the assumptions for ANOVA. The two-way interaction between program organization and experience level was very significant ($F(6,209)=7.3$, $p<.001$). This interaction suggests superior line recall among expert programmers when the program is organized in executable order and also when the chunks are presented in random order. The ANOVA for lines recalled revealed significant main effects for level of experience, ($F(3,209)=66.1$, $p<.001$), and program organization, ($F(2,209)=27.7$, $p<.001$).

Figure 2 shows the number of chunks recalled for each experience group across the two levels of chunk organization. An ANOVA revealed a significant two-way interaction between program organization and experience level ($F(3,155)=9.8$, $p<.001$). This interaction suggests that chunk recall was significantly greater for the Expert and Intermediate experience groups than for the Naive and Novice groups. An ANOVA for number of chunks recalled revealed a significant main effect for experience level ($F(3,155)=188.6$, $p<.001$), but no effect for program organization ($F(1,155)=<1$, $p>.05$).

Table 1 shows the proportion of total chunks which were recalled by each experience group when the chunks were presented in executable and randomized order. When the program was presented in executable order, clear differences in chunk recall arise between the groups. Naive and Novice programmers tend to recall only the first and last chunks presented in the program. As experience level increases, a greater number of subjects recall the additional program chunks. A similar recall pattern appears when the chunks of the program are randomized.

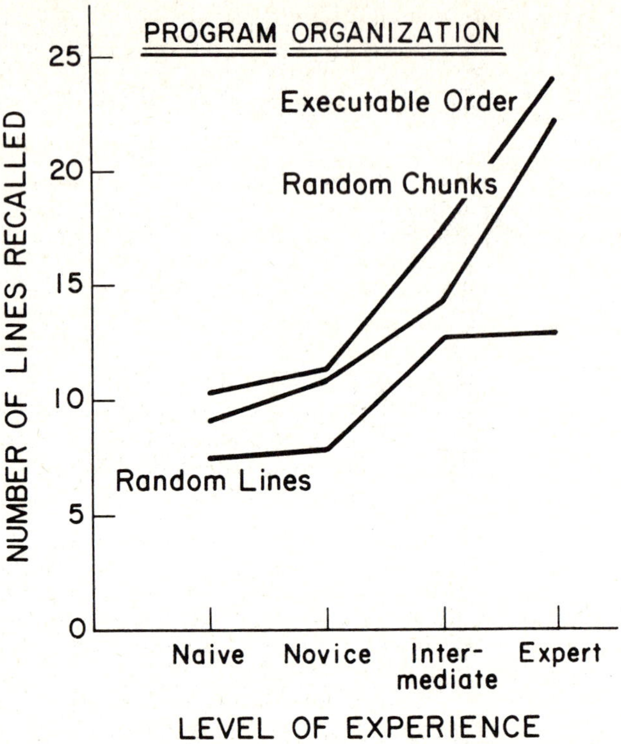

Figure 1. Number of lines recalled across experience level and program organization.

IMPLICATIONS FOR KNOWLEDGE ENGINEERING

The results described above show characteristic differences in program recall for expert versus non-expert programmers. Experts are able to recall significantly more of the program due to their ability to recognize and cluster meaningful groups of statements into chunks. When the lines of the program are completely randomized, the expert's performance drops to the level of the intermediate programmer. Since the mental capacity of the experts is the same as that of the other groups, experts' superior recall in the executable order and random chunk conditions may be due to the process of chunking programming code into meaningful clusters. These clusters, or chunks, appear to ease the burden of memorization and/or recall.

We believe that this ability to chunk information is useful in separating expert and novice performance in complex problem solving domains (see Egan et al., 1979). These results are generally supported by the previous findings of Shneiderman (1976, 1977) and McKeithen et al. (1981).

A number of tentative hypotheses for knowledge engineering can be made based on the results of the above study. First, the ability to chunk meaningful information may be used to distinguish the performance of experts and non-experts in a specific knowledge domain. Second, results may provide us with an indication of how

Expert's organize the information in their knowledge-base. Third, the ability to chunk domain specific knowledge may be used to identify and select candidates for knowledge engineering and fourth, an expert system based on the chunks derived from experts may be compatible with the mental model end-users form of these systems.

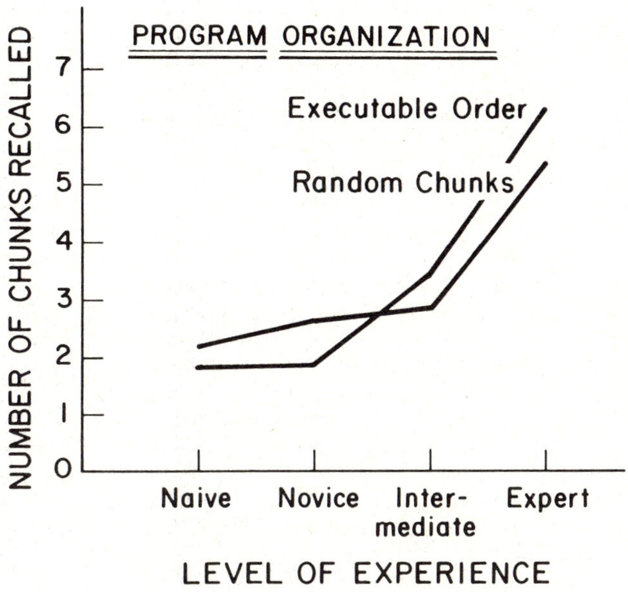

Figure 2. Number of chunks recalled across experience level and program organization.

Acknolwedgement

This study was completed while the first author was an IBM Fellow at the School of Industrial Engineering, Purdue University. Special thanks go to Stuart Weiss and Jack Posey for providing class time for experimentation.

RESEARCH NOTE: A positive correlation ($r=.78$, $p < .01$) was found between chunk recall and program comprehension for the intermediate level of experience group when the program was organized in executable order. There was no significant relationship between chunk recall and program comprehension for the naive and novice groups when the program was presented in executable order. Program comprehension was not tested for any other condition or group.

Table 1. Proportion of total chunks recalled for each experience group across the two levels of chunk organization.

BASIC Program	Executable Order – Experience Level: Expert	Intermediate	Novice	Naive	Randomized Chunks – Experience Level: Expert	Intermediate	Novice	Naive	Random Chunk Organization
10 DIM X(100) 20 INPUT N 30 S=0	100%	78%	71%	79%	71%	21%	39%	28%	(7)
40 FOR I=1 TO N 50 INPUT X(I) 60 S=S+X(I) 70 NEXT I	91%	38%	27%	00%	86%	26%	17%	00%	(3)
80 FOR I=1 TO N-1 90 FOR J=I+1 TO N 100 IF X(J) > X(I) THEN GOTO 140 110 T=X(J) 120 X(J)=X(I) 130 X(I)=T 140 NEXT J 150 NEXT I	100%	69%	14%	00%	71%	21%	05%	0%	(4)
160 M=INT((N+.5)/2) 170 MD=X(M)	82%	31%	02%	00%	71%	32%	44%	28%	(2)
180 FOR I=1 TO N 190 PRINT X(I) 200 NEXT I	82%	54%	02%	00%	86%	38%	05%	05%	(5)
210 A=S/N 220 PRINT A 230 PRINT MD	91%	31%	08%	07%	71%	76%	67%	67%	(1)
240 STOP 250 END	91%	62%	59%	93%	86%	71%	78%	89%	(6)

REFERENCES

Adelson, B., 1981. Problem solving and the development of abstract categories in programming languages, *Memory and Cognition, Vol. 9,* 422-433.

Anderson, J.R., 1983. *The architecture of cognition,* Cambridge, Massachusetts: Harvard University Press, 1.

Boose, J., 1984. A framework for transferring human expertise, in *Human-Computer Interaction,* (Ed. by Gavriel Salvendy), Elsevier Science Publisher: Amsterdam, 247.

Chase, W.G., and Simon, H.A., 1973. Perception in chess, *Cognitive Psychology, Vol. 4,* 55.

Curtis, B., Forman, I., Brooks, R., Sloiwa, Y.E., and Ehrlich, K., 1984. Psychological perspectives for software science, *Information Processing and Management, Vol. 20,* 81.

Davis, J.S., 1984. Chunks: A basis for complexity measurement, *Information Processing and Management, Vol. 20,* No. 1, 119.

Egan, D.E., and Schwartz, B.J., 1979. Chunking in recall of symbolic drawings, *Memory and Cognition, Vol. 7,* 149.

Hamill, B.W., 1984. Psychological issues in the design of expert systems, *Proceedings of the Human Factors Society-28th Annual Meeting,* San Antonio, Texas, 73.

Hopkin, V.D., 1984. Some human-factors implications of expert systems, *Behavior and Information Technology, Vol. 3,* 79.

Kidd, A.L., 1983. Human factors in expert systems, *Proceedings of the Ergonomic Society's Conference,* (Ed. by K. Coombes), Taylor and Francis, New York, 37.

Mayer, R.E., 1979. A psychology of learning BASIC, *Communications of the ACM, Vol. 22,* 589.

McKeithen, K.B., Reitman, J.S., Reuter, H.H., and Hirtle, S.C., 1981. Knowledge organization and skill differences in computer programmers, *Cognitive Psychology, Vol. 13,* 307.

Michie, D., 1980. Expert systems, *The Computer Journal, Vol. 23,* 369.

Shneiderman, B., 1976. Exploratory experiments in programmer behavior, *International Journal of Computer and Information Sciences, Vol. 5,* 123.

Shneiderman, B., 1977. Measuring computer program quality and comprehension, *International Journal of Man-Machine Studies, Vol. 9,* 46.

Wilensky, R., 1983. *Planning and Understanding: A computational approach to human reasoning,* Reading, Masschusetts: Addison-Wesley.

Trends in Ergonomics/Human Factors II
R.E. Eberts and C.G. Eberts (Editors)

A STUDY OF VDT CHARACTER LEGIBILITY AS A FUNCTION OF SOME RELATED PARAMETERS

ABDUL RAOUF

DEPARTMENT OF SYSTEMS & INDUSTRIAL ENGINEERING
UNIVERSITY OF PETROLEUM & MINERALS
DHAHRAN, SAUDI ARABIA

SASAN HATAMI

DEPARTMENT OF INDUSTRIAL ENGINEEERING
GENERAL MOTORS OF CANADA LIMITED
ST. CATHARINES, ONTARIO
CANADA

VDT (Visual Display Terminal) character legibility was studied in a contextual form as a function of character generation method, width to height ratio, and image polarity, by evaluating discriminability among 5 pairs of upper-case characters with similar appearances. A paced reading-searching task was used to simulate text proof reading situations. Method of generation was found to have a significant impact on legibility of some of the tested characters. Legibility was unaffected by image polarity. The effect of character width to height ratio was significant in case of some characters but it was not consistent for all characters.

INTRODUCTION

Recent research about Visual Display Terminals (VDTs) and man-VDT interaction has identified visual work to be the single most important element of almost all VDT tasks. Visual work in the form of reading, searching or scanning is by far the most common element of such tasks. Display legibility is, therefore, one of the most important criteria by which the merits of a VDT are judged. Poor legibility can have negative consequences on the ability of the operator to carry out the work. At the simplest level and provided that they are not too frequent, errors are a source of inconvenience and frustration. At a higher level, such as billing, credit checking, etc., the consequence of errors can become more serious and costly.

The legibility of alphanumeric symbols is usually measured in terms of the confusion - or lack of confusion - that occurs in recognizing and discriminating between individual characters in a display. As a result character sets ability in preserving fine differences in stroke width, curvature, etc., is particularly important in order to avoid similarities among characters.

Factors which may affect character legibility and thereby affect workers visual performance and perhaps comfort in VDT tasks are numerous. Among them to name a few, are character luminance, character font, stroke width, contrast and size (McCormick and Sanders, 1982; Snyder and Taylor, 1979; Vartabedian, 1971a & b). There exists in literature a number of studies pertaining to the study of some of these factors. Snyder and Taylor (1979) concluded that at viewing distances of less than 1.5 meters, character size, dot luminance, and viewing distance have little effect on VDT character recognition accuracy. Vartabedian (1971b) found upright symbols superior to 20 degrees slanted character font. In studying legibility parameters, most researchers have used tachistoscopic recognition technique in non contextual forms where symbols are presented briefly and individually. This technique has the disadvantage that it does not provide the necessary legibility information for VDT related tasks. This is due to the fact that the legibility of alphanumeric characters is affected by the information content of characters when presented in the form of words or character strings, rather than as isolated or unrelated characters (Snyder and Taylor, 1979; Brillouin, 1963).

The significance of symbol legibility, aside from affecting productivity of man-VDT interaction is that the more difficult it is to discriminate among letters, more stress is likely to be exerted upon users eyes. Although some degree of visual discomfort and stress is inherent in any task involving extensive visual work, poor legibility of VDT symbols may also become a contributing factor. Character generation method is an important factor as almost all VDTs use dot matrix method instead of conventional stroke method used in printing on paper. The reasons for popularity of dot matrix, over several other commercial methods of generating symbols lie in being both more economical and more versatile (Vartabedian, 1971).

Dot matrix character generation is the technique which is most commonly used for many VDTs using the raster scanning principle. By this method, the required dot positions for each character are stored in a dot matrix memory which is incorporated in the refresh store and coupled to the scanning system of the VDT. Each character is defined by a set of dots from a rectangular matrix of dots as the electron beam scans through its raster pattern. The resolution of the matrix is defined by the number of horizontal and vertical dots which determine the maximum possible horizontal and vertical extends of the character image. (Cakir et al, 1980)

It is generally believed that character definition (or legibility) improves as the density of the dot matrix increases. There are dot matrices of various densities (5x7, 7x7, 7x9, etc.) for certain characters, such as those having contours (e.g. B or C) or inclined segments (e.g. K or Y), however, lower densities may not provide the necessary legibility. Designing characters with higher resolutions, on the other hand, may result in unnecessary extra costs.

METHOD

Four different displays were used; specifications and description of which are summarised in Table-1.

DISPLAY	MODEL	MANUFACTURER	GENERATION METHOD	APPROXIMATE WIDTH TO HEIGHT RATIO	IMAGE POLARITY
Hard Copy	-	-	Stroke	1/1	+ve
VDT	VT-52	DEC	7x7 Matrix	2/3	-ve
VDT	TV-910	Tele-	7x7 Matrix	1/2	+ve
VDT	TV-910	Video	7x7 Matrix	1/2	-ve

TABLE 1: Specifications (in terms of independent variables of study) of displays used in the study.

10 male university students whose first language was English, in the age group of 22-28 and with no apparent physical handicap participated in these experiments. Legibility was studied through letter substitution and

subjects were required to search for errors in a paced visual search task. The text was in upper case and consisted of 48 pages; each page containing 20 lines. Subjects had 45 seconds for each page. The objective in allowing a limited time was to simulate a systematic proof-reading condition (i.e., a reading or reading-scanning visual inspection strategy) rather than a random search. Therefore the time had to be long enough, on one hand, for subjects to detect and verify the errors and short enough on the other to force them to adopt a reading-searching strategy where errors could be detected on the basis of recognition of words in which errors were embeded rather than close inspection of characters.

Two factors were taken into account in setting 45 seconds as the allowed time for processing each page. First, a reading speed of approximately 320 words per minute obtained by Lang (1969) in reading a lower case material was adopted as the average reading speed for the purpose of the study. The test used in the experiments contained, on an average, approximately 190-205 words per page. Second, since the text was presented in upper case letters more time was allowed as lower case texts are read faster (as much as 13.4%) than that in all capitals (Tinker and Paterson, 1928).

Timing in case of hard copy display was done manually using a digital stop watch and in the case of VDTs (the other three displays) a PDP/11 mini computer was used in displaying and timing each page. Subjects had no prior knowledge of the type of error they would encounter but they were told that only one of the six types of errors, presented in Table-2, could appear in any one page at a time; they were required to record the type of error and number of times it was found in that page on the sheets provided. Each pair was presented in eight different pages and for a total number of 20 of each type of error. The first five pairs were used in discriminability study of like characters. These were chosen on the basis of similarity of their appearances and are shown in Figure-1 as they appeared on the displays used. The last one, named CR (Character Repeated) errors were repetition of one of the characters in a word whose correct spelling would be known to subjects.

TYPE OF ERROR	EXAMPLE
I - 1	T1ME
K - X	EKAMPLE
S - 5	WA5
D - 0	HDUR
U - V	MANVAL
CR	AAND

TABLE 2: Character pairs used as errors in study of discriminability

(CR) errors were included for two reasons: (a) to determine the search strategy adopted by subjects. If the material was to be read, repetition of one of the characters would be easily detectable as its structure and shape would appear different to the reader's eyes. If the material, on the other hand, was to be scanned in search of any of the other types of error, (CR) errors may not be so easily identifiable. (b) Since factors other than character legibility such as inter-character and inter-line spacings may also have some effect on performance, (CR) errors could point out possible differences in performance due to above factors as detection of (CR) errors is independent of their visual appearance or legibility.

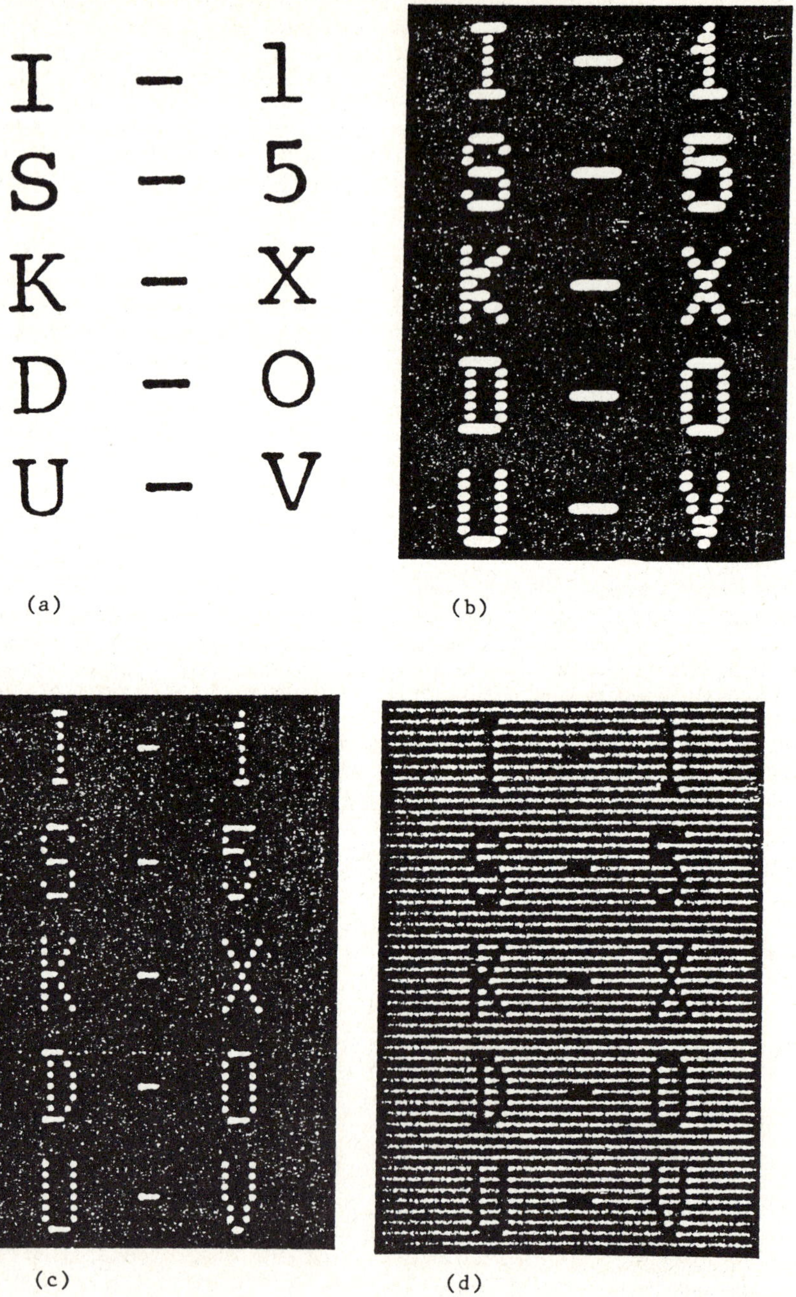

FIGURE 1: Photographs of character pairs on:
a) Hard-copy paper
b) VT-52 VDT
c) TV-910 VDT negative polarity
d) TV-910 VDT positive polarity

Each experimental session lasted for approximately 50 minutes. A 5 minute practice preceded each session so that subjects could adjust their speed within the limited time allowed for each page. to provide an assessment of fatigue as a result of 50 minutes of paced visual search Critical Fusion Frequency (CFF) tests were performed before and after each session. Measurements of CFF values were performed using an OG-Giken (Model CE-1) digital flicker. This device used a light source of 2.5 mm in diameter with a brightness of 120 cd/m^2 on a peripheral background with a brightness of 25 cd/m^2.

ANALYSIS, RESULTS AND DISCUSSION

In statistical analysis of all data, a 5% level of significance is used. Results are reported in two parts; first results of character legibility as indicated by subjects' performance and second results of CFF tests as a measure of induced fatigue.

CHARACTER LEGIBILITY

Obtained data were analysed according to a mixed-nested factorial experimental design with subjects (SUB) effect treated as random and type of error (BCD) nested in display (DISPLAY). Type of error was treated as a nested effect since levels of (BCD) did not cross those of display effect as appearance of characters was unique for each display type. Table-3 shows the results of ANOVA for the dependent variable performance. All main effects and their interactions were found significant. Significance of the factor (SUB) points out differences among subjects in their overall performance in detecting errors. Further analysis of data showed this difference to be significant only when working on VDTs. This can be due to the varying levels of subjects' familiarity with VDT screens' particulars. Significance of the factor (DISPLAY) means that there were significant differences among the four displays in terms of their overall legibility. Similarly, it can be infered from significance of factor (BCD) that types of errors used in the experiment were not of the same level of difficulty (I.E., it was easier to discriminate between certain character pairs than others). Further interpretation of these effects as well as their interactions is provided in analysis of data for each type of error. Duncan's multiple range test was performed and statistical ranking of displays in terms of their overall legibility (based on the 6 types of error tested) is shown in Table-4. Overall legibility of the first three displays were statistically the same, while the last one (i.e., VDT4) which used a positive polarity showed a significantly poorer legibility.

SOURCE	df	SS	F VALUE	FINDING
SUB	9	2.600	3.34	Sig.
DISPLAY	3	2.996	4.31	Sig.
SUB*DISPLAY	27	6.253	2.67	Sig.
BCD	20	42.329	19.20	Sig.
SUB*BCD	180	19.846	1.27	Sig.
ERROR	1680	145.549		

TABLE 3: ANOVA for results of performance evaluation

MEAN (%)	DISPLAY	GROUPING
76.04	PAPER	A
74.81	VDT2	A
72.23	VDT3	A
65.81	VDT4	B

KEY: VDT2: VT-52 VDT, WITH NEGATIVE POLARITY
VDT3 TV-910 VDT, WITH NEGATIVE POLARITY
VDT4: TV-910 VDT, WITH POSITIVE POLARITY

TABLE 4: Statistical ranking of displays for their overall legibility. Means with the same letter are not statistically different at 5% level of signaificance

Duncan's multiple range test is performed for each dispaly and results are shown in order of discriminability of each character pair in Table-5. From results of the statistical ranking of the character pairs two conclusions may be drawn. First, the order of discriminability of character pairs is not the same on all displays and thus suggest significant legibility differences among the four displays. For example the legibility of characters (I) and (1) was very poor on the hard copy display while subjects found it relatively easier (with respect to other types of error) to discriminate between (I) and (1) on VT-52 display. Second, the relative position of (CR) errors in above rankings shows that the objectives of the experiment in simulating proof reading situations is achieved as detection of (CR) errors is not the easiest and thus suggests a reading-scanning visual search strategy.

(a)

BCD	MEAN (%)	GROUPING
S-5	93.76	A
K-X	89.43	A
CR	77.96	B
U-V	77.82	B
D-0	77.58	B
I-1	39.69	C

(b)

BCD	MEAN (%)	GROUPING
K-X	87.49	A
S-5	83.66	A B
CR	76.50	C B
I-1	73.45	C
U-V	72.61	C
D-0	55.11	D

(c)

BCD	MEAN (%)	GROUPING
K-X	91.89	A
S-5	84.30	A
CR	73.38	B
D-O	63.58	C
I-1	61.08	C
U-V	59.16	C

(d)

BCD	MEAN (%)	GROUPING
K-X	90.81	A
CR	78.55	B
S-5	75.34	B
U-V	60.43	C
D-0	53.28	C
I-1	36.44	D

TABLE 5: Statistical rankings of discriminability among character pairs for:
a) Hard copy display
b) VT-52 VDT
c) TV-910 VDT with negative polarity
d) TV-910 VDT with positive polarity
Means with the same letter are not statistically different at 5% level of significance

In analysing significance of the factor display; Duncan's multiple range test, presented in Table-6 shows how the four displays varied from one another in terms of legibility of the tested characters. With the exception of I-1 character pair all other character pairs were found statistically either more discriminable or equally discriminable from one another on the hard copy display. Performance in detecting (CR) errors is shown to be the same on all four displays. As mentioned earlier, detection of CR errors was independent of the characters' appearance and hence it can be infered that factors other than character legibility (such as inter-character and inter-line spacings) had insignificant effects on subjects' performance and thus any variations in subjects' performance in detection of errors on different displays is merely due to the legibility of their characters. Furthermore, as seen from Table-6, comparison of the two different polarities of TV-910 VDT display (i.e., VDT3 and VDT4), indicates a significant difference between the two displays in case of two character pairs. These two displays differed from one another only by the polarity of their images. Specifically, discriminability between I and l, and S and 5 were significantly poorer when working on the positive polarity of the display.

(a)

DISPLAY	MEAN (%)	GROUPING
VDT2	73.46	A
VDT3	61.08	B
PAPER	39.69	C
VDT4	36.44	C

(b)

DISPLAY	MEAN (%)	GROUPING
PAPER	93.76	A
VDT3	84.30	B
VDT2	83.66	B
VDT4	75.34	C

(c)

DISPLAY	MEAN (%)	GROUPING
VDT3	91.89	A
VDT4	90.81	A
PAPER	89.42	A
VDT2	87.49	A

(d)

DISPLAY	MEAN (%)	GROUPING
PAPER	77.83	A
VDT2	72.61	A
VDT4	60.43	B
VDT3	59.16	B

(e)

DISPLAY	MEAN (%)	GROUPING
PAPER	77.58	A
VDT3	63.58	B
VDT2	55.11	B
VDT4	53.28	B

(f)

DISPLAY	MEAN (%)	GROUPING
VDT4	78.55	A
PAPER	77.96	A
VDT2	76.50	A
VDT3	73.38	A

LEGEND:

PAPER:	HARD COPY DISPLAY
VDT2:	VT-52 VDT, WITH NEGATIVE POLARITY
VDT3:	TV-910 VDT, WITH NEGATIVE POLARITY
VDT4:	TV-910 VDT, WITH POSITIVE POLARITY

TABLE 6: Statistical rankings of displays for each character pair of: a) I & l, b) S & 5, c) K & X, d) U & V, e) D & O and f) CR. Means with the same letter are not statistically different at 5% level of significance.

Comparison of results for the two displays of VT-52 and TV-910 (both with negative polarity) VDTs shows that there was a significant difference

between the two displays in the case of I-1 and U-V error types. These displays, both had a negative image polarity and employed a 7x7 dot matrix. They differed, however, from one another in their character width to height ratio. This suggests that character width to height ratio, may affect the legibility of some characters.
It appears, however, that the effects that character width to height ratio may have on legibility of characters depends itself on characters' appearance and shape and may not be a consistent effect. As a result, for example, a different character width to height ratio may improve legibility of certain characters while degrading definition of some others in the same alphanumeric set.

ANALYSIS OF INDUCED FATIGUE

Results of flicker tests were analysed according to a mixed factorial experimental design, with subject's effect treated as random. Table-7 displays ANOVA results. Percent change in mean CFF values, for all displays, and from pre to post experiment was approximately % 2.26.

Subjects showed signficant differences in their CFF readings. This is normal as individuals have varying levels of sensitivitity to flicker frequency (Simonson and Brozek, 1952). Interaction of subjects by the other two main effects of condition (COND) (i.e., before or after experimental values) and type of display (DISPLAY) as well as the three way interaction were significant. These imply that certain subjects found the task (SUB*COND), or certain display(s) (SUB*DISPLAY), or their crossed effect (SUB*DISPLAY*COND) more fatiguing than average. Effects due to factor (COND) were found to be significant and thus imply measurable levels of fatigue as a result of 50 minutes of paced visual work. Levels of fatigue however, for each display were the same for all displays as the contribution of factor (DISPLAY) is insignificant. This therefore suggests that VDT usage may not be necessarily more fatiguing than conventional paper material usage for the same level of involving work and time duration.

SOURCE	df	SS	F VALUE	FINDING
SUB	9	3315.858	213.54	Sig.
COND	1	77.602	14.56	Sig.
SUB*COND	9	47.963	3.09	Sig.
DISPLAY	3	26.442	1.01	-
SUB*DISPLAY	27	235.554	5.06	Sig.
COND*DISPLAY	3	4.853	0.94	-
SUB*DISPLAY*COND	27	151.801	3.26	Sig.
ERROR	400	690.127		

TABLE 7: ANOVA for analysis of CFF readings

CONCLUSIONS

Character generation method is a significant factor in character definition. Some of the tested characters in this study showed to be more legible when generated by stroke method than by a 7x7 dot matrix. A 7x7 dot matrix, however, is suggested to be sufficient for generation of uppercase letters as the difference between the two generation methods in their overall legibility of the tested characters was statistically insignificant.

Character width to height ratio is also a significant factor in character definitiion. two displays with the same polarity and same dot matrix resolutions but with different character wdith to hegith ratios, showed significant differences in legibility of 2 of the character pairs tested. No general conclusion can, however, be stated as the possible effects of this factor appears to depend on the characters shape and pecularities. Raster lines in positive image polarities degrade the legibility of characters. The display with positive polarity in this study showed to be the worst among the four tested displays on the basis of its overall legibility.

As a result of 50 minutes of paced visual work, a small but significant decline in flicker frequency was observed. The difference between displays, however, was not significant. Thus it can be concluded that, for short time periods, reading from a VDT screen may not be necessarily more stressful or fatiguing than that from a paper display.

REFERENCES

[1] Brillouin, L., Science and information theory (2nd ed.) New York, Academic Press, (1963)

[2] Cakir, A., Hart, D.J., and Steward, T.F.M., Visual display terminals, New York: Wiley, (1980)

[3] Lang, A.M. Reasearch report on standards for reading operations. Fairlawn, N.J. : MTM association for standards and research, (1969)

[4] McCormick, E.J. and Sanders, M.S. Human factors in engineering and design (5th ed.) New York: McGraw Hill (1982)

[5] Simonson, E., and Brozek, J. Flicker Fusion Frequency: Background and applications. Physiological Reviews, (1952) 32, 384-397

[6] Snyder, H.L. and Taylor, G.B. The sensitivity of response measures of alphanumeric legibility to variations in dot matrix display parameters. Human Factors, (1979) 21, 457-471

[7] Tinker, M.A. and Paterson, D.G. Influence of type form on speed of reading. Journal of Applied Psychology, (1928) 12, 359-368

[8] Vartabedian, A.G. The effects of letter size, case, and generation method on CRT display search time. Human Factors, (1971) 13, 363-368 (a)

[9] Vartabedian, A.G. Legibility of symbols on CRT displays. Applied Ergonomics, (1971) 2, 130-132 (b)

ACKNOWLEDGEMENTS

The authors acknowledge the financial support provided by the Natural Science and Engineering Research Council of Canada. The writers would also like to thank Mrs. Cheryl Perras for the typing of this manuscript.

Trends in Ergonomics/Human Factors II
R.E. Eberts and C.G. Eberts (Editors)

DEFINING AND MEASURING COMPUTERPHOBIA

Dr. Tim Jay

Department of Psychology
North Adams State College
North Adams, MA, USA 01247

Both popular and professional press have supported the notion that many people have developed an anxiety about computers. While the idea is quite pervasive, few have defined computerphobia or devised an instrument to assess the phenomenon. This research is an attempt to define and measure computerphobia. Identified here are two general factors that represent peoples' attitudes about computers and their impact on society. Variables such as sex, experience, and education are also implicated.

INTRODUCTION

The field of human factors has produced an immense array of data on man-machine systems (MMS) about perceptual, cognitive, or performance aspects therein. We know much less about the affective variables involved. The problem is that some emotional and personality dimensions affect outcomes in these systems. To the degree that goal performance is altered, the researcher needs to control affect. We are becoming more attentive to the nature of affect in man-machine interaction, as recent research indicates (Emmett, 1981; Kiesler, et al, 1984).

It is probably safe to say that one of the most important emotions affecting performance is the dimension of fear-aggression, or some underlying very basic avoidance-attack tendency. Human factors personnel are probably most likely to encounter the aggressive end of this dimension in MMS, when users become frustrated with certain characteristics of the system (see Emmett or Kiesler et al). We are less likely to see fear or avoidance because those who have acquired such tendencies are likely to avoid MMS employment or computer technology altogether. Now, however, we must retrain those with avoidance tendencies because current jobs are being redesigned to include MMS; also new positions are created which include computer architecture. So, where do we search for answers to problems of innovation in job or training demands?

RESISTING INNOVATION

Psychologists know that people develop fear and anxiety over a variety of objects and events in their surroundings. The security that we establish through job, routine, or lifestyle will be threatened by any change in those habits. Threatening security creates anxiety, so many are motivated to maintain status quo and avoid change. What about innovation? Computerphobia is merely an extension of a fear of inventions, especially at work and to a lesser extent with other nonvocational aspects of lifestyle. A similar episode occurred during the industrial revolution, and from pervasive inventions such as automobiles, television, and robotics. Innovation that changes habits is anxiety provoking and engenders negative affect.

The psychological reasons for our need to control job requirements or lifestyle have been elaborated by Maslow (1966). Research on how the electronic revolution will alter job design also appears in a more recent NSF report (1979) and a variety of books, such as Osborne's (1979). There are now emerging many texts on how computers affect workers and more specifically, reports on the impact of computers in instructional settings (Jay, 1984; Martellaro, 1981; or Spuck, 1981, for example). Thus, the problem is recognized and solutions are sought to enable computer and man to work together.

A general conclusion from previous research is that people resist changes that require altering their schedules, values, skills, knowledge, and personalities. Recently, two papers examined the implementation of computers for instruction and teachers' attitudes about such usage (Cunningham, 1981; Moscow & Bolton, 1981). Both studies report affective changes in teachers' attitudes about computers following training and use. Users felt less helpless, uneasiness, frustration, and dehumanized; they (computers) were viewed as more challenging, valuable, and enjoyable as well (Cunningham, 1981). Further, computers were found to be more valuable, interesting, and helpful for students (Moscow & Bolton, 1981). So, now we know that people react affectively and what some of those dimensions of emotionality are. Now for the current research.

The goal of this research is threefold: to define the phenomenon of computerphobia or negative attitudes about computers; to construct a survey to measure computerphobia; and finally to identify the degree to which people possess these negative attitudes.

DEFINING COMPUTERPHOBIA

Based on subjective reports and observations, I previously defined computerphobia (Jay, 1981 a) by enumerating the negative attitudes and behaviors toward technology. Resistance to talking about, thinking about computers, as well as a fear of damaging or being replaced by computers was apparent. These attitudes result physiologically in fear, anxiety or aggression. I further refined these ideas in the context of the computer literacy issue (Jay 1981 b), as one hurdle to be overcome in training settings. The current task is to identify computerphobics by measuring the degree to which people identify with these negative attitudes towards computers.

METHOD

QUESTIONNAIRE. Based on previous operational definitions of computerphobia, a 40 item questionnaire was constructed (see Table 1). The items were to tap both positive ("I enjoy writing computer programs") and negative ("Computers dehumanize people") attitudes.

SUBJECTS. Subjects were 33 females and 17 males enrolled in an introductory psychology class at NASC and were given course credit for participation.

PROCEDURE. All 50 subjects were run in one group. They were given a copy of the questionnaire and told to indicate on a 7 point scale how much they agreed or disagreed with each of the 40 items. Scale values ranged from 1, strongly disagree to 7, strongly agree. The session took about 15 minutes. Data were then analyzed using the SPSS factor analysis program.

RESULTS

The factor analysis procedure resulted in 5 factors that accounted for 50% of the total common variance. Questionnaire items that have a high factor loading, that is near the +/- .50 level are noted in Table 1 for the first two factors. Factor 1 accounted for some 22% of the variance and Factor 2 for about 8%. These two are the most interpretable factors and the discussion will be restricted to them.

Factor 1 contains questions that could be classified as "attitudes toward computers." These questions represent positive affect as the negatively loaded questions were negatively affectively worded. This factor probably reflects what people feel about computers and technology in general.

Factor 2 contains items with high loadings that may describe a recognition of the social-technical impact of computers on society. Rather than indicative of feelings or attitudes, this factor indicates a realization that the computer will have specific outcomes on society.

ITEM	Factor 1	Factor 2
1. I like computers	.688	
2. Computers dehumanize people		.626
3. I talk about computers with my friends		
4. Computers cannot replace people		
5. Modern technology is out of control	-.493	
6. Computers should be used in schools	.494	
7. I would enjoy writing computer programs	.762	
8. Using computers will cause unemployment		.600
9. I am afraid of using computers		
10. Students learn better with computers than from teachers		.545
11. I can use the computer like any tool		
12. People who know a lot about computers impress me	.520	
13. Computers cause alienation among people	-.554	
14. I don't think I'd be good working with computers	-.781	
15. Computers, like CB radios, will be another passing fad		
16. I have no fear of using computers		
17. A person could be replaced by a computer		.468
18. Using computers requires more effort to get some jobs done than not using computers for the same job		
19. I feel like a slave to computer technology	-.538	.470
20. People are responsible for many of the errors blamed on computers		
21. A computer is similar in function to a human brain		
22. I feel I could use a computer to do almost any job	.588	
23. Computers will eventually solve some of our most puzzling everyday problems		
24. If I had the money I would purchase my own computer	.732	
25. I probably have "computer anxiety," they make me nervous		
26. Computers have eliminated many routine and boring tasks	.423	
27. Computers are very simple machines		
28. I would like to know more about computers	.769	
29. Without computers life would be more difficult	.495	
30. I will probably never use a computer	-.689	
31. Computers can be used to make life more fun	.440	
32. I intend to take a computer course in the future	.712	
33. Computers should be used to replace humans in dangerous jobs		
34. Computers cannot think like humans		
35. I understand computers and their role in society	.627	
36. Most of my friends do not understand computers		
37. Most kids understand computers better than their parents		
38. I would entrust my life to a computerized decision	.551	
39. A computer is just a large fast calculator		
40. Students should be required to take a course in computers	.500	

TABLE 1

Questionnaire Items and Factor Loadings

SUBJECT CHARACTERISTIC ANALYSIS

In this study the survey was administered to a second sample of students in order to determine how their personal background was related to attitudes about computers. The reader is urged to use caution in interpreting these results due to the restricted nature of the sample.

METHOD

QUESTIONNAIRE. In addition to the 40 item questionnaire, subject data were recorded with regard to age, sex, experience with computers, both academic and vocational (whether touched a computer or not), as well as major academic course concentration.

SUBJECTS. Subjects were 222 college students (74 male, 148 female) enrolled in a general psychology course.

PROCEDURE. Administration was similar to the initial study. Scale values were analyzed with a BMD computer package for analysis of variance. Independent variables were sex of subject, years of experience with computers, and touched versus never touched a computer. Dependent variables were scale responses to positive items about computers and responses to negative items about them.

RESULTS

Present analysis focuses on the major independent variables under study. A variety of interactions were obtained with specific items on the questionnaire but are not of interest here. A more complete ANOVA table may be obtained from the author. Note that the term significance here refers in all cases to a p value less than .01.

SEX OF SUBJECT. In general males displayed a more positive attitude about computers than females. Males were significantly ($p < .01$) more in agreement with positive statements than were females. Similarly, males disagreed significantly more with negative items than females disagreed.

EXPERIENCE. Those who have TOUCHED a machine are significantly more positive about them than those who have NOT TOUCHED. This trend is stronger for males than females, in agreement with the sex differences. As for YEARS EXPERIENCE, in general the more experience for both males and females, the more positive is the attitude about the machine. These results were indicative of both more positive agreement with positive items and more disagreement with negative statements about computers. Obtained were a main effect for both types of experience and an interaction of experience with sex of subject.

GENERAL DISCUSSION

The goal of this research was to define positive and negative attitudes about computers, design a questionnaire to measure these emotions, and identify subjective differences in attitudes about computer technology. The implications of the results are applied to settings where job design requires the use of computers and to assess the possible impact of job design and training on those people affected. What have we learned?

One can conclude that people do differ in their feelings about computers. A sex effect typical of those found in math and engineering has emerged, that is males being more receptive to the subject than females. However, sampling only college students who are biased toward instruction and computers in a positive manner, is probably different than a population of working adults. These data suggest that with time and experience with computers attitudes become more positive and less negative. In the future we might use this questionnaire and data to assess an individual to identify her particular attitude and design an individualized course of training in order to overcome potential difficulties with the implementation of computerized work requirements.

REFERENCES

(1) Bolton, H., and Moscow, D., Microcomputers in the classroom: A foot in the door, Educational Computer, 1(3) (1981) 34-36.

(2) Cunningham, S., Off to a good start with microcomputers, Educational Computer, 1(3) (1981) 38-39.

(3) Emmett, R. VNET or GRIPENET?, Datamation, Nov (1981) 48-58.

(4) Jay, T.B., What to do about 'computerphobia,' Educational Technology, Jan (1981) 47-48.

(5) Jay, T.B., Are you a computer 'illiterate?' The Computing Teacher 8(4) (1981) 58-60.

(6) Jay, T.B., The cognitive approach to computer courseware design and evaluation, Educational Technology, Jan (1983) 22-26.

(7) Jay, T.B., Psychology and CAI (PSYCAL Press, North Adams, MA, 1984).

(8) Kiesler, S., Siegel, J., & McGuire, T., Social psychological aspects of computer-mediated communication, American Psychologist 39(10) (1984) 1123-1134.

(9) Martellaro, H.C., Why don't they adopt us?, Creative Computing, Sept (1980) 104-105.

(10) Maslow, A.H. The psychology of science (Harper & Row, New York, 1966).

(11) Osborne, A., Running wild: The next industrial revolution (Osborne/ McGraw-Hill, New York, 1979).

(12) Spuck, D.W., An analysis of the cost-effectiveness of CAI and factors associated with its successful implementation in higher education, AEDS Journal, Fall (1981) 10-22.

(13) Technology in science education: The next ten years. Report prepared for the National Science Foundation, July 1979.

Trends in Ergonomics/Human Factors II
R.E. Eberts and C.G. Eberts (Editors)

A PC-BASED GATEWAY FOR INFORMATION RETRIEVAL

William Leigh
Univ. of Southern Mississippi, Hattiesburg, Mississippi, U.S.A.

Ray Souder
Northern Kentucky Univ., Highland Hts., Kentucky, U.S.A.

There are many bibliographic citation databases available to computer terminal equipment over telephone lines. These databases are useful for technological research and the development of bibliographies.

This paper describes the design and use of a user-interface for these information retrieval systems which achieves a common command language for several databases. This user-interface resides in a personal-computer and mediates the researcher's dialog with the online systems.

INTRODUCTION

Since the 1960's, when the enabling computer technology became available, many bibliographic citation and full-text databases covering almost any area of knowledge have been built and made available for access with computer terminal equipment via telephone lines. Figure 1 is a diagram of this arrangement. Lancaster (1979) supplies a good overview of the history and capabilities of information retrieval systems. Salton (1980) offers a survey of technical developments in the field.

Bibliographic databases contain citation entries composed of titles, authors, publication information, and abstracts of articles, books, and reports. Full-text databases contain the complete texts of the documents in addition. There is one entry in these databases for each document. Usually the databases specialize in one area of knowledge. Glossbrenner (1983) is a directory to the most widely available databases, what they contain, and how they are accessed. Table 1 contains a list of some major databases and a description of their contents. Wente (1971) describes one bibliographic database system in detail.

Entries are accessed with keywords. Keywords are supplied for the documents by the author, by a human indexer, or with automatic methods. Figure 2 contains a citation including keywords which might be from a computer science bibliographic database.

Queries are formulated by the researcher to access databases. These queries must be expressed in the special language supported by the specific database. Efforts are underway to standardize these languages across databases. The International Organization for Standardization (ISO) is

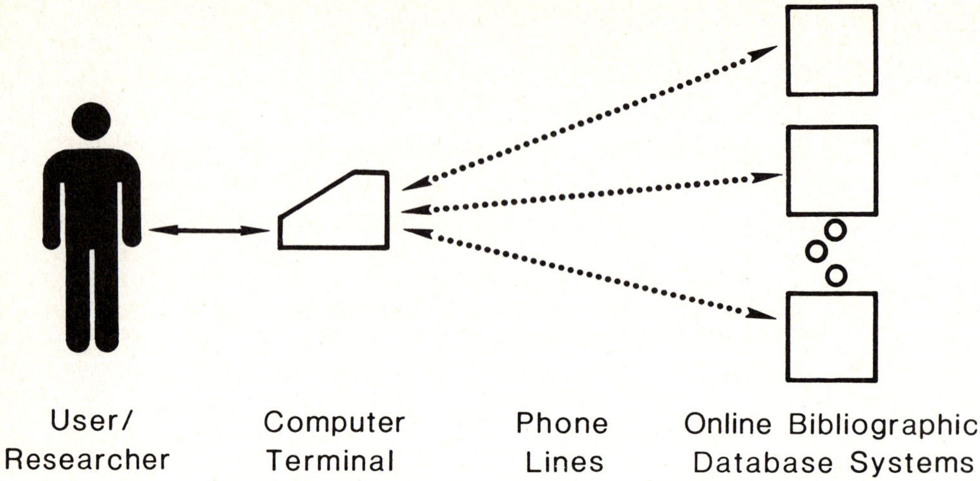

Figure 1
ONLINE INFORMATION RETREIVAL

ABI/INFORM	Business and Economics
AGRICOLA	Agriculture and Related Subjects
COMPENDEX	Worldwide Coverage of Engineering
ERIC	Educational Materials
INSPEC	Physics, Electrotecnology, Computers
MEDLINE	Medicine, Dentistry, Nursing
NASA/RECON	Aerospace technology

Table 1
Some Major Databases and Their Subjects

Title - A Data Structure for Word Processing

Author - Jones, John M.

Date - August, 1982

Source - Journal of Computer Science

Pages - 354-368

Keywords - DATA STRUCTURES, WORD PROCESSING, OFFICE AUTOMATION, TEXT PROCESSING, TEXT EDITING

Abstract - A data structure is presented which has desirable properties in situations requiring automatic justification under conditions of insert and over-write. This situation occurs in screen-oriented word-processing systems. Examples of use and test results are reported.

Figure 2
Citation from a Typical Bibliographic Database

considering one such standard language as described in the "Draft Specification for Commands for Interactive Search Systems" (1984). Figure 3 contains several queries expressed in this draft standard query language. Another solution to the problem of multiple querying languages which does not involve changing the database systems themselves is the "gateway" approach. Marcus (1982) has built such a system which uses a computer to convert queries specified in a standard language to the specific languages of the individual database systems. Figure 4 shows the relationship of the components in the "gateway" approach to online information retrieval.

FIND WORD PROCESSING AND DATA STRUCTURES

FIND (WORD PROCESSING OR TEXT PROCESSING) AND TEXT EDITING

FIND (DATA STRUCTURES OR ALGORITHMS) AND WORD PROCESSING

FIND TEXT EDITING AND DATE 1977

Figure 3
Example ISO Standard Queries

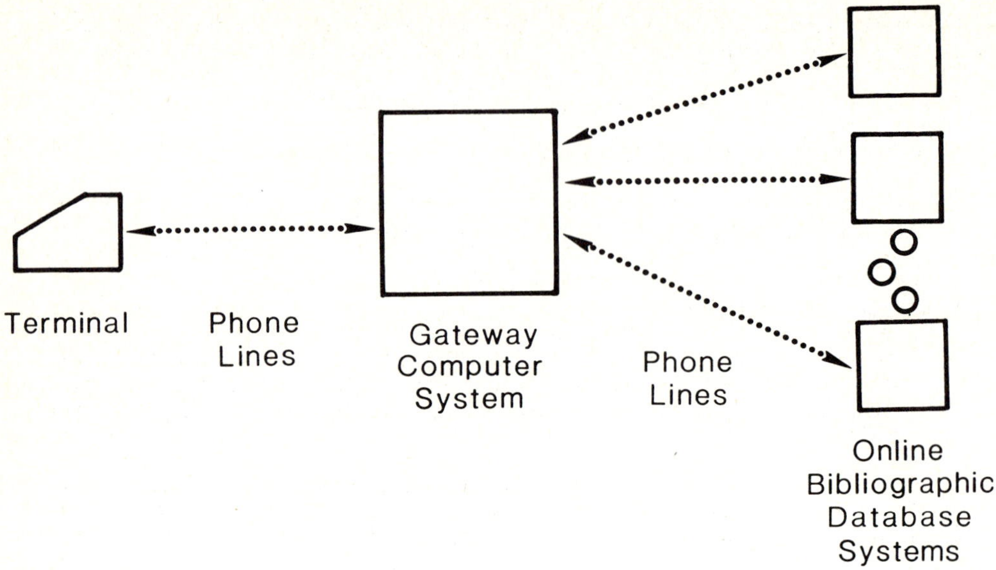

Figure 4
DIAGRAM OF A 'GATEWAY' ARCHITECTURE

DEVELOPING A GATEWAY

The development of a "gateway" is a difficult task. There are many subtleties of using the various databases that are not apparent in the manuals. Developing truly equivalent commands which remain equivalent across several databases may not be possible. However, before searching can be done by non-experts, this must be accomplished.

For example, a first issue to be resolved is the meaning of "basic index". The different databases have different semantics for this term. It is logical that the user be presented with the same basic index on every database. The basic index which makes the most sense includes all searchable terms and stems from title, abstract, and the controlled vocabulary of the system.

This default index should be set up by the "gateway" for each system as the default by the logon--this process being "transparent to the user."

The semantics of the search language for the various systems is often obscure. For example, a single term search template for NASA/RECON:

```
SEARCH searchterm + ATL/searchterm + AX/searchterm
```

An AND search for term 1 AND term 2:

```
SEARCH (searchterm1 + ATL/searchterm1 + AX/searchterm1)
         *(searchterm2 + ATL/searchterm2 + AX/searchterm2)
```

A MENU-STYLE INTERFACE

Our system is based on the ZOG(7) approach to non-machine communication. A network of menu's is presented to the user. That network is designed by an expert in the application domain, in this case computer bibliographic searching.

Indented menu's are available for several databases. (See Figure 5 and 6 for example) The user can select a function with confidence that semantically equivalent commands are transmitted to the subject database system.

```
           MASTER MENU

A     DROLS MENU
B     NASA/RECON MENU
C     EUROPEAN STANDARD MENU
D     POST-PROCESSING MENU
E     MODIFY PCCL
G     PRINT SEARCH TERMS
H     ENTER/CHANGE SEARCH TERMS
X     RETURN TO SYSTEM


--enter your selection

                                      F1-help, F2-end
```

Figure 5
The Master Menu

```
        EUROPEAN STANDARD MENU

A       SIGNON
B       HELP
C       SELECT DATABASE
D       TERMINATE SESSION
E       DISPLAY SEARCH TERMS
F       SEARCH ST1
G       SEARCH ST2
H       SEARCH ST1 AND ST2
I       DOWNLOAD CITATIONS
J       SPECIAL FEATURE
K       ENTER/CHANGE SEARCH TERMS
X       RETURN TO PCCL MASTER MENU


--enter your selection

                                      F1-help, F2-end
```

Figure 6
An example of a search menu.

This menu-style interface is implemented on a personal computer. The menu's and their underlying commands may be coded by the experts or by the user with a screen-oriented text editor. Also, expanded HELP texts may be prepared.

In operation, the user can employ his own communication program to handle the telephone lines between his personal computer and the remote online database system. A commercial "windowing" environment program is used to accommodate the communication program and our menu-style interface simultaneously on the same personal computer. The user can activate our program and select his function. The underlying data base commands may then be passed to the communication program with the Facilities of the "windowing" program and then transmitted to the remote database system.

SUMMARY

There are many bibliographic citation databases available to computer terminal equipment over telephone lines. These databases are useful for technological research and the development of bibliographies. Usually, for a researcher to use these databases, he must seek the assistance of an "expert intermediary" who specializes in the details of online searching and knows the command languages of the different systems.

We have built a user-interface for these information retrieval systems which achieves a common command language for several databases. This user-interface resides in a personal computer which acts as a "gateway" to the bibliographic database systems. This user-interface mediates the dialog between the user and the online systems. This system functions in a menu-based manner. Integral help facilities are present. User extension is allowed.

This interface is a "first-aid kit". Lowering of the syntax barrier makes searching a do-it-yourself proposition. Many engineers wish to be self-sufficient and not dependent an expert. This tool relieves the expert of syntax problems, also. Most importantly, this system provides a convenient tool for structuring and codifying expert searcher's knowledge.

REFERENCES

(1) Glossbernner, A., Personal Computer Communications (St. Martin's Press, New York, 1983).

(2) International Organization for Standardization (ISO), Draft Specification for Commands for Interactive Search Systems (Documnent ISO/TC 46/4/5 N49, 1984).

(3) Lancaster, F., Information Retrieval Systems (Wiley-Interscience, New York, 1979).

(4) Marcus, R., User Assistance in Bibliographic Retrieval Networks Through a Computer Intermediary, IEEE Transactions on System, Man, and Cybernetics, Vol. SMC-12, No. 2 March/April(1982) 116-132.

(5) Robertson, G., McCracken, D., and Newell, A., The ZOG Approach to Man-Machine Communication, International Journal of Man-Machine Studies, 14 (1981) 461-488.

(6) Salton, G., Automatic Information Retrieval,Computer, September (1980) 41-56.

(7) Wente, V., edited by D. Walker, NASA/RECON and User Interface Considerations, in Interactive Bibliographic Search: The User/ Computer Interface (AFIPS Press,Montvale, New Jersey,1971).

Trends in Ergonomics/Human Factors II
R.E. Eberts and C.G. Eberts (Editors)

A TAXONOMY OF USER-ORIENTED DATA PROCESSING FUNCTIONS

James A. Carter Jr.

Department of Computational Science
University of Saskatchewan
Saskatoon, Saskatchewan
Canada

This paper presents a taxonomy of user-oriented data processing functions. The taxonomy is composed of a hierarchy of user-oriented functions generally required in data processing systems plus the corresponding command names for these functions that were most often suggested by a group of potential users. The taxonomy can be used for commercial system design and evaluation and as a basis for further research.

In order to define a computer-user interface one must define both the activities that will take place at the interface and those that will pass through the interface. From the user's point of view, the activities that pass through the interface deal with what the user desires to get done, while the activities taking place at the interface determine how it is done. User-oriented interfaces should provide maximal flexibility in activities taking place at the interface while standardizing those passing through it.

If activities passing through, commonly called functions, are not standardized, confusion will result for single users of multiple systems (4, 8, 12, 16). The large number of different systems that a person may use makes the development and implementation of a set of standards both highly desirable and very difficult (17). This standardization must include parts of the operating system that the user interacts with directly, as well as application programs (11, 15).

These standards should be as natural to users as possible, incorporating both the characteristics and the needs of the users (1). User functions should be based on a conceptual model of how the user will use the system to accomplish tasks. According to Carey (3), "this knowledge begins with the system's representation for information in the task domain (files, records, etc.) and the operations on functions that apply to them (storing, copying, etc.)". This suggests the use of object-oriented systems such as the Smalltalk system (9).

Card et al (2) state that the number of immediate actions should be limited to avoid user confusion. Schneiderman (14) suggested that a synatactic/ semantic model of command formulation proceeds from a user's perception of the hierarchial organization of the task to be performed.

A variety of authors have dealt with the problem of hierarchial organizations or taxonomies of user functions (5, 6, 8, 10, 15). Schvanveldt et al. (15) discuss problems resulting where, "In many computer systems, the user interface has been the result of individual programmers using intuition about the best way to interact with users". They emphasize the need of research into the types of information exchanged and the method of exchange in individual interactions, but did not report such a taxonomy. Cohill (8) identified 10 function areas based on reviewing nine existing systems. Lenorovitz et al (10) identified a three level hierarchy that they refer to as a "User-Input Taxonomy". It contains five main functional areas and 37 nodes.

The most thorough hierarchy is that proposed by Carter and Everett (5) and refined in Carter (6). It contains four levels and 148 nodes, including 100 terminal nodes representing different specific functions generally required by users. This hierarchy was developed by analyzing 42 systems in detail, consisting of 20 application packages, 17 operating systems, and five query languages. Successive sets of consolidations and organizations of the identified functions were used to produce a logically organized description of the various user-oriented functions without redundancies. Further refinements made it compatible with the object oriented approach to data systems.

METHODOLOGY

Operational definitions were developed for each of the terminal nodes in the hierarchy (including nine additional terminal nodes added upon further analysis) to survey typical users of systems to find what command names they would associate with each function. For example, the operational definition developed for function "1.1.1.1 Add a Field" was:

> Suppose I have already told the computer that I am going to be entering some names and some addresses and I have told the computer where I am going to enter them. Now I have decided that I also want to enter telephone numbers at the same time as the names and addresses. What command would I type in to the computer to achieve this?

Surveys composed of groups of operational definitions were administered to 500 students with various backgrounds in computing. Approximately 300 were entering an introductory computer science course for non-majors. Their involvement represents the opinions of naive users. Approximately 100 were entering a management information systems class for commerce students. They had taken the introductory non-majors course two years earlier. Their involvement represents the opinions of users with only slight computing experience. Approximately 100 were in the middle of their third year of majoring in computer science and had completed a course in data processing. Their involvement represents the opinions of users with a moderate level of experience. The ratio of different types was based on the availability of students rather than on other factors. This ratio was maintained in the distribution of surveys. From the average of 100 users surveyed for each function, an average of 72% responded for each function.

The various suggested names were compiled and analyzed. Differences in verb tense and number were resolved into the more common version of the command. Where answers contained two or more command words, they were split into one occurrence of each of the commands. Answers that were inappropriate to the function described in the operational definition were deleted as were answers expressed as computereese such as "ctrl C". An average of 34% of the answers made were deleted accordingly. Twenty operational definitions were found unsatisfactory and were redone and re-surveyed. Two additional functions were identified after the completion of the survey and were added to the hierarchy without suggested command names.

A cross-reference identifying the various occurrences of each suggested name was produced to analyze the hierarchy of functions. Groups of adjacent terminal nodes were identified containing many of the same suggested command names. Two instances of non-adjacent occurrences of primarily the same suggested names were identified suggesting minor modification of the hierarchy.

The taxonomy was developed by grouping adjacent terminal nodes where the combination of their suggested names indicated that they should be treated as various occurrences of a single terminal node rather than as separate terminal nodes. The criteria for agreement was either: 70% agreement between the combined frequency of terms in nodes being combined or 70% agreement between the suggested names in the nodes being combined. In most cases of combination the first of these criteria was applied. An additional criteria was that no more than 5% of the previously appropriate responses from the resulting combination should be found to be inappropriate for the combined node. The application of these combination criteria yielded a hierarchy containing 89 terminal nodes (including 33 combined subnodes) with an average of 10 appropriate command names per terminal node.

The resulting taxonomy, as reported below, contains a hierarchy of functions plus the command names most often suggested for these functions. The criteria for reporting command names was a 10% or greater frequency of suggestion (based on total appropriate responses for a function) and a minimum of five suggestions of the name. Where no suggested command names met these criteria (either due to a large variety of suggested names or due to a low response rate to the operational definition) this is indicated by an * in place of suggested command names. A more thorough report of the taxonomy, including descriptions of these functions, the operational definitions used in the surveys, and the complete listing of appropriate user suggested command names is found in Carter (7).

THE TAXONOMY

A. THE TAXONOMY OF GENERALLY REQUIRED FUNCTIONS

HIERARCHY	SUGGESTED COMMAND NAMES
1 DATA FUNCTIONS	
1.1 ADD OR ADD TO AN OBJECT	
1.1.1 Create an Object	
1.1.1.1 Create a Field	add 48%, enter 33%
1.1.1.2 Create a Record	*
1.1.1.3 Create a File	create 22%, file 18%, new 13% edit 11%
1.1.1.4 Create a Directory	*
1.1.1.5 Create a Form	*
1.1.1.6 Open a Window	split 60%, display 13%
1.1.2 Define an Object's Format	
1.1.2.1 Define Field Format	*

```
  1.1.2.2 Define Record Format       *
  1.1.2.3 Define File Format         format 29%
  1.1.2.4 Define Form Format         *
 1.1.3 Add to an Object
  1.1.3.1 Add Data to a Record       *
  1.1.3.2 Add to a File
   1.1.3.2.1 Add a Record to a File  add 44%, enter 19%
   1.1.3.2.2 Append a File to a File add 46%, combine 35%, append 10%
  1.1.3.3 Link a File to a Directory search 50%
  1.1.3.4 Join (Append) Directories  combine 46%, group 25%, merge 13%
  1.1.3.5 Add Protection to an Object
   1.1.3.5.1 Protect an Object       lock/security 20%, protect 17%
   1.1.3.5.2 Assign a Password       password 76%
   1.1.3.5.3 Grant Access            *
 1.1.4 Copy an Object
1.1.4.1 Copy a Field or a Record     copy 55%, transfer 27%
        Copy a Field
        Copy a Record
1.1.4.2 Copy a File                  copy 70%, duplicate 20%
1.1.4.3 Copy a Window                duplicate 32%, double 25%, split 18%

1.2 CHANGE AN OBJECT FUNCTIONS
 1.2.1 Change a Field                change 38%, replace 17%,
                                     substitute 15%, new 10%
  1.2.1.1 Move a Field               move 40%
 1.2.2 Change a Record
  1.2.2.1 Move a Record              change 40%
 1.2.3 Change A File
  1.2.3.1 Split a File               split 38%, divide 33%, separate 23%
  1.2.3.2 Sort a File                sort 54%, order/rearrange 17%
  1.2.3.3 Translate Characters       substitute 31%, change 23%,
                                      replace 21%, exchange 13%
  1.2.3.4 Encrypt/Decrypt a File     code or decode or encode 53%,
                                      scramble or unscramble 39%
          Encrypt a File
          Decrypt a File
  1.2.3.5 Update one File on Another add/combine 37%, merge 19%
 1.2.4 Change an Object's Name
  1.2.4.1 Change a Field Name        change 50%, substitute 30%
                                      replace 17%
  1.2.4.2 Change a File/Form Name    rename 34%, change 28%
                                      substitute 10%
          Change a File Name
          Change a Form Name
 1.2.5 Change Object Format          change 55%, revise 15%
       Change Field Format
       Change Record Format
       Change File Format
       Change Form Format
 1.2.6 Change Window
  1.2.6.1 Modify Window              change 38%, screen 30%
  1.2.6.2 Link Windows               together 55%
 1.2.7 Change Protection
  1.2.7.1 Change Password            password 45%, change 40%
  1.2.7.2 Change Access              *
```

```
1.3 DELETE OR DELETE FROM AN OBJECT FUNCTIONS
 1.3.1 Delete an Object               delete 53%, cancel 12%, remove 11%
        Delete Field
        Delete Record
        Delete File
        Remove Directory
        Delete Form
        Close Window
        Delete an Index
 1.3.2 Remove from an Object
  1.3.2.1 Empty or Remove an Object   delete 45%, erase 23%, remove 14%
                                       clear 10%
          Empty a Field
          Empty a Record
          Empty a File
          Remove Trailing Blanks
          Remove Link to Directory
  1.3.2.2 Compact a File              condense 30%, compress 19%
                                       compact 17%
  1.3.2.3 Remove Protection
   1.3.2.3.1 Unprotect an Object      open 44%, access/unlock 12%
   1.3.2.3.2 Remove Password          delete 38%, cancel 34%
   1.3.2.3.3 Remove Access            change 41%, lock 15%

1.4 GET AN OBJECT FUNCTIONS
 1.4.1 Find an Object
  1.4.1.1 Find a Field/Record         find 49%, search 22%
           Find a Field
           Find a Record
           Find a File
  1.4.1.3 Find Protection/Access
   1.4.1.3.1 Find Protection Status   *
   1.4.1.3.2 Find Access              access 38%, list 21%
  1.4.1.4 Obtain Data Item Statistics*
  1.4.1.5 Count Words                 count 49%, size 20%, word count 13%
  1.4.1.6 Compute Two or More Fields  *
  1.4.1.7 Find Object Type            *
 1.4.2 List All Objects
  1.4.2.1 List All Field Names        *
  1.4.2.2 List File Names/Directories files 57%, list 19%
           List All File Names
           List All Directories
  1.4.2.3 List All Access Permissions list 21%, recall/who 14%
 1.4.3 Output an Object
  1.4.3.1 Output a Field              *
  1.4.3.2 Output a File/Record        list 30%, display 25%
           Output a File
           Output a Record
  1.4.3.3 Output a Directory          files 34%, list 33%, names 18%
 1.4.4 Specify an Object
  1.4.4.1 Select a File               *
  1.4.4.2 Activate A Directory        recall 40%, get 14%
  1.4.4.3 Designate an Index Key Field   read 33%
```

Function	Terms
1.5 COMPARE AN OBJECT FUNCTIONS	
1.5.1 Compare Fields/Records	compare 40%, check 31%, equal 14%
Compare Fields	
Compare Records	
1.5.2 Compare Files	*
1.5.3 Compare Directories	compare 52%, difference 35%
2 TASK EXECUTION FUNCTIONS	
2.1 SIGNING ON	login 29%, start 26%
2.2 SIGNING OFF	end 17%, logout/quit 14%, stop 11% bye/exit 10%
2.3 CHANGE SIGNON PASSWORD	password 61%, change 26%
2.4 STARTING EXECUTION FUNCTIONS	
2.4.1 Execute Now	run 27%, do 22%, send 19%
2.4.2 Execute in the Future	at 33%, delay 15%, run/time 11%
2.4.3 Execute Again	repeat 81%
2.4.4 Process Batch	"JOB NAME" 38%
2.5 STOPPING EXECUTION FUNCTIONS	
2.5.1 Leave Task Normally	end 35%, quit 19%
2.5.2 Terminate Task Abnormally	stop 39%
2.5.3 Interrupt Task	hold 35%, pause 24%, break 19%, stop 11%
2.5.4 Continue Task	resume 38%, continue 25%
2.5.5 Stop batch Request Before Start	cancel 59%, kill 10%
2.5.6 Undo Previous Processing	*
2.6 EXECUTION STATUS FUNCTIONS	list 30%, tasks 22%, jobs 15%
Obtain Status of Current Processes	
Obtain Status of Batch Requests	
2.7 PRIORITY FOR EXECUTION FUNCTIONS	
2.7.1 Change Priority	priority 91%
2.7.2 Obtain Queue Information	*
3 INFORMATION FUNCTIONS	
3.1 USER ASSISTANCE FUNCTIONS	
3.1.1 Assistance with Commands	
3.1.1.1 List Commands	help 48%, list 24%, commands 10%
3.1.1.2 Explain Command Description	define 39%, explain 20%, help 13%
3.1.2 Assistance with Errors	
3.1.2.1 List Errors	error 35%, help 17%
3.1.2.2 Explain Error Type	explain 38%, error/help 13%
3.1.3 Provide Tutorial	help 64%, instructions 18%
3.1.4 Provide Help Screens	
3.1.4.1 Display Next Help Screen	help 29%, information 24%, explain 17%
3.1.4.2 Display Previous Help Screen	help 26%, option 18%
3.1.4.3 Return From Help	return 39%, resume 21%, continue 14%
3.1.5 Provide Keyboard Assistance	keys 36%, functions/list 16%, keyboard 13%
3.2 PROVIDE OTHER INFORMATION FUNCTIONS	
3.2.1 Info. on System Users	users 47%, list 29%, names 11%
3.2.2 Info. on Last Signon of Users	users 50%, who 21%
3.2.3 Info. on Where a User Is	file 41%, name 32%
3.2.4 Info. on Last Command Executed	*

B. ADDITIONAL SPECIAL FUNCTIONS
(are identified only as a high level hierarchy, they were neither expanded further nor surveyed)

4. PERIPHERAL FUNCTIONS
 4.1 TERMINAL FUNCTIONS
 4.2 DISK FUNCTIONS
 4.3 OTHER PERIPHERAL CONTROL FUNCTIONS

5. SPECIAL FUNCTIONS
 5.1 USER TO USER COMMUNICATIONS FUNCTIONS
 5.2 DATE AND TIME FUNCTIONS
 5.3 WRITING AIDS
 5.4 ARITHMETIC OPERATIONS
 5.5 SYSTEM UTILIZATION ACCOUNTING

C. APPLICATION DEPENDENT FUNCTIONS
(not further described here)

6. BIBLIOGRAPHIC FUNCTIONS

7. SPREADSHEET FUNCTIONS

8. GRAPHICS

9. TEXT EDITING FUNCTIONS

USES OF THE TAXONOMY

The taxonomy can be used in a variety of ways both for the development of commercial systems and as a basis for further research into the human factors involved with computer-user interactions.

The taxonomy can provide various immediate benefits to the development and evaluation of commercial systems. The hierarchy within it can be used to analyze individual systems for completeness or to analyze groups of systems for compatibility. Similarily the suggested command names can be used to evaluate the ease of use of command names in an application program. In systems analysis the hierarchy can provide a check list to remind the analyst to look for all of the logical needs that might correspond to the various functions. In systems design the taxonomy can be used to standardize the design of the physical implementations of various logical systems, to provide checks for compatibility and consistency of functions and function command names between various systems and to suggest command names for new systems that do not require compatibility with existing systems.

The hierarchy can provide a basis for further research into the design of user-oriented functions, the interactions between them, and the naming of commands to perform them. Experiments are planned to evaluate the suggested command names for their accuracy and efficiency of use and surveys are planned to evaluate them for their user preference and their semantic properties. The results of this further research should provide a basis for proposing a user-oriented standardized vocabulary of command names.

REFERENCES

[1] Benbasalt, I., The Impact of Cognative Styles on Information System Design, MIS Quarterly, 2(2),(1978), 43-54.
[2] Card, S.K., Moran, T.P., and Newell, A., The Psychology of Human Computer Interaction, (Lawrence Erlbaum Associates Publishers, Hillsdale, N.J., 1983).
[3] Carey, T., User Differences in Interactive Design, Computer, (November 1982), 14-20.
[4] Carroll, J.M., The Adventure of Getting to Know a Computer, Computer (November, 1982), 49-58.
[5] Carter, J.A. and Everett, H.J.M., A Hierarchy of User-Oriented Functions, in: Gotlieb, L. (ed.), Proceedings of the Canadian Information Processing Society Session '84, (Calgary, Alberta, 1984), 464-469.
[6] Carter, J.A., User-Oriented Structured Evaluation and Design of Data Processing Applications, in: Hendrick, H.W., and Brown, O. (eds.), Human Factors in Organizational Design and Management, (Elsevier Publishers B.V., Amsterdam, 1984), 259-268.
[7] Carter, J.A., A Taxonomy of User-Oriented Functions, Technical Report 8501, (Department of Computational Science, University of Saskatchewan Saskatoon, Saskatchewan, 1985).
[8] Cohill, L.F., A Taxonomy of User-Computer Interface Functions, in: Salvendy, G. (ed.), Human-Computer Interaction, (Elsevier Science Publishers B.V., Amsterdam, 1984), 125-128.
[9] Goldberg, A. and Robinson, D., Smalltalk-80: The Language and Its Implementation, (Addison-Wesley, Reading, MA., 1983).
[10] Lenorovitz, D.R., Phillips, M.D., Ardrey, R.S., and Kloster, G.V., A Taxonomic Approach to Characterizing Human-Computer Interface Functions in: Salvendy, G., (ed.), Human-Computer Interaction, (Elsevier Science Publishers B.V., Amsterdam, 1984), 111-116.
[11] Morland, D.V., Human Factors Guidelines for Terminal Interface Design, Communications of the ACM, 26(7), (1983), 484-494.
[12] Scapin, D.L., Computer Commands in Restricted Natural Language: Some Aspects of Memory and Experience, Human Factors, 23(3), (1981), 365-375.
[13] Schneider, M.L., and Thomas, J.C., The Humanization of Computer Interfaces, Communications of the ACM, 26(4), (1983), 252-253.
[14] Schneiderman, B., Software Psychology: Human Factors in Computer and Information Systems, (Winthrop Publishers, Cambridge, MA., 1980).
[15] Schvanveldt, R., Cooke, N., Durso, F., Onorato, L., and Bailey, G., A Taxonomy of Human-Computer Interactions, in: Salvendy, G. (ed.), Human-Computer Interaction, (Elsevier Science Publishers B.V., Amsterdam, 1984), 121-124.
[16] Thomas, J.C. and Carroll, J.M., Human Factors in Communication, IBM Systems Journal, 20(2), (1981), 237-263.
[17] Treu, S., Uniformity in User-Computer Interaction Languages: A Compromise Solution, International Journal of Man-Machine Studies, 16, (1982), 183-210.

Trends in Ergonomics/Human Factors II
R.E. Eberts and C.G. Eberts (Editors)

APPLICATION OF INEXPENSIVE PERSONAL COMPUTERS TO HUMAN FACTORS RESEARCH

Evan P. Rolek and Mark S. Crabtree

SYSTEMS RESEARCH LABORATORIES, INC.
2800 Indian Ripple Road
Dayton, Ohio 45440
U.S.A.

Inexpensive personal computers have been developed to the point where they are frequently preferable over larger systems for many applications. The tendency to rely exclusively on large, costly systems continues to make much human factors research costlier and less timely. This paper examines uses of inexpensive personal computers in human factors research. Several specific applications are presented as a demonstration of the utility of these economical devices as research tools.

INTRODUCTION

In spite of technological advances with computer systems, many barriers still exist that prevent the conduct of cost-effective research. Experimenters have the capability to break some of these barriers. Take, for example, the "barrier" formed by the words human factors, computer, research, and $200. Quite a few avenues are closed by this roadblock--or are they?

It is not the intent of the authors to elaborate on the use of computers for conducting human factors research. Most of us have probably had many encounters with large (mainframe) computer systems throughout our careers. But have the mainframes held most of the territory too long? Refinement of the microprocessor over the past few years has made available to everyone computer power that would have taken several hundred thousands of dollars to acquire 10 years ago.

In this paper, we will describe some current trends and associated problems that develop from adhering to the policy of using mainframes exclusively. It will be shown that an alternative does exist if one is willing to believe that personal computers "under $200" are also "under utilized." Not only will general applications be discussed, but specific examples of some ongoing research will be presented as a demonstration that there are some cost-effective approaches to conducting human factors research that are not widely recognized.

GENERAL USE OF COMPUTERS IN HUMAN FACTORS RESEARCH

Whether basic or applied, most human factors experiments require some amount of preexperimental activities following the definition of the problem. Furthermore, hypotheses are often derived from computer-driven mini-experiments that provide preliminary data that demonstrate the effect of particular variable manipulations. The mini-experiment is very critical to the design of more complex experiments. It is desirable, however, for the computers that drive mini-experiments to be very flexible and easily programmed so that a large development effort is not required for changing experimental parameters and procedures.

Generally, computers have the capability of displaying stationary and moving letters, numbers, patterns, cursors, targets, colors, etc., which are necessary for eliciting subject responses, with speed and convenience that surpass any other means of presentation. In the past, experimenters have had to sketch visual stimuli or otherwise describe auditory stimuli to programmers who would attempt to develop software to the experimenter's specifications. Investigators and their programmers can use computers to help design the stimuli that are displayed. They can evaluate a variety of stimuli in mini-studies, and then select the stimulus and range of treatment levels that give the desired result. Experimenters that are versed in computer languages can often design their own stimuli and control scheme and make changes to suit their needs. In some cases, computers can be programmed to develop stimuli automatically, given the constraints imposed by the experimenter. Some experimenters are entering into the computer the necessary formulas and algorithms that will permit the computer itself to develop the stimuli automatically.

Since many computers contain built-in random number generators, they can be used to choose random presentations of stimulus sequences. The sequences can be stored in arrays and the computer can present the stimuli in the predetermined sequences. Alternatively, the stimuli can be chosen in real-time within specific constraints.

Computers can help with the selection of appropriate summary statistics since the preliminary data collected in ministudies can be examined and the best statistic(s) selected. The computer can be programmed to give a quick-look or summary of preliminary data. When important characteristics (i.e., mean or median, standard deviation, etc.) are known, appropriate descriptive statistics can be chosen. The same preliminary data can be examined for important characteristics (e.g., variability, amplitude, frequency, etc.) that can help in choosing appropriate statistical tests.

Because the procedures, techniques, variables, etc., that are chosen during preexperimental activities are so interdependent, it is extremely important for a cost-effective experiment to have the choices made based on an adequate amount of data. Preliminary experiments assisted by computers is the best way to collect the necessary data. Even if the actual experiment is a field study or is performed in an applied environment, much of the aforementioned preliminary work can be completed in the controlled environment that the laboratory provides. Therefore, procedures and techniques can be refined far in advance of the actual experiment.

Computers are also used for formal training of subjects on the experimental task. If adaptive training is used, the computer can keep track of the subjects' training schedules and whether or not the subjects are busy with another task. If paced training is used, the computer can determine optimum pacing rates and make certain that the necessary stimuli are presented in accordance with the pacing schedule.

Most human factors experiments involve measuring some aspect of human performance, subjective experience, and/or physiological/neurological response to stimuli under a specified set of conditions, or in a specific environment. The response itself, such as its accuracy or amplitude, and the amount of time taken to produce the response (latency or response time) are often important dependent variables. Many experiments require only a simple button press, turn of a knob, or force applied to a stick. Signals from these devices may be analog or digital and are processed according to preprogrammed algorithms. Responses and reaction times are sampled at the desired rate and are then stored in the computer's RAM or dumped directly to a mass storage device (tape or disk, typically) for reduction and analysis. Since many experimenters are familiar with programming languages such as BASIC or FORTRAN, they can make necessary modifications to the canned data analysis programs or to the summary statistics used in the programs.

THE $200 ALTERNATIVE TO THE MAINFRAME

We will now present some observations concerning increased use of "under $200" personal computers. These observations emphasize areas that can be improved with today's technology.

Although small personal computers are becoming more like appliances, with word processing and spread sheets being two examples, there are still some apprehensions about bringing low cost computers into the hands of the researcher for more of the research tasks (i.e., the personal computer is still not being used as effectively as it could be as a research appliance or tool). Frequently, there is a communication gap between the vocabulary and thinking of the behavioral researcher and that of the programmer. Hence, the lack of software and/or good documentation in the past has hindered computers from being used extensively during each research activity except by professional programmers.

There are many attitudes present today that limit the application of some cost-effective approaches for conducting behavioral research. While these attitudes might not be universal, they are common enough that large sums of money can be seen being spent less efficiently than necessary. In the areas of man-in-the-loop simulation and training, there is tremendous concern for physical fidelity and a lower concern for functional fidelity; some are content that "it looks like it should."

When conducting research with human subjects, the computer might be only part of the research apparatus with operator consoles and other devices also being needed. Maximum use of subjects' time requires that large functional blocks of the apparatus be available when the subject is present; prolonged delays require retraining of the subjects and motivational problems may arise.

There is also an attitude that selection of a computer should not be based on its capabilities so long as it can be expanded at a later date for additional cost. In many cases, the cost of the expansion is more than the total cost of another entire system with that capability.

It is not too difficult to find organizations that standardize by requiring the use of a specific $1500 plus computer. It is also not difficult to find that many of these applications have more computer power available than they need; yet, administrators are more than willing to pay the price of limiting the number of machines available other than consider using many inexpensive personal computers.

What are the consequences of some of these observations? On the surface, some of them seem positive--having rigid ideas about the use of computers simplifies many decisions because one doesn't have to be bothered with examining alternatives for the most cost-effective solution. It is typical to spend more time and money preparing for an experiment using a mainframe computer than running it. As a result, preexperimental work, mini-studies, and part task investigations are frequently eliminated. For many applications, preventive and corrective maintenance are a significant percentage of the research budget and highly skilled technical specialists are required. It is tempting for a researcher to become a programmer or hardware expert. To do so, however, may take a toll on the researcher's primary area of specialization.

There is sometimes a tendency to avoid selling a computer, even through its useful life has long since passed. In today's market where you can get more computer for the dollar as time passes, a used computer is often sold for much less than the owner would like because prospective buyers can get more capability, for any given price, in a new computer than the used computer. Inexpensive personal computers, on the other hand, can be considered to be "disposable."

Personal computers, even "under $200 models," offer many benefits for applications where mainframes are not really needed. Some of the desirable results include a shorter development time for software, developing initial software specifications as communications tools to avoid "I didn't think it would work like that," and avoiding the inconvenience of having the "computer down." The time and money invested in hardware and highly skilled personnel for mainframes can then be used more efficiently for those cases where they are actually needed. The low cost personal computer is ideal for part task research or training where total system capability is not needed nor is of interest. For applications where "testing in the field" is necessary, the mainframe is usually inappropriate. Availability of systems is frequently enhanced by using several totally independent computers and associated apparatus that are in widely different locations. It is easy to use one system for development for the next experiment while the other system is running the current experiment. It is sometimes cheaper to replace an entire personal computer for a cost that is a fraction of the parts to fix a mainframe. Now we will consider how low cost personal computers can be used for improving human factors or behavioral research.

The ease of setup and operation of the low cost personal computer lends itself to the conduct of mini-experiments or preexperimental work. Refining the range of independent variables, changing the experimental design, etc., are easy to accomplish. One of the largest benefits with using the personal computer to develop the stimulus presentation is that the experimenter can show programmers the final product to be implemented on the mainframe rather than going through iterative development cycles between experimenter and programmer.

Early personal computers were not suitable for presenting stimuli for a variety of reasons including limited memory, difficulty with programming, poor graphics and sound, etc. This is not true today with models that include excellent graphics, 16 bit processors, large choice of peripherals, etc., which make these machines suitable for real-time applications in the human factors laboratory. If one visualizes the CRT and computer as a "poor man's computer generated graphics" system, it is easy to recognize the potential of the personal computer as a stimulus presentation device. When programming these machines, it is not always necessary to use machine language. Using BASIC and compiling the program is more than adequate for many applications. Even if the final product will be used on a mainframe, the personal computer is an excellent method of demonstrating intermediate progress for briefings and discussions.

Subject training is another area where the inexpensive personal computer can greatly assist the researcher. Whether the actual experimental task or a related task is used, the personal computer and related apparatus can be used to screen, test, maintain performance, and familiarize subjects with the experimental task without disrupting the mainframe and its apparatus.

Depending on the sample rates and duration of experimental trials, the personal computer may be more than adequate for data collection. There are cases, however, where high sampling rates and many dependent variables necessitate the use of the mainframe. Here again, if the "under $200" personal computer will suffice, it will require a much smaller investment initially and avoid the high cost of maintenance if failures do occur.

Unless one has special analysis requirements, it is likely that one of the currently available analysis programs will be adequate. However, personal computers allow the experimenter to compute many different summary statistics and determine the results for the data base without the delays associated with using a centralized data processing facility.

When preparing the findings for a report, it is often desirable to change the format of the presentation. Changing the scales on a plot, using different symbology for encoding data points, altering the number of curves on a plot, etc., enable the experimenter to provide the graphics support staff with a product that is complete, only it needs to be made "print ready." Even computers in the "under $200" price class can be used for word processing when preparing a report. Given that one is organized enough to have a detailed outline in hand, inexpensive computers and their respective word processors are more than adequate for preparing a draft report. If one writes then organizes, it may be necessary to use a more sophisticated word processor which may or may not be available when needed. Here, too, the individual must decide if it is necessary to produce the final copy when a professional publications staff is available. Doing their job for them only reduces the experimenter's time for conducting research.

EXAMPLES OF CURRENT APPLICATIONS

The previous discussion is not merely a list of potential applications for the "under $200" personal computer, but it is illustrative of current applications that are familiar to the authors. Following is a list of some applications where personal computers are providing cost-effective approaches to some research that would not be done if there were total reliance on mainframe computers.

Several studies have been conducted at Wright State University to look at the effects of such variables as workload level, nature of the loading task, duration of time interval between experienced workload and reported workload, and the nature of the task intervening in the time interval, on the subjective workload reported by subjects. All workload tasks were presented by the Commodore VIC 20 computer. The computer recorded responses, determined accuracy of response and reaction time, calculated summary statistics, and generated stimuli for each task prior to the experimental trial by using various algorithms furnished by the experimenter (Eggemeier, Crabtree, and LaPointe (1983) and Eggemeier, Melville, and Crabtree (1984)).

A few commercial firms have developed batteries of tests that identify specific brain disorders. The tests consist of a variety of stimuli that patients or clients must respond to in some fashion. Depending on the response and the reaction time, some types of brain dysfunction can be diagnosed. The Commodore VIC 20 is being used to administer and score these tests in some doctors' and psychologists' offices.

Systems Research Laboratories, Inc. (SRL), has developed a field portable test system from a small computer for assessing human cognitive performance in various real-world environments. These units are totally self-contained with the necessary software residing in EPROMS and data being stored in nonvolatile memory until it can be dumped to tape or disk. The units have built-in LCDs and are battery powered. They can be connected to larger computers easily through built-in ports.

A variety of secondary tasks for human operator workload assessment have been developed by both the Air Force Aerospace Medical Research Laboratory, Human Engineering Division (AFAMRL/HEG), and SRL for use with Commodore computers, including the 64, the VIC 20, and the PET. The tasks have been used very successfully in both laboratory and field settings, and they feature a high degree of user-friendliness and simplicity of operation. They require the operator to perform an additional task (i.e., "secondary task") besides his primary, most important task. His performance on this "secondary" task is an index of his "reserve mental capacity." All tasks include stimuli presentation, response recording, and preliminary data analysis capabilities (Shingledecker and Crabtree (1982); Crabtree, Bateman, and Acton (1984); and Crabtree and Rolek (1983)).

AFAMRL/HEG has developed a set of tasks called the "Criterion Task Set," which measures human cognitive perfomance at each stage of information processing. The battery is currently in use in many different geographical locations (Hawaii, Amsterdam, Germany, Paris, Washington, D.C., Brooks Air Force Base, University of Oklahoma, Wright-Patterson Air Force Base, and others). The entire battery, which consists of 12 tasks, was written in BASIC and then compiled for the Commodore 64. The battery features self-administration, multiple menus, and built-in data analysis. Over 2500 hours of experimentation has been conducted with this battery (Shingledecker, Crabtree, and Acton (1982)).

Research is being conducted by SRL concerning the effects of various types of radar degradation on crew performance. With a very limited budget, data are being collected to assist with making decisions concerning the direction of future research in this area. A Commodore 64 personal computer costing under $200 was the only means available for providing these data within the cost and time constraints.

A Commodore 64 is also being used by AFAMRL and SRL to demonstrate how data collected on a man-in-the-loop mainframe simulation can be used in the cockpit to lower pilot workload and improve performance of his mission. The organization sponsoring this work has a very limited budget for this task and rapid turn-around time is a requirement. No other alternatives could be found to using an inexpensive personal computer to perform certain segments of this work. The mainframe couldn't handle the graphics requirements and the personal computer couldn't handle some of the real-time tasks. Together, they fulfilled the required specifications.

SUMMARY

This paper has demonstrated that a complete research apparatus using a personal computer costing under $200, with necessary peripherals costing perhaps another $500 to $600, definitely has a place in human factors research. The large computers which were typical of the '60s and '70s are no longer required for every research application. In fact, such systems may actually be detrimental to research facilities that desire to maintain maximum productivity and cost-effectiveness. Yet, there are many attitudes still supporting exclusive use of large systems and in opposition to small, efficient systems.

In essence, there is no real need to favor one approach over the other exclusively. There are still many requirements where only large systems make sense. There are also many applications in which both large and small systems can be used together effectively. And finally, there are those activities in which the flexibility, ease of use, and other outstanding features of the "under $200" personal computer make it the system of choice for many applications, such as preexperimental activities, testing of tentative concepts, developing experimental procedures, and perfecting test paradigms before they are applied in real-world settings. More extensive use of the inexpensive personal computer in human factors research could increase the efficiency at which we scientists conduct research and solve man-machine problems.

REFERENCES

[1] Eggemeier, F. T., Crabtree, M. S., and LaPointe, P. A., The effect of delayed report on subjective workload ratings, Proceedings of the Human Factors Society 27th Annual Meeting (1983) 139-143.

[2] Eggemeier, F. T., Melville, B. E., and Crabtree, M. S., The effect of intervening task performance on subjective workload ratings, Proceedings of the Human Factors Society 28th Annual Meeting (1984) 954-958.

[3] Shingledecker, C. A. and Crabtree, M. S., Subsidiary radio communications tasks for workload assessment in R&D simulations: II. Task

sensitivity evaluation, AFAMRL-TR-82-57, Air Force Aerospace Medical Research Laboratory, Wright-Patterson Air Force Base (September 1982).

[4] Crabtree, M. S., Bateman, R. P., and Acton, W. H., Benefits of using objective and subjective workload measures, Proceedings of the Human Factors Society 28th Annual Meeting (1984) 950-953.

[5] Crabtree, M. S. and Rolek, E. P., Using the personal computer in research environments, Proceedings of the IEEE 1983 National Aerospace and Electronics Conference (1983) 1002-1004.

[6] Shingledecker, C. A., Crabtree, M. S., and Acton, W. H., Standardized tests for the evaluation and classification of workload metrics, Proceedings of the Human Factors Society 26th Annual Meeting (1982) 648-651.

Trends in Ergonomics/Human Factors II
R.E. Eberts and C.G. Eberts (Editors)

An Interdisciplinary Approach to Human Factors in Computer System Development

Bonnie Kaplan
Departments of
Quantitative Analysis and Information Systems
and
Pathology and Laboratory Medicine
University of Cincinnati
Cincinnati, OH 45221
U.S.A.

Concrete implementation of human factors engineering principles has been neglected in medical computer software systems, in part due to ignorance of how to determine environmental features that bear upon user acceptance. A case history of designing and implementing an on-line, interactive computer system in a large metropolitan hospital illustrates methods by which such features were identified and incorporated into computer systems design.

INTRODUCTION

Despite considerable expenditures of time and money for medical computer software systems, Dowling[1] reports an estimated 45% failure rate of computer-based medical information systems that were technically sound. One reason is the failure to take the context of the system into consideration sufficiently. Human factors engineering principles have been neglected, in part due to ignorance of how to determine those features of the environment that bear upon user acceptance.

Users may not be able to articulate their needs or describe the relevant features of the environment into which a system must fit, and there are few techniques for testing and refining a system for its fit with the users' environment. The development of a computer system for clinical laboratory test order entry and results reporting illustrates how usual systems techniques were supplemented to help solve these problems.

BACKGROUND

A computer system to support the Pediatric Biochemistry Laboratory of Michael Reese Hospital and Medical Center, a 1000-bed teaching hospital in Chicago, was undertaken by the Department of Medical Physics at the request of the laboratory director in late 1976. An initial order entry and results reporting system was installed when, in October 1978, the laboratory approved a minimal trial system. Users complained that the system was too cumbersome and slow, and that it interfered with their other responsibilities. An upgraded system was installed in January 1979. It was in use, with repeated upgrading, until late 1981, when the functions of the Pediatric Laboratory were incorporated into the main biochemistry laboratory, which was using a different computer system.

The system has been described elsewhere.[2,3] It provided 24-hour support to a laboratory with approximately the same work load for each shift. It was interactive, on-line, and table driven. Each item was checked immediately after data entry, and the user could not continue until a valid entry was made.

This paper describes how the initial system was modified to be more acceptable to the laboratory technologists who were its primary users.

METHODS

> Because data and information management deals in abstracts, and projects are often conceived and initiated at a level of management removed from direct operations, it is not unusual for the basic development of processing and results to be done in terms of the idealized picture.[4]

User involvement and the usual methods of system analysis were not sufficient to produce a system that users would use. Attention to non-obvious system needs also was important because some system requirements are latent and cannot be stated explicitly by users.[5] Initial systems analysis discovered primarily manifest and idealized, rather than latent, system needs. Therefore, other methods were used to characterize laboratory work-load, data flow, and system requirements. We later realized that two key techniques used to supplement the earlier systems analysis techniques were adaptations of the social science methods of participant observation and unobtrusive measures.

Participant Observation

Participant observation, or direct fieldwork, allows the researcher to study social situations or complex social organizations by collecting data in terms meaningful to the subjects of study. Participant observation employs such techniques as observation, interviewing, document analysis, and participation, but requires repeated, durable social interaction by a researcher participating in the daily life and activities of the people comprised by the system under study.[6]

Prior to joining the Pediatric Biochemistry project, one of the designers had been a clinical chemist in the laboratory. Although her previous first-hand experience in the laboratory was an advantage, the use of participant observation does not depend upon it. Participant observation is often used by researchers to gain the knowledge that this kind of experience provided.

Participant observation was used to observe laboratory activities and actual system use. System developers worked in the laboratory, using the system to enter data, to correct software problems, and to assist users.

Unobtrusive Measures

Unobtrusive (or non-reactive) techniques assess behavior and outcomes without intervening directly in the social system.[7] Hard (printed) copy data entry sessions were examined to determine system usage patterns and to check system functioning. These hard-copy records enabled errors and problems to be traced, and users' behavior to be "observed" in the absence of system developers. In addition, system software was written to record the frequency and type of input errors. Both of these methods documented data patterns and input procedures.

RESULTS

Participant observation proved the key to determining latent system needs, with unobtrusive measures providing additional important information. Through designers' observations and use of the programs, data entry was streamlined. These changes, and others,[2,3] increased the accuracy of the data entered by decreasing the difficulties of data entry and by standardizing data. Ultimately, the number of key strokes was reduced by two-thirds. Moreover, the presence of system developers in the laboratory had further benefits: (1) it enables users to bring their problems and concerns to developers' attention, (2) it improved rapport between developers and users, and (3) it helped developers gain users' interest and cooperation in adapting the initial system to the needs of laboratory staff members. Using these methods resulted in changes in data entry, error handling, and user acceptance.

Data Entry

By entering data under real conditions in the laboratory, developers learned that data entry programs were not well adapted to entering data from laboratory test requisition slips, or to the pattern of test orders. With the guidance of users' suggestions, programs were changed to match data patterns and users' skills better.

Menus, which had been included to guide laboratory staff members in system use, proved time-consuming for experienced users. Consequently, users were given the options of eliminating menu text display and of entering common groups of tests as units instead of as individual tests.

Users also did not take advantage of features intended to speed entry but which instead required them to interrupt data entry flow. Default options, for example, were not used; users found it simpler to answer all questions than to remember which could default and what the default values would be. Explanations of defaults and data entry paths were included in the users' manual. However, when users did not interrupt data entry to refer to this manual, these explanations were printed at the terminal along with data entry program queries in order to assist the user. One user, who found these messages annoying, suggested that they be printed only if "problem" data were entered, and this were done.

Initially programs were designed to request patients' names rather than identification numbers in order not to depersonalize both patients and users. However, as one technologist told us, technologists preferred entering identification numbers instead of names because they were more skilled at numeric data entry. The programs were changed to take advantage of users' abilities, of which neither developers nor users had been explicitly aware previously.

At users' suggestions, data entry formats were matched to those used on test requisition slips, and changed each time these slips were changed. However, some requisition slips were handwritten, which made data entry more difficult because of the variety of formats used and the illegibility of data. Data verification routines consequently were changed to include some of these non-standard entries, or to allow free format entries from which new standards could be developed.

Users' styles of input were found to vary. Some users conformed to expectations unintentionally built into the programs by sorting requisition slips before entering them. Others, who entered information directly from unsorted slips, found the programs more difficult to use. The programs were changed so that either data entry method was convenient.

Error Handling

Frequency and type of input errors, as recorded in hard copy data entry sessions, were studied. Programs were modified by increasing latitude in acceptable input, and by ignoring specific "irrelevant" errors. These changes decreased the number of times users were interrupted by error messages. Where such "errors" could not be eliminated, programs were changed so that free format input was used, and these answers were studied to discover formats for later incorporation into data validation standards.

Unanticipated input styles also necessitated changes in error handling. Some users who were familiar with the system typed ahead of the program. Sometimes, the data that had been typed ahead was used by the program as input for a question other than the one it was intended to answer. When an error was detected, the user was therefore unable to determine where it had occurred. Modifying the programs to ignore all entries made after an error was detected solved the problems caused when users typed ahead.

User Acceptance

Participant observation, because it involved becoming part of the social system of the laboratory, increased rapport and improved communication between laboratory staff members and system developers. Participant observation and studying the hard copy sessions also provided evidence of problems that otherwise might have gone undetected. Designers' presence in the laboratory then made it easier to explain to users how "malfunctions" were due to their misunderstanding of system functions, or because of previous users' errors.

Participation also reduced suspicion or uneasiness and increased mutual trust. Laboratory staff members made suggestions for improving the system, and were further encouraged when their ideas were incorporated into system changes. Participant observation also provided an opportunity for laboratory staff members to observe system development and use, thus giving them informal training and reinforcing that the system was under human control.

DISCUSSION

Because medical problems are not well-defined, and "standard" practices are subject to human variability (of both patient and practitioner), it is helpful for system developers to work within the users' frameworks and to provide flexible systems that can be changed easily when users' needs have not been met. Kaplan[8] previously described the importance of users' frameworks in medicine and the need for developing good rapport with users in order to gain first-hand knowledge of what these frameworks are. Therefore, analysis, design, and implementation should incorporate more than manifest system requirements that can be determined through interviews and through study of data flow and work loads. Latent system requirements--such as skills users consider so commonplace that they do not know they have them--requirements that are not stated or recognized explicitly, must be met in order for a system to succeed. As has been reported, social science techniques are helpful in determining and meeting these requirements.[9,10]

The combination of participant observation, unobtrusive measures, and software engineering techniques that made the programs easily adaptable, enabled us to implement Kaplan's suggestions for taking the medical environment into account through (1) flexible software that allowed for users' skills and idiosyncracies, and for lack of standard medical problem definition, (2) ease of system use, and (3) encouraging a sense that the user is in control. These goals were met by (1) establishing good rapport with the users, (2) understanding each user's model, and (3) providing user-oriented software.[8]

These techniques may seem to require a large commitment of developers' time, but this need not be the case. In the four years between the laboratory director's initial request for a computer system and the last version of the system, only during the three months following installation of the initial minimal system were these methods used intensively. We felt that unless this effort was made, the users would reject the system. These three months changed a technically sound system into one that was better accepted by the users. After installing a new version of the system, designers' interaction with the laboratory was reduced, but the rapport remained, allowing further improvement of the system through observation and discussion with the users.

CONCLUSIONS

Three important general lessons were reinforced by this experience: (1) the value of all users as full partners in system development, (2) the impossibility of fully specifying a system in advance, and (3) the importance of human factors in insuring acceptance. Supplementing usual systems analysis methods by techniques adapted from the social sciences enabled the data entry programs to be customized for the needs of laborator technologists. In particular, the following approaches to determining user requirements were especially helpful:

(1) using scientific methods to determine latent system needs and user skills;

(2) designing to take advantage of users' skills and the laboratory environment;

(3) iterative systems development that incorporated human factors findings into each cycle.

We believe such approaches can be productive in reducing the high cost of system failure.

REFERENCES

[1] Dowling, A.F., Do hospital staff interfere with computer system implementation? Health Care Management Review 5 (1980) 23-32.

[2] Nobel, D. and Kaplan, B., User-oriented approach to developing an interactive computer system for a laboratory, in: Proceedings, MEDCOMP'82, the First IEEE Computer Society International Conference on Medical Computer Science/Computational Medicine (IEEE Computer Society, Silver Spring, Md., 1982).

[3] Kaplan, B. and Nobel, D. An interdisciplinary approach to user oriented interactive laboratory computer system development, in: Proceedings, First Annual Conference, the American Association for Medical Systems and Informatics (American Association for Medical Systems and Informatics, Bethesda, MD., 1982).

[4] Thompson, G. and Handelman, I., Health Data and Information Management (Butterworth's, Woburn, Mass., 1979).

[5] Merton, R.K., Social Theory and Social Structure (The Free Press, Glencoe, Ill., 1949).

[6] McCall, G.J. and Simmons, J.L. (eds.), Issues in Participant Observation: a text and reader (Addison-Wesley, Reading, Mass., 1969).

[7] Webb, E.J., Campbell, D.T., Schwartz, R.D., and Sechrest, L., Unobtrusive Measures (Rand McNally, Skokie, Ill., 1966).

[8] Kaplan, B., The context of medical computer software systems development: medical user needs and expectations, in: Dunn, R.A. (ed.), Proceedings of the Third Annual Symposium on Computer Applications in Medical Care (IEEE Computer Society, Silver Spring., Md., 1979).

[9] Brown, B., Harbort, B., Kaplan, B., and Maxwell J., Guidelines for managing the implementation of automated medical systems, in: Heffernan, H.G. (ed.), Proceedings of the Fifth Annual Symposium on Computer Applications in Medical Care (IEEE Computer Society, Silver Spring, Md., 1981).

[10] Kaplan, B., et al., Systems analysis, project management, and evaluation: are quantitative methods enough?, in: Dayhoff, R.E. (ed.), Proceedings of the Seventh Annual Symposium on Computer Applications in Medical Care (IEEE Computer Society, Silver Spring, Md., 1983).

Trends in Ergonomics/Human Factors II
R.E. Eberts and C.G. Eberts (Editors)

TOWARD A MODEL OF AMBIGUITY

William Leigh
Univ. of Southern Mississippi, Hattiesburg, Mississippi, U.S.A.

Ray Souder
Northern Kentucky Univ., Highland Hts., Kentucky, U.S.A.

Perhaps the central problem in natural language understanding by computer is the lack of an adequate model of syntactic and semantic ambiguity. Such a model should: a.) identify the presence of ambiguity in natural language statements; b.) enumerate the alternate interpretations possible; c.) rank the possibilities as to likelihood of being correct. In this paper we examine the conventional model for grammar, the transition network, in the light of these requirements. We propose a type of model for this purpose which is rarely so considered, discrete optimization.

INTRODUCTION

One approach to natural language interpretation is to regard natural language as a formal language with a large grammar. If such a grammar can be obtained, then natural language may be interpreted using the compiler mechanism developed in computer science for the translation of higher-level programming languages to lower-level machine languages.

The compiler may be described as a program which analyzes an input string by submitting it to a sequence of distinct phases of analysis: lexical, syntactic, and semantic.[1] Each phase in the analysis usua-ly arrives at a unique interpretation according to its area of knowledge before the next phase begins.

The lexical phase classifies tokens (or words) into lexical categories. These categories might be noun, adjective, verb, etc., for natural language; or variable, operator, etc., for a programming language. The lexical determination is accomplished using only the form and content of the token with no consideration of the context surrounding the individual token.

The syntactic phase is what is conventionally called the "parsing" phase. It is here that larger structures such as noun phrase, verb phrase, clause, and sentence are recognized in natural language. Statement, expression, etc. are the larger structures for a programming language. Much research in formal grammar techniques has gone into making the syntactic phase efficient.

The semantic phase synthesizes the target language code from the structure perceived by the lexical and syntactic phases. Default conventions are used to supply missing information; the final association of references to concepts is made (symbols to addresses for programming language), and errors are detected.

The partitioning of the task into these distinct phases is a "divide and conquer" strategy which demands assumptions about the input string to be interpreted. Backtracking to a previous phase of analysis is not allowed. Ambiguity must be resolved within a phase and not between phases.

The syntactic requirements on the input string are specified by requiring that the input string be a member of a language generated by the rules of a formal grammar. Such a grammar may be alternatively specified as a transition network.[2] An example transition network for a sentence:

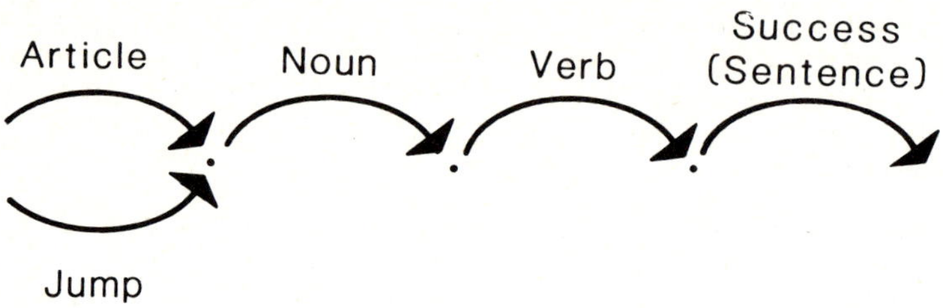

The example transition network accepts sentences such as "The pitcher pitches."

Many phenomena in language are generative in nature and require a recursive transition network:

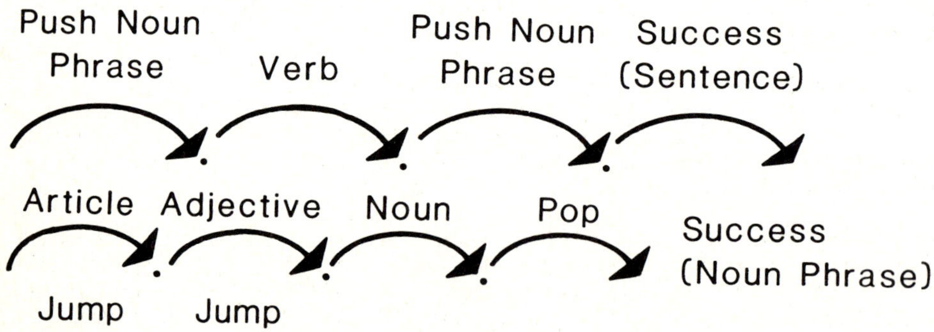

The addition of the "Push" and the "Pop" capabilities gives the recursive transition network the power of recognizing language generated by context-free grammars.[3] The example network can accept sentences such as "The tallest player caught the ball."

A compiler-model interface processes a query such as "How many trucks are in New York?" in distinct passes. The first pass is that of the lexical phase. The lexical process looks up each word in a dictionary and assigns a part of speech:

Adverb Adjective Noun Verb Preposition Noun

The syntactic phase determines the structure of the query by matching the lexical assignment against a recursive transition network. The semantic phase uses the path through the network found by the syntactic phase to guide the generation of a computer code for answering the question. The semantic phase has available to it code generation rules associated with the arcs in the syntactic transition network. These rules specify how the code is generated in the semantic phase.

The use of the compiler as a model for a natural language interface allows only a subset of English for command or query specification. This is so that the size of the grammar may remain in manageable bounds. Also, a complete grammar for English has never been defined. The approach of using the compiler as a model with a subset of English has been used in some experimental systems.

One of the earliest natural language interfaces for database question-answering was BASEBALL (Green (1961)). This program answered questions concerning the month, place, teams, and scores of baseball games. The system was patterned after the compiler with separate lexical, syntactic, and semantic phases. Within its very constrained domain of knowledge, the system was successful. However, the type of information in the database was limited and structured to facilitate question-answering. The methods used are not generally portable to other domains of information.

TQA, "Transformational Wuestion Answerer," is a recent compiler-model question-answering system (Plath (1976)). A context-free grammar recognizer is used in the syntactic phase. The semantic phase uses rules which can transform the syntactic parse into a "canonical" form. This form is easily translated into code. This system is currently being evaluated on a long-term basis for querying land-use data in a city in New York.

Shortcomings in the compiler-model approach have been encountered. There is low tolerance for ambiguity. It is difficult to arrive at a suitably complete grammar for English. Such systems have little ability to handle fragmentary, incomplete, or ungrammatical input such as is commonly received from human users (Hayes and Mouradian (1981)).

Simon (1969) observed that syntax ans semantics are nearly decomposable subsystems and that there is a small, fixed set of possible interactions between them. The compiler-model approach capitalizes upon the decomposability. However, a solution to the problem of natural language interpretation may require more emphasis on the interaction.

ALTERNATIVES TO THE COMPILER-MODEL APPROACH

Regarding natural language as a formal language with a large grammar requires the identification of such a grammar. In practice this has been found to be difficult. Also, the requirement that any ambiguity be resolvable within a single type of knowledge source, or phase, may be too restrictive in practice. There may be a unique solution meeting all the criteria of the various phases in total, but in many cases that solution can be determined only if the knowledge in the phases is combined to analyze the input string. Some alternatives to the strict compiler-model approach have been developed to exploit an interaction between syntactic and semantic sources of knowledge.

One such alternative is based on the Augmented Transition Network (ATN) formalism (Woods (1970)). The ATN allows semantic meanings to be attached to the arcs of a recursive transition network. Also, whereas compiler-model formal language algorithms require arc choices to be unique or resolvable within a look-ahead of finite length, the ATN approach allows non-unique choices for arc branches with a "most-likely" ordering assigned.

An example ATN using the Noun Phrase transition network shown previously:

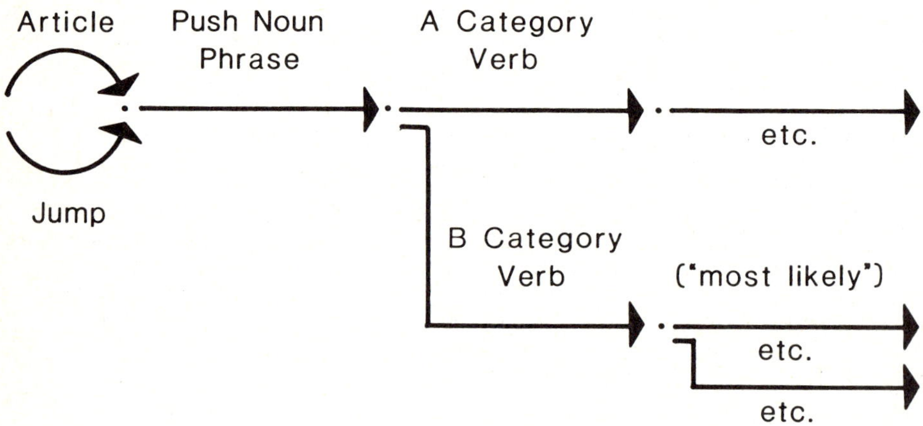

This ATN allows verbs of two semantic categories, A and B. For example, A category verbs might be those concerning travel to flying, so the succeeding network would involve objects such as "airplaes," etc. B category verbs might concern travel by water, so "boats" would be subsequently suggested. Notice that two branches are allowed for B Category verbs. One is marked as the "most likely." The semantic significance of this might be that travel by water in steamships is more likely than in sailing ships.

ATNs were employed successfully in the LUNAR database question-answering system (Woods,Kaplan and Nash-Webber (1972)). This program served as an interface between NASA geologists and a database of information on samples of moon rock. Like BASEBALL, LUNAR exploited unique characteristics of the information in its database. This information was simply structured and possessed attributes making it amenable to natural language manipulation. A large vocabulary or grammar was not required. LUNAR was one of the more successful natural language experiments. Another mechanism employing ATNs is ROBOT, a natural language database question-answering system which has been offered commercially (Harris (1977)).

The ATN is an effective formalism for natural language interfaces. However, transferring an ATN-based system to another domain is difficult since much domain-dependent knowledge is built into the ATNs. Also, the application must be well-structured and small in scope in order to keep the ATN grammar within a reasonable size. Efficient algorithms for recognizing this type of grammar have not yet been developed, and parses can become lengthy.[4]

The ATN allows syntactic and semantic knowledge to coexist in an interpreting network. Some experimenters have noticed that most of the power in interpretation in this type of mechanism seems to come from the semantic information.[5] They have developed approaches that use semantic considerations only, with no reference to the usual syntactic categories of noun, verb, and so forth. In practice this has involved developing various semantic grammars which use more semantically meaningful categories. The semantic grammars have either been general, applying to common and elementary concerns of humans such as "going," "getting," and "doing," or specific, applying to a narrow domain. Those experiments in developing grammars of the first category are basic research and are not meant to result in working database interfaces. Those of the second type, the domain-dependent semantic grammars, have resulted in some of the most successful systems, and this is the method of choice at the present (1983) state of the art of natural language interfaces to databases. PLANES (1975) and LIFER (1978) are recent systems based on semantic grammar approaches.

PLANES allowed Navy flight-maintenance mechanics to query a database of information about airplanes (Waltz (1978)). The PLANES semantic grammar is expressed as a set of "concept case-frames." Each concept case-frame is a set of data attributes which are likely to be associated in a query. For example, the query, "What maintenances were performed on plane 48?" would be matched to a concept case-frame:

QWORD MAINTTYPE MAINTACTION PLANETYPE

A program accepts the parameters of the slots in the concept case-frame and searches the database for the proper information. The order of the semantic references in the query was found to be not significant in this experiment. The concept case-frames also held default information to be used in the database search if the slots were unfilled explicitly by the query. (PLANES is used as a benchmark system for evaluation in this work.)

LIFER lets the user enter his own semantic grammar sentence templates (Hendrix (1978)):

WHAT IS THE (ATTRIBUTE) OF THE (SHIP)?

Along with the template, the user must supply a formal query language statement which, when supplied with the (ATTRIBUTE) and (SHIP) values, will process the database to provide the answer. If the user's subsequent natural language questions matches the template, then the values are supplied to the formal query language statement for execution.

PLANES and LIFER are domain-dependent because the programs require modification for application to a different type of database. They both imply an enumeration of query types before they can process natural language from "casual users." The grammar resulting from this approach becomes large and cumbersome if the domain is not restricted.

The alternatives to the compiler-model approach such as ATNs and semantic grammar approaches are efforts to bring specific, semantic knowledge to bear on the interpretation of the natural language. The objective is to assemble enough knowledge so as to make only one unique interpretation possible. Theoretically, such a constraint set can be assembled. However, in practice the knowledge required to arrive at a unique interpretation is sometimes not completely present. In these cases the system should be able to generate alternate interpretations and rank them as to likelihood.

The ATN handles this in a greedy manner selecting the "most likely" interpretation for each arc. This is not an optimization, but a greedy, "satisficing" approach. ROBOT, however, generates multiple queries for such an ambiguity and processes them all against the database. The reply returning a non-null answer is selected as "most likely." Multiple non-null answers are a problem.

Another available knowledge source is the human user. Humans often ask for clarification in conversation. RENDEZVOUS asks the human for such clarification whenever an ambiguity is encountered, and look-ahead is not employed (Codd (1974)).

COPING WITH AMBIGUITY THROUGH OPTIMIZATION

The hopelessness of ever having a complete syntactic grammar of English is generally accepted.[7] The continuing presence of ambiguity, even when operating with a syntactic grammar adequate for a restricted domain of usage, is recognized.[8] These problems prompted the development of the semantically-oriented parsing methods as alternatives to the syntax-driven compiler-model approach.

In recent work the possibility of resolving natural language ambiguity on completely semantic bases is questioned.[9] The necessity of integrating syntax and semantics along with any other knowledge sources is being recognized. From an initial preoccupation with syntax in the 1960s to a reliance on exclusively semantic methods in the 1970s, computational linguistics is now developing integrated methods. However, an integrated method has not been tried in an operational environment, but only in test parsers.[10]

If it is accepted that it is not usually practical to limit the domain of application sufficiently and/or to develop a syntactic and semantic constraint set complete enough so as to insure always having a unique parse, then ambiguity must be admitted. Indeed, one definition of "natural language" cites the necessary presence of ambiguity as a consequence of open-endedness of the application domain (Kummel (1979)).

Examining the compiler-model approach and its alternatives from the optimization point of view highlights the inherently greater power of optimization models. The compiler-model approach exploits no coupling of the constraint sets and does not allow multiple feasible solutions (ambiguous parses). ATNs employ a limited coupling of syntax and semantics. ATNs allow multiple feasible interpretations but choose between them on a greedy basis at the arc level, with no global optimization at all. The "concept case-frame" and the semantic grammar rely on totally enumerating the solutions so as to apply a special constraint to each. There is no organized method for determining the "goodness of fit" of the case-frames or the grammars. In general, it is not specified what to do if several solutions fit equally well. ROBOT consults the database, and RENDEZVOUS, the user.

There must be some way to evaluate comparative "goodness of fit" to resolve the ambiguity. This implies some type of objective function. Systems have endeavored to maximize a "semantic density" but have only used greedy methods (Wilke (1975)). Other system have used a greedy "distance" metric on the database schema. Our research examines whether an integer programming framework can couple syntax with these and other dimensions of language and more efficiently guide the search for the "best fit."

The main objective of our research is to develop a model which considers the complications of natural language interpretation more completely than the compiler-model approach or its alternatives. The compiler-model approach is a unified and economical, though simplistic, framework. The alternatives such as the ATN and the semantic grammar begin to integrate the syntactic and semantic knowledge sources so as to profit from interaction. However, these alternatives only use that interaction in a greedy fashion. We propose a model which will more clearly define such interaction: optimization.

A restricted subset of the grammar implicitly used in our work, expressed in the phrase structure formalism, is of the form:

S abS
S ab
S cdS
S cdeS
S cde
S efgS
S (null)S
S (null)

Such a grammar can be used to parse a query:

a b c d e f g ?

(1) a b
c d
e f g

OR

(2) a b
c d e
(null)
(null)

A parsing scheme which works from left to right selecting the longest fitting rule will arrive at parse 2 above. Parse 1, however, may be better on several counts. Parse 1 would be picked by a parsing method which picked the parse including the fewest rules. The example of a parse which results in parse 1 is the example which is used in this research.

The parsing method first maps and fits grammar rules to the query. Each rule is fitted as a column:

<u>Query</u>

a	a				(null)						
b	b					(null)					
c		c	c				(null)				
d		d	d					(null)			
e			e	e					(null)		
f				f						(null)	
g				g							(null)

This fitting matrix is converted to a 0-1 matrix:

1	0	0	0	1	0	0	0	0	0	0
1	0	0	0	0	1	0	0	0	0	0
0	1	1	0	0	0	1	0	0	0	0
0	1	1	0	0	0	0	1	0	0	0
0	0	1	1	0	0	0	0	1	0	0
0	0	0	1	0	0	0	0	0	1	0
0	0	0	1	0	0	0	0	0	0	1

This 0-1 matrix is the constraint matrix for an integer program. Each row in the 0-1 matrix is set equal to 1. This may be interpreted to mean that each word in the query must have some rule or part of a rule assigned to it. The variables in the integer program are 0-1 corresponding to the columns. A solution to the integer program is a list of the columns which are used to parse the query.

The rules used in this work have the general form:

S aSbSc.

The flexibility of the method used in fitting these rules is inconsistent with conventional phrase structure grammer practice. The context rules are allowed to overlap:

query:	a	b	c	d	e	f	?
parse:	a		c	d			
		b			e	f	

Variable length gaps are allowed in the fitting of the context rules. The possibility of overlapping fittings greatly increases the number of phrase structure rules necessary for an equivalent conventional representation.

Semantic fittings and the syntactic fittings result in submodels in the integer program. The submodels are coupled with coupling constraints. (Those coupling constraints would require additional context-sensitive rules to model in the phrase structure grammar formalism.) The resulting constraint matrix is of the form:

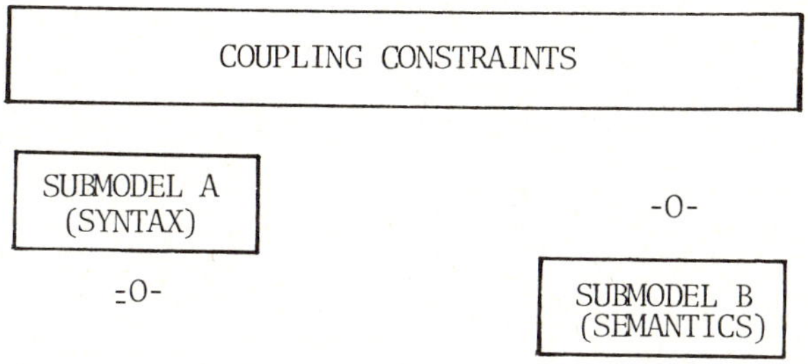

The method of parsing queries employed in this research differs from conventional methods:

1. In that the rules used for expressing the grammar are of a different formalism than that of usual phrase structure rules or the equivalent ATN rules. This new type of rule is difficult to model equivalently in the phrase structure formalism and requires in many cases context-sensitive rules, which are not conveniently handled in conventional systems.

2. In that syntactic and semantic rules are integrated into one coupled model. Conventional practice is to handle these two processes separately.

3. In that the method employs a global optimization rather than a greedy selection of the parse.

Casting the parsing and interpretation problem in the optimization format permits the appealing concept of ambiguity as the presence of alternate optima or near optima. Ambiguity is now a density concept, with more ambiguity implying more feasible solutions in the neighborhood of the optimum. This suggests a metric for ambiguity.

FOOTNOTES:

(1) There are many textbooks on compiler design. For instance, see Peter Calingaert, Assemblers, Compilers, and Program Translation (Computer Science Press, Inc., 1979).

(2) For a transition network explanation of formal language methods (as opposed to the equivalent phrase structure grammar presentation), see Harry Tennant, Natural Language Processing (New York: Petrocelli, 1981).

(3) The formal models of grammar can be divided into classes corresponding to the type of automata (formal models of computing machines) which can recognize languages generated by them. The simplest class of grammar is the "regular grammar," and it can be recognized by a "finite automaton." The next more general class in the hierarchy is the context-free grammars. This class requires the more powerful "push-down automaton" for its recognition. Programming languages are usually of this grammar class. The next class of grammar, of increased complexity and generality, is the "context-sensitive" grammars. These require a "linear-bounded automaton." The most general class is that of the unrestricted grammars, which is all those that require the full power of the "Turing Machine" for their recognition.

(4) This problem is addressed by M. P. Marcus, A Theory of Syntactic Recognition for Natural Language (Camgrigce, Massachusetts: The MIT Press, 1980). This book also contains a critical review of natural language methods.

(5) The work of Schank is expecially associated with this point of view. See Tennant or R. C. Schank, Conceptual Information Processing.

(6) Algorithms are considered "greedy" when they proceed by selecting alternatives to include in a solution based on optimization criteria applied locally. For many problems, this results in solutions which are sub-optimal by the same criteria applied globally.

(7) See Hayes and Mouradian, "Flexible Parsing," American Journal of Computational Linguistics, 7 (October-December 1981). This article describes a parser developed for handling fragmentary input and also provides an overview of the limitations in the current state of art in natural language processing.

(8) P. Dell'Orco, V. Spadavecchia, and M. King, "Using Knowledge of a Data Base World in Interpreting Natural Language Queries," in Information Processing 77, ed. B. Gilchrist (North Holland,1977). This article suggests the use of the database schema information in conjunction with a semantic grammar for interpreting queries. A demonstration of the problem of determining references syntactically is supplied as a motivation for the work.

(9) See Marcus, A Theory of Syntactic Recognition for Natural Language (Cambridge, Massachusetts: The MIT Press, 1980). This book also contains a critical review of natural language methods.

(10) M. P. Marcus, The test parser is one of Marcus, called PARSIFAL, A Theory of Syntactic Recognition for Natural Language (Cambridge, Massachusetts: The MIT Press, 1980).

REFERENCES:

(1) Codd, E.F., edited by J. W. Klimbie and K. L. Koffema, Seven Steps to Rendezvous with the Casual User, in Data Base Management (North Holland, 1974)

(2) Green, B., et al., Baseball: An Automatic Question Answerer, Computers and Thought (New York: McGraw-Hill, 1961).

(3) Harris, L.R., User Oriented Data Base Query with the ROBOT Natural Language Query System, International Journal of Man-Machine Studies 9 (1977) 697-713.

(4) Hayes, P.J. and Mouradian, G.V., Flexible Parsing, American Journal of Computational Linguistics, 7 October-December (1981).

(5) Hendrix, G.G., et al., Developing a Natural Language Interface to Complex Data, ACM Transactions on Database Systems 3 (2) June (1978) 105-47.

(6) Kaplan, S.J., On the Difference Between Natural Language and High Level Query Languages, Proceedings of the ACM 78, (1978) 136-147.

(7) Kummel, Paul, Formalization of Natural Language (Springer-Verlag, New York, New York, 1979).

(8) Plath, W.J., Request: A Natural Language Question-Answering System, IBM Journal of Research and Development 20 (6) July (1976) 326-35.

(9) Simon, H.A., The Sciences of the Artificial (The MIT Press, Cambridge, Massachusetts, 1969).

(10) Waltz, D.L., An English Language Question Answering System for a Large Relational Database, Communications of the Association of Computing Machinery, 7 July (1978).

(11) Wilks, Y., An Intellegent Analyzer and Understander of English, Communications of the Association of Computing Machinery, 18 (5) May (1975) 264-74.

(12) Woods, W.A., Transition Network Grammars for Natural Language Analysis, Communications of the Association of Computing Machinery 13 (10) October (1970) 591-606.

(13) Woods, W.A., Kaplan, R.M., and Nash-Webber, B., The Lunar Sciences Natural Language System: Final Report (NTIS N72-28984, 1972).

Trends in Ergonomics/Human Factors II
R.E. Eberts and C.G. Eberts (Editors)

ERROR CAUSE ANALYSIS

Joseph S. Papa
Julie Dolyak

AT&T Communications
White Plains, NY

A mechanized system of Error Cause Analysis is proposed to aid the Human Factors Engineer and related professionals, in designing effectively for error control. The system will serve as an evaluative tool for new and existing large computer applications. As an evaluative tool the system may go beyond the data made available by the current labor intensive manual analysis and provide savings in both time and money. A further advantage of the system is to increase user satisfaction and performance by enabling the Human Factors Engineer to assess the effectiveness of training, documentation, screen format design, and edit message design.

INTRODUCTION

In todays automated office environment, the Electronic Data Processing (EDP) system is generally considered to be a reliable tool for the efficient achievement of work goals. In this sense, the connotation of reliability is that EDP systems are static instruments until replaced by the next generation of technology. Humans on the other hand, although reliable, are inherently a dynamic lot by nature. A users job performance can be affected by variables such as: the deprivation of the basic needs, relationships, motivation levels, environmental factors, and the user-friendliness of computer systems. One indication of affected job performance can be the making of errors when working with computer systems. However, errors cannot be solely attributed to users alone; large EDP systems also have their own idiosyncratic error sources, such as hardware or software bugs. It is reasonable to state that errors, whether computer or user based, tend to degrade the total efficiency and effectiveness of a system. From a behavioral standpoint, as the error rate increases and approaches a maximum level, the system may in fact be rendered unusable. Furthermore, the cost to correct errors increases in direct proportion to the elapsed time from when the error occurs to the point where the error has been remedied. Thus, the cost to an organization in both user time and money to treat errors can be quite expensive.

TREAT THE CAUSE NOT THE SYMPTOM

The last step in the development of an EDP system is to perform an evaluation of total system performance. Typically, the objectives of this analysis are to provide evaluative findings concerning: system performance, the interactive performance of the system and the user working conjunctively, the development effort, user attitudes about the "friendliness" of the system, and the bottom line economic impact. During the data collection phase of the system performance analysis, much time is often spent on determining what data should be collected, the sources of data, and the method of data collection. These steps are essential because the sample data for analysis must be representative of the population from which it was derived. However, equal attention must also be given to thinking about how to utilize the collected data.

One of the major measurements that is evaluated during a system performance analysis is the overall error rate. Errors are defined as "wrong or omitted data fields that occur as direct result from incorrectly performing an activity. Another error cause, is the introduction of an activity to a system which should not have been performed at all" (Bailey, 1983)*. In short, our research efforts center on developing a mechanized Error Cause Analysis (ECA) methodology by utilizing error data/statistics.

The existing error statistics limit the Human Factors Engineer to minimal data such as acceptance and rejection rates. These statistics do not provide the additional information necessary to fully evaluate the types and sources of errors. Today, the methodology responsible for providing the additional information is a labor intensive manual analysis. Although the manual analysis goes beyond the current statistics in providing the information useful in error control, it is too labor intensive to collect. For example, a manual analysis of a data dump showing all transactions made within one day, for only 2 types of screens, by multiple users on a large EDP system, required approximately 16-20 hours of work. Additionally, due to the lack of some relevant information such as user identification and time delays between data transactions, this method remains inefficient.

With a mechanized Error Cause Analysis system its efficiency can be readily defined in terms of large labor reduction costs and speed. This system should also be "portable" so it can be utilized by most EDP systems thereby maximizing its efficiency. The mechanized ECA system for example, would first direct the Human Factors Engineer to a statistics screen displaying overall error rates summarized by a screen, edit message number, or total number of transactions for some period of time. The investigator would then have the additional capability of windowing to another screen depicting a further analysis of the selected error summary. The concept of this system is to provide the investigator with an increased progression of detail or breakdown concerning system errors to accurately identify error causation and treat accordingly. For example, this type of data breakdown might indicate that a screen or edit message was responsible for a large percentage of the rejected transactions, directing the Human Factors Engineer to assess the "user friendliness" of these human/computer interfaces. If the error breakdown indicated that 40 of 50 errors were committed by a single user, the investigator could focus on that one user rather than on the population of users.

CONCLUSION

The result of our research is the proposal of a on-line system to provide for a methodological approach to the analysis of error cause which is more effective than the present labor intensive and costly manual analysis.

Such a system may enable the Human Factors Engineer to assess the effectiveness of: training, documentation, screen format design, and edit message design. Once the effectiveness of each of these components has been assessed, the investigators will have a better understanding of which direction to take in designing for the prevention of errors. By aiding investigators to determine the appropriate error preventive measures the ECA system may limit expenses, provide for a more user-friendly system, and increase user satisfaction and productivity. Thus, a system concerned with the causes and sources of system errors has the increased potential to operate at an optimal level by directly treating error preventive measures.

REFERENCES

*1. Bailey, Robert W., Human Error In Computer Systems. (Prentice-Hall, Inc., N.J., 1983)

2. Bailey, Robert W., Human Performance Engineering. (Prentice-Hall, Inc., N.J., 1982)

3. McCormick, Ernest J., Job Analysis: Methods and Applications. (Amacom, N.Y., 1979)

4. Shneiderman, B., Software Psychology - Human Factors In Computer And Information Systems. (Winthrop Publishers, Inc., Mass., 1980)

Trends in Ergonomics/Human Factors II
R.E. Eberts and C.G. Eberts (Editors)

IN DEFENSE OF THE TRADITIONAL (NON-ICON) INTERFACE: A POSITION BASED ON A SELECTIVE REVIEW OF THE LITERATURE

Debra M. Switchenko, Ph.D.

Burroughs Corporation
Plymouth, MI
U.S.A.

The purpose of this presentation is to provide some thoughts and comments relevent to the issue of interface "style;" particularly with reference to the use of icons as an interface technique. These thoughts are derived from readings within a variety of disciplines (referred to below). They are presented here in hopes of providing some impetus to the long range goal of developing adequate criteria for selecting (user, task, and functionally) appropriate interface techniques in computer system design. Additionally, there exists the short range goal of alterting readers to the shortcomings of the literature base on which present-day design decisions are being made. It is hoped that this review and analysis will stimulate further research, and ultimately provide the human factors community-at-large with more information for design recommendations.

Statement of the Problem

Since the introduction of the Xerox Star, and especially since the marketing success of Apple's Macintosh, an increasing amount of enthusiasm has been generated for the incorporation of icons into the human/computer interface. Enthusiasts promote a variety of assertions regarding icons, including: (1) that icons are more quickly recognized and their meaning more easily conveyed than analogous verbal (printed) labels; (2) that icons communicate a greater amount of information in a smaller amount of space than verbal labels; (3) that they provide an effective mechanism for introducing the novice user to the complexities of computer assisted tasks. Thus, primarily because of their assumed efficiency and simplicity, icons are widely touted as the principle components of an idealized user-friendly system.

The position taken in this paper is that the industry trend towards the implementation of icons (or minimally, their acceptance) as a "user-friendly" interface technique is moving well in advance of published literature regarding the efficacy of this approach. This position will be defended with the citation and evaluation of selected published reports from three domains: the evaluation of traffic symbols, the cognitive processing of picture/word symbols, and finally, the (limited) available human factors evaluation of iconic interfaces.

1. Traffic Symbols

An understandable interest in the relative efficacy of printed and symbolic (e.g., pictographic) messages has existed in the traffic directives literature for some time. The early published studies have the virtues of simplicity of design along with readily controllable relevent contextual factors. Various measures of performance, including glance legibility, distance legibility, and speed of comprehension, have been taken. Although, as in many experimental disciplines, the data have been equivocal, there is general recognition within this field that icons are superior to printed commands under degraded viewing conditions, and may generally be associated with speedier driver responses.

Again, as in many experimental disciplines, this rendering of the issue is clouded with some problems concerning experimental task requirements. For example, Dewar and Ells(1) and Dewar, Ells and Mundy(2) reported that printed verbal messages were responded to faster than pictographic messages in a recognition task; latency of response between tachistocopic presentation of an isolated symbol to vocal naming response was measured.

Later, Ells and Dewar(3) acknowledged that the naming response may have artifactually facilitated responses to the printed stimulus relative to the pictographs, because of the high degree of compatibility between the verbal stimulus and the verbal response. (Subjects in this condition may have simply read the word.) In their modified test, subjects peformed a same-different judgement to the two types of messages. An experimenter read aloud a traffic message; this was followed by the tachistocopic presentation of a single message stimulus; subjects verbally responded "yes" or "no" to the question "Are these two items (i.e., the projected message and the antecedent referent message) the same?"

With this modified test, responses to pictographs were speedier than to printed messages. Whitaker and Stacey(4) confirm this finding. However, the modified task may have produced a different type of artifact. In this case, competition between two inconsistent verbal items - the reading of the stimulus word and the required "yes" or "no" response - may have inflated response latencies in the word condition relative to the pictographic condition.

The Whitaker and Stacey report poses some additional problems for the extrapolation of this research into the domain of human/computer interface. Their investigation employed an arrow as the principal stimulus representative for the pictograph condition. There is some suggestion by Rohr and Keppel(5) that an arrow may behave untypically with respect to other computer-based icons. Specifically, in their report of user ratings of various icon features, only the arrow produced significantly high values for functionality and manipulative information. More than other stimulus attributes, arrows easily convey the meaning they represent. Their high meaningfulness value may render them inappropriate as stimuli on which to determine performance with computer based icons.

There is a final caution which should be levied in the consideration of this literature base. There are considerable differences between the tasks and the stimulus environments of the driver in a car and the computer user. Consider that the driver of a car is motorically active within a three dimensional stimulus space. In addition, the visual distinction between the natural environment (with its perceptible dimensions of shape, texture, shading, and the full range of natural colors) and the pictographic cue (limited to the stimulus properties representable in two dimensions) can be dramatic. While it could be argued that the user of a keyboard (or any other interface device) is "motorically active," it must be recognized that the computer users' task is primarily an abstract one (i.e., dealing with representations on the screen) and their "actions" are performed at the interface of the two dimensional stimulus area, and not integrated within it. Because of these task and environmental differences, the roles that icons play in conveying information to the driver/user, may themselves be different.

In sum, the contribution that the traffic directive literature can make to our understanding of iconic performance must be evaluated with caution; this because the literature is equivocal with respect to the rapid comprehension measure of symbol types, because of possible differences between the typical traffic and computer based pictographs, and finally, because of recognizable differences between the drivers' and the computer users' task environment.

2. Cognitive Psychology

Cognitive psychologists have long been interested in processing differences which may exist between various representational forms. Data from this domain bear indirectly on the issue at hand. Most studies in cognitive psychology are performed with the goal of modeling the unseen psychological processes pertinent to various symbol types. Rarely is a strict performance comparison taken (for example, merely asking, "Under what conditions is performance faster?") unless important theoretical information is obtained as well.

Nonetheless, research from this domain does provide food for thought concerning various human/computer interface issues. Hemenway(6) has discussed some psychological issues which are important considerations in the use of icons.

As she points out, graphic representations can be made more visually distinctive from each other than are printed words. This provides for a possible performance benefit: the ability of users to discriminate between items and locate a target item more quickly for icons than words.

Further, she suggests that graphical representations may save screen space, because of their ability to represent, in one symbol, information which would require more than a few words to describe. She illustrates this point with her presentation and description of a graphics package menu, meanwhile providing comment that graphic packages are particularly well-suited to benefit from this possible advantage. (Their ability to save screen space for concepts which are less easily represented by an icon needs to be considered.)

Hemenway discusses various attributes of icons which may influence their initial comprehension and subsequent retention. One important consideration is the "directness of the link between what's depicted and the command". The more straightforward the link (e.g., a graphic representation of a telephone to represent phone services), the greater the presumed benefit of the icon. Icons that rely on analogies in order to convey meaning, or arbitrarily shaped or coded icons, suffer from a decreasing lack of "directness."

Hemenway's presentation poses many interesting hypotheses. She encourages research on these questions, accurately stating, "to a large extent the effects of icons on users' performance with a system are unknown."

Switchenko(7) has reviewed the basic psychological literature relevent to the evaluation of iconic interfaces. She examines several widely known findings which are potentially interpretable as favoring the use of icons.

First, there exists a large literature base documenting differences in the retention of information presented in a picture versus comparable information presented in a printed verbal form. In both recall and recognition, performance with pictures is superior to performance with words(8). Presumably, graphic representations can share in this advantage to the extent that they are more picture-like than the words they represent.

Secondly, there is a memory advantage for concrete concepts over abstract concepts, as demonstrated in a wide variety of memory tasks(9). This is relevent to the icon issue, since it is widely assumed that by providing a pictorial representation, icons "concretize" the concepts they represent.

Finally, there has been evidence that non-orthographic representations, like pictographs, favor an immediate access to the meaning of the representation. Printed English words, in comparison, may require a "sounding-out" strategy which takes time, and comes between the visual representation itself and the meaning for which it stands. From this, it could be assumed that icons provide a cognitively easier and speedier mechanism for the representation of meaning than do printed words.

Unfortunately, the icon advantages suggested by these findings are undermined by the following issues:

1. Computer generated icons do not possess the richness in detail and vividness in characterization of color, depth and texture that is present in pictures. This is important since it has been suggested that just such stimulus "richness" is responsible for the picture superiority effect in memory. In support of this, Park and Arbuckle(10) have demonstrated that the memory advantage for pictures is not matched by memory for ideograms (meaningful symbols which are more abstract than pictures). Hence, the more iconic the representation, the less the memory benifit.

2. Related to this, the relative simplicity in the detail of icons may increase their potential for confusion. To illustrate, icons may share more features such as "solid black line" than do pictures. Searching for a menu item in an alphabetically ordered list of words may be easier than distinguishing between visually similar graphical elements.

3. It is not clear to what extent abstract concepts will "behave" cognitively as concrete objects, simply by virtue of an iconic representation. Although research to date demonstrates superior memory for pictures over words, with somewhat reduced memorial benefits for ideograms over words, this research has primarily dealt with concrete concepts. If verbs are less concrete than nouns, then verb-like software functions may not benefit from iconic representation. In support of this contention, Rohr and Keppel(5) report data (discussed more fully below) which indicate superior performance with words over icons for the representation of editing functions.

4. Finally, there are a growing number of reports which do not support the position that ideograms are more simply processed than alphabetic representations. The most recent evidence indicates that an internal "sounding-out" translation mediates the visual symbol and the access of meaning for ideaograms as well as words. If this is true, it is not safe to assume that the use of icons represents a non-verbal process. Similarly, it may not be safe to assume that the use of a mouse and icons "eliminates unnecessary spatial-linguistic conversions"(11), even in a graphically based task.

In sum, the literature base in cognitive psychology provides mixed support for the use of icons. While certain widely held findings seem to support the position that icons are a preferred representation for software functions, there is reason to be cautious. There do exist a number of findings and considerations which are simply at odds with this conclusion and many others which indirectly generate relevent contradictions. What is strongly suggested by this literature is the need for adequate, comprehensive evaluative (i.e., human factors) research before a final judgment on icons can be reached.

3. Human Factors Psychology

Evaluations of iconic interfaces and the experimental testing of icon-related issues have only recently begun to appear in the published human factors literature. Thomas(12) discussed in a general way the human factors implications of the picture superiority effect, but provided no empirical results. Flamm and Marshall(13) measured recognition memory for various types of pictoral stimuli. They reported several significant effects: performance was worse for line drawings than for photographs, was worse for black and white than colored representations; for illustrations, performance was worse for complex than simple object representations. While these data have obvious implications for the design of icons, no icon-word comparison was made.

Bewley, Roberts, Schrod and Verplan(14) described the human factors testing which contributed to the development of the Xerox "Star" user interface. Among many other tests, they collected information on the relative familiarity, recognizability, distinguishability, and user opinions of four alternative icon sets. Their data indicate that, once learned, their alternative sets did not differ in their recognition value. Significant differences in the distinguishability of icons was measured; those which were most distinguishable resulted in improved performance with a mouse pointer. However, no indication was given as to the preferred dimensions for icon representations. Again, no direct comparison between icons and words was reported.

The relative effectiveness of pictorial, printed and (auditory) voice formats was a concern to Hawkins, Reising, Lizza and Beachy(15). Their study incorporated these various representations into the emergency procedures for experienced aircrew members in simulated combat missions. The investigators hypothesized that a pictorial representation would prove to be more effective than either text or voice. Their data proved them to be wrong in their predictions. For these users, there was no significant difference between text, voice and pictorial representations, as measured by flying task performance and task completion times.

Recently, Rohr and Keppel(5) of IBM presented evidence directly applicable to the issue of computer generated icons. Their report is timely and relevent for two reasons: (1) they describe a system for prototyping and evaluating the use of icons in software interfaces; (2) they directly compare the use of empirically derived icons to traditional verbal commands under several conditions using several dependent measures.

The Rohr and Keppel study deals directly with many of the assumptions regarding the presumed benefits of iconic interfaces, and the authors do a fine job of making those assumptions explicit. In particular, they focus on the relationship of icons to a user's underlying mental model of the system. They hypothesize that icons do a better job than verbal commands at imparting to the user the actions and inferences which together constitute the conceptual system model, i.e., the model from the user's standpoint. Additionally, they were interested in identifying the components of the visual representation which are most important for conveying meaning to the user.

In their first experiment, the investigators relied upon naive subjects to evaluate a set of 72 icons for a contrived word processing application; these icons were constructed on the basis of 16 elementary features and their combinations. As stated previously, only the element "arrow" was interpreted as conveying any consistently meaningful information. Additionally, subjects rated types of icons according to their word processing function. Subjects preferred "file handling" functions to be represented by more pictorial symbols; they preferred "editing" functions to be represented by more abstract elements.

Using the information from experiment one, the investigators created two sets of icons for testing in an interface experiment. The same word processing functions were also represented by a verbal command set. Three groups of naive subjects were assigned the task of learning to use the prototypical word processing system. The subjects were required to perform three sets of operations, or "states" (general, file handling, and editing states); each state had a different HELP system associated with it. Dependent measures included task performance time, number of faults per operation, and number of help requests.

Of interest here are comparisons between the verbal and iconic interfaces. No significant difference between symbol sets was obtained for either the time or error measures.

The authors report a symbol set x state interaction which was significant for number of help requests. Of considered importance to them was the finding that the number of help requests for words was greater than for either of the icon sets in one of the state conditions. Because this state condition was designed to represent the most general kinds of functions, the authors conclude that "the verbal command group had more difficulty acquiring a mental model of the system...(Icons) impart a better mental model of the system structure to be worked on."

It is difficult to evaluate the authors' interpretation of their data, since necessary statistical information is not reported. (The authors were most likely constrained by space limitations in this report.) In particular, no between-the-means test following the significant interaction was reported. This is problematic since the interpretation of these data depends upon the pattern of differences obtained between the icon and word symbol sets for the three conditions. The differences between the word and icon symbol sets appear (from their graph) to be larger in the other two conditions; further, they are in the opposite direction. In other words, there are fewer help requests for the word stimuli in two of the three conditions tested. Without this information, it could be argued that these data support the opposite conclusion: That verbal commands are more effective overall in imparting information to naive users, as demonstrated by relatively smaller number of requests for help in two out of three conditions.

Because the Rohr and Keppel study is one of the human factors "ground-breakers" in terms of confronting important issues related to iconic interfaces, it is hope that their work continues and finds public release in a more comprehensive form.

Conclusion

This presentation was an attempt to evaluate the industry trend toward the acceptance of iconic interfaces as a user-friendly standard, with an appeal to relevent literature bases. While disparate in nature, the fields selected share an interest in understanding performance and/or processing differences with respect to various symbol types. The review undertaken leads to the following conclusions:

(1) The traffic directives literature does not provide a solid literature base from which to make inferences regarding human/computer interaction. Particularly because of task and environmental differences between the car driver and the computer user not captured in the published experiments, as well as some problems in the interpretation of published data, these studies are not entirely applicable. Therefore, no real support for the implementation of computer-generated icons comes from this field.

(2) The literature in cognitive psychology emphasizes the modeling of unobservable processes and not simply the observable performance differences desired by human factors specialists. Still, this complicated literature base has bearing on many icon-related issues,

particularly those with respect to the acquisition of the mental models of systems. In general, there is mixed support for the use of icons from this domain; however, there are generated enough counter-arguments concerning their use that some caution regarding oversimplification of the psychological issues related to icon processing seems necessary.

(3) The human factors literature is just beginning to tackle the issues involved. The published literature base is presently weak; more studies like that of Rohr and Keppel(5) are called for. Because of the limited information available, no clear support for icons can be culled from this field.

In short, the general recommendation to be drawn from this review must be for a continuation of the empirical investigation of the issues and assumptions involved in the use of icons. The scientific understanding of behavior and cognition at the iconic interface is just beginning.

References

(1) Dewar,R., and Ells,J., A comparison of three methods for evaluating traffic signs, Trans. Res. Rec, 503 (1974) 38-47.

(2) Dewar,R., Ells,J., and Mundy,G., Reaction time as an index of traffic sign perception, Hum.Fact., 18 (1976) 381-392.

(3) Ells,J., and Dewar,R., Rapid comprehension of verbal and symbolic traffic sign messages, Hum.Fact., 21 (1979) 161-168.

(4) Whitaker,L., and Stacey,S., Response times to left and right directional signs, Hum.Fact., 23 (1982) 447-452.

(5) Rohr,G. and Keppel,E., Iconic interfaces: Where to use and how to construct, in: Hendrick, H.W. and Brown, O.,Jr. (eds.), Human Factors in Organizational Design and Management (North-Holland, Amsterdam, 1984).

(6) Hemenway,K., Psychological issues in the use of icons in command menus, Proc.Hum.Fact.in Comp.Sys., (1982) Gaithersburg, Md., 20-23.

(7) Switchenko,D., A psychologist's critical appraisal of the use of icons ot represent software functions, Burroughs Corp. Technical Paper, Burroughs Corp., Plymouth, Mi., (October 1984).

(8) Paivio,A., Imagery and Verbal Processes (Holt, Rinehart and Winston, New York, 1971).

(9) Paivio,A., Mental imagery in associative learning and memory, Psych.Rev., 76 (1969) 241-263.

(10) Park,S., and Arbuckle,T., Ideograms versus alphabets: Effects of script on memory in "biscriptal" Korean subjects, J.of Exp.Psych:Hum. Learn. Mem., 3 (1977) 631-642.

(11) Foley,J., and Sibert,J., How to design user-computer interfaces, Proc. CHI'83, Tutorial 1, 53.

(12) Thomas,E., Visual superiority effect in memory (Pictures are worth a thousand words), Proc. 27th Ann. HFS (1983) 714.

(13) Flamm,L., and Marshall,C., Picture recognition: A human factors study of four illustration types, Proc. 27th Ann. HFS, (1983) 87.

(14) Bewley,W., Roberts,T., Schrod,D., and Verplan,W., Human factors testing in the design of Xerox's 8010 "Star" Office Workstation, Proc. CHI'83, 72-77.

(15) Hawkins,J., Reising,J., Lizza,G., and Beachy,K., Is a picture worth 1000 words---written or spoken? Proc. 27th Ann. HFS, 1983, 970-972.

VI

DESIGN

Trends in Ergonomics/Human Factors II
R.E. Eberts and C.G. Eberts (Editors)

HUMAN ERROR AND ITS CONSIDERATION IN SAFE PRODUCT DESIGN

Joseph P. Ryan

National Standards Technology, Inc.
28840 Southfield Road, Suite 142
Lathrup Village, Michigan 48076

The design of safe products is necessary to conform to public policy with respect to both social and legal requirements. Mere conformance to Standards may not be viewed by the Courts as meeting consumer needs for safety. This paper discusses the application of human factors principles, and the critical role of perceptual impairment and its causal relationship to the safe use of products. Human error and its issues as it relates to hazardous product design is also presented.

INTRODUCTION

The design of safe consumer and industrial products is necessary to conform to public policy with respect to both social and legal requirements. Engineers have been made aware of the need to satisfy consumer demands in safe consumer product design by the persistent efforts of the Human Factors Society in forming the Consumer Products Design Division in 1972, and by the inclusion of numerous Consumer Product Design sessions in the Annual Meetings since 1968.

Further, numerous papers have presented both engineering and legal criteria on the subject of safe product design. Safety in product design can be achieved, with some degree of assurance, by complying with acceptable Safety Standards, authoritative Regulations, and special Codes. However, from a judicial perspective, mere conformance to these criteria may not, in the Courts view, meet the legal requirements for safety to the product user. This principle of liability to the product designer and manufacturer is frequently referred to as the unreasonably dangerous defect doctrine (Cartwright, 1979; Eginton, 1977; Restatement (2nd) of Torts, 1965).

When a product is held to be unreasonably dangerous to the user, it is also held to be defective in its design and manufacture. A perplexing problem to both designers and attorneys is the fact that the Courts have refused to define the "unreasonably dangerous" term, and left that burden on the shoulders of the jury. In effect, it is only necessary to prove that the product was defective in its design and manufacture which was the proximate cause of the injury or loss.

PRODUCT DESIGN AND MANUFACTURE DEFECT

In general, consumer and industrial products are defective as the design and manufacture of the product relates to the following:

A. Design Defect
   1. The product was not provided with devices for safeguarding to prevent user injury.
   2. The material used in the product was inadequate in strength, and foreseeable wear and tear for its reasonable use and life.
   3. The product was not provided with proper packaging, warning labels, and proper instructions for safe use, and mis-use.
   4. The product was unreasonably dangerous beyond the consumer's expectations.

B. Manufacture Defect
   1. The product was not manufactured with proper care in process and assembly to reduce or eliminate hazard.
   2. The product was not subjected to quality control with respect to raw materials used, and final assembly.
   3. The product was not subjected to proper testing with respect to laboratory and commercial development for safety and reliability.

The defects in product design and the award for damages resulting from these defects are the subject of Tort Law. The extension of Tort Law into the theory of Product Liability Law as it is practiced today is actually the progressive application of social justice in the Courts. A rudimentary reading of this law is recommended for all human factors engineers (Noel & Phillips, 1981; Weinstein, et al, 1978).

HUMAN AND SYSTEMS RELIABILITY

The study of human error and its issues begins with an understanding of human performance. By human performance we mean the overt behavior or response that a person makes when using some system, machine or device (Van Cott, 1984). It is widely accepted that the capacities and limitations of human performance can be best integrated into product design by systems task analysis (Miller, 1953).

Currently, our understanding of human error rate lacks any scientific basis for predictability and, unfortunately, there is little research in this critical area of human performance. Numerous papers have been published in attempts to quantify human error research data based on S-R testing into predictive expressions. As Adams (1982) points out, most of these attempts have failed to formulate expressions since there has, to date, been no direct link between systems reliability expressions and human error in task performance.

The application of reaction-time equations, e.g., Fitt's Law,

Power Law of Practice, etc. (Card, 1981), as a measure of human performance is questionable and only confirms the fact that the processing of a stimulus takes time, and that time varies with each person being tested. Other attempts to quantify human performance by the application of probabilistic theory used for system component reliability analysis have also failed to reach an acceptable expression for human error rate prediction (McCormick, 1981).

In a recent paper, Swain (1982) used an expression based on the binomial distribution to determine the significance level of human error in a systems task. The method used to determine the probability of human error in the task was subjective consensus of three analysts who reviewed the task. This cannot be viewed as a prediction of human error in task analysis.

A review of the literature also presents other methods for predicting human performance and error (Salvendy, 1975; Rook, 1962; Danaher, 1980; Dannhaus & Halcomb, 1975). However, these methods are not, in fact, predictive, but rather expressions of single-cause errors in a task analysis. As pointed out by Embrey (1976), human error is too complex to be directly applied to component failure reliability equations as described in the foregoing. More research in this area is needed before we can develop predictive expressions.

HUMAN ERROR AND PERCEPTION

It is recognized that human performance degradation leads to human error. Adams (1982) discussed the problems relating to the issue of behavioral units for reliability analysis and the machine-man interface multidimensional complexity. He also discusses the promise of synthesizing human and systems reliability by use of Monte Carlo simulation modeling. The Monte Carlo technique is based on an analogous stochastic process in which the man-machine system is observed, and the results are tabulated and treated as if they were experimental data representing the actual problem. An approach which offers promise into quantifying human error in systems design is to examine the variables in human performance degradation as shown in Table 1.

Since human error is the direct result of incorrect behavior caused by perceptual impairment, it is possible further research into perceptual impairment will give us better insight into the causes of human error. This is especially critical in any hazardous machine-man system in which the operator error may lead to catastrophic trauma. Further, it is of critical importance for the human factors engineer to understand how perceptual impairment affects product design with respect to reasonable and foreseeable use and mis-use of the product.

A review of the literature indicates our knowledge of human performance is based on data compiled from the aviation and aerospace industry (Pickrel & McDonald, 1964). As a result, human factors engineers have leaned heavily from this source, rather than develop our own set of operating principles for human error. In effect, this approach is somewhat indirect; however, the study of human behavior is still evolving and, therefore, safe product design requires application of known safety principles.

The search for methods of analyzing and quantifying human error led to the development of a classification of behaviors (Berliner, et al, 1964, modified by Rabideau, 1964), as shown in Table 2.

As applied to human performance, Berliner's taxonomy of behavior provides an organized structure in task analysis for unimpaired perceptual processes. If we are to make any progress in the area of human error analysis, we must temporarily put aside our search for quantative expressions of human error and concentrate on the perceptual factors which lead to performance degradation with resultant error.

ERROR ANALYSIS

Human error takes place in product or system design at any level, including conception, design, manufacturing, testing, and marketing. However, it is at the man-machine interface that most errors result. To be considered an error, the human behavior of the operator must potentially have a degradative effect on the product or system performance, and result in personal or property loss. Meister (1964) defined several ways in which error could occur:

1. Performance of a required action incorrectly.
2. Failure to perform a required action.
3. Performance of a required action out of sequence.
4. Performance of a non-required action.

It should be noted that human error in skilled performance is statistically small and complex. Several sources of human error empirical data has been developed that are useful to product and systems designers: Data Store developed by the American Institute for Research (1962), and the Sandia Human Error Rate Bank (Rigby, 1967). A word of caution should be mentioned in using these empirical data--it has not yet been shown there is any validity between human error probability and system reliability analysis. One of the objectives of this paper is to encourage more meaningful research in human error and its causes.

PRODUCT DESIGN, SAFETY, AND PERCEPTION

The place to attack potential human error is in product design. We must continue the search for human error analysis, and it is possible the following discussion on the relationship between perceptual impairment and product design will help us design safe products.

Despite the existence of the U.S. Consumers Product Safety

Commission since 1972, and the existence of thousands of Safety Standards and Codes for over 60 years, the current increasing rate of injuries and deaths caused by defective products requires a re-evaluation of product design methods. Obviously, there is something wrong in the way we design and manufacture consumer and industrial products. A review of the growing number of lawsuits in products liability litigation indicates that the major problems exist in the basic design of the product (ATLA, 1984). As discussed in the foregoing, defective product design includes inadequate safeguarding, safety devices, warning labels and instructions, and human factors testing prior to marketing.

Some research has been started in the area of perceptual impairment and its relationship to personal injury in product use (Etherton, et al, 1979); Blignaut, 1979; Rethans, 1980; Ryan, 1980). In all of these publications, the authors were in general agreement that operator error in the performance of his task was the result of perceptual impairment. In this context, perceptual impairment is the failure of the operator to correctly respond to a received stimulus resulting in personal injury or death, or a near-miss. Attributes of perception contributing to incorrect response in hazardous machine operations are shown in Table 3.

To design a product safely requires an appreciation of operator behavior and the consequences of human error in the reasonable and foreseeable use and mis-use of the product. Since there are not reliable product-risk methods or error-rate equations to assist in product design, we must rely on both laboratory and field reliability testing. By using the above perceptual attributes relating to human error action, it is possible to critique all the components of the product or system design and compare the results to human reasonable and foreseeable use.

HUMAN FACTORS IN SAFE PRODUCT DESIGN

There are now many handbooks, texts, and papers published on human factors engineering useful in product design and manufacturing. All have a common objective which is to provide human factors applicable to product design which satisfies the average consumer with respect to comfort, fatigue, skill, efficiency, long-life, and esthetics. What is missing from these sources is criteria for safe use of the product. To design a product with hazard present when released to the marketplace is inviting a lawsuit unless the product has been provided with proper safeguarding of the hazard. The place to start is in the design stages.

One design engineer cannot assume the total burden for safety. He should be part of a team effort to design safety into the product. A suggested method to critique the product in the initial stages before the prototype model is built is as follows:

A. DESIGN

1. The design team should be staffed with competent and experienced engineers. One engineer should be designated Product Manager, and supervise the team.

2. One engineer should be designated Safety Engineer and coordinate safety criteria from all sources. Any required training in product safety, as it relates to the user of the product, should be given to the staff as needed.

3. A complete log book, records of meetings, design changes, inter-office communications, schedules, costs, etc. should be maintained and legible for photocopying.

4. Safety devices, safeguarding, warning labels, etc. should be tested for long-life, wear and tear, abuse, and mis-use early in the design stages.

5. The state-of-the-art in this type of product should be made by review of Patents, similar product in the market-place, and a comparative analysis of safety features and technology available in similar products.

B. HUMAN FACTORS DESIGN EVALUATION

1. After initial design concept has been approved, the user of the product should be identified and described in terms of age, skill and other factors required for safe product use.

2. User simulation testing both in the laboratory and in the field is necessary, from a safety and legal perspective.

3. Where safety devices fail to provide that degree of reasonable safety expected by the user, then the product must be redesigned to provide that safety.

4. Reliance on existing Safety Standards or Codes is not the ultimate, but rather the minimum requirements for product safety. Other criteria may be a design challenge, but if it is technically and economically feasible to provide the required safety device, then it is expected by the user and the Courts.

5. The design team should hold separate meetings for design and safety. Outside personnel with design safety experience should attend the product safety meetings for objectivity.

C. USER CONSIDERATIONS

1. Has the user been properly identified? Should only men use the product with respect to safe use? How can you eliminate mis-use of the product?

2. Does long use of the product induce fatigue (chain saws, circular saws, etc.)? Has this been covered in warning labels fixed to the product, stated in the Owner's Manual, and labeled to the package?

3. If the hazard is high-risk and difficult to adequately safeguard, a separate Safety Instruction Booklet should be prepared.

4. Has the mis-use of the product been tested to the extent that the questions of foreseeability and reasonableness have been properly answered?

5. Have both the laboratory and field testing confirmed the assurance of adequate simulation with positive results?

The opportunities for human factors engineers to apply these valuable principles in safe product design is immeasurable. Further, besides the satisfaction of common sense use of these principles, there is also the reward of contributing to society by reducing the unnecessary injuries and deaths today.

TABLE 1
Factors Affecting Human Performance

Factors	Variables
Perception	Stimuli, Processing, Cognition & Response
Information Processing Patterns	Preference in Selection & Presentation
Physiological	Capacity & Limitation
Environmental	Imposed Conditions
Skilled Performance	Skill & Intelligence Limitations

TABLE 2
Behavior Classification

Process	Activities	Behaviors
Perceptual	Search for/Receive	Detects, Processes, & Responds
Mediational	Information Processing	Categorizes & Computes
Communication	Decision Making	Informs & Directs
Motor Control	Simple/Complex	Activates, Adjusts, & Tracks

TABLE 3
Perception & Attributes in Man-Machine Hazard

Perceptual Impairment	Operator Response Error
Lack of Attention	Permits machine element to move beyond limit; Over/Under cutting; Incomplete Monitoring; Loss of Control
Lack of Vigilance	Misses critical stimuli; Induces Confusion; Loss of sequence; Loss of Tracking Deviation
Psychological Set	Expects similar stimuli to require identical action; Causes Cognition Degradation, Induces Omission Errors
Expectation	Confusion of proper S-R action; Imperfect Learning Curve Application
Information Processing Limitation	Channel overload causes improper task cognition; Memory failure; False communication.
Hazard Recognition Limitation	Induces high-risk taking; Creates false safety mode; Mis-matches skill performance level required.

REFERENCES:

1. Adams, J.A., "Issues in Human Reliability", Human Factors, 24,(1),1982
2. Blignaut, C.J.H., "The Perception of Hazard", Ergonomics, 22,(11),1979
3. Card, S.K., "The Model Human Processor: A Model for Making Engineering Calculations of Human Performance," Proc. of the Human Factors Society, 25th Annual Meeting, 1981.
4. Cartwright, R. E., Esq., "Considerations in the Design Defect Case", TRIAL, November, 1979.
5. Danaher, J.W., "Human Error in ATC System Operations", Human Factors, 22, (5), 1980
6. Dannhaus, D. M. and C. G. Halcomb, "Methodology for the Prediction of Complex Skill Performance", Proc. of the Human Factors Society, 25th Annual Meeting, 1981
7. Data Store (Index to Equipment Operability), American Institute for Research in the Behavioral Sciences (AIR), 1962.
8. Eginton, W. W., Esq., "From Winterbottom to Magnuson-The Developing Law of Product Liability", Proc. of the Products Liability Prevention, 1977
9. Law Reporter, American Trial Lawyers of America, Washington, D.C.
10. McCormick, E.J., Human Factors in Engineering and Design, New York: McGraw-Hill Book Co., 1981
11. Meister, D., "Methods of Predicting Human Reliability in Man-Machine Systems", Human Factors, 6, (6), 1964.
12. Miller, R. B., "A Method for Man-Machine Task Analysis", Wright-Patterson AFB, Ohio, 1953
13. Noel, D. W. and J. J. Phillips, Products Liability, St. Paul: West Pub. Co., 1981
14. Pickrel, E. W. and T. A. McDonald, "Quantification of Human Performance in Large, Complex Systems", Human Factors, 6, (2) 1964
15. Rabideau, G. F., "Field Measurement of Human Performance in Man-Machine Systems", Human Factors, 6, (6), 1964
16. Rethans, A., "Consumer Perceptions of Hazard", Proc. of Products Liability Prevention, 1980
17. Rigby, L. V., "The Sandia Human Error Rate Bank" (SHERB), Sandia Corp., Albuquerque, N.M., 1962
18. Rook, L. W., "Reduction of Human Error in Industrial Production", Sandia Corp., Albuquerque, N.M., 1962
19. Ryan, J. P., "Human Factors Design Criteria for Safe Use of Consumer Products", Proc. of the Human Factors Society, 17th Annual Meeting, 1983
20. Ryan, J. P., "The Perception of Hazard and its Causal Relationship to Personal Injury in Machine Operations", Unpublished Dissertation, California Coast University, 1984
21. Salvendy, G., "The Prediction of Operator Performance on the Basis of Performance Tests and Biological Measures", Proc. of the Human Factors Society, 19th Annual Meeting, 1975
22. Swain, A.D., "A Note on the Accuracy of Predictions Using THERP", Human Factors Society Bulletin, 25 (4), 1982.
23. Van Cott, H. P., "From Control Systems to Knowledge Systems", Human Factors 26, (1), 1984
24. Weinstein, A. S., A. D. Twerski, H. R. Piehler, and W. A. Donaher, Products Liability and the Reasonably Safe Product, New York: John Wiley & Sons, 1978

Trends in Ergonomics/Human Factors II
R.E. Eberts and C.G. Eberts (Editors)

EVALUATING WORK STANDARDS IN INDUSTRY: TOWARD A SCIENTIFIC DEFINITION OF FAIR AND REASONABLE

Frank E. Gomer and Denise M. Yanko

Research and Development Division
General Physics Corporation
201 South Main St., Suite 810
Dayton, Ohio 45402 (USA)

We describe an independent evaluation of a new work standard affecting machine-paced mail sortation. The framework adopted for the evaluation was that increased productivity and cost savings, key factors in management's decision to implement a new work standard, must not be associated with undue levels of job stress. Primary considerations in developing the methodology included the indentification of a suitable control group, the selection of valid measures of job performance and stress, and the use of nondisruptive data collection procedures.

INTRODUCTION

Contracts agreed to by management and labor often have articles which specify that new work standards must be fair and reasonable. In accordance with the current contract between the U.S. Postal Service (USPS) and the American Postal Workers Union, AFL-CIO (APWU), the APWU is permitted through qualified representatives to conduct its own evaluation of a new work standard if it has concerns with the adequacy of the test conducted by the USPS. Potential increases in productivity and reductions in cost clearly influence the decision to implement a new work standard, whether this work standard relates to the operation of new equipment or to a modification in the procedures which are followed when existing equipment is operated. However, several recent review articles have emphasized the significance of such factors as the work environment, equipment design, job content, and work practices as causes of occupational stress (Cohen, 1981; Smith, 1981; Sharit and Salvendy, 1982). Therefore, an acceptable framework for evaluating a work standard should include the explicit provision that desired gains in productivity and cost savings not be associated with undue levels of job stress (cf. Karger and Hancock, 1982).

Work Standards For Machine-Paced Mail Sortation. Most of the first-class mail handled by the USPS is sorted by keyboard operators assigned to multiple position letter sorting machines (MPLSMs). These machines consist of 12 consoles for coding unsorted mail, arranged in a row along the front, and 9 adjacent modules of letter receptacles (or bins), positioned along the back. Each console is equipped with a numerically labelled keyboard containing 10 upper and 10 lower piano-type keys (see Figure 1).

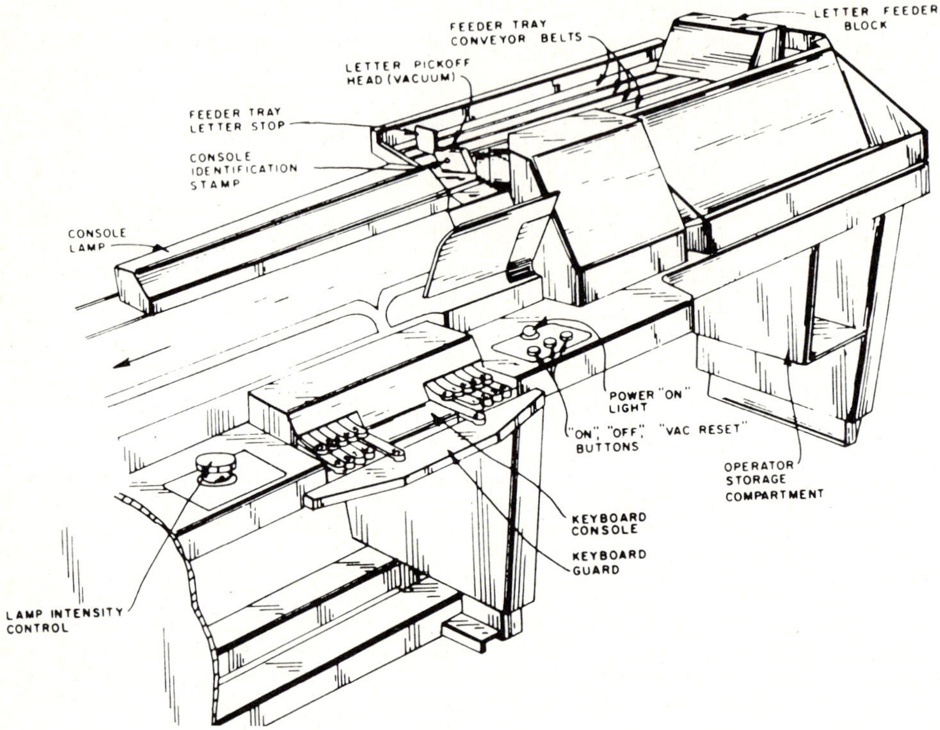

Figure 1. MPLSM console used for machine-paced mail sortation.

Unsorted letters are placed on the consoles and advanced automatically to the pickoff position, where single letters are removed by the vacuum arm and dropped to the level of the viewing window. Here, individual letters move at machine-paced intervals into the viewing window directly above the keyboard. When the letter arrives at the viewing window, it is stopped momentarily to allow the operator to search for and then read the relevant address information. After the letter moves from the viewing window toward the machine, the operator enters a numeric code by pressing in sequence the appropriately designated keys. Errors occur if an operator is unable to enter a code during the keying cycle, or if an operator enters a code that has not been programmed into the machine's memory. In addition, operators at times will enter a code that is recognized as legitimate by the computer, but nonetheless is not appropriate for the particular letter being processed.

Operators assigned to MPLSMs must rotate between 3 separate machine functions: console ledge loading, bin letter removal, and keyboard operation. MPLSMs are used for 3 types of mail distribution in most large cities. For the first, outgoing primary distribution, mail is distributed to regions outside the city and to separate zones in the city on the basis of the ZIP Code designation. For the second, box mail distribution, mail is distributed to certain businesses, colleges, hospitals, etc. on the basis of the post office box number which appears on the envelope. For the third, secondary distribution, mail which previously has been distributed to separate zones in the city and to suburban communities receives a "finer" sortation to individual homes or businesses on the basis of the street address.

Keep in mind that a code can be entered only after the envelope passes from view. This means that an operator, after deciding what code to enter, must store that code in short-term memory until the keying cycle commences. The principal keyboard tasks, then, are: (1) visually search the envelope; (2) translate the address information on the envelope to the proper numeric code and store that code in short-term memory; and (3) execute the key presses to enter the code.Since the keyboard function is machine-paced, there are finite time periods available to search an envelope and to enter a code. Figure 2 schematically depicts these time periods for mail sortation performed at rates of 60 and 55 letters per minute (lpm). In the top half of the figure, we show that 3-digit codes must be entered at a mail sortation rate of 60 lpm. If we begin with the first unsorted letter in the stack, we can see that an operator has 600 msec to visually search the envelope while it is stationary in the viewing window. As indicated in the figure, it takes 400 msec for the next letter to appear in the viewing window. However, as discussed in greater detail later, the mean response time for entering a 3-digit code is approximately 560 msec. Thus, on average, for about 160 msec of each 600 msec viewing period, an operator is completing code entry for the previous letter. A premium is placed on time-sharing ability when performing the keyboard function; an operator must begin to search an envelope and memorize a new code while recalling the code and completing the key presses for the previous letter.

A work standard which is applied to a particular type of mail distribution must address at a minimum: (1) the number of operators comprising a full crew; (2) the number of operators assigned at any time to each machine function; (3) the rotation schedule followed in performing the three machine functions; (4) the accuracy level demanded of keyboard operators to avoid disciplinary action and/or loss of job; (5) the number, spacing, and duration of breaks; (6) the rate at which the mail will be sorted; and (7) the maximum number of key presses required to enter a code.

METHOD

Although we examined loading and letter removal performance, the work standard issues presently being contested involve only

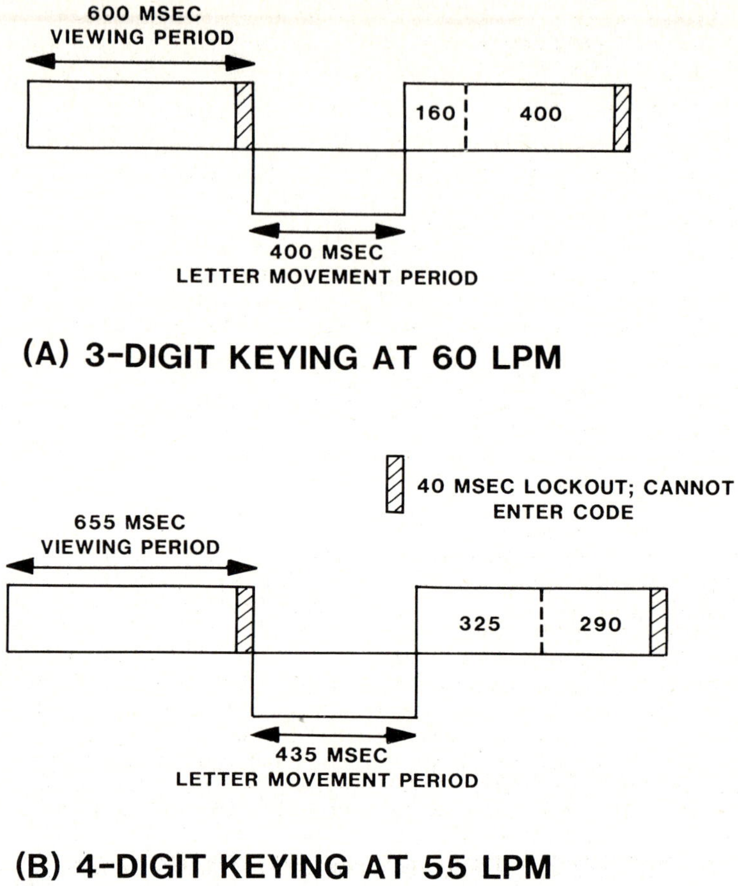

Figure 2. Schematic representation of machine-paced intervals for visual search and code entry.

the performance of keyboard operators.

<u>Crews Studied.</u> Crews performing 4-digit coding of box mail were studied on site in each of the 3 cities where the USPS presently uses this method for box mail distribution. Crews in the same 3 cities performing 3-digit coding of outgoing primary mail (a method of mail distribution acceptable to both the USPS and the APWU) served as control subjects. The groups did not differ with respect to these potential causes of job stress - the work environment, equipment characteristics, type of training, time of shift, overtime requirements, and machine- versus self-pacing. Moreover, the box mail and outgoing primary crews in each city were matched quite well on the basis of age, female-to-male ratio, and experience as MPLSM operators. Visual search demands and rotation and break schedules also were very similar for the 2 work standards.

Rather, treatment and control crews differed primarily in terms of the coding requirements - 4-digit keying at a rate of 55 lpm versus 3-digit keying at a rate of 60 lpm.

Design. Work-related data (see below) were collected at different time periods within the work day (i.e., beginning, middle, and/or end of shift). The beginning of the shift was defined as 0.5 to 2.0 hours into the shift. The middle of the shift was defined as 3.5 to 5.0 hours into the shift. Finally, the end of the shift was defined as 6.5 to 8.0 hours into the shift. Crews were studied on two days during the work week. Because of scheduled days off, the composition of some crews changed slightly from the first to the second test day.

Data Collection and Reduction. Consistent with the requirements of a multifaceted research strategy (Salvendy, Smith, Morgan, Sen, Triggs, and Haider, 1981), these types of data were collected: work-related, behavioral, physiological, and subjective.

In terms of work-related data, we examined both accuracies and response times in entering codes. With respect to coding accuracies, 25-letter edits were conducted at the beginning, middle, and end of shift. On day 1, the groups included 30 4-digit and 46 3-digit operators. On day 2, the groups included 33 4-digit and 46 3-digit operators. A high-resolution JVC video camera equipped with a time base generator was used to record keying-related response times. Each successive video frame corresponded to a 30 msec increment. Because of the time required to move and then carefully position the video recording equipment at the consoles, these data were collected from a maximum of 10 operators per crew at both the beginning and end of shift. Beginning with the second minute of the recording period, response times were extracted from the video tapes for 25 consecutive 3- or 4-digit code entries. On day 1, the groups included 24 4-digit and 30 3-digit operators. On day 2, the groups included 21 4-digit and 30 3-digit operators.

Repetitive keying motions of the kind observed during machine-paced sortation place some operators at risk for developing cumulative trauma disorders, particularly carpal tunnel syndrome (Armstrong, 1984; Armstrong and Chaffin, 1979). Moreover, keyboard operators often complain of muscle fatigue in the forearms, hands, and fingers. We selected 2 behavioral measures sensitive to local muscle fatigue in the forearms - hand-steadiness and sustained grip strength. With respect to the former, a Lafayette Instrument hand-steadiness tester was used. The device had 7 apertures of decreasing diameter. Operators were instructed to position the stylus in the center of each aperture and maintain that placement for 30 seconds. No forearm support was provided, and the dominant hand was used during the test. We recorded the number of contacts that the stylus made with the sides of each aperture. With respect to sustained grip strength, a Lafayette Instrument hand dynomometer was used. Data were collected from the dominant hand for 30 seconds while the operator exerted maximal grip strength. To allow a valid pre- and post-work comparison, the 2 tests had to be administered to operators before the shift began and immediately after their last keyboard assignment. USPS management agreed to this protocol only in City 1. Therefore, these data were collected from a total of 10 4-digit and 10 3-digit operators. The time constraints permitted testing on one day only.

We also recorded surface electromyographic (EMG) activity from the right forearm muscle of operators engaged in machine-paced keying. The displayed record of surface EMG activity usually is referred to as an interference pattern; it reflects the summation of individual motor units. With increased tension and effort there is greater recruitment of individual motor units. These motor units discharge asynchronously. However, during muscle fatigue, one observes increasing synchrony in the discharge of individual motor units. Therefore, if the interference pattern recorded during muscle fatigue is subjected to frequency analysis, one can see an increase in power for low frequency bands (e.g., 1-40 Hz) and a decrease in power for high frequency bands (e.g., 81-120 Hz) (Chaffin, 1968; Goodgold and Eberstein, 1977). Bipolar recordings were taken over the flexor digitorum. Signals were fed to a Colbourne Instrument amplifier. From the amplifier, signals were fed to a 16 channel A/D converter. Sampling rate was 280 Hz. An Apple IIE was programmed to record and then write to floppy disc during alternate seconds. EMG activity was recorded from operators in the training facility before the shift began and immediately after their last keyboard assignment. The same 20 operators who participated in the behavioral testing of muscle fatigue participated in the physiological testing. The data were collected during 6.5 minute periods of either 3-digit or 4-digit coding. First, the data were subjected to fast Fourier transformation to determine coefficients which defined 3 Hanning digital filters (1-40, 41-80, 81-120 Hz) with 12 dB per octave slopes. Then, relative power in the 3 frequency bands was computed for each second in the record.

A questionnaire consisting of self-report inventories and rating scales was administered to obtain subjective estimates of: (a) the workload associated with the keyboard function; (b) job satisfaction; (c) job-related anxiety; (d) general health; and (e) physical discomfort. We tailored a general workload scale recently validated by Wierwille and Casali (1983), a ten-point rating with scale descriptors in the form of a decision tree, to permit assessments of the workload experienced: (1) in visually searching envelopes; (2) in storing and then retrieving codes from memory; and (3) in keying codes within the allotted time. The scale allowed operators to assign ordinal values to easily handled, difficult but manageable, objectionable, or impossible levels of workload. A portion of the Job Descriptive Index, copyrighted by Bowling Green State University, was included to assess job satisfaction. Operators were asked to use a five-point scale when rating how well individual adjectives (e.g., challenging, creative, endless, etc.) described their impressions of the job. The State-Trait Anxiety Inventory, copyrighted by Consulting Psychologists Press, was administered to assess job-related and general anxiety. Operators were asked to use a five-point scale when rating how often they felt a particular way (e.g., at ease, calm, tense, etc.) immediately after work and when at home. Health questions probed the use of sick days, eating habits, drinking patterns, sleeping patterns, caffeine intake, smoking patterns, and types of diagnosed illnesses. Self-reports of physical discomfort and trauma associated with the keyboard function were grouped according to type. Five-point rating scales were used to obtain estimates of the extent of: (1) discomfort in fingers, hands, wrists, arms, and shoulders; (2) discomfort in the legs; (3) discomfort in the neck; (4) discomfort in the back; (5) headaches; and (6) visual discomfort. The percentages of operators returning completed questionnaires were 76% for the box mail group and 77% for the outgoing primary group.

RESULTS AND DISCUSSION

Coding Accuracies. Accuracy data were analyzed in two ways. First, we calculated the average accuracy for each crew (by pooling across individual operators) on each of the two test days at the beginning, middle, and end of shift. This resulted in 18 measures of crew accuracy for the two groups (3 crews X 2 days X 3 measurement periods). For the box mail group, crew accuracies ranged from 76.0 to 96.0%, with a mean accuracy of 90.1%. For the outgoing primary group, crew accuracies ranged from 93.9 to 98.5%, with a mean accuracy of 96.6%. An operator must maintain a coding accuracy of 95.0% to avoid disciplinary action and/or loss of job. Therefore, we compared the two groups on the basis of the number of occasions (out of 18) on which crew accuracy equalled or exceded the criterion level of 95%. For the box mail group, this occurred 1 out of 18 times. For the outgoing primary group, this occurred 17 out of 18 times. A chi square test showed this difference to be highly significant ($X^2 = 24.8$, df = 1, $p < 0.001$). Next, we calculated the proportion of individual operators on a crew who achieved a 95% accuracy in coding. For the box mail group, the proportions ranged from 0.18 to 0.83, with a mean proportion of 0.48. For the outgoing primary group, the proportions ranged from 0.56 to 1.0, with a mean proportion of 0.82. By pooling across crews, we determined the proportion of operators in each group achieving the criterion level of accuracy for each measurement period of each test day. The six chi square tests applied to these data revealed that a significantly smaller proportion of box mail operators was able to meet the demanded accuracy level in each time period of each test day (X^2s > 5.95, df = 1, ps < 0.02).

Response Times in Entering Codes. We knew that response times would be longer in general for 4-digit as opposed to 3-digit codes. However, we wanted to determine whether the groups differed reliably with respect to increases in response times by the end of the shift. Sharit and Salvendy (1982) have summarized the literature supporting the position that an increase in the time required to complete a well-practiced task can be a sensitive measure of job stress. Thus, average response times were computed for each operator tested on the two days at both the beginning and end of shift. On day 1, mean response time for the box mail group was 686 msec at the beginning of the shift and 761 msec at the end of the shift; comparable response times for the outgoing primary group were 566 msec and 564 msec, respectively. On day 2, mean response time for the box mail group was 760 msec at the beginning of the shift and 822 msec at the end of the shift; comparable response times for the outgoing primary group were 553 msec and 556 msec, respectively. For each test day, average operator response times were entered into a 2 X 3 X 2 analysis of variance, representing group, city, and time within shift factors, respectively. Each analysis of variance revealed a highly significant group by time within shift interaction; Day 1 $F(1,48) = 35.6$, $p < 0.001$ and Day 2 $F(1,45) = 14.2$, $p < 0.001$. Simple tests showed that operators entering 4-digit codes exhibited a substantial increase in response time at the end of the shift on both test days ($ps < 0.001$). No such increase was found on either test day for operators entering 3-digit codes. This finding is very important from the perspective of group differences in job stress. Also, consider the premium that such protracted response times place on efficient time-sharing during machine-paced coding. Figure 2B shows that at a mail sortation rate of 55 lpm, an operator has 655 msec to visually search each envelope and memorize the appropriate code. Moreover, the time interval between successive viewing periods is only 435 msec. Thus, if we assume a mean response time of 760 msec for entering a 4-digit code, we can see that an operator, on average, actually is completing the code for the previous letter during the initial 325 msec of each viewing period. This leaves only 290 msec within each viewing period when the demands of searching the envelope and memorizing a code <u>are not</u> competing with the demands of recalling a code and executing key presses. Figure 2A shows that a much smaller overlap exists for 3-digit keying at a sortation rate of 60 lpm.

<u>Behavioral Measures of Local Muscle Fatigue.</u> Difference scores (post-work minus pre-work) for hand steadiness were entered into a 2 X 7 analysis of variance, representing group and aperture size factors, respectively. The analysis of variance revealed a highly significant group by aperture size interaction, $F(6,108) = 3.82$, $p < 0.002$. Simple tests of the change in hand steadiness on the two smallest apertures showed that operators who had entered 4-digit codes during the shift exhibited substantially more tremor after their last keying assignment than did operators who had entered 3-digit codes ($ps < 0.01$). However, in terms of the grip strength testing, the two groups did not differ significantly with respect to any measures of sustained force over time.

<u>EMG Measures Of Local Muscle Fatigue.</u> We first computed percent relative power in each of the 3 frequency bands during successive minutes of the 6-minute test. These values were entered into a 2 X 2 X 6 X 3 analysis of variance, representing

group, pre- vs. post-work testing, time interval (minutes 1-6), and frequency band, respectively. The analysis of variance revealed a highly significant 4-way interaction, $F(10,180) = 2.45$, $p < 0.009$. For those operators who entered 4-digit codes, Figure 3 shows that there were changes from pre- to post-work testing in percent relative power within both the low and high frequency bands. Across the 6-minute test conducted after the last keying assignment in the shift, we see <u>increases</u> in percent relative power for the low frequency band and <u>decreases</u> in percent relative power for the high frequency band. As described earlier, this pattern is consistent with increasing synchrony in the discharge of individual motor units, a clear indication of local muscle fatigue. For those operators who entered 3-digit codes, there were no changes from pre- to post-work testing, i.e., percent relative power in both bands remained relatively stable across the 2 6-minute tests.

<u>Self-Report And Rating Scale Results.</u> Mann-Whitney U tests were applied to operator ratings of: (1) visual workload; (2) mental workload; and (3) keying-related workload. Given the large samples (box mail group = 28, outgoing primary group = 43), U values were converted to z scores. Operators entering 4-digit codes reported significantly higher and more objectionable levels of workload for all 3 categories. Mean ratings and z scores were: (1) visual workload, 3-digit coding = 4.4 and 4-digit coding = 5.7, $z = 1.75$, $p < 0.04$; (2) mental workload, 3-digit coding = 3.9 and 4-digit coding = 5.1, $z = 1.85$, $p < 0.03$; (3) keying workload, 3-digit coding = 4.2 and 4-digit coding = 6.9, $z = 4.18$, $p < 0.001$.

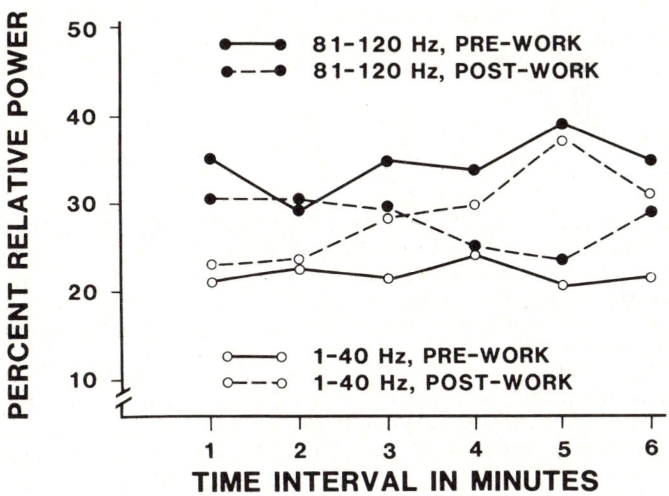

Figure 3. Average percent relative power in low and high frequency bands of the EMG activity recorded from operators entering 4-digit codes.

No significant differences were found between the groups in terms of operator ratings for either job satisfaction or general and job-related anxiety. Moreover, the groups did not differ with regard to self-reports of general health and diagnosed illnesses. It is worth noting that operators considered the 3 machine functions - letter removal, loading, and keyboard operation, when estimating job satisfaction and job-related anxiety. Perhaps it is not surprising that operators assigned to both types of machine-paced mail distribution would find their jobs to be similarly dissatisfactory and anxiety producing.

Mann-Whitney U tests were applied to aggregate ratings of each type of physical discomfort (i.e., the ratings were summed across individual questions in each of the 6 categories). Operators entering 4-digit codes reported significantly greater discomfort for 2 of the 6 categories. Mean aggregate ratings and z scores were: (1) discomfort in the fingers, hands, wrists, arms, and shoulders, 3-digit coding = 29.9 and 4-digit coding = 42.4, $z = 1.71$, $p < 0.05$; and (2) visual discomfort, 3-digit coding = 5.6 and 4-digit coding = 7.9, $z = 1.74$, $p < 0.04$.

CONCLUSIONS

For all classes of data that we collected, the new work standard differed significantly from the existing work standard agreed by both parties to be fair and reasonable. Bear in mind that we only studied operators on active status, and that we could not address such long-term consequences of job stress as the incidence of diagnosed cumulative trauma disorders. Our conclusion, then, was that 4-digit coding of box mail at a rate of 55 letters per minute is unfair and unreasonable from a scientific standpoint.

REFERENCES

(1) Armstrong, T.J. Evaluation and design of jobs for control of cumulative trauma disorders. Proceedings of the University of Michigan Conference on Cumulative Trauma Disorders of the Upper Extremities, Ann Arbor, Michigan, 1984.

(2) Armstrong, T.J. & Chaffin, D.B. Carpal tunnel syndrome and selected personal attributes. Journal of Occupational Medicine, 1979, 21, 481-486.

(3) Chaffin, D.B. Electromyography - A method of measuring local muscle fatigue. Journal of Methods - Time Measurement, 1968, 14, 29-36.

(4) Cohen, A. Human factors methodology for stress evaluation in machine-paced jobs. In G. Salvendy & M.J. Smith (Eds.), Machine Pacing and Occupational Stress. London: Taylor and Francis, Ltd., 1981.

(5) Goodgold, J., & Eberstein, A. Electrodiagnosis of Neuromuscular Diseases. Baltimore: Williams and Wilkins Co., 1977.

(6) Karger, D.W., & Hancock, W.M. Advanced Work Measurement. New York: Industrial Press, Inc., 1982.

(7) Salvendy, G., Smith, M.J., Morgan, B.B., Sen, T.K., Triggs, T.T., & Haider, M. Current status and research needs in occupational stress with special reference to machine-paced work. In G. Salvendy & M.J. Smith (Eds.), Machine Pacing and Occupational Stress. London: Taylor and Francis, Ltd., 1981.

(8) Sharit, J., & Salvendy, G. Occupational stress: Review and reappraisal. Human Factors, 1982, 24, 129-162.

(9) Smith, M.J. Occupational stress: An overview of psychosocial factors. In G. Salvendy and M.J. Smith (Eds.), Machine Pacing and Occupational Stress. London: Taylor and Francis, Ltd., 1981.

(10) Wierwille, W.W., & Casali, J.G. A validated rating scale for global mental workload measurement applications. Proceedings of the 27th Annual Meeting of the Human Factors Society, Norfolk, Virginia, 1983.

Trends in Ergonomics/Human Factors II
R.E. Eberts and C.G. Eberts (Editors)

A SUGGESTED SUBMAXIMAL TEST FOR JOB DESIGN AND PRE-EMPLOYMENT SCREENING TESTS

A. M. Genaidy*, S. S. Asfour*, T. Vinciguerra**, and A. M. Abdel-Salam***

*College of Eng.
University of Miami
Coral Gables, Florida
U.S.A.

**Ind. Safety Division
Frank Hall and Co.
Coral Gables, Florida
U.S.A.

***School of Medicine
University of Miami
Miami, Florida
U.S.A.

The main objective of the present paper is to devise a new submaximal test for job design and pre-employment screening tests. This test is based on the determination of the anaerobic threshold. A complete discussion of the various factors affecting the anaerobic threshold is provided.

INTRODUCTION

For job design and pre-employment screening tests, maximum oxygen uptake (VO_2 max) has long been used by industrial physicians as an objective measure of an individual's maximal physiologic capacity to perform hard physical work. However, there are problems associated with the direct measurement of VO_2 max in terms of the excessive physical stress placed upon the subjects, as well as limitations imposed on many researchers and practitioners in stressing human subjects to their limits. The main objective of the present paper is to devise a new submaximal test based on the determination of the anaerobic threshold.

DEFINITION

Anaerobic threshold (AT) is the oxygen consumption value attained during work above which the rate of lactate production exceeds the rate at which it can be catabolized, thus causing arterial lactate to increase (Wasserman et al. (1981)).

FACTORS AFFECTING THE ANAEROBIC THRESHOLD

1. Determination of Anaerobic Threshold

Formation of lactic acid during exercise permits oxidation to proceed anaerobically and accounts for most of the oxygen debt which accumulates during heavy exercise. Consequently, measurement of accumulated lactate invasively has been used as an index of anaerobic metabolism (Wells et al. (1957)). With the introduction of reliable rapidly responding gas analyzers and on-line computer processing Wasserman et al. (1973) identified AT non-invasively by the point of (1) non linear increase in minute ventilation, (2) non linear increase in volume rate of carbon dioxide production, (3) an increase in end-tidal oxygen without a corresponding decrease in end-tidal carbon dioxide, and (4) an increase in respiratory quotient as work load is increased during an incremental exercise test. Reinhard et al. (1979) reported that the ventilation equivalent, which is the minute ventilation divided by the oxygen consumption, is the best estimator for AT. The findings of Caizzo et al. (1982) were also in favor of the ventilation equivalent as the best predictor of AT in a comparison of gas exchange indices used to detect the anaerobic threshold.

The AT detection from gas exchange measurements has some limitations. The subjective determination of AT from ventilatory measures plotted versus time presents the major limitation. Davis et al. (1976) reported that estimation of AT can rarely differ by more than one minute among all observers. Orr et al. (1982) developed a computer algorithm that modeled the ventilatory response to exercise using multisegment linear regression. AT was reported as the first break point in that model. The second limitation is the reproducibility of the AT measurement. Davis et al. (1976) found a correlation coefficient of 0.95 between the gas exchange parameters and AT determined from repeated serial venous lactate samples.

2. Effect of Sex and Age

The findings of Reinhard et al. (1979) showed higher workload levels and oxygen consumption at AT in males, and a significant negative correlation to age. Moreover, there were no differences with regard to sex in AT for weight corrected work rates. In contrast to the findings of Reinhard and coworkers, Wasserman et al. (1973) reported that the lower limit values for AT appeared to be the same for all age groups and both sexes.

3. Duration of Workloads

Wasserman et al. (1973) found that the increase in respiratory quotient is greater during 1-min incremental test than 4-min incremental test. For this reason, and the obvious advantages of the shorter test, Wasserman and coworkers used a 1-min incremental work on the bicycle ergometer as a standard work test. They also suggested the use of a warm-up period of 4 minutes of pedaling on an unloaded bicycle ergometer as the initial work rate, then followed by the 1-min step increment in workload. Studies by David et al. (1976), Davis et al. (1979), Reinhard et al. (1979), and Orr et al. (1982) used the 1-min increment test previously described by Wasserman. Stamford et al. (1978) designed a discontinuous exercise protocol in which each progressive increment was performed 3-min followed by a 2-min rest period.

4. Type of Exercise

Davis et al. (1976) reported that when AT is expressed as a percentage of VO_2 max, no significant difference was found between the leg exercise modes (leg cycling and treadmill walk-running). However all arm (arm cranking) versus leg comparison were significantly different. Stamford et al. (1978) found that while absolute differences existed at given work levels, blood lactate concentration, minute ventilation, respiratory exchange ratio, heart rate and cardiac output were similar during one versus two legged cycling when expressed as relative percentage of VO_2 max. According to the authors, this indicated that the size of the exercising muscle mass is not necessarily an influential factor in determination of AT which is in contradiction with the results of Davis and coworkers.

A study by Intaranont (1983) indicated that there was no significant difference between AT determined by leg work (leg cycling and lifting from floor to knuckle height). The same result was found in the case of arm work (arm cycling and lifting from knuckle to shoulder height). The frequency of lift (6, 7.5, 9 lifts/min) had no significant effect on the AT. However, the height of lift (floor to knuckle and knuckle to shoulder) had an effect on AT.

5. Training

Few studies were conducted to study the effect of intensity and length of training on AT. Davis et al. (1979) found a 44% rise in AT expressed as absolute oxygen consumption value and by 15% expressed relative to VO_2 max compared to a 25% increase in VO_2 max after 9 week program (4 days per week, and 45 min/day). Ready and Quinney (1982) followed the same training program described by Davis and coworkers except that each training lasted for30 min and each subject was trained at 80% of VO_2 max, a value which was above the mean AT for all subjects. Following the 9 week training, elevations of 70.4% and 19.4% occurred in AT expressed as absolute oxygen consumption value and percent of VO_2 max respectively. In contrast, Skinner et al. (1977) found no change in relative AT after 8 weeks of endurance training, 3 times per week for 30-45 min at around 65% VO_2 max.

DISCUSSION

The reduction of the number and severity of over-exertion injuries in heavy industry could be achieved by either pre-employment screening to match job demands with individual capacities or by redesigning jobs to match human capabilities. A screening program separates the fit from the unfit, the normal from the unhealthy (Ayoub (1982)).

Maximal oxygen uptake has been used over the years by numerous investigators as an objective measure of the upper limit to the capacity of the cardiorespiratory system to transport oxygen to the muscles. However, the complexity of the direct measurement of VO_2 max, and the excessive physical stress placed upon the subjects have led to the development of indirect methodologies for estimating VO_2 max. The main objective of this paper is to suggest using the anaerobic threshold as a submaximal test for job design and pre-employment screening tests. The anaerobic threshold can be regarded as a measure of the maximum work rate that can be performed totally aerobically so that there is no lactic acidosis. Therefore, AT is an important assessment of the ability to perform endurance type work. The higher the AT value obtained, the more fit the person is.

Many cardiology, pulmonary and exercise laboratories include the measurement of AT as a part of their standard protocol. Since the AT is reached before maximum exercise, the stress and duration of the exercise required of patients is reduced. Thus, it is suggested that AT be applied as a part of pre-employment screening programs to assess the cardiorespiratory capacity. The test should be as follows: (1) allow a 4-min warm up period on an unloaded bicycle ergometer, then (2) increment the load every minute by 15 W (Figure 1). The ventilation equivalent is plotted versus time, and the minimum value is considered as the AT (Reinhard et al. (1979)). Research is needed to provide AT norms in order to assess the physiologic capabilities of industrial workers. It is felt that the use of AT as a submaximal test will help designers of industrial tasks to assess human physiologic capacities without exerting workers at maximal levels.

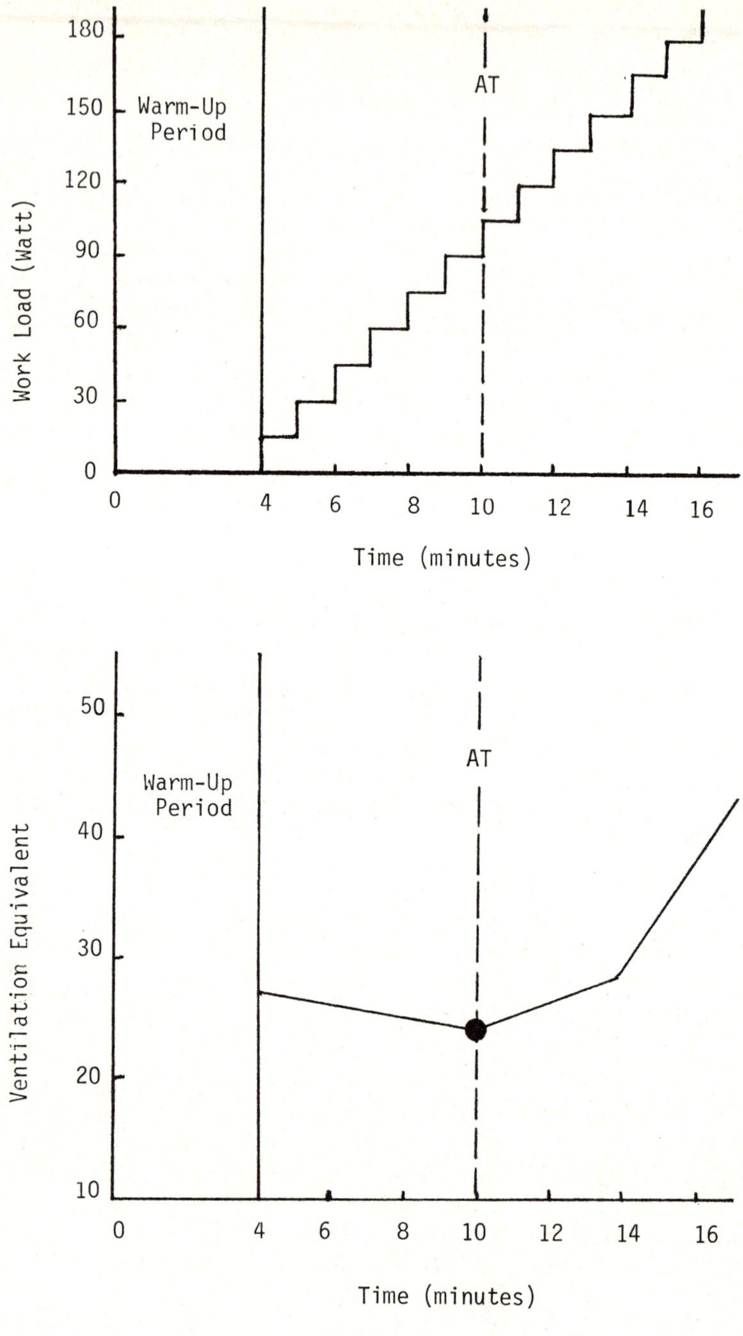

Figure 1

1- min Incremental Work Test for Detection of Anaerobic Threshold (AT)

REFERENCES

[1] Ayoub, M. A., Pre-employment screening programs that match job demands with worker abilities, J. Ind. Eng., 14 (1982) 41-48.

[2] Caizzo, V. J., Davis, J. A., Ellis, J. F., Azus, J. L., Vandagriff, R., Prietto, C. A., and McMaster, W. C., A comparison of gas exchange indices to detect the anaerobic threshold, J. Appl. Physiol., 53 (1982) 1184-1189.

[3] Davis, J. A., Frank, M. H., Whipp, B. J., and Wasserman, K., Anaerobic threshold alterations caused by endurance training in middle-age men, J. Appl. Physiol., 46 (1979) 1039-1046.

[4] Davis, J. A., Vodak, P., Wilmore, J. H., Vodak, J., and Kurtz, P., Anaerobic threshold and maximal aerobic power for three modes of exercise, J. Appl. Physiol, 41 (1976) 544-550.

[5] Intaranont, K., Evaluation of anaerobic threshold for lifting tasks, Doctoral Dissertation, Dept. of Ind. Eng., Texas Tech. Univ. (December 1983).

[6] Orr, G. W., Green, H. J., Hugh, R. L., and Bennett, G. W., A computer linear regression model to determine ventilatory anaerobic threshold, J. Appl. Physiol., 52 (1982) 1349-1352.

[7] Ready, A. E., and Quinney, H. A., Alterations in Anaerobic Threshold as the result of endurance training and detraining, Med. Sci. Sports Exercise, 14 (1982) 292-296.

[8] Reinhard, V. Muller, P. H., and Schmulling, R. M., Determination of anaerobic threshold by the ventilation equivalent in normal individuals, Respiration, 38 (1979) 36-42.

[9] Skinner, J., Lemieux, G., and Taylor, A., The effects of endurance training on the kinetics of oxygen intake at absolute and relative work loads, Med. Sci. Sports, 9 (1977) 68.

[10] Stamford, B. A., Weltman, A., and Fulco, C., Anaerobic threshold and cardiovascular responses during one - versus two - legged cycling, Res. Q., 49 (1978) 351-362.

[11] Wasserman, K., Whipp, B. J., and Davis, J. A., Respiratory physiology of exercise: metabolism, gas exchange and ventilatory control, in: Widdicombe, J. C. (ed.), Respiratory Physiology III (Univ. Park, Baltimore, Maryland, 1981), 149-211.

[12] Wasserman, K., Whipp, B. J., Koyal, S. N., and Beaver, W. L., Anaerobic threshold and respiratory gas exchange during exercise, J. Appl. Physiol., 35 (1973) 236-243.

[13] Wells, F. G., Balke, B., and Von Fassau, D. D., Lactic acid accumulation during work. A suggested standardization of work classification, J. Appl. Physiol., 10 (1957) 51-55.

Trends in Ergonomics/Human Factors II
R.E. Eberts and C.G. Eberts (Editors)

THE IMPACT OF AUTOMATION ON WATER PLANT OPERATOR PERFORMANCE

Richard L. Shell, Nicholas A. Damachi, O. Geoffrey Okogbaa

Department of Mechanical and Industrial Engineering
University of Cincinnati
Cincinnati, Ohio 45221
U. S. A.

Water utilities increasingly adopt various levels of automation in water treatment plants to enhance operational efficiency and performance. The automation efforts feature advanced instrumentation and digital computer technology and impacts on the duties and responsibilities of water plant operators. It demands the acquisition of a new set of skills by the water operator. Automation in water plants involves the redesign of water operator tasks, proper allocation of functions between humans and machines, and raises the issues of humanization of work, operator workstation design and a host of other factors. This paper examines the automated work environment of water operators, the impact of automation on their tasks and functions, and offers some guidelines for the redesign of tradiitonal water operators' tasks.

INTRODUCTION

Water utilities, particularly large ones, adopt various levels of automation in utility operations in their continued strive to provide quality service to customers and to meet stringent Environmental Protection Agency regulations on water quality and OSHA standards. Automation of water utility plant operations generally involve advanced instrumentation and digital computers to monitor flow and process variables in order to optimize chemical treatment of water in the plants while conserving energy. The trend toward increased water utility automation impacts on the duties and responsibilities of water operators. Traditionally, water operators have been a valuable human resource and are vital to smooth water plant operation. As automation continues to impinge on their activites, a need arises for a close examination of their changing job requirements. Other concerns that arise in an automated water plant environment include human-machine interface, the role of the operator as the ultimate controller of the water plant, increased skill level requirements, water operator work station design, job design, job satisfaction and other motivational considerations. This paper examines the automated work environment of the water operator, and offers some guidelines for job redesign.

To facilitate an understanding of the role of a water operator, a typical water plant configuration is briefly presented. A typical water system is composed of a water supply unit, treatment facilities and a distribution unit. Raw water obtained from sources that may include dams, reservoirs, rivers or ground wells, is conveyed into the treatment plant using low service pumps, or gravity flow. At the plant, the water undergoes treatment. Although the treatment process varies at individual water plants, the successive steps at some plants are generally screening, coagulation using liquid alum, flash mix, softening with line and soda ash or caustic soda, one or two stages of flocculation, carbonate stabilization with carbon dioxide, pre-and post chlorination using liquid chlorine, flouride adjustment, taste and odor control with activated carbon or potassium permaganate, stabilization with polyphosphate and filtration. The treated water is conveyed into the distribution system using high service and booster pumping stations.

The traditional water supply plant operator has the primary responsibility for the operation and maintenance of the water supply works and for the day-to-day delivery of safe, clear, and sufficient water to the community. He or she performs any related work assignment as required, checks the delivery of water from wells and conduits, maintains a constant check on the efficiency of the plant, monitors the water distribution system, and supervises work of subordinate employees, [5].

OPERATORS" TRADITIONAL SKILL LEVEL REQUIREMENTS

The skill levels traditionally requried by operators to run a water plant vary with the size, complexity and level of sophistication of the plant. Table 1 shows typical skills, abilities and knowledge requirements needed to run a small, medium, and moderately large sized waterwork systems, which correspond respectively with classes I, II and III Water Supply Works (Ohio Department of Health Classification). In addition to a knowledge of the principles, practices and methods of water treatment, storage and distribution, operators are required to possess a knowledge of occupational hazards and safety, public health needs as well as minor maintenance skills.

AUTOMATED WATER PLANT WORK ENVIRONMENT

In automated treatment plants, electronic instrumentation and digital computer system provide the capability for total automated control of treatment and pumping operations; based on the level of automation. In fully automated plants, computer systems monitor the flow of water, pace all chemical feed systems to plant flow, and automatically adjust the chemical treatment of water in the plants. In semi-automated plants, chemical feed systems may be manually and remotely set at a central master control console, but automatically paced to plant flow. There is provision for closed loop control and monitoring of some treatment processes. In an automated plant environment, a central control computer receives data from other computers within the plant, which is analyzed by resident algorithms; performs a myriad of complex calculations requried for supervisory and regulatory control and the generation of reports. It maintains controlled surveillance over all process control systems. The data is

Small Sized Waterwork	Medium Size Waterworks	Moderately/Large Sized Waterworks
Knowledge of operation, maintenance and repair of mechanical and electrical equipment. Ability to: Check and detect flaws in operation of equipment. Make minor repairs to equipment. Use hand tools. Keep neat records and reports. Make routine test. Give clear and understandable instructions. Good physical condition.	Knowledge of: Principles, practices and methods of water treatment. storage, and distribution. Operation of mechanical and electrical equipment. Occupational hazards and safety. Proper methods of lubricating, maintaining and adjusting equipment. Ability to: Maintain, repair, and adjust pumps, chemical feeders, instruments and other water works equipment. Check operation of mechanical equipment, detect flaws and defects, and make corrections. Take accurate readings, keep routine records, and make reports. Plan, schedule and supervise of subordinates. Advises municipal officials on present and future requirements to maintain safe and adequate water supply.	Knowledge of: Equipment, instruments, materials and operating practices for water treatment plant and distribution system. Occupational hazards and safety. Water chemistry, bacteriology, laboratory techniques. Public health needs. Practices, methods, materials and tools of mechanical trades. Methods of training and instructing employees. Ability to: Analyze complex maintenance and repair need. Supervise maintenance and repairs in plant and system. Plan, layout and inspect work of personnel. Establish and maintain systems of cost accounting, bookkeeping, stock keeping and record keeping. Supervise and handle personnel effectively. Skill in communications and public relations.

Table 1
Skills, Abilities and Knowledge Requirements for Operators

also stored for later mathematical calculations that leads to optimization of water quality and cost on a continuous basis. Typically, all locations in the water system communicate with each other over microwave or radio telemetry network links; and critical information is telemetered to the supervisory control center for system-wide coordination [3]. Also, the water operator is generally provided with manual backup controls for computer control override. The operator is provided with a fully equipped chemical and bacteriological laboratory facilities and sampling systems are designed to supply samples to the laboratory on continuous basis. Automatic analyzers are provided for the analysis of parameters of interest. Security check is accomplished through the use of TV cameras strategically located throughout the plant and to monitor desired operations.

OPERATOR'S RESPONSIBILITIES AND INTERFACE

The utilization of advanced instrumentation technology and comptuer based control techniques to enhance overall water plant opreation and performance presents the traditional operator with a new physical work environment and demands the acquisition of new sets of skills. The trend toward increased automation is usurping traditional mental processes performed by operators. However, in an automated water plant environment, the operator must ultimately exercise discretion and supervise the plant. As discussed, plants are generally regulated by sets of controllers which compare measurement signals with set points, and issue final commands to final actuators [1]. The human operator/supervisor gets informed about plant conditions through displays. Based on the information displayed, the operator decides whether intervention is needed and plans appropriate control strategies for coping with any malfunction. Some of the numerous displays and report available to the operator include:

- System summary displays: Show current operating conditions such as station pumpage, pressure district consumption, total system consumption, water supply information, equipment out of service and alarm information.

- Data Point Information Displays: Provide identification limits, and current values for all real-time data points monitored by the system.

- Water supply summary displays: Show impounding reservoir levels, river levels, lake releases and any other data related to water supply.

- Pressure district Summary displays: Show current operating conditions by pressure district, including station flow and pressure, pumps on, pumps out of service, storage elevations and rates of change and alarm conditions.

- Alarm summary conditions which list all current alarm conditions.

- Maintenance request summary displays: List all maintenance requests handled by the operator on the previous or current days.

- . Outstanding maintenance requests displays: Summarises all maintenance requests which have not been resolved.

- . Previous Day Summary Displays: Show total values for previous day operational information. These displays may be available for any day during a specified previous period.

- . Indices or list of available displays and system functions to aid the operator in utilizing the system.

In an automated plant environment, the operator stations provide the facility for operator interface with the plants. The operator interface is generally equipped with CRT screens and alphanumeric keyboard, low speed printing device which prints an exception log, and an audible alarm annunciator. The operator uses the standard keyboard for manual entry of data which is pertinent to the plant and/or distribution system operation, but which is not monitored by real-time sensors. Typical examples of such data include hydrological information regarding impounding reservoir levels and valves positions of manually operated valves. The operator can make manual entries to change the status of pump stations or individual pumping units, change operational limits, record request for maintenance services, correct data items, etc. The functional keyboard is also used to retrieve information through color displays, which may be in the form of X-Y plots, schematic presentations or alphanumeric displays. Examples of typical displays include both real-time data and ambient background information such as headings, schematics and labels. The operator-machine interface should be designed to allow active operator control. Complex human-machine interface would inhibit the control capability of operators.

IMPACT OF AUTOMATION ON OPERATOR TASK: ADVANTAGES AND LIMITATIONS

1. Increased skill level requirements:
The automation of water plant operations does not eliminate the waterworks operator; however, it places the need for additional skill requirements on the operator and reduces the number of workers under his or her direct supervision. The areas for additional skill development and training include basic knowledge of computers and familiarity with alphanumeric keyboard as well as the numerous functions, displays and charts available to him and some basic knowledge about the digital instrumentaiton used for plant automation. A knowledge of the key functions allows the operator to know which ones to use, and how to locate and interpret displays. The skill mix required to successfully operate automated water plants would include electronic technicians, computer analyst, and system maintenance personnel.

2. Job design considerations:
Increased skill levels requried of an automated water plant operator should be matched by adequate financial compensations. Factors relating to the physical environment of the operator workstation such

as safety, health, comfort and attractiveness are typically well taken care of in automated water plants.

Job enrichment and enlargement: should be considered in assigning traditional water operators to run automated plants. Generally, the good operators have considerable experience in their fields and plants. Their knowledge and familiarity is so high that they become an integral part of the plant with senses well tunned to the plant functions. In an automated plant, provision should be made for areas of decision making which the operator can call his or her own. Since most functions are automated in computer controlled plant environment the operators spends considerable time on plant inspection, checking and ensuring that plant operations are running smoothly; for instance that valves are open/close; wells are turned on/off; monitoring maintenance needs e.g. pumps that require lubrication, filters that require change on any piece of equipment, and undertaking any other tasks that are not automated, depending on the level of plant sophistication.

The operator may be requried to perform other additional tasks, for instance, grass-cutting, painting, janitorial services and undertake light maintenance work on pumps, motors, valves, meters and other mechanical equipment. A fraction of time is spend on bacteriological laboratory analysis to satisfy EPA requirements and to verify automated continuous on-line chemical analysis results.

Career path: automation of water plants demands additional skill acquisition by the operator thereby offering him/her learning and personal development opportunities as well as opportunity for career advancement. Some individual autonomy and self regulation should be designed into the operator's job.

3. As a result of automation, some water plants operate extended shifts, particularly when there is increased water demand; and reduced work week.

4. As previously mentioned, automation reduces the number of personnel needed to operate a water plant and expands or changes the skill mix. Overall, automation results in less personnel dollar expenditure while it increases the expenditure for equipment.

5. Automation of water plant creates the need for more research in human operator - machine interface, and social operator; mental work output and fatigue effects, and motivational and social needs as increasing levels of automation are implemented in the water industry. There is also the need to establish guidelines for allocating functions between humans and machines.

6. Automation may facilitate the attainment of stringent EPA regulations on water quality due to improved monitoring and measurement of relevant parameters of interest. Improved measurements tend to result in improved control and subsequent improved quality for any process.

7. Automation improves overall water utility management by improved management reporting systems.

8. Some water operators accustomed to traditional water plants tend to resist efforts toward automation of water operations.

CONCLUSION

This paper has examined the changing work environment of the water operator with the trend toward increased automation in water plants. The skills, abilities, and knowledge requirements of traditional water operators are identified as well as additional technical and basic computer skills requried to function effectively in an automated water plant environment. Some guidelines are offered for job re-design for water operators which is necessitated by automation.

REFERENCES

[1] Beaverstock, M. C., Stassen, H. G., Schneider, H. W., Modelling the operator: a Needed Approach to Interface Design in Tech, April 1979 49-52.

[2] Guarino, C. F., Radziul, J. V., Hess, A., Cairo, P. R., Day, R. E., Smith Jr., J. R., Philadelphia's Water System Automation Plan, Journal of the Environmental Division, Oct. 1976, 953-968.

[3] Damachi, N. A., and Barnes, S., Telemetry for a Utility Computer Network: A Cost Effectiveness Study. Institute of Industrial Engineers Annual Conference Proceedings, 1983.

[4] Shell, R. L. and Damachi, N. A., Improving Water Utility Management Water Engineering and Management 2 (1984) 42-44.

[5] Shell, R. L., Damachi, N. A., Souder, H. R., Leigh, B., An Analysis of water utility operator task, in: Mital, A. (ed) Trends in Ergonomics/Human Factors Vol. I (North-Holland, Amsterdam 1984).

[6] Shell, R. L. and Okogbaa, O. G., The effect of mental fatigue on knowlege worker performance. Institute of Industrial Engineers, Fall Conference Proceedings, Toronto, Canada, 1983.

[7] Shell, R. L., The impact of automation on work measurement, Institute of Industrial Engineers, Fall Conference Proceedings, 1982.

Trends in Ergonomics/Human Factors II
R.E. Eberts and C.G. Eberts (Editors)

A HUMAN FACTORS ANALYSIS OF AMBULANCE DESIGN

Robert E. Shook III
Department of Psychology
Miami University
Oxford, Ohio 45056

Philip F. Spelt[1]
Department of Psychology
Wabash College
Crawfordsville, Indiana 47933

Ambulance design suffers from a paucity of Human Factors considerations, especially for the driver's compartment. Data are presented which suggest that noise levels and control positions could contribute to decreased performance by the driver during high-speed emergency runs, and remedies are suggested. Soundproofing is the greatest need. Control relocations are also suggested.

When an ambulance is used for an emergency run, there are always stressful conditions from the high speed and the very nature of the emergency. In addition, there may also be unfavorable weather and road conditions present. Indeed, under severe weather conditions helicopter transfers are not possible and ground transportation is the only means left to move a critically ill patient to an acute care facility that may be an hour or two away. Even under optimal conditions, the operator is still exposed to stresses from noise and excess highway speed. Despite these problems, the only human factors considerations present in current ambulance design are those common to the automobile industry at large, which do not take into account the high stress conditions under which ambulance operators work. The little research that has been done on human factors in ambulance design relates to patient comfort and physical arrangement of the patient compartment. (Heuse, _et al_, 1975; Snook, 1972).

The analysis which follows will show that there clearly need to be improvements made in ambulance design. The needed changes fall into three categories: soundproofing of the driver's compartment (cab); redesign and relocation of controls peculiar to the unit's operation as an ambulance; and relocation of some driving controls. The discussion will center on problems and solutions in these three areas.

I. Noise

One major contributor to operator stress is background noise in the cab; this noise is comprised of engine and road noise, the siren, radio traffic, and often a two-way intercom between the technician and the driver.[2] A certain amount of trade-off is necessary in this area: the siren must be loud enough to warn other drivers, and the ambulance driver must also be able to hear other emergency equipment responding to the same call. Moreover, if the ambulance has the siren on, the radio must be turned up so that it can be heard above the increased noise in the cab, and the driver and technician must be able to communicate. That these sources of noise contribute to driver stress is supported by data we gathered from three ambulances in wide use in America, one based on the 1975 Dodge Tradesman van, and the other two on Ford Econoline vans with Vanbulance (1977 model) or Wheeled Horse (1983 model) modification packages. Measures were made with a Realistic Sound Level Meter (Radio Shack #42-3019) at several vehicle speeds and operating conditions. The meter was set on the A scale and Slow setting to give an average sound level reading. The meter was held by the person riding in the passenger's seat, pointed at the windshield, since pointing the meter in various directions did not seem to affect the readings. All measures were made with the windows rolled up. These data are summarized in Figure 1, which shows driver's compartment noise levels with and without siren for three ambulances[3] running at four different road speeds.

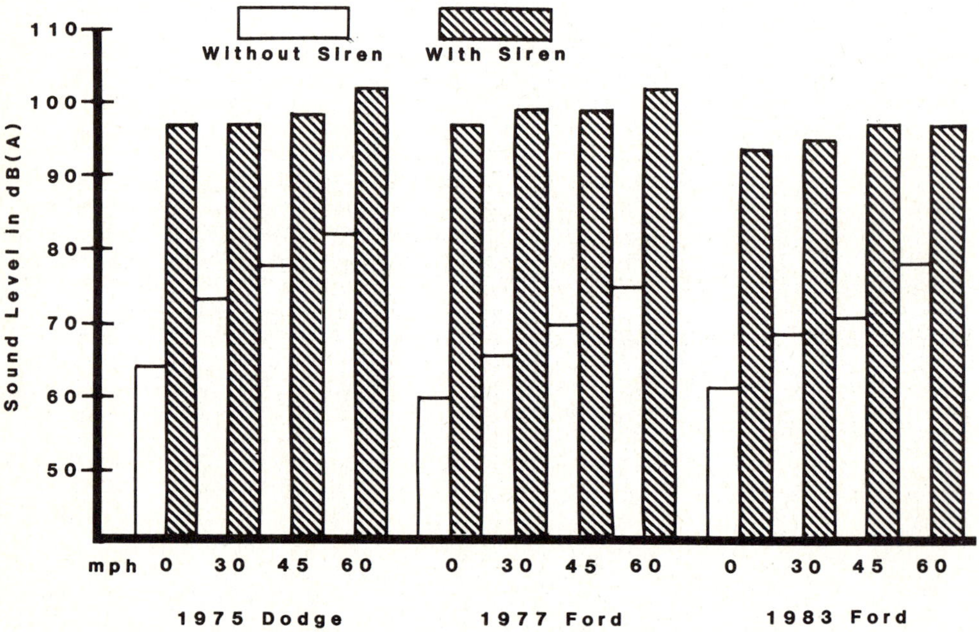

Figure 1. Cab noise levels in dB(A) as a function of speed for three ambulances, with and without siren.

As can be seen in Figure 1, noise levels without siren rise as a function of road speed (open bars). The oldest vehicle was subjectively the noisiest, a fact reflected in the higher bars for it than for the two newer ones. In addition, the older Ford was "clearly" noisier (subjectively) to those riding in it than the newer one, although the meter readings do not reflect this. Some of the noise increase of the older vehicles may have come from the rattling of doors and medical equipment which were more securely fastened in the 1983 model. For reference, the average sound level in a quality passenger car is about 65 to 70 dB (Snook, 1972).

With the siren on, average noise levels in all three vehicles rose to between 96 and 102 dB, completely masking vehicle noise. It should be noted that these are average levels, and the peak levels were momentarily higher. Although it does not show in the graphs, the subjective sound level in the newest Ford was subjectively lower than in the other two ambulances. Speaking and radio traffic added another 7 to 10 dB to the general noise level without siren and about 5 dB with siren. In one of the vehicles, there is an intercom which was either "on" or "off" for the driver, so that if the technician does not speak, the driver hears an amplification of the background noise from the patient compartment. While this intercom did not add anything to the measured noise level, it did add to the variety of noises to which the people in the cab were subjected.

Recent evidence presented by Ayers and Hughes (1984) indicates that these cab noise levels may actually reduce visual acuity, in addition to contributing to the general stress of the driver and medical technician. These authors measured human visual performance on three tasks under two levels of rock music (70 dBA and 110 dBA). They found a significant reduction in visual acuity reflected in subjects' detection of a small gap in a computer-generated square. There was no deterioration for either visual search or for tracking on a moving target.

As our data indicate, noise levels in the cab need to be attenuated by soundproofing. Indiana regulations state that "The patient compartment shall be heavily insulated to minimize conduction of heat, cold, or external noise entering the vehicle interior."[4] This regulation neither specifies an acceptable level of noise in the patient compartment nor calls for any soundproofing in the cab. Snook (1972), in discussing the detrimental effects of excessive sound on the patient, has also called for more engine soundproofing. His recorded sound levels in the patient compartment were lower than the levels reported here for the cabs. Proper soundproofing should reduce both vehicle and siren noise to levels comparable to those in quality passenger cars (about 65 to 70 db). To quiet the intercom, a call button could be installed, or a voice-activated switch could be set to turn on the intercom when the sound reaches a certain threshold (i.e., a voice above conversation level). The doors on the unit (both cab and patient compartments) should fit tightly to reduce external noise and rattles.

Part of the radio noise comes from unwanted radio channels used by other service vehicles (police, fire, etc.). It would be a benefit for the driver to be able to <u>easily</u> select which frequencies are monitored on the radio, thereby reducing the extraneous radio traffic and making it easier for the driver to detect "important" traffic that may be coming across.

II. Ambulance Controls

Controls that are used by the operators also show very little consideration for human factors design principles. The guiding philosophy of the current arrangement was that the controls be equally accessible to both people in the driver's compartment. Thus, most of the switches that control major functions (including lighting, heating/air conditioning, vacuum pump, auxiliary power, loading lights, warning lights, and the generator) are located above and to the right of the driver's head, in the center of the cab. The function of the switch is written above it (and backlit at night), but all the surfaces feel identical and the driver must take his/her eyes off the road to locate the proper switch before using it. There are two master switches located within 10 cm. (4 inches) of each other; one is the master for the red lights, the other is the main electrical system cut-off for everything including the ignition. It is far too easy to shut down the entire unit when reaching up to shut off only the red lights.

The controls for the siren and outside speaker are located to the lower right of the driver, on the engine cover. The selector is a multi-position switch that is rotated to select one of five desired options. These options are illuminated by a green light, but the unit is located far enough from the driver that selection must be done by trial-and-error, day or night, since the lettering is so small. Also, one must step through all the intervening options to get to the desired one. There is a single "hit" button located on the left of the unit (used to give just one cycle of the siren or to give a quick burst of a higher frequency when the siren is on) but the driver must have his/her right hand off the steering wheel to use it.

The radio, another essential piece of equipment, is located on top of the dashboard, in the center. Day/night options are available which change the brightly-lighted daytime display to backlit selections and a dimly-lighted display at night. The microphone is within easy reach of the driver, and automatically moves the radio to the dispatch frequency when the microphone is removed from its clip. The channel selector is a touchpad that "beeps" when used, although the operator must remember which service/frequency is located on which number selector.

The controls that are currently above and to the right of the driver's head must be relocated. While they are within the reach envelope for anyone using the unit, it is still unsafe to have to take one's eyes off the road to locate the proper control. There are two possible options for relocation: 1) A re-designed steering wheel with a command center in the middle or occupying the bottom half; 2) A small panel protruding from the dashboard, near the gear selector, in easy reach and sight of the driver.

The steering wheel option would keep the controls within very easy reach and sight of the driver, although they would not be accessible to the technician if s/he were riding in the front passenger seat. The switches should still be backlit and illuminate when activated. The switches themselves could be redesigned to fit a smaller panel (especially if they are more within the sight range of the driver), although they should still feel different from each other.

The siren controls should be altered to have them set as single-select push button switches that remain depressed when the given option is selected. The "hit" switch should be moved to the center of the steering wheel control panel for easy access, although it should not be located in a common "horn" position.

The radio should be designed with an alphanumeric display that shows what service is currently called up not just by channel number, but by name, like "IHERN"[5], "FIRE", etc. The touchpad could be designed with transparent plastic tabs to go where the numbers are to again describe that selection by name, not number. The touchpad keys should also be made larger.

III. Driving Controls

Driving controls in the units we tested were in the locations the manufacturer had placed them. The foglight switch is under the driver's seat (1977 Ford) out of reach, and the headlight dimmer foot-switch is difficult for the driver to find. Foglights should be activated from the same area as the headlights, and the bright/dim control should be put on the directional signal as it is in most newer cars.

Seating is uncomfortable on long trips, and the vinyl upholstery is excessively hot in summer and cold in winter. The seats need to be redesigned to distribute weight more evenly across the backs of the legs, and the seats should be made of cloth to reduce driver discomfort in temperature extremes.

The "kill" switch that disconnects the entire electrical system should be made more difficult to use; it is generally used only when shutting the unit down back at the garage after a run. There should be a cover over the switch to prevent it from accidentally being hit, or else it should be moved to a remote area of the dashboard. Chapanis and Kincaid (1972, cited in Kantowitz and Sorkin, 1983) list seven ways to prevent accidental operation of a control: recessing the switch, relocating it, orienting the direction of activation away from one of accidental activation, covering, locking (2 motions to activate), operation sequencing, and increasing resistance to operation.

There are many aspects of ambulance design which need improving. The safe arrival of the patient and the crew should be regarded as the prime objective of all ambulance design work. This safety is presently compromised by the faults described above. It can be assured by thorough application of the principles of sound human factors design, taking into consideration the increased stress and less-than-optimal working conditions that face ambulance drivers.

FOOTNOTES

1. Reprint requests should be sent to Dr. Philip F. Spelt at this address.

2. In Indiana, both people working on an ambulance are usually state-certified Emergency Medical Technicians (EMTs) or paramedics. For the purposes of this paper, "driver" refers to the person operating the ambulance, and "technician" refers to the medical technician working with the patient in the rear patient compartment.

3. The authors wish to express their thanks to Statewide Transfer Ambulance and Rescue (STAR Ambulance) and to Crawfordsville Emergency Medical Services, both of Crawfordsville, Indiana.

4. State of Indiana Official Rules and Regulations, 836 IAC 1-3-3.k.3.

5. Indiana Hospital Emergency Radio Network

BIBLIOGRAPHY

--------, "State of Indiana Official Rules and Regulations for Operation and Administration of Emergency Medical Services, Pursuant to IC 16-1-39 Acts of 1974, Public Law 55, Section 1." Indianapolis: Indiana EMS Commission, 1982.

Ayres, T.J. and Hughes, P., Amplified music and visual task performance. Applied Psychology, Clarkson Univ. (1984, submitted for publication).

Heuse, A., Dewaay, C., Demul, C., and Degueldre, C., Etude ergonomique des voitures ambulances, Travail Humain, 38, 2 (1975), 373.

Kantowitz, B.H. and Sorkin, R.D., Human Factors: Understanding People-System Relationships (John Wiley and Sons, New York, 1983).

Snook, R., Medical Aspects of Ambulance Design, British Medical Journal, 3 (1972) 574-578.

Trends in Ergonomics/Human Factors II
R.E. Eberts and C.G. Eberts (Editors)

A MICROCOMPUTER-BASED SYSTEM FOR MATCHING INDUSTRIAL WORKERS TO PHYSICAL TASKS IN FUZZY ENVIRONMENT

Waldemar Karwowski and Nancy Mulholland
Department of Industrial Engineering
University of Louisville
Louisville, Kentucky
U.S.A.

A procedure that allows matching of workers to physically demanding jobs in presence of imprecise information about the worker's characteristics and task requirements is proposed. The qualitative descriptors of the worker's characteristics are expressed as linguistic variables. Similarity relations between the desired and actual values of the domain sets are used for construction of the index of match. Such an index is then calculated as an intersection of the respective similarity measures, and its maximum value is selected as a criterion for the worker/task assignment.

INTRODUCTION

Despite the recent progress in automation and mechanization of work, it is still necessary in many industrial settings to perform physical tasks which are biomechanically and physiologically demanding. Furthermore, as indicated by Chaffin (1982), as many as 33% of industrial jobs require potentially hazardous physical effort on an occasional basis during a work day. Therefore, it has been recommended that an assessment of the physical capabilities of the workers being considered for such tasks be performed. This is often a case when redesigning of the hazardous jobs is neither technically possible nor economically justifiable.

The performance of the worker, and the chances that s/he will safely accomplish the given task depend on such factors as muscular strength and endurance, aerobic work capacity (or maximal aerobic capacity), age, safe behavior, and experience on the job. The above characteristics, as well as estimates of the job requirements, should be used then in developing procedures for a worker/task assignment.

Unfortunately, not always are the requirements of the task and characteristics of the workers well known. Instead, only imprecise and qualitative information about relevant characteristics is available. Furthermore, such knowledge is primarily based on human judgment and experience with job assignment procedures. The supervisor responsible for assignment can only specify that the potential candidates for a given task should have "medium" <u>experience</u> (relevant to the nature of the task), be "more or less" <u>strong</u>, and of "average" <u>physical condition</u> (fitness status).

The applied human judgment and intuition are not easy to formalize, and do not allow for straightforward quantification of matching criterion. In view of the above, the matching process is taking place in a fuzzy environment (Karwowski and Ayoub, 1984), and as such, it requires fuzzy treatment to allow for vagueness of the verbal descriptions (Karwowski and Mital, 1984). This paper outlines the matching procedure that facilitates an intuitive assignment process in presence of imprecise information about worker's characteristics and task requirements.

ASSUMPTIONS AND MODELING TECHNIQUES

It is assumed here that the assignment process takes place on a daily (or weekly basis), and the time available for making an assignment decision is limited. The decision must be made in a situation where there exists a group of workers with different characteristics (called profiles) of the selected domains of interest. Also, a set of distinct tasks with different physical demands is to be performed on a daily basis.

Since only qualitative data (verbal descriptions) regarding the relevant task demands and workers' characteristics is readily available, the matching procedure should incorporate and facilitate human judgment and allow manipulation of vague information. In order to satisfy, at least partially, the above conditions, the concepts of fuzzy linguistic variables (Zadeh 1975) and similarity relations (Zadeh, 1973) is used to model the ill-defined categories, and to develop an index of match for the worker/task assignment.

LINGUISTIC VARIABLES AND SIMILARITY RELATIONS

In linguistic approach (Zadeh, 1975), the linguistic variables with words (phrases) of a natural language are being used as values. The linguistic values can be interpreted as fuzzy restrictions on the possible values of some base variable (for example young is a value of the linguistic variable age). These fuzzy restrictions are represented by the appropriate compatibility functions. Each such function associates with each value of the base variable a number in the interval [0,1], representing its compatibility with the fuzzy restriction.

Typical values of linguistic variables contain primary terms, called here domain values, hedges, and fuzzy connectives such as and, or and negation not. The hedges and connectives can be used as fuzzy modifiers of the domain values (for example: very young).

In our case, the linguistic values are used to model the verbal descriptions of both, the desired profile (having its basis in task demands), and the actual profile of the worker. Three domain sets of the desired and actual profiles are considered. These are muscular strength (S), physical fitness status (F) and relevant experience (E). The domain values for strength (S) are: weak, low, fair, average, good and strong, and for fitness (F) are poor, average, and excellent. The linguistic values for experience (E) are: low, fair, average, good, and high. The derivation of these values is explained later.

The domain values can be modified by the fuzzy modifiers: more-or-less, minus, plus, and very. The meaning of such modifiers can be explained as follows. Assume that strength is equal to fair. Then, one could also say that strength is very fair, or more-or-less fair.

In terms of the relevant compatibility function f(u) for low, this can be written as follows (Zadeh, 1975):

$$\text{more-or-less low} = \text{low}^{0.5} <-> f^{0.5}(u)$$

$$\text{minus low} = \text{low}^{0.75} <-> f^{0.75}(u)$$

$$\text{plus low} = \text{low}^{1.25} <-> f^{1.25}(u)$$

$$\text{very low} = \text{low}^{2} <-> f^{2}(u)$$

In order to evaluate similarity between the desired and actual profiles of the worker, the concept of similarity relations is utilized (Zadeh, 1973), and appropriate similarity matrices are developed.

A similarity relation S (Buckles and Petry, 1983) is defined over the domain values d of the domain set D, for either desired or actual profile:

$$S: \; D \times D \dashrightarrow [0,1]$$

and d_{ij} denotes a set of domain values for a domain set (i-desired, j-actual); while s_{ij} is a similarity measure between particular values of the desired and actual profiles.

The above similarity relation is reflexive, i.e.: S (low, low) = 1, but non-symetrical, i.e.: S (low, average) # S (average, low). The latter property ensures that the assignment not only provides for the necessary safety of the worker, but it also attempts to maximize the efficiency of assignment, and makes a full use of the worker capacities where these are required.

In other words, the attributes of a worker with a high level of experience or strength would be wasted if he was assigned to a task requiring only an average experience or strength. Therefore, the similarity index between the desired and actual values will be higher than the one applied in a situation where the desired level of experience is high and the actual experience of the worker is average.

DERIVATION OF SIMILARITY RELATIONS FOR THE DOMAIN SETS

Similarity measures for the desired versus actual values of experience and strength were derived on a purely subjective basis. In fact, the user has an option of changing these values at the beginning of the program.

Although no structured procedure was used to derive the respective similarities, care was exercised in order to ensure that some safety margin is incorporated to compensate for a possible error in actual strength assessment. It should be emphasized that if the experimental data referring to actual static or dynamic strength of the workers is available, the similarity matrix should be altered, and more exact analysis of the match between strength requirements and actual capacities should be performed. The similarity matrix for experience is identical to that of strength, and is given in Table 1.

TABLE 1

SIMILARITY MATRIX FOR STRENGTH OR EXPERIENCE

	DESIRED STRENGTH (EXPERIENCE)					
	WEAK	FAIR	AVERAGE	GOOD	STRONG	
WEAK	1.0	0.8	0.6	0.3	0.1	
FAIR	0.9	1.0	0.7	0.5	0.3	**ACTUAL STRENGTH (EXPERIENCE)**
AVERAGE	0.8	0.9	1.0	0.7	0.5	
GOOD	0.7	0.8	0.9	1.0	0.8	
STRONG	0.6	0.7	0.8	0.9	1.0	

General fitness status is used here as an indirect description of the worker's aerobic capacity, and his ability to meet energy demands likely to be imposed by the specific task. Physical fitness allows a comparison of the estimated energy demands of the task and energy available for work that can be produced by a worker.

The total energy available to an individual per day depends upon his age and his maximal aerobic capacity. If the aerobic capacity is unknown, as it is in our case, it can be estimated (Mital and Shell, 1984). Instead of using the aerobic capacity as a category for matching purposes, the domain of general physical fitness with three values: poor, average and excellent is introduced. These categories correspond to the qualitative estimates of physical conditions of low, average and high, respectively.

The amount of energy available for work to an 18 year old male with an unknown aerobic capacity can be estimated for different categories of physical conditions (National Academy of Sciences, 1980). In order to account for the effect of age, a decline of one percent per each year of age above 18 years can be used (Mital and Shell, 1984). The estimated values for the total amount of energy available for work for the 18, 32 and 48 year old males are given in Table 2.

The ratios of energy available for work between different classes of physical condition, i.e.: low/average, average/high and low/high were calculated and used to develop similarity values for actual versus desired fitness status. The above ratios were equated with the corresponding ratios for fitness, giving a value of 0.7 for poor/average, 0.52 for poor/excellent, and 0.75 for average/excellent. These were the average values calculated for different age categories. The similarity values for fitness status are given in Table 3.

TABLE 2

ESTIMATED TOTAL ENERGY AVAILABLE PER DAY (Kcal) FOR MALES AS A FUNCTION OF AGE AND PHYSICAL CONDITION

	Physical condition				
Age	low	fair	average	good	high
18*	2100	--	3050	--	4000
32	1785	--	2590	--	3600
48	1470	--	2035	--	2800

*based on Mital and Shell (1984)

TABLE 3

SIMILARITY MATRIX FOR FITNESS

	DESIRED FITNESS			
	poor	average	excellent	
poor	1.0	0.7	0.52	ACTUAL FITNESS
average	0.8	1.0	0.75	
excellent	0.65	0.85	1.00	

In addition to the domain sets described above, two age groups: young (18-33) and middle (34-48 years old) were proposed to account for the effect of age on the actual amount of energy available for work. The respective ratios of the total energy available for both age group categories at their upper age bounds (33 and 48 years old) were then calculated for all fitness categories.

The average ratio of 0.8 was used as the modifier of similarity values for the middle age category to account for a decreased potential for energy expenditure. For example, if the similarity values for the fitness (with respect to the same task) were 0.75 and 0.52 for the middle and young age workers, respectively, the fitness for the middle age worker would be modified to 0.75 x 0.8 = 0.6, due to the reduced amount of energy available to him in comparison with the young worker.

MATCHING PROCEDURE

After the similarity measures between the desired and actual characteristics of the workers are derived, these should be modified by the appropriate

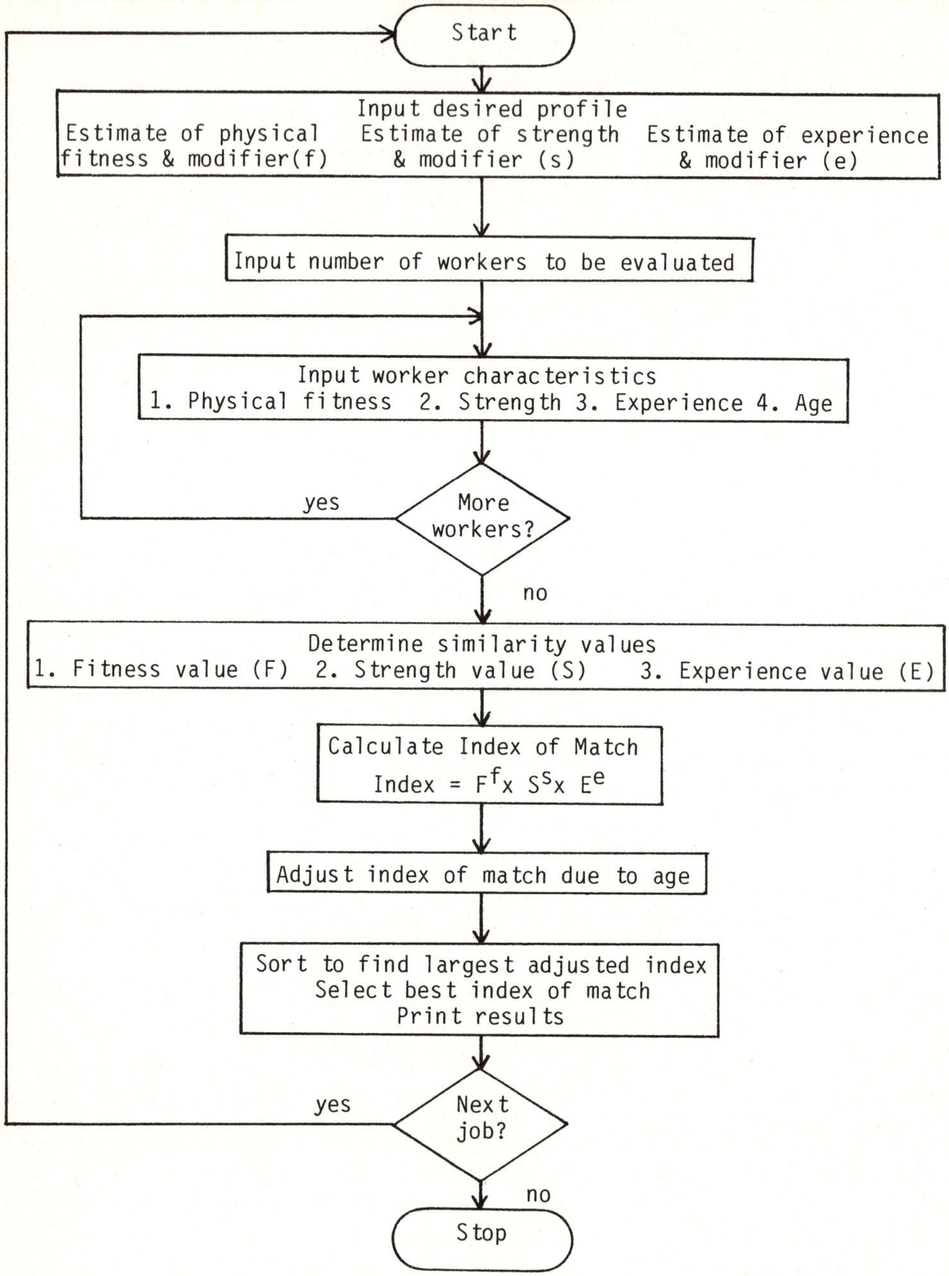

Figure 1
Flow Chart for the Matching Procedure

TABLE 4

MATCHING PROCEDURE - EXAMPLE

WORKER 6

▷ DESIRED PROFILE
▶ ACTUAL PROFILE

PHYSICAL FITNESS

1. POOR
▷ ▶ 2. AVERAGE
3. EXCELLENT

CHOOSE THE LEVEL OF PHYSICAL FITNESS THAT BEST DESCRIBES THIS WORKER (1 TO 3)

STRENGTH

▶ 1. WEAK
2. FAIR
3. AVERAGE
4. GOOD
▷ 5. STRONG

CHOOSE THE LEVEL OF STRENGTH THAT BEST DESCRIBES THIS WORKER (1 TO 5)

EXPERIENCE

▶ 1. LOW
2. FAIR
▷ 3. AVERAGE
4. GOOD
5. HIGH

CHOOSE THE LEVEL OF EXPERIENCE THAT BEST DESCRIBES THIS WORKER (1 TO 5)

AGE

WHAT IS THE AGE OF THIS WORKER? 30

SIMILARITY VALUES

WORKER	FITNESS	STRENGTH	EXPERIENCE	INDEX
1	1	.3	.9	.44
2	.85	.3	.7	.17
3	.7	.8	.8	.4
4	.7	1	.7	.34
5 *	.7	.5	1	.49
6	1	.1	.6	.11

WORKER 5 HAS THE BEST INDEX OF MATCH

linguistic modifiers specified for the domain values. The next step involves derivation of the index of match (I) defined as a product of the modified similarity values. An optimal assignment will be one with the highest value of I. Figure 1 shows the flow chart for the proposed matching procedure. The user is asked to specify the desired profile of the worker based on the estimated task requirements, and to assess actual characteristics of the potential candidates. An example involving six workers as candidates for the task is shown in Table 4.

CONCLUDING REMARKS

Although the example shown above is a simple one, one may imagine a situation in which more than three domain sets (worker characteristics) need to be specified, and a large group of workers need to be evaluated. Under such circumstances the choice for an assignment decision becomes extremely difficult, and human intuition and judgement must give up to the complexity of the problem. The outlined procedure for the worker/task assignment in a fuzzy environment requires further work and empirical verification.

REFERENCES

[1] Buckles, B. P. and Petry, F. E., Information-theoretical characterization of fuzzy relational databases, IEEE Transactions on Systems, Man and Cybernetics, Vol. SME-13, (1983) 74-77.

[2] Chaffin, D. B., Functional assessment for heavy physical labor, In: M. H. Alderman and J. J. Hanley (ed.), Clinical Medicine for the Occupational Physician, (Marcel Dekker, New York, 1982).

[3] Karwowski, W. and Mital, A., Soft descriptions and classifications in human factors engineering, In: A. Mital (ed.) Trends in Ergonomics/Human Factors I (Elsevier, Amsterdam, 1984) 325-330.

[4] Karwowski, W. and Ayoub, M. M., Fuzzy modelling of stresses in manual lifting tasks. Ergonomics, Vol. 27 (1984) 641-649.

[5] Mital, A. and Shell, R. L., Determination of rest allowances for repetitive physical activities that continue for extended hours, In: Proceedings of the 1984 Annual International Industrial Engineering Conference (1984) 637-645.

[6] Recommended Dietary Allowances, Food and Nutrition Board, National Academy of Sciences, National Research Council, Washington, D.C., (1980).

[7] Zadeh, L. A., The concept of a linguistic variable and its application to approximate reasoning. Information Sciences, Part I, Vol. 8 (1975).

[8] Zadeh, L. A., Similarity relations and fuzzy orderings. Information Sciences, Vol. 3 (1971) 177-200.

Trends in Ergonomics/Human Factors II
R.E. Eberts and C.G. Eberts (Editors)

MEASURING MOVEMENT AT ERGONOMIC WORKSTATIONS

Leonard S. Mark, David C. Vogele, Marvin J. Dainoff, Scott Cone, & Kermit Lassen

Department of Psychology
Miami University,
Oxford, Ohio 45056

A graphical method for measuring changes in body position over long time periods is used to assess worker movement at computer workstations. Comparisons of six people working in "best case" and "worst case" environments indicate that over several hours there is considerably less movement in properly adjusted ergonomic furniture.

A well-designed ergonomic workstation should assist a person in maintaining a relatively fixed (stable) working posture for prolonged periods of time (Dainoff, 1982). But even under ideal circumstances, movement is required to avoid muscolo-skeletal discomfort (e.g., eschemia) and other health problems (Branton, 1969; Pustinger, Dainoff, & Smith, 1985). On the other hand, large and frequent amounts of movement are undesirable because they involve expenditure of energy and interrupt work, as indicated by measures of productivity (Pustinger et al., 1985; Secrest & Dainoff, 1984). How successful are ergonomic workstations at minimizing the amount of body movement and postural adjustments required to maintain a fixed working posture?

To examine this question, a method is needed for measuring the amount and range of body movements and postural changes over long periods of time. While electromyography (EMG) can monitor muscular effort and fatigue involved in postural adjustments, EMG is a costly and highly invasive method for studying activity of individual muscles. More importantly, anatomical, mechanical and physiological indeterminacies in the action produced by a muscle (Bernstein, 1967) make it difficult, at best, to determine the overall amount and trajectory of body movement from EMG recordings of specific muscles. For this reason EMG is impractical for measuring the amount of movement. The current study documents an inexpensive "graphical" technique for measuring variability in body position over a long periods of time. This graphical representation is then used to look at differences in the amount of movement at "best case" and "worst case" computer workstations.

THE METHOD OF SUPERPOSITION

A procedure was developed for graphically representing body movement by superposition of successive frames of an event.[1] Six body landmarks (lips, shoulder, elbow, wrist, hip, knee) were identified on a photograph of a person working at a video display terminal (VDT). The joint positions were chosen because the actual distance between successive joints is nearly constant over time, thus permitting the user to infer limb position if desired. The lips were selected as a landmark on the head because they could be reliably identified from the photographs with which we worked. As a whole, these landmarks were intended to provide a representation of an individual's position at a given moment. (Although it would have been highly desirable to have also used a landmark below the knee, such as the ankle, the room in which the study took place was too small to permit the use of a wider angle lens without serious distortion. However, there is nothing to preclude the use of additional points in future work.)

The six body landamrks identified above were digitized using a Radio Shack Color Computer and X-Pad (256 X 192 resolution). To allow these landmarks to be superimposed at a later time, two reference points on the edge of the workstation were also digitized. These reference points were selected because at each workstation, they remain fixed relative to the position of the camera. The computer reoriented and scaled each picture, superimposing them on the two workstation landmarks. Thus, superposition of successive frames provided a graphic representation of positional variability of each of joint over time (Figures 1 & 2).

HYPOTHESIS AND PURPOSE OF THE PRESENT STUDY

If a subject makes a large amount of movement, then there should be a larger dispersion of at least some body landmarks than if the amount of movement is small. Variance in the position of each body landmark over time is taken as a index of the amount of movement. If ergonomic workstations help to minimize the amount of movement needed to maintain a relatively fixed posture for prolonged periods, it should be the case that: For each worker, there will be a smaller range (variance) of movement in a so-called "best case" work environment compared to a "worst case" or less optimal environment. That is, over a period of several hours, the dispersion of each body landmark should be smaller in the properly adjusted workstation than it is in the improperly adjusted work environment. The current study examines this prediction.

PROCEDURES FOR EXAMINING THE METHOD OF SUPERPOSITION

Our initial test of this graphic representation utilized video tape records from a study at the National Institute of Occupational Safety and Health (NIOSH), which attempted to assess the effect of workstation characteristics on productivity, perceived comfort and muscle exertion (measured by EMG). (Secrest & Dainoff [1984] describe details of a similar study run prior to this one.) Pairs of typists performed a data-entry task for 6 hours a day for 5 days (Monday through Friday). One member of each pair spent the first four days at a "best case" (properly adjusted) ergonomic workstation and the fifth day at a "worst case" (improperly adjusted) workstation[2]. For the other subject, the order of conditions was reversed. Each subject worked at the same workstation for all five days. On the fifth day, however, the workstations were readjusted to the condition opposite to the previous four days. To facilitate identifcation and digitization of body landmarks (above), reflective tape was placed on each subject's joints prior to the start of each session. Subjects were told not to wear loose fitting clothes which might cause the position of the tape to move with the apparel rather than the joint.

We used videotapes from the first three weeks of the NIOSH study. The two workstations were recorded successively; a 30 sec recording was made at one workstation, after which the computer would switch the video recorder to the other workstation.[3] Composite drawings of each subject were made from photographs taken at half hour intervals from the videotapes. The only criterion for selecting a video "frame" to photograph was that subjects had to be working on the task (i.e., hands on the keyboard). These pictures were digitized as discussed above. A total of 30 composite pictures were made for the six subjects. The composites were either of the morning or the afternoon work sessions. These composites permitted us to make 17 best case/worst case comparisons, with the restriction that composites of the morning session were compared to morning composites and afternoon composites were compared to afternoon composites.

RESULTS

In analyzing and evaluating the composite pictures, our concern is with positional variability of an <u>individual</u> worker in the best case workstation relative to the worst case. Thus, for each worker, we want to compare the amount of movement in the best case to the worst case. Figure 1 and 2 show examples of composites for four subjects.

From a <u>graphical</u> perspective, it appears that the best case composites exhibit less variability in joint position than the worst case composites. This is true regardless of whether the subject went from best case to worst or worst to best. But while an important goal of this work is to develop a graphical method that allows observers to "see" the relative amounts of movement, this variability must be quantified in order to examine our observations of Figures 1 and 2.

Toward this objective, a mean location was computed for each landmark based on the positions of that landmark in the composite. The average distance of each point from the mean provides a measure of the dispersion of each landmark around the mean. For convenience, we refer to this measure as the Mean Distance for a given landmark. As a coarse measure of the overall dispersion of points, we computed for each composite an Overall Mean Distance (OMD), which is defined as the average of the Mean Distances for each of the six body landmarks. For every subject, we plotted the OMD in the worst case (OMD_{worst}) against the OMD in the best case (OMD_{best}). Figure 3 shows that for each of the 17 best case/worst case comparisons, OMD_{best} was less than OMD_{worst} ($\underline{t}[16] = 9.67$; $\underline{p}<.001$). These data confirm our observations of the composite pictures (Figures 1 & 2): There is significantly less variability in overall body position for subjects working in a properly adjusted workstation compared to an improperly adjusted one.

However, do all landmarks contribute equally to the OMD or are particular landmarks subject to more positional variability than others? An analysis of the point dispersion indicates that the variance at each landmark is significantly less in the best case than in the worse case

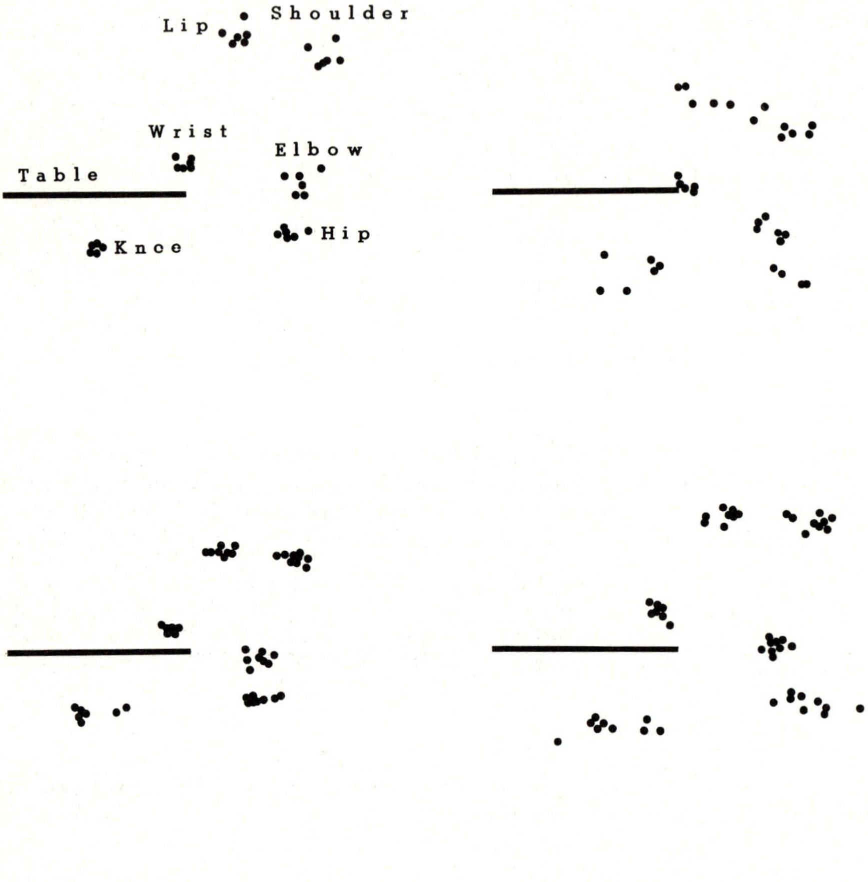

Figure 1

Composite drawings for two subjects (S1, top; S2, bottom) who worked at the "best case" workstation during the first four days (left) and "worst case" on the fifth day (right). The dispersion of each landmark in the "best case" composite is significantly less than in the "worst case" composite.

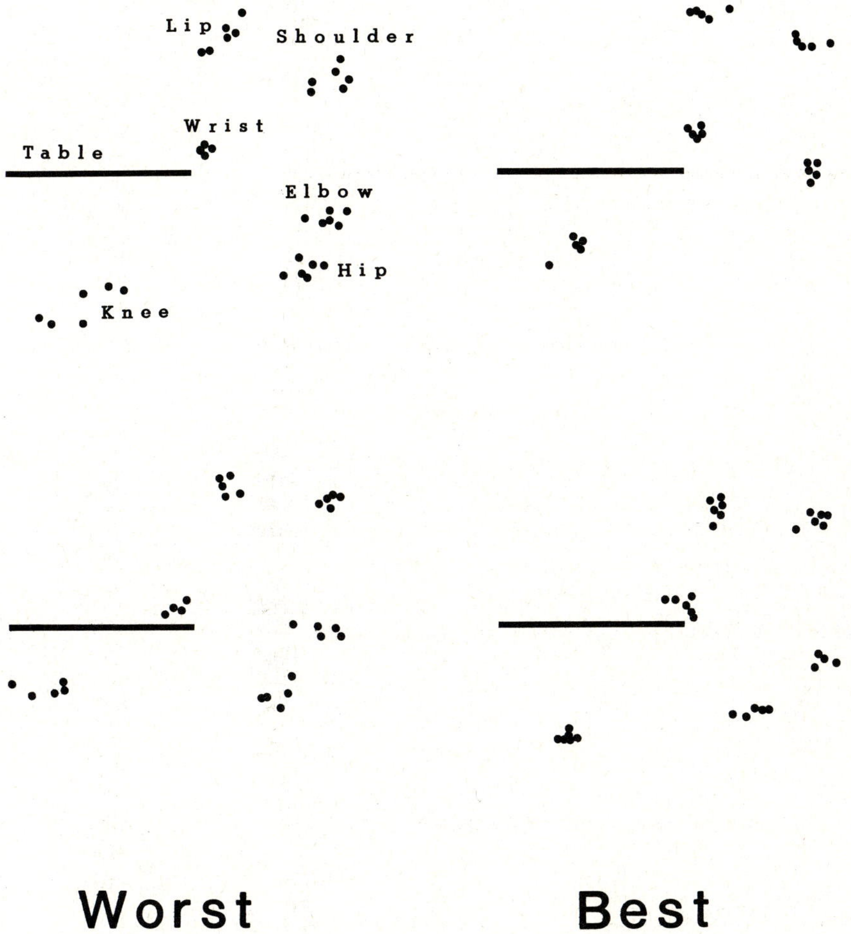

Figure 2

Composite drawings for two subjects (S1, top; S2, bottom) who worked at the "worst case" workstation during the first four days (left) and "best case" on the fifth day (right). The dispersion of each landmark in the "best case" composite is significantly less than in the "worst case" composite.

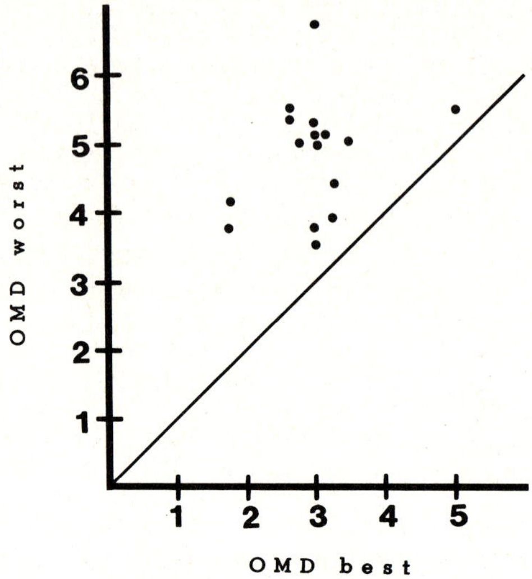

Figure 3

The Overall Mean Distance (OMD) for the best case composite (x-axis) and worst case composite (y-axis) are plotted for the 17 best case/worst case comparisons. The diagonal line shows OMD_{best} equal to OMD_{worst}. [OMD_{best} is less than OMD_{worst} for all 17 comparisons.]

Table 1

Landmark	Mean Difference between Worst Case minus Best Case*	t value (16 df)	pr>|t|
Lips	0.95	3.69	.002
Shoulder	1.60	5.91	.001
Elbow	0.96	2.19	.043
Wrist	1.37	2.47	.025
Hip	2.27	5.81	.001
Knee	3.39	7.93	.001

* In the coordinate system of our computer, one screen unit corresponds to roughly 0.85 cm.

(Table 1). But on close examination, the largest difference between conditions occurs at the knee, hips and shoulder -- the latter two are subject to considerable discomfort and fatigue from working at a fixed posture (Secrest & Dainoff, 1984). It also appears that differences at the wrist and elbow are smaller than for the other landmarks. This result is not surprising, since a criterion for taking a photographic record from the video tape is that subjects must have their hands on the keyboard, which remains in a fairly constant position relative to the two reference points on the workstation that are used to standardize the data; this constrains variation in the position of the wrist and, consequently, the elbow, since the two joints are separated by a single linkage.

DISCUSSION

In summary, the data from our method of superposition are consistent our prediction that <u>optimality</u> of the workstation configuration has a marked effect on worker movement. But are the small differences (Table 1) in the amount of movement really important? After all, 1 unit in screen coordinates is equal to roughly 0.85 cm. Even the largest mean difference (knee = 3.39) would correspond to slightly less than a 3 cm difference in point dispersion. Does this seemingly small difference have significant implications for the worker and for the evaluation of workplace designs?

We are cautious in evaluting the significance of the current findings of movement differences between best case and worst case conditions, since this study used a limited sample of data. However, the outcome of a recent study indicates that movement is inversely related to productivity. Pustinger et al. (1985) report a negative correlation (r = -0.60) between movement and worker productivity on a word processing task. This finding points toward the need for better measures for documenting movement over periods of at least several hours. For this reason, we believe the present work warrants further studies using the method of superposition as a tool for measuring worker movement and, ultimately perhaps, evaluating one behavioral effect of ergonomic furniture.

FOOTNOTES

[1]Variations of this technique have been used by anthropolgists, biologists and clinicians for studying the kinematics of growth (Moyers & Bookstein, 1980), and by human factors researchers analyzing movements of manual laborers (e.g., Gilbreth & Gilbreth, 1919).

[2]Each typist worked in a separate room. Each room contained a DEC VT100 VDT, an adjustable 2 level VDT workstation (either IBM Synergetics or Wright Line ETS 2000), and a chair. At the beginning of each week, one room was designated as "best case" and the other as "worst case". In the best case room, a Steelcase Concentrix chair, with lumbar support and height adjustmment, was employed; both the chair and the workstation were properly adjusted to the anthropometric dimensions of the typist. In the worst case room, a standard issue non-adjustable secretarial chair without lumbar support was used. In conjunction with

the standard chair, the workstation was maladjusted to simulate a single unit (monitor attached to keyboard) VDT on a high table. This maladjusted workstation was not atypical of which might be found in actual work environments.

[3]Due to a variety of computer malfunctions, one or both channels were not always recorded throughout the week. As a result, there was not always a sufficient amount of data (at least six pictures) at a given period of the day (morning or afternoon) to permit us to make a composite representation of movement. However, we were able to make at least one "best case" and one "worst case" composite for each of the six workers during the three weeks, thereby allowing us to examine the effect of workstation adjustment on the amount of movement.

REFERENCES

[1] Bernstein, N. The coordination and regulation of movement (Pergammon press, London, 1967).

[2] Branton, P. Behavior, body mechanics and discomfort. Ergonomics. 12, (1969) 316-327.

[3] Dainoff, M. J. Occupational stress factors in VDT operation: A review of empirical literature. Behavior and Information Technology. 1 (1982) 141-176.

[4] Gilbreth, F. B. and Gilbreth, L. M. Fatigue study (MacMillan, New York, 1919).

[5] Moyers, R. E. and Bookstein, F. L. The inappropriateness of conventional cephalometrics. American Journal of Orthodontics. 75 (1979) 599-617.

[6] Pustinger, C., Dainoff, M. J., and Smith, M. VDT workstation adjustability: Effects on worker posture, productivity and health complaints. Mid-Central Human Factors/Ergonomics Conference. (Purdue, 1985).

[7] Secrest, D. and Dainoff, M. J. Performance changes resulting from exposure factors in VDT data-entry workstation design. in Mital, A. (ed.) Trends in Ergonomics/Human Factors 1 (North-Holland, Amsterdam, 1984).

Trends in Ergonomics/Human Factors II
R.E. Eberts and C.G. Eberts (Editors)

WORK ENVIRONMENT DEMAND CHARACTERISTICS AND USAGE OF PROCEDURAL DOCUMENTATION

Greta L. Myers and Lorraine K. Tankersley

Human Factors Engineering
AT&T Communications
Cincinnati, Ohio

Structured interviews were conducted with 91 users of procedural documentation from 3 AT&T-C regions to investigate relationships between demand characteristics of the work environment and user response to and use of procedural documentation. Results are discussed relative to the 3 major variables found to influence the amount and method of procedural documentation use: book referencing skills, supervisory support, and type of training.

INTRODUCTION

Work environments are rarely stable, consistent, and unchanging. Instead, most work environments evidence dynamic characteristics, including lateral and vertical shifting of employees among task responsibilities, changes in work group size and composition, and changes in the work procedures themselves. Such procedural changes may result from shifts in the work group's responsibilities, the streamlining of job functions, the development of upgraded manual and mechanized processes to support overall corporate goals, or even changes in the composition of the organization as a whole.

Whatever the cause of such procedural change, the task of communicating both procedures and updates to all affected employees is of critical importance. The communication of work procedures by word of mouth or from incumbent to trainee without the support of documented procedures can be expected to result in the eventual generation of grossly inaccurate and highly inconsistent information. By contrast, company-wide user training programs, in conjunction with well-documented work procedures, can be expected to promote a common user knowledge base and to improve both work accuracy and efficiency of job performance. It is the purpose of this paper to propose how certain key environmental and user characteristics can mediate the effectiveness of training efforts and the usability of procedural documentation.

The type of procedural documentation referred to specifically in this paper is the Methods and Procedures (MAPs) manual. The MAPs were developed to provide users of AT&T computer systems with accurate, job-related procedural documentation. The manuals support the training effort for the inexperienced person and serve as a reference tool for the experienced system user.

The MAPs were so named for their intended logical resemblance to a road map. The MAPs route their users from the receipt of an input to the creation of an output. Both the manual and computer-supported components of the MAP user's job are included. Thus, each job is conceived of, trained, and documented as an integrated whole.

The MAPs consist of six components. The Overview section provides a description and workflow detailing the relationship between the job that is documented and the overall team. The Contents section enumerates the specific tasks documented for the work position. The Workflow section depicts each job task in logic tree format and lists those sections in the MAP which provide specific additional information. The Reference section provides the "how to" information to support the Workflow tasks. Unfamiliar terms can be clarified using the Glossary, and the Index provides an additional referencing tool.

The MAP documents were designed to have utility for both novices and experienced employees. While the novice may consult the MAP almost continuously in order to complete his or her job tasks, the expert may utilize key sections of a particular Workflow or Reference in the performance of critical or infrequent tasks.

METHOD

Relationships between demand characteristics of the work environment and user response to and use of procedural documentation were studied by means of a structured interview methodology. Ninety-one MAP users from three regions of AT&T-C were interviewed individually. The interview protocol was developed with the combined goals of obtaining a better understanding of our user population, determining methods of procedural documentation usage, establishing the perceived usability of the MAPs in particular, and understanding the environmental conditions mediating procedural documentation use.

Interview questions were generated by an interdisciplinary research team. The multiple choice format questions address current job level, tenure on the job and with the company, length of experience with the MAP documents, frequency of MAP use, perceived usability of the various MAP sections, ease of information retrieval, usage of retrieval aids such as the Contents and Index sections, format and perceived effectiveness of MAP training, and perceived level of supervisor support for MAP use. In addition, each user was asked how he/she might reorganize or redesign the MAP in order to facilitate use.

Interviews were conducted individually in 30-minute sessions by the authors. Subjects were encouraged to ask questions or offer any additional comments.

RESULTS

The data from the three regions were analyzed and compared relative to the three major variables found to affect user orientation toward the use of procedural documentation in the work environment: general book referencing skills, supervisor support, and type of training. Figure 1 illustrates these findings.

Type of training and user book referencing skills substantially influenced how efficiently information in the MAPs was located. Across regions, it was found that for those users who had received applied training (training

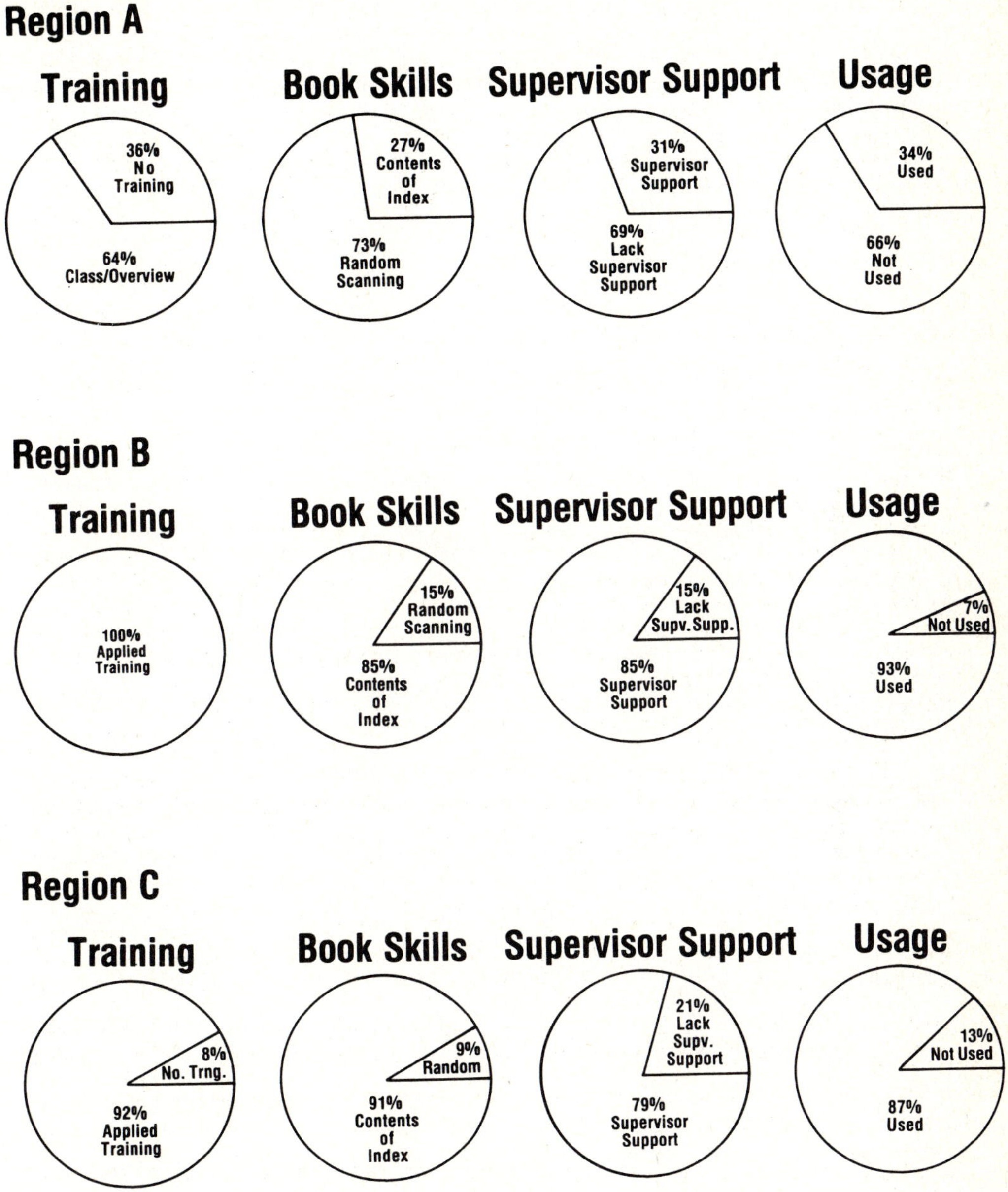
Region A
Training
36% No Training
64% Class/Overview
Book Skills
27% Contents of Index
73% Random Scanning
Supervisor Support
31% Supervisor Support
69% Lack Supervisor Support
Usage
34% Used
66% Not Used
Region B
Training
100% Applied Training
Book Skills
15% Random Scanning
85% Contents of Index
Supervisor Support
15% Lack Supv. Supp.
85% Supervisor Support
Usage
7% Not Used
93% Used
Region C
Training
8% No. Trng.
92% Applied Training
Book Skills
9% Random
91% Contents of Index
Supervisor Support
21% Lack Supv. Support
79% Supervisor Support
Usage
13% Not Used
87% Used

Figure 1

that included instructions on how to use the Table of Contents, Index, and other search aids), use of the Contents or Index for initial help in locating information was significantly more likely than for those who received no such training. Eighty-five percent of the people in Region B used the Contents or Index to help locate information; 100% of those in Region B received applied training, focusing in part on effective search strategies. In Region C, 91% used the Contents or Index; 92% received applied training. By contrast, in Region A, only 27% used the Contents or Index; 36% of this group had no MAP training and 64% were given only an overview of the purpose and content of the MAP; search techniques were not trained. When type of training and book referencing skills were compared to MAP usage in each region, usage was found to be much higher in regions where users received applied training. Ninety-three percent of the people in Region B and 87% in Region C reported using their MAP for reference or to solve problems, as compared to only 34% in Region A.

Supervisor support for the use of procedural documentation was also found to be a critical variable affecting user attitude toward the use of MAPs. In Region A, 69% of the users reported a lack of supervisor support; 66% of these users also reported that they did not feel comfortable using their MAPs on the job. Eighty-five percent of the users in Region A and 79% in Region B, however, reported that they had supervisor support; usage in these regions was reported at 93% and 87%, respectively.

CONCLUSIONS

Certain critical characteristics of both the user and the work environment may determine whether procedural documents actually achieve their intended purpose. Specifically, users lacking book referencing skills cannot make effective use of procedural documentation without remedial training. This clearly presents an important consideration for most training programs: the rudimentary skills necessary for timely and effective documentation use must first be identified and then assessed in the user group. Once user skills are assessed, training can be developed and applied in a manner that is consonant with user needs and that promotes optimal training effectiveness.

User response to the question, "If you could reorganize or redesign this manual, how would you do it to best suit your needs?", provided insight into specific user needs that ultimately improved our document design. It was found that even in regions where users were trained to use the Table of Contents or Index, many users still found it difficult to locate information. User comments indicated that the Table of Contents was not organized in a manner that facilitated timely retrieval of information. Specifically, users requested that the Table of Contents be structured to reflect the sequential steps of their job process. This new structure has since been instituted and has received a favorable user response. Thus, once we determined this particular user requirement relative to interfacing with our product, a more effective design resulted.

In addition, the more experienced users tended to "customize" their manuals; frequently-used material was extracted and organized into a smaller manual representing only the information the user perceived to be pertinent to the job process. Further study of these "customized" manuals is needed to determine a more efficient documentation design for the experienced user.

With regard to the work environment itself, a number of factors converged to yield two different scenarios. All regions were impacted by tremendous pressures to produce. Users in those regions where information retrieval strategies were not trained and supervisors were unsupportive of documentation use viewed the MAPs as a hindrance to effective job performance. By contrast, users in those regions where applied MAP training was provided and supervisors encouraged documentation use viewed the MAPs as a critical link in the meeting of job demands. Clearly, supervisor support has a major impact on user response to procedural documentation.

Trends in Ergonomics/Human Factors II
R.E. Eberts and C.G. Eberts (Editors)

VDT WORKSTATION ADJUSTABILITY: EFFECTS ON WORKER POSTURE, PRODUCTIVITY, AND HEALTH COMPLAINTS

Carole Pustinger[1]
Department of Environmental Health
University of Cincinnati
Cincinnati, Ohio

Marvin J. Dainoff
Department of Psychology
Miami University
Oxford, Ohio

Michael Smith[2]
Chief, Motivation and Stress Research Section
Applied Psychology and Ergonomics Branch
Division of Biomedical and Behavioral Sciences
National Institute for Occupational Safety and Health
Cincinnati, Ohio

Subjects alternated between two ergonomic workstation configurations while performing a word processing task. In the Adjustable condition, subjects were instructed how, but not why, to adjust the terminal stand and chair. In the Fixed condition, no adjustment was possible. Postural and performance analyses were carried out and subjective health complaints were assessed. Results indicated less motion and fewer health complaints under the Adjustable condition. In addition, motion was negatively correlated with performance.

INTRODUCTION

The introduction of video display terminals (VDTs) has caused a major revolution in the nature of office work. A growing body of literature suggests the possibility of health costs resulting from the intense use of poorly designed VDT equipment within poorly designed environments. As a result, many governmental and private sector organizations are generating guidelines and standards requiring the use of "ergonomic" equipment, with a goal of minimizing musculoskeletal and visual demand on operators. Many of these guidelines are predicated on the model of a static operator posture in which back and head are erect, the lumbar region of the spine is concave and supported, the forearms and thighs parallel to the floor, and the feet flat on the floor (see, for example, Cakir, ref. 1). Given the anthropometric variability of humans, it seems clear that such posture cannot be achieved without workplace furniture which is adjustable in several attributes.

However, it is not at all clear the model posture indicated above is, in fact, used by operators even if adjustable furniture is available.

Preliminary evidence (Grandjean et al., ref. 2) suggests that many operators prefer a backward leaning posture, with worksurface heights which are far from the theoretical optimal. The present study represents a laboratory investigation of these issues, in which task performance, subjective health complaints, and body movement are examined as a function of the opportunity to self-adjust the parameters of the workstation.

METHOD

Design

The study employed a repeated measures design using fourteen subjects; 12 women, 2 men. Subjects, who served as their own controls, worked for 3 hours without a break on each of two consecutive days; one in an "adjustable" workstation condition, and one in a "fixed" condition. Conditions were counterbalanced such that 7 subjects started in the adjustable mode and 7 in the fixed, with subjects assigned randomly to modes.

Subjects

Fourteen subjects with no prior experience working at adjustable VDT office furniture were recruited from an employment agency specializing in temporary clerical workers. They were required to have been tested for basic keyboard skills (at least 40 wpm) and received a physical examination and ophthalmologic evaluation prior to participation.

The physical requirements were:

1. General good health with no history of serious musculoskeletal disorders.
2. No bifocals, although glasses were permitted if vision was corrected within 20:18 to 20:22.
3. No history of hypertension.
4. Be willing and able to work for 3 hours without a break.
5. May not smoke during the work period.

Workstation

An effort was made to provide a model workstation environment in which the only variable would be furniture configuration. A 275-cm by 550-cm room in which temperature, humidity, and lighting could be controlled was employed. The furniture consisted of an IBM Synergetix adjustable terminal stand and chair. The table had separately adjustable keyboard and screen positions. The chair had a 5-leg, swivel-wheeled base and was moderately upholstered with a cloth surface. The backrest and seat were slightly contoured. The chair had a pneumatically assisted seat height adjustment and a lumbar supporting backrest which adjusted in both height and tension. A DEC VT-100 terminal was used, with screen controls kept at default settings. Additional glare control was provided by fitting the VDT with a Polaroid CP70 polarized glare filter. Wrist-rest and adjustable document holder were not provided because it was thought that individual adjustment of these accessory items might be a confounding variable. The entire work session was videotaped using a Panasonic video camera/recorder. The task, which was self-paced, consisted of typing manuscript text. This was considered to be a moderately visually demanding work task with high musculoskeletal and postural demands.

Measures

Posture. The subject's posture during the 3-hour work period was recorded on a videotape system which recorded the time visually on the tape. Postural changes by type, number, and time of occurrence were characterized over the entire work period utilizing the "pause" or "freeze-frame" capability of the system to allow accurate scoring. The categorical rating developed for this project is similar to, and expands upon, that used by Laville (ref. 3). Body and limb components which were scored were:

I. Manipulative System Changes
   - A. Hand/finger (flexion)
   - B. Wrist (bending/rotation)
   - C. Arms (stretching/flexion, left, right, both)

II. Postural System Changes
   - A. Head/neck (tilt forward, backward/lean sideways)
   - B. Shoulder (lean forward/press into chair/shrugging, slumping)
   - C. Trunk inclination (lean from hips)
   - D. Trunk rotation (shoulders only/hip swiveling, hip position change/hip to hip weight shift)

III. Leg/foot System Changes
   - A. Legs (vertical/forward/backwards/sideways)
   - B. Feet (flattened/crossed/elevated/on chair base)
   - C. Ankle (flexion)

These movements were scored as episodes of change by classifying them by system and type as follows:

1. Simple movement in one system; recorded by system.
2. Multiple movements in one system; recorded by system.
3. Combination movements involving more than one system; components named.

Furniture. Adjustments to chair, keyboard, screen table, and copy table were also recorded by type, number, and time of action for the entire work period.

Health complaints. Checklists were given to each subject at the beginning and end of the work period to examine subjective reports of symptoms of irritation, fatigue, pain, visual function difficulties, and general systemic complaints. The list used was the same as that reported by Dainoff et al. (ref. 4) and included three questions on the visual system, six on the musculoskeletal system, and one on headache.

Productivity. Productivity was evaluated by computer. The terminal was linked to a PDP 11/20 computer, and each subject's copy was compared to a previously entered master file. A program was written which timed the 3-hour work session from the first keystroke and signaled the subject when the time was completed. During the work period the program also tracked number of keystrokes, elapsed time, corrections, uncorrected errors, and backspaces.

Procedure

On the first day, the subject's height (without shoes) was measured at head, shoulder, elbow, hip, and knee. A Titmus Vision Tester was utilized to conduct visual screening of each subject on each day before and after the work period. Measurements taken included near- and mid-range acuity and vertical and lateral phoria. (Mid-range acuity was obtained by adding a -1.25 diopter lens to the far mode.) A ten-item, paper-and-pencil, Health Complaints Questionnaire was also given before and after each work period.

If the subject began at the fixed station, the keyboard and screen tables were set at a standard desk height of 750 mm (29.5 inches). The chair height was fixed at 470 mm, with the back raised to its highest position, thereby approximating a standard, nonadjustable chair. The subject was instructed not to change any feature of the workstation.

When the workstation was in the adjustable mode, subjects were able to vary keyboard and screen supports separately, and could control screen height, angle, viewing distance, keyboard height, horizontal distance, as well as adjust the seat height and backrest height and tension. At first, the investigator demonstrated the full range of adjustability for each component. Then subjects were led through the entire procedure again while making the adjustments themselves. This was followed by a few minutes of practice to become accustomed to the controls. Subjects were instructed as to how to adjust the furniture, but not as to why, or what one would expect to accomplish by the change. Comfort was not mentioned, only that the subject should feel free to arrange the various elements in a manner which allowed her/him to best perform the work, and that readjustments would be allowed at any time during the work period.

On each day the subject was allowed a 5-minute warm-up typing session. A bathroom break was then given and the subject returned and typed for 3 hours. The adjustable mode furniture was repositioned at the break so that the keyboard table was at its lowest (660 mm), the screen table at its highest (820 mm), and the chair seat height adjusted to the individual's knee height. It was explained that this was done so that everyone started the same but that the subject should feel free to self-adjust. The videotape was also started during the break to assure no loss of recording of adjustment.

In both modes the subject was instructed in the mechanics of the VDT and the typing task. The workstation was measured at the end of the session. Additional anthropometric measures taken at this time included elbow height, eye height, and eye-to-screen distance.

RESULTS AND DISCUSSION

Posture

The total number of all episodes of motion was greater in the fixed than in the adjustable condition (Wilcoxon one-tailed test, $p < .05$, $N = 14$).

Health Complaints

For these analyses, only the six musculoskeletal items were employed. Response scales for each item, in terms of intensity of discomfort, ranged from 0 to 52. Difference scores (after work-before work) were combined across all six items for each subject. Results indicated greater mean difference scores in the fixed condition than in the adjustable condition (Wilcoxon one-tailed test, $p < .05$, $N = 14$).

Productivity

No significant differences were found. All subjects showed some improvement on the second day regardless of workstation condition, indicating a probable practice effect.

Table 1. Median episodes of motion, before-after differences in health complaints, and keystrokes per minute by day and condition sequence: adjustable-fixed or fixed-adjustable

	Day 1	Day 2
Motion		
Adjustable-fixed	218	227
Fixed-adjustable	594	426
Health complaints		
Adjustable-fixed	10.9	9.6
Fixed-adjustable	14.3	8.6
Keystrokes per minute		
Adjustable-fixed	147.42	153.78
Fixed-adjustable	149.06	162.33

Correlational Analysis

There was a significantly large negative correlation ($r = 0.605$) between keystrokes per minute and total motion episodes ($p < .01$). Evidently increased movement resulted in an adverse effect on productivity.

There were also significant correlations obtained between motion episodes and the specific health complaint item relating to lower back pain. Obtained $r = 0.59$ for fixed; $r = 0.58$ for adjustable ($p < .05$).

None of the other health complaints were significantly correlated with movements, and there were no significant correlations of health complaints and productivity.

Order Effects

In addition to these basic findings, an order effect pattern emerged from the data. In the postural movement data, the subjects who began in

Day One/Fixed (FA) showed higher median levels of movement, greater variability, and greater rates of change than the Day One/Adjustable (AF) group. This was most evident on Day One/Fixed but true for both days. This group of subjects also made more furniture changes in the adjustable condition and raised the seat height and keyboard height. At the end of the session group FA had adjusted to a mean keyboard table height of 69.5 cm and a mean seat height of 48.9 cm versus Group AF's mean table height of 66.6 cm and mean seat height of 44.6 cm. These differences are not likely to be the result of anthropometric factors since the two groups were virtually identical. Mean standing height was 163.6 cm for AF and 166.5 cm for FA. It should be noted that although Group FA adjusted to some 4 cm higher than Group AF, the seat level to keyboard distance means were 28.98 cm for AF and 27.56 cm for FA. Both fall within the 25-35 cm optimal range described by Hunting et al. (ref. 5).

The order effect was also observed in the health complaint data, Day One/Fixed always showing the greatest effect and Day Two/Adjustable rarely being as good as Day One/Adjustable.

CONCLUSIONS

The results of this study indicate better subject performance under the adjustable condition in terms of fewer health complaints and fewer episodes of motion believed to be caused by adverse loads on worker musculoskeletal systems. In addition, while no direct difference in keystroke productivity was measured, there was a clear inverse relationship between number of keystrokes per minute and frequency of episodes of motion. Beginning in the fixed furniture condition had an effect on how the worker adjusted the furniture when allowed complete freedom of adjustment without instruction as to 'expected' settings. The fixed-first group changed the furniture more on the adjustable day and chose higher table and seat heights. This group also displayed higher levels of movement, with greater variability and rate change over the 3-hour work periods. Further study with a larger group and longer work period and alternative training methods would be useful. Further analyses of the existing data are continuing.

ACKNOWLEDGMENTS

This study was conducted at the National Institute for Occupational Safety and Health, Division of Biomedical and Behavioral Science, Cincinnati, Ohio.

Special thanks are given to Ron McFarland of Dynalectron Corp., who wrote the computer program for the productivity evaluation.

FOOTNOTES

[1]Now at the Department of Psychology, Washington University, St. Louis, MO.

[2]Now at the Department of Industrial Engineering, University of Wisconsin, Madison, WI.

REFERENCES

[1] Cakir, A., Hart, D. J., and Stewart, T. F. M., Video Display Terminals (Wiley, New York, 1976).

[2] Grandjean, E., Hunting, W., and Piderman, M., VDT workstation design: preferred settings and their effects, Human Factors 25 (1983) 161-175.

[3] Laville, A., Postural reactions related to activities on VDU, in: Grandjean, E., and Vigliani, E. (eds.), Ergonomic Aspects of Visual Display Terminals (Taylor and Francis, Ltd., London, 1980).

[4] Dainoff, M. J., Fraser, L., and Taylor, B. J., Visual, musculoskeletal and performance differences between good and poor VDT workstations: preliminary findings, in: Proceedings of the Human Factors Society 26th Annual Meeting (Human Factors Society, Santa Monica, California, 1982).

[5] Hunting, W., Laubli, T., and Grandjean, E., Constrained postures of VDU operators, in: Grandjean, E., and Vigliani, E. (eds.), Ergonomic Aspects of Visual Display Terminals (Taylor and Francis, Ltd., London, 1980).

Trends in Ergonomics/Human Factors II
R.E. Eberts and C.G. Eberts (Editors)

EVALUATION OF SUBJECTIVE MEASURES OF CHAIR COMFORT

Raymond A. Daley, Brenda Voisard, and Marvin J. Dainoff
Department of Psychology
Miami University
Oxford, Ohio
U.S.A.

Three chairs were compared using 4 subjective measures of comfort and a performance measure. The chairs could not be differentiated on comfort or performance for subjects as a group, but a best case chair could be identified for 89% of the subjects. Performance was the only variable to interact with order, indicating a practice effect. Subjects' responses to specific body areas and chair features show that the rating scales allowed subjects to identify critical aspects of chair comfort after a relatively short time span.

INTRODUCTION

Reports of postural pains and discomfort during seated work tasks have led to an increase in ergonomic research and the production of ergonomically designed chairs. Assessment of these ergonomic chairs has been based on existing anthropometric recommendations, operator performance, and measures of comfort. Although Dainoff (ref. 1) has shown performance differences between well-designed and poorly designed workstations, comfort has become the criterion most widely used (ref. 2). Reports that ergonomic recommendations based on anthropometric data do not match with actual postural adaptations of workers (ref. 3) nor predict seat comfort (ref. 4) have led Shackel, Chidsey, and Shipley to call subjective measures of assessment "the ultimate criterion of comfort against which other more convenient and more objective methods may be validated" (ref. 4, p. 303).

Several different measures have been used to assess comfort. The General Comfort Scale (GCS) was developed by Shackel et al. (ref. 4) to measure overall comfort. Although the GCS has been used to obtain ratings of individual chairs, it does not allow a direct comparison of target chairs, but rather requires independently obtained ratings. Alternatively, Chair Rank Ordering provides a more direct comparison. This measure is obtained by asking subjects to rank order designated chairs on overall comfort. Two additional measures, the Chair Feature Checklist (CFC) and Body Area Discomfort Map (BAD), have been used to provide a more refined analysis of chair comfort. The CFC was designed to direct attention to 10 specific aspects of chairs (e.g., seat height, slope of seat, and position of backrest), while the BAD is designed to direct attention to specific body areas.

These subjective measures have been used in several studies, but the amount of time that a rater must be exposed to a chair in order to obtain valid and consistent ratings is uncertain. For instance, Barkla (ref. 5) found that 5-minute ratings of seat comfort were unstable and not consistent with 30-minute ratings by the same subjects. However, Drury and Coury (ref. 6) found ratings over a 2-1/2-hour period to be stable. A related question is consistency across subjects. Shackel et al. (ref. 4) and Christensen, Casali, and Kroemer (ref. 7) were not able to distinguish comfortableness of chairs when analyzing data averaged across subjects.

This study is the first in a series aimed at exploring the issues of consistency in chair assessment. In this study 4 subjective comfort measures and a clerical performance measure were used to compare 3 task chairs using 15-minute evaluation periods. Further studies will utilize longer evaluation periods.

METHOD

Subjects

Sixteen young adults from a midwestern university participated in the study. The sample consisted of an equal mixture of males and females.

Subjective Rating Scales

1. The General Comfort Scale (GCS) is a single-item, 11-point scale (from 1, "I feel completely relaxed" to 11, "I feel unbearable pain") which is considered to approximate an interval level of measurement (see ref. 6).

2. Chair Rank Order (CRO) is obtained by asking subjects to rank order a group of chairs on overall comfort after they have sat in each chair for a fixed period of time. Subjects were not told that they would be asked to rank order the chairs until the conclusion of the experimental session.

3. The Chair Feature Checklist (CFC) is composed of a list of 10 chair features and rating scales appropriate for the characteristics of each feature (see ref. 6). The CFC used in the present study was a slightly modified version of the Drury and Coury (ref. 6) edition. (Specific individual features will be presented in the Results section.)

4. The Body Area Discomfort Map (BAD) consists of a diagram of the front and back views of a human body with specific body areas outlined. In order to complete the BAD, subjects were asked to rate the discomfort of each specific body area through the use of a five-point scale with anchors of (1) slightly uncomfortable to (5) extremely uncomfortable. Ratings were made on lines next to the mannequin diagram that pointed to the specific body area.

Performance Measure

A letter cancellation task was used to measure clerical performance. Subjects were presented with nonsequential pages of a research manuscript and asked to cross out all the letter e's they could find in each line. The total number of e's cancelled while at each work station was calculated as the performance score.

Workstations

The workstations in which the tasks were completed varied only through the type of chair. The stations consisted of a task chair and a table (29 in. high, 32 in. wide, and 30 in. deep) surrounded to the front and sides by blank walls. The chairs themselves were structurally quite different. Although none of the chairs had arms, the predominant features of the chairs, a student work chair with hard plastic seat and back rest (chair #1), a wooden office chair (chair #2), and a chair with padded seat and back rest (chair #3), were quite distinctive (see Figure 1). Seat #2, unlike the others, allowed for swivel and backward tilt. However, the work posture dictated by task and table height did not allow the tilt feature to be utilized.

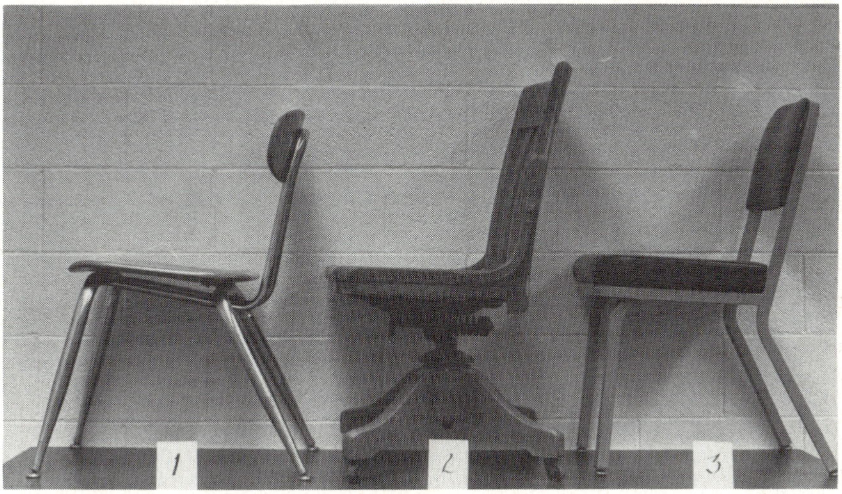

Figure 1
Task Chairs Which Were Evaluated

Procedure

Three subjects participated simultaneously in the same room with three different orders of presentation maintained during the study. The subjects were seated at workstations and commenced the letter cancellation task. At fifteen-minute intervals, subjects stopped work, stood up and filled out the rating scales in the following order: General Comfort Scale, Chair Feature Checklist, and Body Area Discomfort Map. After completing these rating scales, the subjects were moved to the next workstation and continued with the letter cancellation task. At the end of the three 15-minute trials, subjects reported the Chair Rank Order.

RESULTS

Comparison of Chairs

Univariate analyses of variance were computed for each of the dependent measures in 3 x 3 (Seat x Order) designs. Chair, order of presentation, and their interaction served as the independent variables in these analyses. For each of the measures, except performance, the greater the rating, the more negative the response was.

General Comfort Scale

Recall that the General Comfort Scale ranges from 1-11 with 11 being the least favorable. Although the chairs were structurally quite different (plastic, wooden, and padded), results of the analysis did not find differences among them. No main effect for seat was found (F = .17, df = 2, $p < .84$), nor was there any interaction with order of presentation. The mean ratings and standard errors of the mean for seat #1 were 4.62, 0.52; for seat #2, 4.25, 0.50; and for seat #3, 4.50, 0.56, respectively.

Chair Feature Checklist

Nine items of the CFC were applicable to the chairs in this study. Six of these items--seat height above floor, seat length, seat width, slope of seat, position of backrest, and curvature of back support--required ratings that were bidirectionally suboptimal (e.g., seat height: too high/ adequate/too low), while the other three--clearance for feet under chair, seat shape, and moulded chairback--were unidirectional (e.g., seat shape: good to poor). In both cases a 13-point rating scale was used. In the bidirectional case, the scale ranged from -6 to 0 to +6, and in the unidirectional case the scale ranged from 1 to 13. The difference in response characteristics for these two sets of items precluded their combined analysis.

The differences between mean scores for the three chairs on the unidirectional items was negligible and no further analyses were performed. Absolute values of the ratings of the 6 bidirectional items were used in an analysis, in which the seat variable approached significance (F = 3.23, df = 2, $p < .056$). The mean ratings and standard errors of the mean for seat #2 were 5.81, 1.24; seat #2, 4.88, 1.15; seat #3, 8.56, 1.41, respectively. The mean frequency of suboptimal responses to Chair Feature Checklist items is shown in Table 1.

Body Area Discomfort Map

The BAD item responses were combined in an additive fashion following the procedure of Shackel et al. (ref. 4). The chairs were not able to be distinguished on this measure. No significant main effect was found for chair (F = .29, df = 2, $p < .75$). The means and standard errors of the mean for chair #1 were 31.06, 7.35; chair #2, 34.25, 7.65; and chair #3, 35.38, 8.60, respectively. The mean frequency of suboptimal response to body areas combined across chairs is indicated in Table 2.

Table 1. Frequency of suboptimal response per chair on items of Chair Feature Checklist

Chair feature	Frequency
Seat height	.56
Seat length	.40
Seat width	.31
Seat slope	.58
Position of backrest	.60
Curvature of backrest	.67

Table 2. Frequency of suboptimal response per chair for body areas

Body area	Frequency
Head, neck, and shoulders	.59
Arms, wrists, and fingers	.30
Up/mid back, chest, and abdomen	.36
Lower back	.60
Buttocks, thighs, and legs	.20

Chair Rank Order

No significant effects were found for chair rank order. The mean CRO for chair #1 is 2.19, chair #2, 1.81; and chair #3, 2.00.

Performance

The data for the performance measure consist of the total number of e's cancelled at each work station. No differences were found between seat conditions (F = 1.08, df = 2, $p < .35$), with the mean for chair #1, 588.19; chair #2, 543.69; and chair #3, 583.69. A Seat x Order interaction was significant (F = 7.86, df = 4, $p < .0003$). Not surprisingly, performance rates increased as the sessions progressed. Means for successive 15-minute periods, combined across chairs, were 489.00, 586.62, and 639.94.

Summary--Analyses by Chair

The grouped data indicate no significant differences between chairs, although the Chair Feature Checklist approached significance. A Seat x Order interaction on the dependent variable of performance is consistent with an interpretation of practice effects. Order of seat condition was not found to interact with any of the other dependent measures.

Individual Analysis

To examine the data on a more individual basis, seat rank orders were computed for each rating scale. The percentages of subjects who rated either seat #1, #2, or #3 as best on each rating scale are presented in Table 3. These rank orderings were also examined for subjects individually. If we use as a criterion for selection for a best chair for a given individual that individual's ability to rate the <u>same</u> chair as best on 3 out of 4 measures of comfort, then 14 out of 16 subjects could be said to have defined a best chair. The percentages of subjects for whom chair #1, #2, and #3 were each rated as best are given in Table 4. This confirms the group data indicating the lack of a consistent preference for any one chair.

In the 14 cases where a best chair could be defined, there were 5 instances in which a chair was rated best on all 4 measures. This left 9 cases in which one measure was discrepant. General Comfort Scale and the Chair Feature Checklist were each discrepant once, the Body Area Discomfort measure twice, and the Chair Rank Order five times.

Table 3. Proportion of subjects who ranked a specific chair best on each rating scale

Rating scale	Chair #1	Chair #2	Chair #3
General Comfort Scale	.30	.30	.40
Chair Rank Order	.19	.50	.31
Chair Feature Checklist	.32	.47	.21
Body Area Discomfort	.35	.35	.29
Mean	.29	.40	.30

Table 4. Proportion of subjects who rated a specific chair best on 3 out of 4 measures

Chair #1	Chair #2	Chair #3	No choice
.25	.38	.25	.12

CONCLUSION

One of the goals of chair evaluation is the ability to make a choice from among several alternatives. In this study, a single chair could not be identified as a best choice for our subjects as a group on any rating scale. Using similar measures, Shackel et al. (ref. 4) found clusters of as many as 6 chairs that could not be differentiated and Christensen et al. (ref. 7) failed to find significant differences between two chairs. One possible reason for these findings is the homogeneity of the chairs. Some chairs might be too similar to differentiate, suggesting that subjective measures of evaluation are not precise enough to be used for this task. If this is the case, then it would be difficult to choose, based on comfort, one out of a set of homogeneous chairs for a group of workers.

However, our data suggest that this conclusion does not necessarily hold for individual workers. Although the analyses do not define a best chair for the group, observation of rank ordered ratings of individual subjects shows that a single best chair could be distinguished for 14 out of 16 subjects. It appears that individual differences in the ratings of chairs (i.e., different subjects rate different chairs as best) precludes the averaging of data to define an overall best chair.

It is of interest to take a more detailed look at the Body Area Discomfort

Map and the Chair Feature Checklist. The frequency of response to items on the BAD indicates that the "head, neck, and shoulder" and the "lower back" receive the greatest response. Thus, the BAD allowed subjects to distinguish areas of discomfort that are most commonly reported in the literature. This is also true for the CFC. Concern for seat height, slope of seat, and backrest is indicated by the frequency of suboptimal responses. These are dimensions of seats that have again concerned researchers. Thus, in a relatively short amount of time, the CFC and BAD elicited responses from subjects that identified chair features and body areas that have been of concern to researchers and designers.

Although a single best case could not be defined after 15-minute trials, critical aspects of the chair and body during seating were able to be identified and optimal chairs could be identified for individuals. Whether this same pattern of results would be found in a longer term sitting trial is the next problem that will be addressed in this research program.

REFERENCES

[1] Dainoff, M. J., Visual display terminals: the relationship between ergonomic design, health complaints and operator performance, Occupational Health Nursing, December (1983) 29-33.

[2] Leuder, R. K., Seat comfort: a review of the construct in the office environment, Human Factors 25 (1983) 701-711.

[3] Grandjean, E., Hunting, W., and Piderman, M., VDT workstations design: preferred settings and their effects, Human Factors 25 (1983) 161-175.

[4] Shackel, B., Chidsey, K. D., and Shipley, P., The assessment of chair comfort, Ergonomics 12 (1969) 269-306.

[5] Barkla, D. M., Chair angles, duration of sitting and comfort ratings, Ergonomics 7 (1964) 297-304.

[6] Drury, C. G., and Coury, B. G., A methodology for chair evaluation, Applied Ergonomics 13 (1982) 195-202.

[7] Christensen, C., Casali, J. G., and Kroemer, K. H. E., A methodology for work chair assessment, Proceedings--First Mid-Central Ergonomics/ Human Factors Conference (University of Cincinnati, Ohio, 1984), 339-344.

VII

SAFETY

Trends in Ergonomics/Human Factors II
R.E. Eberts and C.G. Eberts (Editors)

HAZARD PERCEPTION AND RISK ESTIMATION IN ACCIDENT CAUSATION

Bernhard Zimolong

Institut of Psychology
Ruhr-Universität Bochum
4630 Bochum, Postfach 102148

A study was carried out to suggest additional measures of safety in the field of accidents caused by falling. Methodology is briefly described and results are reported. Risk estimation was measured with a new pictorial technique. Results show a significant mis-match between estimates and objective risks. They are explained in terms of 'bounded rationality'. Proposals are derived to improve rational decision making under occupational hazards.

INTRODUCTION

Hazard perception and risk estimation seem to be important factors in accident causation. A great deal of recent experimental effort has been devoted to understanding hazard cognition, including hazard perception and hazard assessment (Hoyos 1980), and risk taking behavior. Dunn (1972) compared subjective estimates of risk made by chainsaw operators with an objective risk distribution derived from reported accident frequencies. He found very little correlation between the subject's estimates and the objective distribution.

Zimolong (1979) studied subjective risk estimates of shunters. Subjects estimated relative frequencies and degrees of seriousness of accidents on a 5 point scale for a variety of work places and work activities. Little correspondence was found with objective accident frequencies and distribution of minor, severe and fatal injuries. Lichtenstein et al. (1978) demonstrated that individuals do not have valid perceptions about the frequency of hazardous events in other fields, too. As an example, floodplain residents heavily misjudged the probability of floods as compared to empirical calculations (Slovic et al. 1974).

Klebelsberg (1973) and Wilde (1978) studied decision making in hazardous situations. They hypothesized on the basis of empirical investigations, that an individual will always maintain a constant margin of safety, irrespective of any changes in skill level, supporting equipment or other technical aids. The accepted risk level is established as a result of previous experiences and cognition. If the perceived hazard will exceed the accepted risk level, the person will improve his/her precautions.

Recent laboratory experiments in the field of decision making under uncertainty have provided strong support for the concept of 'bounded rationality' (Simon 1957). This research indicates that people systematically violate the principles of rational decision making when judging probabilities, making predictions or otherwise attempting to cope with probabilistic risks. Frequently, these violations can be traced to the use of judgmental heuristics or simplification strategies (Tversky & Kahneman 1974). Slovic (1982) summarizes several types of judgmental bias investigated in natural settings as well as violations of rational decision making when attempting to integrate probabilities and values into the selection of some course of action.

It is obvious then that the variability of decision making in hazardous situations may arise at least from three sets of factors:

- o those that affect perception and estimation of risk
- o those that affect the criteria upon which decision are based, and
- o those that affect the rationality of decision making.

These factors imply for accident prevention, that attempts to control risk taking must not be based wholly on (1) establishing a code of acceptable and non-acceptable levels of risk, but that an attempt must be made (2) to induce accurate subjective risk estimates. An incorrect estimate of a risk inherent in a situation and the use of inadequate simplification strategies will result in a poor decision, no matter how accurate the decision criteria are.

STUDY: ACCIDENTS CAUSED BY FALLING

A study[1] was performed to develop and suggest additional safety measures in the field of accidents caused by falling (Segger & Zimolong 1982). After briefly describing the methodology and some basic results, I will mainly focus on results of the hazard cognition part of the study.

Two kinds of methods were employed to study accidents and near miss situations due to falling. First, officially available accident statistics were examined and accident reports were categorized according to technical, organizational and personal accident factors. Secondly, investigation procedures focussing on normal and hazardous situations were carried out for occupational groups most affected by accidents caused by falling.

Results of accident analysis are shown in Table 1. Serious injuries are those resulting in monthly financial compensations for labour related accidents (erstmals entschädigte Unfälle).

Table 1: Occupational accidents and accidents caused by falling from 1977-1979 in West Germany

	Occupational Accidents	Accidents due to Falling	Percentage
Total number	4,347,746	75,988	1.75
Serious Injuries	108,441	12,345	11.38
Fatal Injuries	3,899	839	21.52

1 This research was granted by the Bundesanstalt für Arbeitsschutz und Unfallforschung in Dortmund 1981.

The fatal accident risk of occupational groups was determined by a risk quotient, calculated from the number of fatal accidents (A), divided by the number of employees (N) and multiplied by 10,000: RQ = 10,000 (A/N). Table 2 shows the results of the classification.

Table 2: Occupational fatal accidents due to falling from 1977-1979 in West Germany

Occupational Groups	Accidents	Percentage	Risk Quotient	Rank
Air and waterway transportion	71	8.46	5.04	1
Carpenter, tile-layer (on the roof), scaffolding assembler	123	14.66	2.55	2
Building construction workers (temporary)	44	5.24	1.15	3
Building construction workers (e.g. brick-layers)	99	11.80	0.75	4
Auxiliary building trade	22	2.62	0.58	5
Steel construction	100	11.92	0.41	6
Painter and lacquerer	29	3.46	0.36	7
Transport Worker	49	5.84	0.34	8
Plumber	24	2.86	0.30	9
Electrician	40	4.77	0.21	10
Surface transportion (e.g. truck driver)	20	2.38	0.07	11
Total	621	74.02	0.48	

Information from accident reports is of limited value as far as the identification of accident causes is concerned. Accident causes are incompletely reported, shaped by an individual bias of the investigator or are even intentionally faked (Schwerdtfeger & Zimolong 1973). Therefore additional information about accident causing factors was gained by investigating working conditions and personal behavior in hazardous situations. Study design methods are summarized in Table 3.

The critical incident technique was used to sample information about type and frequency of near miss fall situations. Questionnaires were employed to record objective and personal stress factors at work, and the individual assessment of various accident causes, including organizational factors as well as climate and social conditions.

The risk estimation techniques was used to measure the estimates of relative frequencies of accidents due to falling. Pictorials of places of work were presented on a metal board, including ladder (l), scaffolding (sc), roof (r), building under construction (c), lorries, bridges and platforms (lo), steps and stairs (st), and openings in walls, ceilings and floors (o).

Subjects were asked to allocate 25 magnets to the 8 pictorials according to the supposed accident frequency distribution. Each magnet represented 10 accidents. The ranking exercise was performed twice: for all accidents (minor injuries) and for accidents with serious/fatal injuries.

Table 3: Study Design Methods

Object	Method
Personal Factors	
o hazard perception	Critical incident technique
o risk estimation	
o knowledge, motivation and skills related to accident prevention	Risk estimation technique Questionnaire for weighing accident causes
Job Characteristics	
o job demands	Observations
o stress factors (physical, organizational, personal)	Interviews Group discussion Questionnaire

Six occupational groups with high risk quotients were selected for closer examination. Subjects of the study were 153 employes of mainly small and medium size companies of the building industry and auxiliary building trade. Table 4 summarizes occupational groups and number of participants.

Table 4: Occupational groups and number of participants

Occupational Group	Worker	Supervisor	Manager
Carpenter	20	3	3
Tile layer (on the roof)	15	2	3
Scaffolding assembler	15	1	3
Construction worker	30	5	5
Painter	20	3	2
Steel Construction worker	20	2	1
Sum	120	16	17

An objective measure of risk distribution was derived by analyzing official accident reports of the years 1977-1979 (Abt 1979). Types and frequencies of accidents were allocated to occupational groups and places of work.

Fig. 1 shows the differences of the estimates and the objective frequencies in percentage of all injury accidents and serious/fatal accidents respectively. Chi-square statistics were calculated to test the independence of subjective estimates and objective frequencies. As the figures in Table 5 indicate, the hypothesis of independence between subjective and objective

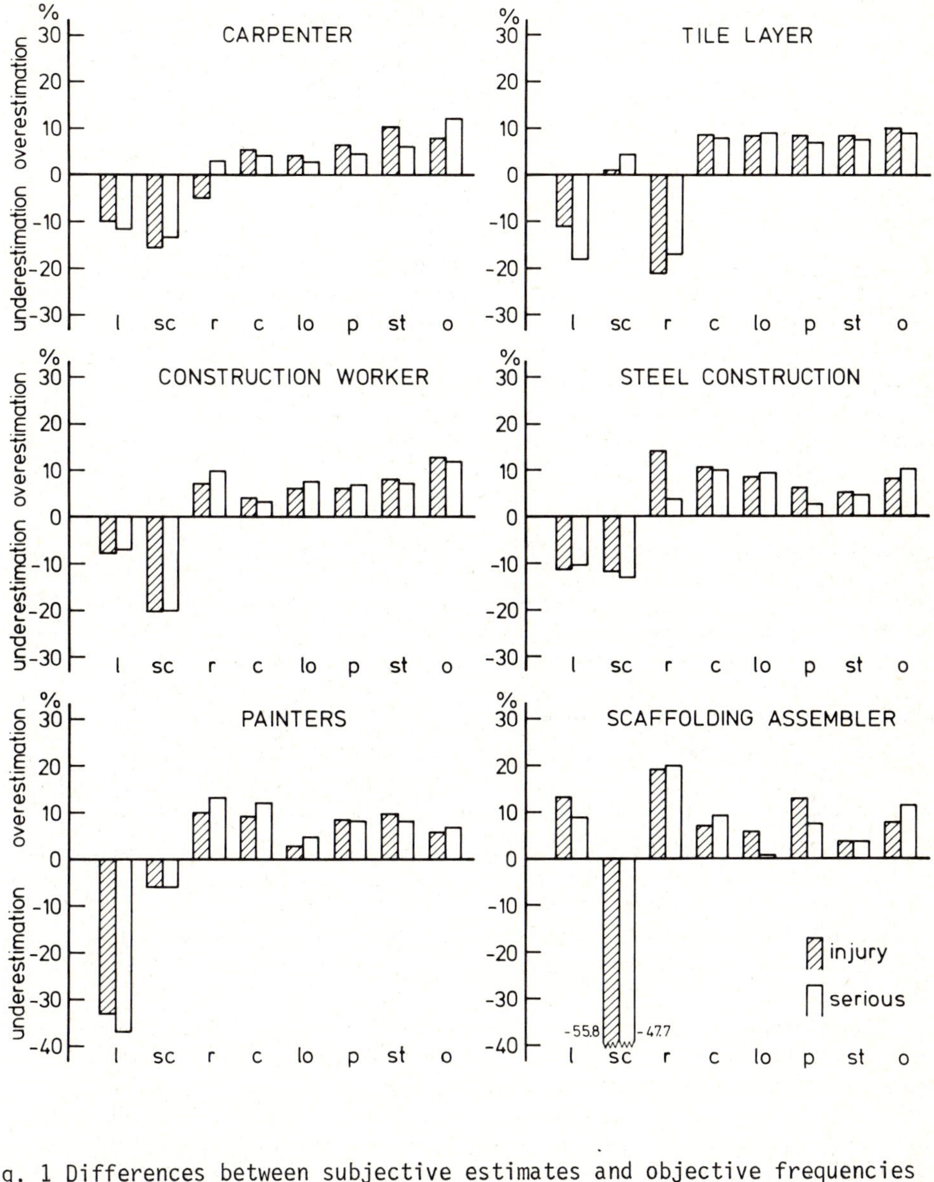

Fig. 1 Differences between subjective estimates and objective frequencies of accidents due to falling.

frequencies could not be rejected on the 1% level of significance. A closer analysis and an extended representation of data can be found in Segger & Zimolong (1982).

Table 5: Chi-Square statistic for testing the independence of subjective estimates and objective frequencies of accidents ($df = 7$, $p \leq 0.001 \geq 24.32$)

Occupation	Injury Accidents	Serious/fatal Accidents
Carpenter	37.36	29.71
Tile layer	48.46	46.34
Scaffolding assembler	89.96	68.66
Construction worker	29.23	36.82
Painter	53.12	61.09
Steel Construction worker	41.94	34.59

DISCUSSION

As far as the results from the estimation study are concerned, three major conclusions can be drawn initially:

(1) There is a significant agreement amongst the employees of each group about which places of work are the most likely to be injured.

(2) There is a mis-match between the risk as perceived by the subjects and the risks as objectively measured. Liability to injury is underestimated in situations with high risk probabilities and vice versa.

(3) Employees are more likely to underestimate high risk situations, if they had a long time experience of these hazards.

The results of this study support the hypothesis, that there is a mis-match between the subjective risk estimates and objective risks. This is not surprising since the subjects' experiences of different types of accidents are based on little evidence and upon a great deal of informal group communication, second hand information and hearsay.

Some misfits between subjective and objective risk distribution seem to be due to special problems of the applied technique. Slovic et al. (1978) have shown that independent from the object under study high risks generally are underestimated and low risks overestimated. They called this phenomenon 'scale compression'. In this study however, underestimation varies systematically with the occupational groups. That means, familiar work situations with high accident risks are more underestimated than unknown situations with comparable risk probabilities.

Similar results from other studies support this point of view. Shunters with long exposure times to hazards underestimated accident risks of familiar work situations more than unfamiliar situations with comparable risk probabilities (Zimolong 1979). Estimates were measured on a five point rating scale, subjects rated frequencies and seriousness of injury accidents.

Probably estimation data reflect the outcome of different processes of the decision making under uncertainty. As far as the 'availability bias' is concerned (Tversky & Kahneman 1973), misfits between estimates and objective data may be caused by little evidence and knowledge of subjects. On the other hand, underestimation of hazards may reflect violations of principles of rational decision making in order to cope with occupational hazards. Available risk information is adjusted and subjectively redefined according to the accepted risk level (Mc Grath 1976).

For the purpose of occupational safety, misfits between subjective risk estimates and objective risks are calling for adjustments and improvements. Traditionally this was thought to be covered by safety posters, safety publications, especially accident reports, and activities of safety experts. It appears that this problem should be attacked on three sides.

(1) An attempt is needed to make hazard perceptions an accurate representation of the risks existing in the real world. That means as far as instructional programs are concerned, they should improve the employee's ability to detect all kinds of hazards at his place of work and to estimate their relative importance.

(2) Criteria of personal acceptable levels of risk must be establ ished to define general acceptable risk levels at work. Beside safety rules and regulations, this is most influenced by model behavior of supervisors and management.

(3) Application of rational principles in decision making can be improved by offering and training techniques and procedures how to cope with hazards due to falling. This includes behavioral training programs how to perceive different types of hazard, how to use safety techniques and how to take personal precautions against accidents due to falling.

REFERENCES

Abt, W., Unfallanlyse 1979. Schriftenreihe des Hauptverbandes der gewerblichen Berufsgenossenschaften e.V. Bonn, 1979.

Dunn, J.G., Subjective and objective risk distribution. Occupational Psychology, 46, 1972, 183-187.

Hoyos, C. Graf, Psychologische Unfall- und Sicherheitsforschung. Stuttgart, Kohlhammer, 1980.

Klebelsberg, D., Risk Taking Behavior. Pharmacopsychiatry, 6, 1973, 83-90.

Lichtenstein, S., Slovic, P., Fischhoff, B., Layman, M. & Combs, B., Judged frequency of lethal events. Journal of Experimental Psychology: Human Learning and Memory, 4, 1978, 551-578.

Mc Grath, J.E., Stress and Behavior in Organization. In: Dunnette, M.D. (Ed.) Handbook of Industrial and Organizational Psychology. Chicago, Rand Mc Nally, 1976, 1351-1395.

Schwerdtfeger, W. & Zimolong, B., Technische und psychologische Bedingungen der Unfallverhütung im Straßenverkehr. Zeitschrift für Verkehrssicherheit, 19, 1973, 143-162.

Segger, H.R. & Zimolong, B., Möglichkeiten zur Verhinderung von Absturzunfällen. Forschungsbericht 314, Bundesanstalt für Arbeitsschutz und Unfallforschung, Dortmund 1982.

Simon, H.A., Models of Man: Social and rational. New York, Wiley, 1957.

Slovic, P., Kunreuther, H. & White, G.F., Decision process, rationality and adjustment to natural hazards. In: G.F. White (Ed.), Natural hazards, local, national, and global. New York, Oxford University Press, 1974.

Slovic, P., Fischhoff, B. & Lichtenstein, S., Accident probabilities and seat belt usage: a psychological perspective. Accident Analysis & Prevention, 10, 1978, 281-285.

Slovic, P., Toward Understanding and Improving Decision. In: W.C. Howell & E.A. Fleishman (Eds.), Human Performance and Productivity: Information Processing and Decision Making. Hillsdale, New Jersey, Erlbaum, 1982.

Tversky, A. & Kahneman, D., Availability: A heuristic of judging frequency and probability. Cognitive Psychology, 5, 1973, 207-232.

Tversky, A. & Kahneman, D., Judgment under uncertainty: Heuristics and biases. Science, 185, 1974, 1124-1131.

Wilde, G.J.S., Theorie der Risikokompensation der Unfallverursachung und praktische Folgerungen für die Unfallverhütung. Hefte zur Unfallheilkunde, 130, 1978, 134-156.

Zimolong, B., Risikoeinschätzung und Unfallgefährdung beim Rangieren. Zeitschrift für Verkehrssicherheit, 3, 1979, 109-114.

Trends in Ergonomics/Human Factors II
R.E. Eberts and C.G. Eberts (Editors)

PUBLIC AWARENESS OF HAZARDOUS WASTE IN THE MIDWEST AND PROPOSED SOLUTIONS

C. Saran, David Begeman, David Mosby, Jim Smith, and Mitch Warmbrodt

Department of Industrial Safety and Hygiene
Central Missouri State University
Warrensburg, Missouri 64093
U.S.A.

This paper reports on surveys conducted to assess the public awareness of hazardous waste in the Midwest of the United States. Rationale for conducting the study is developed. Methodology of conducting the survey, the instrument used, and the analysis of results are detailed.

RATIONALE

The dumping of hazardous waste has been going on for many years but the problems associated with it were not recognized or seriously dealt with until recently. Latent and long term effects--from soil, water, and air contamination to genetic defects--are not always obvious. Disposal of waste is dependent upon the knowledge of its toxicity, applicable regulations, available technology, and public sensitivity at the time of disposal. Public education of these effects has been largely through newspaper and television reports. Recent discoveries of toxic waste sites contaminating entire communities have resulted in controversy and confusion. While the generators and regulators of these sites engage in endless court battles, the public suffers.

Obviously, the residents of the toxic waste dumpsites like Times Beach, Missouri and Love Canal area are concerned and informed. Does the rest of the country care? How personal are the effects to them? What are they willing to do about the problem? In order to answer these and other questions, it was decided to evaluate the public awareness of hazardous waste in the Midwest.

Recent Legislation:

Since the mid-70's governmental regulations have attempted to make the generators of toxic waste responsible for its safe disposal. Paying for the clean-up or abandonment of communities at or near past dumpsites has been controversial. Everyone seems to be shirking responsibility. Absence of immediate, evident, short term effects on humans and the environment complicates the situation. This absence facilitates illegal disposal of hazardous waste for profit. It is estimated that 90% of the waste is disposed in an irresponsible and environmentally unsafe manner.[2] Residential and commercial sources generate solid waste at the rate of nearly 132 million metric tons per year.[4]

The Environmental Protection Agency advises state to develop their own hazardous waste programs. This agency regulates current and future waste practices through the Resource, Conservation and Recovery Act (RCRA) of 1976. Cleanup of old wastesites is conducted through the Comprehensive Environmental Response, Compensation, and Liability Act (CERCLA). RCRA sets standards and guidelines for owners and operators of facilities that treat, store, or dispose of hazardous waste. The primary objective of RCRA is to change the means of managing solid waste so as to ensure the protection of public health through nationwide conservation and protection of land, as well as water and air resources.[3]

Generators of hazardous waste are responsible for safety and control during generation, transportation, treatment, storage, and disposal of hazardous waste. Generators are also required to determine if its waste is hazardous by consulting the list of hazardous wastes contained in the regulation or, if the waste is not listed, by determining if it possesses any one of four characteristics established in the regulation. Some other requirements for the generator are as follows:

1. obtain an E.P.A. identification number
2. obtain a facility permit if waste is accumulated on the generator's property more than ninety days
3. use appropriate containers and label them properly for shipment
4. prepare a manifest for tracking hazardous waste
5. assure, through the manifest system, that the waste arrives at the designated facility
6. submit an annual summary of activities

Although multiple regulations and standards have been put into effect for controlling waste, toxic waste sites are frequently discovered. The discovery of Times Beach, Missouri site has caused great tribulation and grief for the people involved and the government. Hopefully, this grief will cause generators and controllers of toxic waste to be more stringent with their procedures.

The 1974 Safe Drinking Water Act (SDWA) provides for the safety of public water systems. The focus of SDWA is on the development and enforcement of national drinking water regulations, whereby public water supplies must not exceed recommended maximum containment levels.[3]

The purpose of the Clean Water Act of 1977, in general, was to establish regulations that require industries that are "direct discharges" (i.e., that discharge effluents directly into a stream) to treat wastes. Industries that "indirect discharges" (i.e., that discharge effluents to a publicly owned treatment work) are required to pretreat wastes.[3]

Methodology:

Two hundred forty-five residents within 200 miles radius of Warrensburg, Missouri were interviewed in the spring of 1983 and the winter of 1983-84. Four interviewers included one graduate and three undergraduate students from the Department of Industrial Safety and Hygiene at Central Missouri State University. The subjects ranged in age from 15 years to 70 years. About one-third had a college education. Half of the population interviewed resided in rural areas. The survey instrument used and the results are tabulated in Figure 1.

Results and Discussion:

The results have been grouped into discussion items: awareness and reaction, future generations, wildlife, cleanup, and solutions.

Awareness and Reaction: Most of the respondents were aware of the recently discovered Missouri dumpsites and of their hazards to the population near these sites. They were generally unaware of the exact toxic waste dumped at these sites. Most of them did not feel that their own health was affected by these dumpsites yet they were aware of their long time effects on health, birth defects, and environment.

Recent discoveries made them insecure about where they lived. They expected their feelings to be of helplessness, anger, and fear if a dumpsite was discovered in their neighborhood.

Future Generations: They were not very clear if their children would be affected by toxic dumpsites. Yet they felt that they have an obligation to future generations to make these dumpsites safe.

Wildlife: Overwhelming majority (99%) recognized that these dumpsites are hazardous to wildlife and would result in genetic problems and deaths. They also felt obligated to protect the wildlife from these sites. These reactions are not unexpected in view of the population's preference for outdoor activities and sports.

Cleanup: Most of the respondents felt that the people who created these dumpsites should be responsible for their cleanup.

Solutions: Stricter regulations and better legislations could help solve toxic dumpsite problems. Most preferred petitions and protests. About 80% would prefer to leave if a toxic dumpsite was discovered on their property. Most people did not want the government to buy out the property but preferred those responsible to pay. Respondents were evenly split on the government's responsibility to invest more money in research to discover other dumpsites. However, most felt that more toxic waste dumpsites will be discovered in the future. Most felt that the government has covered up toxic waste dumpsites and has kept the public uninformed about potential side effects. It may be noted that most of the local government officials contacted were either unaware of or reluctant to provide information on dumpsites in Johnson County, Missouri.

The respondents felt that future dumpsites would be better regulated. Stricter regulations and heavier penalties were preferred by the majority.

Conclusion:

This study indicates that the public is generally in favor of stricter legislation and its enforcement. It wants the cost to be borne by those who create these dumpsites. This is an understandable feeling but many of the creators of these dumpsites have either gone out of business, cannot be identified, or do not have the money to pay for the cleanup.

There is a feeling that the government has covered up the discoveries of toxic dumpsites and has been less than candid in informing the public about their ill effects. Since the respondents were generally aware of the discoveries of toxic dumpsites and their ill effects, it is obvious that the media has filled up the information gap.

Figure 1: Hazardous waste dumpsite questionnaire and responses.

1. Are you aware that 26 toxic waste dumpsites have been found in Missouri?
   yes 173; no 72

2. Do you know what types of toxic wastes were dumped at these sites?
   yes 65; no 180

3. Do you believe these dumpsites are dangerous to people living near them?
   yes 232; no 12

4. Do the recent dumpsites that have been discovered make you feel insecure about where you live?
   yes 137; no 108

5. What would your reaction be if a toxic dumpsite was found near your home?
   a. anger 19
   b. fear 17
   c. helplessness 52
   d. all of the above 187

6. What long-term effects do you think the dumpsites will have?
   a. health problems; b. possible birth defects; c. environmental problems; d. other.

   a and c - 150
   a, b, c - 72
   c only - 26
   d only -10

7. How do you think the problem of toxic waste dumpsites should be solved?
   a. stricter regulations - 102
   b. removal of toxic waste - 22
   c. better legislation - 57
   d. all of the above - 102

8. What do you think you can do, as a citizen, to stop the dumping of toxic wastes?
   a. petitions - 24
   b. protests - 12
   c. lobbying - 45
   d. all the above - 64
   e. a and b - 101

9. Do you feel the government has made an attempt to cover up toxic waste sites, to keep the public uninformed about the potential side effects?
   yes 87; no 8

10. What type of punishment should be placed on these companies?
    a. fines; b. clean-up costs; c. loss of waste disposal license;
    d. other

b only - 22
a and b - 88
a, b, c - 127

11. Do you think that dumpsites will be better regulated in the future?
    yes 178; no 67

12. What future regulations should be enforced?
    a. stricter regulations; b. heavier penalties for violations; c. no new regulations? d. other
    a only - 43; b only - 21; a and b - 173

13. Do you think towns such as Time's Beach, Missouri should be bought out by the government with taxpayer's money?
    yes 48; no 137

14. If you do not think the government should buy the land, what should be done?
    a. those responsible should pay 202
    b. land should not be bought by anyone 24
    c. other 19

15. If you lived in an area such as Time's Beach, Missouri would you be willing to leave, or would you want to stay on your property?
    stay 13; leave 82

16. Do you think more toxic waste sites will be discovered in the future?
    yes 92; no 3

17. Do you feel the government has the responsibility to invest more money in research to find other potential sites where toxic wastes have been dumped?
    yes 47; no 48

18. Do you believe your children will be affected by the toxic dumpsites?
    yes 122; no 107; other 16

19. What effects might these dumpsites have on future generations?
    a. birth defects - 7   a, b - 118
    b. cancer - 10   b, c - 14
    c. others - 72   a, b,c - 26

20. Do we have an obligation to future generations to make these dumpsites safe?
    yes 240; no 5

21. Do you think your health has been effected by toxic wastes?
    yes 14; no 131

22. Do you think toxic sites create hazards to wildlife?
    yes 242; no 3

23. What effects might we see in wildlife?
    a. genetic problems; b. death; c. no effect; d. other

a only - 9
b only - 7
d only - 5
a and b - 74
a, b, d - 141

24. Do you think the public has an obligation toward wildlife to keep them out of waste disposal areas?
yes 65; no 30

25. If yes, who should be responsible for the cleanup?
a. government 10
b. owners of the property 15
c. people who created the dumpsite 64
d. other 6

References:

1. Cargill, Oscar, William Charvat, and Donald A. Walsh. The Publication of Academic Writing. New York: MLA, 1966.

2. Epstein, Samuel; Lester O. Brown, and Carl Pope. Hazardous Waste in America. San Francisco: Sierra Club Books, 1982.

3. Pojasek, Robert B. Toxic and Hazardous Waste Disposal. An Arbour, Michigan: Ann Arbour Science, 1980.

4. U.S. Environmental Protection Agency. Your Guide to the U.S. Environmental Protection Agency. OPA 212, Washington, D.C. September 1982.

Trends in Ergonomics/Human Factors II
R.E. Eberts and C.G. Eberts (Editors)

THE NEUROBEHAVIORAL EFFECTS OF PRENATAL AND EARLY POSTNATAL LEAD EXPOSURE

Kim N. Dietrich, Kathleen M. Krafft, Rakesh Shukla,
Robert A. Padich, and Paul A. Succop
Department of Environmental Health
University of Cincinnati Medical Center
Cincinnati, OH 45267-0056

Abstract

The neurobehavioral effects of prenatal and early postnatal lead exposure was examined in 102 lower socioeconomic status infants. Prenatal lead exposure was unrelated to Bayley outcome variables at 3 and 6 months. 3 month blood Pb levels were inversely related to 6 month Bayley outcome variables. However, this relationship held only for the small subsample of white infants in the study. These results were interpreted in terms of genetic differences in the rate of early sensorimotor development.

INTRODUCTION:

High doses of lead (Pb) can carry serious neurological, hematological, renal, gastrointestinal, and reproductive sequelae (Chisolm & O'Hara, 1982; Lynman, Piantanida, & Cole, 1981; Singhal, & Thomas, 1980). At low to moderate levels of Pb intoxication (e.g., 10-49 ug/dl whole blood) hematopoietic sequelae include aminolevulinic acid-dehydratase inhibition and increased erythrocyte protoporphyrin (Moore, Meredith, & Goldberg, 1980). There are also data which suggest a depression of 1,25 dihydrocholicalciferol at Pb levels under 30 ug/dl whole blood (Mahaffey, Rosen, & Russel, 1982).

The neurobehavioral effects of low to moderate lead exposure are not as well known. As indicated, these so-called asymptomatic levels of exposure sometimes carry hematopoietic effects. However, the effects of low to moderate exposure on the central nervous system and behavior of children remain very controversial (Bornschein, in press; Yule, & Rutter, in press; Bornschein, Pearson, & Reiter, 1980). Even less is known concerning the effects of prenatal and early postnatal lead exposure on the central nervous system and behavioral development on infants. Bellinger, Needleman, Leviton, Waternaux, Rabinowitz, and Nichols (1984) reported a significant negative relationship between prenatal Pb exposure (umbilical cord blood Pb) and 6 month Bayley Mental Development Index scores. However, cord blood Pb levels were quite low ($\bar{X}$ = 6.6 ug/dl, SD = 3.2) and the parameter estimate for cord blood Pb on Bayley scores in the low, medium and high groups was of questionable clinical significance (Beta =-2.6, SE = 0.918).

The purpose of the present longitudinal investigation was to replicate and extend the findings of Bellinger, et al., with more frequent assessments of Pb exposure and neurobehavioral development from conception to the sixth month of postnatal life.

METHODOLOGY:

The sample consisted of 102 predominantly black (83%), lower socioeconomic status infants and their mothers. These families resided in identified lead-hazardous areas of a midwestern inner-city. Infants born at serious biological risk (e.g., severe prematurity, congenital malformation, genetic disease) were excluded. Mothers who were substance abusers, the mentally retarded, epileptics, and diabetics were excluded from prenatal recruitment.

Blood samples were taken prenatally after the mother was recruited into the study. For infants, blood samples were taken at 10 days, 3 months, and 6 months. Samples were analyzed for Pb by anodic stripping voltammetry using an extraction method (Morrell, & Giridhar, 1976). Blood Pb values were log transformed to produce a more normal distribution.

The Bayley Scales of Infant Development (Bayley, 1969) were administered at 3 and 6 months. Both the Mental Development Scale and Infant Behavior Record were given. Care was taken to insure that the infant was not noticeably sick, hungry, or fatigued at the time of testing. The examiners were trained on the Bayley scales by an experienced developmental psychologist. Intertester reliability for the Bayley Mental Development Scale was .96, and at least 85% agreement was maintained for each item of the Bayley Infant Behavior Record (a 30 item, behaviorally anchored rating scale which assesses qualitative aspects of the infant's performance).

Childhood Pb exposure frequently occurs in the context of other social and biological factors which can mimic or obscure its neurotoxic effects. Therefore, 7 potential confounders were chosen a-priori because of their suspected association with both the independent variable (blood Pb) and dependent variable (behavioral development). These were: (1) Obstetrical Complications; (2) Postnatal Complications; (3) Gestational Age; (4) Birthweight; (5) Home Environment; (6) Socioeconomic Status, and (7) Race. These variables were then pretested for their confounding potential according to the method of Dales & Ury (1978). Of these 7 variables, only postnatal complications (as measured by the Littman-Parmelee Postnatal Complications Scale), birthweight, socioeconomic status (Hollingshead 4-Factor Index), and race were potential confounders. These potential confounders and their interactions with each Pb variable were included in every regression analysis of the effect of exposure on early sensorimotor development.

RESULTS AND DISCUSSION:

Table 1 presents arithmetic means and standard deviations for prenatal and postnatal blood Pb values during the first year. Postnatally, mean blood Pb values increased from 5.1 ug/dl at 10 days to 7.7 ug/dl at 6 months. Variability in blood Pb concentration was evident at all ages.

TABLE 1

Age	Mean Blood Pb(ug/dl)	Standard Deviation	Low	High
Prenatal	8.2	3.6	1	21
10 Days	5.1	2.9	1	15
3 Months	6.0	3.7	1	18
6 Months	7.7	5.4	1	36

Table 1. Arithmetic means, standard deviations, and low-high blood Pb for each age.

Table 2 presents means and standard deviations for the Bayley Mental Development Index (MDI). Although some infants in the investigation scored rather poorly (e.g., MDI <80), the vast majority presented a normal to superior pattern of early sensorimotor development.

TABLE 2

Age	Mean Bayley MDI	Standard Deviation	Low	High
3 Months	101.5	10.1	71	125
6 Months	112.0	16.1	78	150

Table 2. Means, standard deviations, and low-high Bayley MDI scores for each age.

A principal axis factor analysis of the Bayley Infant Behavior Record data set was executed to reduce the large number of rating variables to a few meaningful factors. Factor scoring coefficients were then used to calculate factor scores for individual subjects. The 3 month Infant Behavior Record factors were: (1) Attention/Motor Maturity; (2) Positive Mood; (3) Active Engagement in Test, and (4) Sensory Interest. At 6 months, the Infant Behavior Record factors were: (1) Attention/Motor Maturity; (2) Positive Mood, and (3) Active/Energetic.

Due to the large number of independent and dependent variables in this study, a multivariate regression procedure (SAS, 1982) was used to control Type I error. These analyses included all independent variables (blood Pb, confounders, all interactions) and dependent variables (MDI, Infant Behavior Record Factors) at each wave.

The test for the multivariate linear hypothesis of an overall Pb effect on dependent variables was significant only for the lag relationship between 3 month blood Pb and 6 month Bayley outcomes (Wilks' Criterion = 0.8929, $F(4,91) = 2.73$, $p < .04$). In addition, there was an overall significant 3 month blood Pb by race interaction for Bayley outcome variables at 6 months (Wilks' Criterion = 0.8687, $F(4,91) = 3.44$, $p < .02$).

To determine which dependent variables were affected by Pb, a univariate multiple regression analysis was executed including 3 month blood Pb with all confounders, significant interactions, and covariates. The results of this analysis are presented in Table 3.

TABLE 3

Dependent Variable	Parameter	Estimate	Standard Error	T	p
MDI6	PbB3	-22.321	11.051	-2.02	0.04
	PCS	0.206	0.163	1.26	0.21
	BW	0.009	0.003	2.66	0.01
Model R^2=0.25	GA	2.050	0.925	2.22	0.02
	Race	-14.848	11.087	-1.34	0.18
	SES	- 0.030	0.322	-0.09	0.92
	PbB3XRace	11.000	5.960	1.85	0.06
Attention/Motor Maturity	PbB3	- 1.712	0.583	-2.93	0.004
	PCS	0.013	0.008	1.57	0.11
	BW	0.005	0.001	2.80	0.006
Model R^2=0.29	GA	0.093	0.048	1.92	0.05
	Race	- 1.374	0.585	-2.35	0.02
	SES	- 0.006	0.017	-0.41	0.68
	PbB3XRace	1.010	0.314	3.21	0.001
Positive Mood	PbB3	- 1.356	0.640	-2.12	0.03
	PCS	0.008	0.009	0.92	0.36
	BW	0.000E	0.000E	0.25	0.80
Model R^2=0.15	GA	0.037	0.053	0.71	0.48
	Race	- 0.655	0.642	-1.02	0.64
	SES	0.021	0.018	1.17	0.24
	PbB3XRace	0.737	0.345	2.14	0.03
Active/Energetic	PbB3	- 1.059	0.735	-1.44	0.15
	PCS	0.010	0.359	0.92	0.35
	BW	0.0002	0.0002	0.83	0.40
Model R^2=0.09	GA	0.082	0.061	1.34	0.18
	Race	- 1.242	0.737	-1.68	0.09
	SES	0.003	0.021	0.15	0.87
	PbB3XRace	0.698	0.396	1.76	0.08

Table 3. Univariate multiple regression results for 3 month blood Pb and 6 month Bayley outcome variables: PbB3= 3 month blood lead; PCS = Postnatal Complications Scales; BW = Birthweight; GA = Gestational Age; SES = Socioeconomic Status.

Lead exposure at 3 months was significantly related to 6 month MDI, Attention/Motor Maturity, and Positive Mood. The 3 month blood Pb by race interaction was also statistically significant for MDI at 6 months, and the first two Infant Behavior Record factors.

To determine the nature of the 3 month blood Pb by race interaction on 6 month sensorimotor development, separate bivariate Pearson correlations were calculated for white and black infants. Table 4 presents the results of this analysis.

TABLE 4

6 Month Bayley Variables	White Infants (r) N=17	Black Infants (r) N=85
Mental Development Index	-.52*	-.05
Attention/Motor Maturity	-.62**	.18
Positive Mood	-.50*	.05

Table 4. Pearson correlations between 3 month blood Pb and 6 month Bayley variables for white and black infants.
$^{*}p < .05$
$^{**}p < .01$

Clearly, the inverse relationship between blood Pb at 3 months and 6 month Bayley outcome measures was largely contained in the small but highly influential sample of white infants. To further test this finding, another multivariate regression analysis was executed. In this analysis, however, race and the blood Pb by race interaction was excluded from the full model. Recall that, in the full model, the parameter estimate for 3 month log transformed blood Pb on 6 month Bayley MDI was -22.321. This was a substantial effect. However, with racial factors excluded from the model, the test of the multivariate linear hypothesis of an overall 3 month Pb effect on 6 month Bayley variables was nonsignificant (Wilks' Criterion = 0.9186, $F(4,93) = 2.06$, $p = .10$). Further, in a univariate multiple regression analysis of the reduced model, the parameter estimate for 3 month log transformed blood Pb on 6 month MDI was reduced to a non-significant -2.76. Parameter estimates for 3 month log transformed blood Pb on IBR factors were also nonsignificant in the reduced univariate multiple regression model.

The hypothesis that prenatal exposure to Pb results in delays in early sensorimotor development was not confirmed in this study. This contradicts the findings of Bellinger, et al. (1984) who found an inverse relationship between cord blood Pb and 6 month MDI. Postnatally, only blood Pb levels at 3 months were associated with delayed development, and only for 6 month Bayley outcome variables. Further, Pb-related sensorimotor delays were largely contained in the relatively small subsample of white infants in the study.

There are at least two possible explanations for the racial differences in this investigation. First, the blood Pb by race interaction may only have been a chance finding. This is a possibility that cannot be dismissed given the small number of white subjects in the sample. Second, the interaction may have been due to genetically determined differences in the rate of early sensorimotor development in black and white infants. Relative to American and European whites, American and African black infants display a faster rate of sensorimotor and motor development during the first year (Ainsworth, 1977; Bayley, 1969). It may be that the early development of black infants is more canalized, buffering them from mild neurotoxic effects.

To summarize, prenatal and early postnatal low level Pb exposure appears to pose no threat to a normal course of sensorimotor development in black infants during the first half year of life. On the other hand, white infants may be more vulnerable to early low level Pb exposure. However, due to the small number of white infants in this sample, such a finding can only be regarded as tentative. As this ongoing longitudinal study proceeds, the present investigators intend to confirm and extend these findings on a larger sample of both white and black infants.

ACKNOWLEDGEMENT:

This research was supported by grants from the National Institute of Environmental Health Sciences: Program Project Grant USPHS ES 01566-04, and Center for the Study of the Human Environment Grant USPHS ES 00159. The authors are grateful for the assistance of Ms. Leslie C. Harris, Susan MacDonald, and Mariana Bier in data collection, and Ms. Nancy Knapp who typed the manuscript.

REFERENCES:

(1) Ainsworth, M.D.S., Infant development and mother-infant interaction among Ganda and American families, in: Leiderman, P.H., Tulkin, S.R., and Rosenfeld, A. (eds.), Culture and Infancy: Variations in Human Experience (Academic Press, New York, 1977).

(2) Bayley, N. Bayley Scales of Infant Development (Psychological Corporation, New York, 1969).

(3) Bellinger, D.C., Needleman, H.L., Leviton, A., Waternaux, C., Rabinowitz, M.B., and Nichols, M.L., Early sensorimotor development and prenatal exposure to lead, Neurobeh. Tox. Terat. 6 (1984) 387-402.

(4) Bornschein, R.L., Design considerations for prospective studies of lead's effect on behavioral and cognitive development, in Mahaffey, K. (ed.), Health Implications of Typical Levels of Lead Exposure: Dietary and Environmental Sources (Elsevier, Amsterdam, in press).

(5) Bornschein, R.L., Pearson, D.T., and Reiter, L., Behavioral effects of moderate lead exposure in children and animal models: Part 1, clinical studies, CRC Crit. Rev. Toxic. 8 (1980) 43-100.

(6) Chisolm, J.J. and O'Hara, D.M. (eds.) Lead Absorption in Children (Urban and Schwarzenberg, Baltimore-Munich, 1982).

(7) Dales, L.G. and Ury, H.K., An improper use of statistical significance testing in studying covariables, Int. Jrnl. Epidemiol. 7 (1978) 373-375.

(8) Lynman, D.R., Piantanida, L.G., and Cole, J.F. (eds.) Environmental Lead (Academic Press, New York, 1981).

(9) Mahaffey, K.R., Rosen, J.F., and Russell, W.C. Association between age, blood lead concentration, and serum 1,25-dihydroxycholecalciferol levels in children, Am. Jrnl. Clin. Nutr. 35 (1982) 1327-1331.

(10) Moore, M.R., Meredith, P.A., and Goldberg, A., Lead and heme biosynthesis, in: Singhal, R.L., and Thomas, J.A. (eds.), Lead Toxicity (Urban and Schwarzenberg, Baltimore-Munich, 1980).

(11) Morrell, G., and Giridhar, G., Rapid micromethod for blood lead analysis by anodic stripping voltammetry, Clin. Chem. 22 (1976) 221-223.

(12) SAS User's Guide (SAS Institute Inc., Cary, North Carolina, 1982).

(13) Singhal, R.L., and Thomas, J.A. (eds.) Lead Toxicity (Urban and Schwarzenberg, Baltimore-Munich, 1980).

(14) Yule, W., and Rutter, M., Effect of lead on children's behavior and cognitive performance, in: Mahaffey, K. (ed.) Health Implications of Typical Levels of Lead Exposure: Dietary and Environmental Sources (Elsevier, Amsterdam, in press).

Trends in Ergonomics/Human Factors II
R.E. Eberts and C.G. Eberts (Editors)

NEUROBEHAVIORAL EFFECTS OF LOW TO MODERATE LEAD EXPOSURE IN INFANTS: MEASURES OF ACTIVITY AND ATTENTION AT 18 MONTHS

Robert A. Padich, Kim N. Dietrich, and Douglas T. Pearson
Department of Environmental Health
University of Cincinnati Medical Center
Cincinnati, OH 45267-0056

Attention and activity level in infants at high risk for lead exposure were assessed with age-graded problem solving tasks. No significant correlations were observed between blood lead measures and the infants' ability to attend to and purposefully manipulate task materials. Contrary to numerous reports that lead may cause hyperactivity and attentional deficits, a small decrease in movement about the room was observed. The effect appeared to be related to iron status rather than lead per se.

INTRODUCTION:

Despite the fact that an increasing number of women are working in non-traditional jobs, little attention has been given to the effects of exposure to neurotoxicants on the developing child. Infants may receive undue exposure to chemicals that are inadvertantly brought home on clothing, or from the mother's body during pregnancy and nursing. In fact, some chemicals have become so common in the environment of industrialized countries that many infants and children are exposed to potentially dangerous amounts simply by virtue of living in the inner city, or near manufacturing plants. Lead (Pb) is one such element. It is widely used in industry as a constituent of solder, rust resistant paints, and storage batteries. Children living in housing that was built before World War II may be at risk for Pb exposure due to peeling paint or water pipes made from Pb.

A link between Pb exposure and hyperactivity was first reported by Byers and Lord (1943) who studied a group of children that had been treated for clinical Pb poisoning. A number of investigators, relying primarily upon subjective rating scales, have reported elevated activity levels and attentional problems among children with much lower Pb exposure (David, Hoffman, Clark, Grad, & Sverd, 1983; Gittelman & Eskenazi, 1983; Needleman, Gunnoe, Leviton, et al. 1979). However, other investigators using the same or similar rating instruments, have found that when even such a crude covariate as social class is taken into account, the relationship between Pb and ratings of hyperactivity either disappears (Ernhart, Landa, & Schell, 1981; Smith, Delves, Lansdown, Clayton, & Graham, 1983), or that Pb accounts for very little of the variance in ratings of children's behavior (Winneke, Kramer, Brockhaus, et al. 1983).

All of the studies to date have used cross-sectional and retrospective designs. As a result, they have been handicapped, not only by the use of subjective ratings scales which are prone to problems of reliability and validity (Rapaport & Benoit, 1975; Klein & Gittelman-Klein, 1975), but also by reliance on *post hoc* assumptions about Pb exposure history during the early years of life. Clearly, prospective studies are needed which carefully document Pb exposure, and which directly measure both the child's behavior and important covariates, such as caregiving style, in a variety of settings and situations over time.

RATIONALE:

In the Cincinnati prospective study, three problem solving tasks were used to assess the infants' patterns of activity level and attention in a structured laboratory setting. An important aspect of these tasks is that the caretaker is present during the observation session. At the very young age of 18 months, it would be unethical to leave the infant alone in an unfamiliar clinical setting. Moreover, the caretaker's behavior and very presence are of such importance in the infant's everyday life and functioning, that exclusion of the caretaker would destroy the "ecological validity" of the context of measurement. An empirically derived rating scale (Dietrich, Stewart, & Padich, 1984) was used to record the caretaker's efforts to support and teach the infant during the session. This allowed the variation in caretaking style to be included in multivariate analyses of the data. It was hypothesized that low to moderate lead exposure would produce observable alterations in the infant's patterns of activity and attention. These would be manifested by infants with elevated Pb levels spending less time visually oriented to and manipulating task materials, and more time moving about the room.

METHOD:

The subjects of this prospective longitudinal study were infants who were born and resided in predominantly black, impoverished neighborhoods of Cincinnati, Ohio. The current report is based on a sample consisting of all 125 infants who had reached the age of 18 months by July, 1984 (53% male; 90% black). Blood lead (PbB) levels and other hematopoietic measurements were obtained at 10 days of age, and every three months thereafter. Prenatal PbB (maternal blood) was sampled at the time the mother was recruited into the study.

The problem solving tasks (adapted from Matas, Arend, & Sroufe, 1978), were administered as part of a battery of tests lasting between 45 and 60 minutes. The first two tasks required the infant to first use one long stick, and then put two shorter sticks together, in order to push a packet of sweet cereal out of a clear plastic tube. The third problem required the infant to use a block to counterweight a lever in order to elevate the reward from inside a large clear plastic box. Use of the block was necessary because the infant could not simply hold the lever down and still reach the reward. At the beginning of each task, the caretaker was instructed to allow the infant to work independently at first, but to provide whatever help was necessary should the infant be unsuccessful or become overly frustrated. A maximum of 7 minutes was allowed for each of the tube tasks; 10 minutes for the lever task.

The infant's behavior was scored from videotapes using three behavioral categories. *On Task* was the amount of time the infant was observed to be visually and physically oriented to the tube or lever apparatus. Room Movement was the amount of time the infant spent moving about the room. Active Exploration was the amount of time the infant spent purposefully manipulating task materials. Videotapes were scored using a microprocessor.

To accommodate different completion times for each task, the data on each category were collapsed across tasks and reported as the percentage of the total time it took the individual infant to complete all three tasks. No psychometric significance was attributed to the solution of the tasks, per se. Success or failure were considered unimportant relative to the process of problem solving, and not analyzed. Interscorer reliability for the measures of infant behavior averaged +.97. The caretaker's behavior and interactions with the infant were observed and rated on the Caretaker Behavior Rating Scale (Dietrich et al., 1984). Interrater reliability was +.80.

All experimenters were blind as to the individual Pb exposure history of the infants. No infants were excluded from data analysis for any reason. Fluctuation in degrees of freedom in the analyses reflects missing data, primarily due to laboratory problems (e.g. inadequate or clotted samples).

RESULTS:

All blood measures were corrected for hematocrit, and the values transformed to their natural logarithms to produce a more normal distribution. Table 1 presents the geometric means, standard deviations, and ranges for PbB values at different ages.

TABLE 1

Pb Index	Mean (ug/dl)	S.D.	Minimum	Maximum
Prenatal PbB	7.9	1.6	1.0	27.0
PbB 10 Day	4.2	2.0	3.0	23.0
PbB 18 Months	16.8	1.6	6.0	49.0

Table 1. Geometric Mean, Standard Deviation, Minimum, and Maximum Blood Pb Values.

Cumulative PbB (CumPbB) was used as an additional index of exposure, since it represents an estimate of the infant's total exposure history which is less susceptible to the effects of transitory, acute increases at any point in time. CumPbB is derived from a calculation of the area under the fitted curve of each infant's PbB profile (composed of age-concurrent PbB values) from 10 days to the target age of 18 months.

Table 2 summarizes the raw score performance of the infants on the three behavioral variables selected for observation. Due to non-normality of the distributions of these performance measures, the percentages were arcsine transformed for purposes of analysis.

It was the authors' impressions that none of the infants evidenced patterns of attention or activity level that could be clinically consistent with a diagnosis of attention deficit disorder with hyperactivity. In addition, the tasks appear to be sufficiently engaging and to generate sufficient variability (i.e. no "floor" or "ceiling" effects), to be useful as one measure of attention and activity in our battery.

TABLE 2

Behavior	Mean	S.D.	Minimum	Maximum
On Task	61.4%	17.6%	26.3%	96.9%
Room Movement	6.7%	4.4%	0.1%	21.8%
Active Exploration	26.6%	12.0%	3.0%	67.0%

TABLE 2. Behavioral performance of 18 month old infants on the three problem solving tasks. To accommodate different completion times for different tasks within individual infants, the data on each category were collapsed across tasks and converted to the percentage of the total time it took the individual infant to complete all three tasks.

The Caretaker Behavior Rating Scale was subjected to principal axes factor analysis for purposes of data reduction. Of the resulting four factors, the first two appeared to most purely tap variation in caretaking style. Consequently Supportiveness and Teaching Style were regressed on infant activity and attention variables to produce a data set of residuals for further analysis.

Room Movement was essentially independent of the two caretaking factors (R-Square=.02; $F(2,124)=1.25$; p=.29). Small, but statistically significant, amounts of the variance were removed in On Task (R-Square=.14; $F(2,124)=9.95$; p=.0001) and Active Exploration (R-Square=.09; $F(2,124)=5.89$; p=.0036) by the two caretaking factors. Backwards Stepwise Regression analysis of the residual variances was used to evaluate the impact of Pb on each of the measures of infant behavior, after the caretaker's contribution was statistically controlled. Pb exposure variables were selected for the model *a priori*. Prenatal PbB and 10 day PbB were included to assess possible long term effects of prenatal exposure. Concurrent PbB (CpbB18) and Cumulative PbB (CumPbB18) at 18 months were selected to assess the effects of acute and long term Pb exposure respectively.

No effect for any of the Pb measures was found on either On Task or Active Exploration (all Pb measures were eliminated from the model). For Room Movement, the backward elimination procedure retained only CumPbB, indicating a small, but statistically significant, relationship between this index of long term exposure and activity level (R-Square=.03; $F(1,113)=3.98$; p=.048). Contrary to prediction, the parameter estimate was -.004, the negative sign indicating a *reduction* rather than an increase in activity level.

Lethargy, or low energy level is sometimes observed in infants with iron deficiency (Nutrition Reviews, 1983). Therefore, subsequent analyses were run, incorporating free erythrocyte protoporphyrin (FEP) into the model as an index of exposure. FEP becomes elevated as a result of both anemia and Pb exposure. Both conditions inhibit a key enzyme which attaches the iron molecule to protoporhyrin, forming heme, the oxygen carrying pigment in the blood. FEP levels are highly correlated with PbB levels.

The backwards elimination procedure eliminated all of the original Pb measures from the model, retaining FEP as the best predictor of Room Movement (R-Square=.10; $F(1,107)=11.63$; p=.0009). Again, the parameter estimate was negative, and the amount of variance predicted small, though statistically significant.

To further evaluate the relative contributions of CumPbB and FEP on activity, both variables were simultaneously regressed on Room Movement. CumPbB yielded a nonsignificant parameter estimate of -.001, while FEP yielded an estimate of -.097 ($F=7.49$; $p=.007$). Thus the analysis suggested that the suppression of activity level was most directly related FEP level. However, Room Movement was not related to other available measures of concurrent hematological status (Hematocrit, Serum Iron, Percent Saturation, or Total Iron Binding Capacity). Consequently, iron deficiency does not appear to be an adequate explanation for the suppression of activity level either.

DISCUSSION

Prior to the evaluation of any Pb effects, multivariate techniques were used to remove the caretaker's behavior as a source of variance in the infant's performance. This was done to reduce the possibility that subtle effects due to lead might be masked by other influences on performance during the observation session. The Pb variables were not found to be related to the amount of time the infant spent visually oriented to and purposefully manipulating task materials. Thus the hypothesis that Pb would disrupt the infant's ability to maintain attention was not confirmed.

Contrary to prediction, the data suggested that with increasing Pb exposure, the infants tended to have a lower activity level rather than being hyperactive. Lethargy is sometimes observed in infants with iron deficiency. This possible explanation was explored by incorporating FEP into the multivariate model. The small decrease in Room Movement did appear to be related to FEP more directly than to the index of lead exposure history (CumPbB). However, none of the other available measures of hematopoietic function were related to this measure of activity level. Thus, the precise nature of the Pb-FEP-Behavior relationship is unclear. If dimunition of activity level had been accompanied by decreased attention and active engagement with the task materials, that was also related to Pb or FEP, one might be tempted to speculate that FEP is serving as an index of individual sensitivity to Pb effects. Given that this was not the case, and in light of the small size of the effect on activity, the data suggest that these domains of behavior are not being impaired, in a practical sense, by low to moderate Pb exposure in 18 month old infants.

These results must be interpreted with a great deal of caution. The subjects of this report constitute about one half of the projected total will provide the statistical power necessary to permit the use of more complicated statistical models incorporating more covariates and outcome measures related to activity and attention (e.g. observations of the infant and caretaker at home; ratings of the infant's behavior by the caretaker). Also, the fact that this is a prospective longitudinal study will permit assessment of the same infants over time. To achieve this end, additional age graded problem solving tasks have been developed for use with these children at later ages.

The use of multiple measures over time presents real statistical and logistical challenges. However, given the complexity of individual differences in response to toxic exposure, in addition to the individual differences that occur in normal development, a multivariate approach is clearly necessary. In addition, experience with Pb has suggested that traditional psychometric techniques may either lack sensitivity or fail to adequately define specific impairments due to low level toxicant exposure (Bornschein, Pearson, & Reiter, 1980). Consequently, additional fine grain measures of behavior must be developed. This is an endeavor in which the skills of the human factors specialist in measuring environmental influences on performance could make important contributions towards improving the safety and quality of life.

The need for intensive effort in this area of research is underscored by the fact that although there are an estimated 100,000 chemicals used in industry today, "safe" exposure levels for healthy adult workers have been established for less than 1,000. So far, only Pb, mercury, and polybrominated biphenyls have attracted any attention as to their potential effects on children. It is a sad note that research on these compounds is often begun only after clinical poisoning of epidemic proportions are recognized. Much work remains to be done on the effects of low level exposure to toxicants in both adults and children.

ACKNOWLEDGEMENT

This research was supported by grants from the National Institute of Environmental Health Sciences: Program Project Grant USPHS ES 01566-04 and Center for the Study of the Human Environment Grant USPHS ES 00159.

REFERENCES

(1) Byers, R.K. and Lord, E.E., Late effects of lead poisoning on mental development. Am. J. Dis. Child., 66 (1943) 471-494.

(2) Bornschein, R., Pearson, D. and Reiter, L., Behavioral effects of moderate lead exposure in children and animal models: Part 1, clinical studies. CRC Crit. Rev. Toxicol. (November, 1980) 43-98.

(3) David, O., Hoffman, S., Clark, J., Grad, G. and Sverd, J., The relationship of hyperactivity to moderately elevated lead levels. Arch. Environ. Hlth. 36(6) (1983) 341-346.

(4) Dietrich, K.N., Stewart, C.B. and Padich, R.A., The Cincinnati Lead Program Project Activity Level and Caretaker-Child Interaction Protocol and Coding Manual. (1984).

(5) Ernhart, C., Landa, B. and Schell, N., Subclinical levels of lead and development deficit -- a multivariate follow-up assessment. Pediatr. 67 (1981) 911-919.

(6) Gittelman, R. and Eskenazi, B., Lead and hyperactivity revisited. Arch. Gen. Psychiat. 40 (1983) 827-833.

(7) Klein, D.F. and Gittelman-Klein, R. Problems in the diagnosis of minimal brain dysfunction and the hyperkinetic syndrome. Int. J. of Mental Hlth. 4 (1975) 45-60.

(8) Matas, L., Arend, R. and Sroufe, L.A., Continuity of adaptation in the second year: The relationship between quality of attachment and later competence. Child Develop. 49 (1978) 547-556.

(9) Needleman, H., Gunnoe, C., Leviton, A., Reed, M., Peresie, H., Maner, C. and Barret, P., Deficits in psychological classroom performance of children with elevated dentine lead levels. N. Engl. J. Med. 300 (1979) 689-695.

(10) Rapoport, J. and Benoit, M. The relation of direct home observation to the clinic evalutions of hyperactive boys. Child Psychol. Psychiatr. 16 (1975) 141-147.

(11) Smith, M., Delves, T., Lansdown, R., Clayton, B. and Graham, P., The Institute of Child Health/Southhampton Study. Devel. Med. Child Neurol 25: Supplement No. 47 (1983).

(12) Winneke, G., Kramer, U., Brockhaus, A., Eweres, U., Krujek, G., Lechner, H. and Janke, W., Neuropsychological studies in children with elevated tooth-lead concentration: II Extended studies. Int. Arch. Occup. Environ. Hlth. 51 (1983) 231-252.

Trends in Ergonomics/Human Factors II
R.E. Eberts and C.G. Eberts (Editors)

SAFETY CONSIDERATIONS FOR WATER DISTRIBUTION MAINTENANCE CREWS

Nicholas A. Damachi, Richard L. Shell, O. Geoffrey Okogbaa, H. Ray Souder

Department of Mechanical and Industrial Engineering
University of Cincinnati
Cincinnati, Ohio 45221
U. S. A.

The working conditions and potential for exposure to hazards of water distribution crews is of major importance as a result of the vital role they assume in the maintenance, upkeep and efficient operation of a water utility system. The need for closer examination of the activities of water crews is further stimulated by the deteriorating conditions of the underground water distribution network of many older urban water systems which result in energy waste and create a potential for public health problems. This paper examines the activities, working conditions, potential for exposure to hazards, and identifies safety guidelines for water distribution crews.

INTRODUCTION

Water utilities generally share a common objective regardless of their sizes, which is to operate a water system effectively and efficiently in order to deliver an acceptable level of service to their customers. The water distribution unit and its crews play a vital role in the overall smooth operation of a water utility, and as a result, the working conditions, potential for exposure to hazard, and performance of water distribution maintenance crews assume significant importance. The need for closer examination of the activities and performance of water crews is further stimulated by the deteriorating conditions of many older urban water systems, treatment facilities, pumping stations, and in particular, underground water distribution systems, which result generally in energy waste and create a potential for public health problems (1,3). Deferred water distribution systems maintenance is identified as a serious problem to some extent by most water utilities in the United States (3).

As water systems continue to deteriorate, water distribution crews have the primary responsibility for the maintenance and upkeep of these systems. They undertake a wide range of duties which include minimization of water loss (unaccounted-for water) through leakage, unauthorized use, and meter over and underregistration which have significant cost effects on water systems. This paper provides an in-depth examination of the activities, working environment, potential for exposure to hazards, and safety guidelines for water distribution maintenance crews.

WATER DISTRIBUTUION CREWS: WORK TASKS

To provide an understanding of the role of the water distribution unit and its crews in a utility, it is perhaps useful to view the distribution unit as a subsystem of the entire water utility system. Figure 1 shows the water distribution subsystem along with other major subsystems of a water utility, identified in the same sequential order as water flows through the utility's physical facility. They include water supply and treatment, revenue, and management subsystems. Briefly, the water supply and treatment subsystem comprised of pumping stations and treatment facilities is primarily responsible for pumping raw water into the treatment facilities using low service pumps, treatment of water as needed, and pumping of finished water into the distribution system.

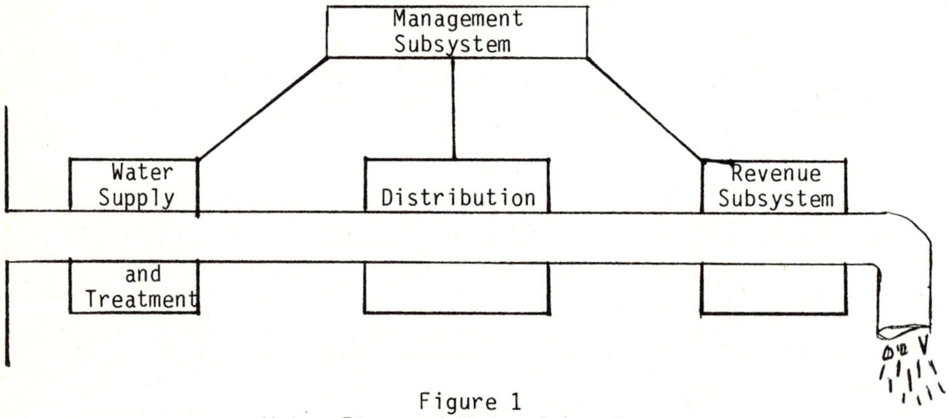

Figure 1
Water Flow and Major Subsystems

The water distribution unit ensures proper functioning of the distribution system to provide acceptable and quality level of service to customers. As previously mentioned, distribution crews undertake a wide range of duties and tasks which include maintenance of all water lines and service taps in the distribution network; maintenance of all fire hydrants and services from water main lines to the curb stop and side feeder valves. In addition, water crews work main line valves and replace broken and frozen valves as they are found. They are also responsible for making new taps in existing mains for new water customers, and laying new main lines of specified maximum dimensions. Table 1 provides the range of tasks undertaken by water distribution maintenance crews.

Metering	Main/service Construction and Maintenance	Fire Hydrants	Plumbing	Valves	Inventory	Communications
In-field installation of new meters of various sizes	Install main lines Install new service lines	Installation Check Repair	Maintenance of plumbing needs in treatment facilities, public fountains, streets and parks	Installation	Fire Hydrants Valves, water pipes	Dispatching and maintenance needs
Field and shop testing of meters (under/over registration)	Repair main and service breaks	Replace old hydrants		Maintenance	Large bends	Employee Performance
Repair and rebuild old meters (if economically feasible)	Renew or put in shape service boxes	Flow test Paint	Constructing and troubleshooting heating systems		Fittings	
Maintain New meter inventory	Cut off old Services					
	Work Valves					
	Respond to emergencies					

Table 1

Work Task for Water Distribution Crews

POTENTIAL HAZARDS AND SAFETY CONSIDERATIONS

Before focusing on the potential hazards and safety issues of water distribution units of water utilities, a brief review and discussion of the water industry's safety performance would be useful and should provide some background for the need for increased industry-wide emphasis on safety. Van Dyke (6) observed that the number of severity of on-the-job accidents in the water utility exceed the figures for industry as a whole and renders safety programs increasingly important in terms of cost savings as well as employee morale. According to recent safety data that include the number of accidents that results in lost time injuries and the number of days lost compiled by the American Water Works Association from many participating utilities in the United States and Canada, there was a total fo 5040 lost time accidents reported resulting in 122,629 lost days. This provides an average of more than 24 days per accident and over one month of productivity lost per person. About 3008 water utilities employing 69,844 people responded to the survey. The projected estimates of the cost of accidents to water utilities in 1981 was placed at $73 billion (6). While some may question the validity of this projection, it however, reflects the scope of magnitude of the cost of water utility accidents. There appears to be potential for increased safety work and cost effective safety programs in the water industry. An examination fo the working environment of water maintenance crews reveals that the following safety related conditions tend to prevail -

1. A significant portion of time of work crews is spent working in the midst of traffic flow. Working in the middle of traffic poses a number of hazards which are potential causes of work-area traffic related injuries to workers, motorists and pedestrians. The following hazards are identified:
   - failure to provide advance warning to motorists.
   - failure to provide effective signals.
   - failure to provide effective guidance patterns.

2. Water crews are frequently engaged in digging trenches and breaking up of concrete pavements using a variety of tools and equipments - backhoes, air hammers, shovels, etc. The use of backhoes in the middle of traffic at some sites necessitates traffic control measures. The physiological impact of some power tools that vibrate while in use deserves analysis.

3. Some work tasks are performed within trenches and ditches dug by water crews. They include cutting and tapping of water main lines, laying and bolting of fittings. While in the trenches, there is the potential hazard of cave-ins.

SAFETY GUIDELINES

In view of the traffic flow conditions that characterize the work environment of water distribution maintenance crews, and the nature of the tasks that they perform, there is need for safety guidelines to minimize hazards and accidents to the work crews and the public. A summary of guidelines is provided: (2,7)

1. Use of effective traffic control measures. These measures assists in guiding traffic directly affected by the job of water crews safely around and through the work areas. In addition, effective traffic control measures provide protection for water crews from potential injuries that may be caused by vehicle or pedestrian traffic. The public is also protected from injuries that may be caused by water utility operations. Adopting effective traffic control measures also ensures compliance with local, state and federal laws and regulations.

2. Traffic warning devices should be used and kept clean, in good condition, and placed where they are not hidden by trees, shrubbery, or parked or moving vehicles. They include traffic signals, traffic cones, flags, pavement markings, barricades, warning lights and signs. Individual cones, signs, and barricades should be equipped with reflectors.

3. A crew member should be assigned the responsibility for traffic control procedures at the work site. It is suggested that such a worker should personally travel throught the work area before setting up controls. It is also advisable to drive thorugh the work area after placing controls. This assures that the devices do not obstruct the view or cause a problem for cross-traffic that will not be directly involved in the work zone. It also eliminates possible misinterpretation of conditions as they appear to drivers, which may be the cause of serious and costly accidents. Table 2 shows the National Safety Council guidelines and recommendations for traffic safety in public work areas.

4. Individuals who are responsible for the safety training of water utility personnel should acquaint themselves with the relevant portions of <u>Uniform Traffic Control Devices for Streets and Highways</u>, a manual of the American National Standards Institute that specify uniformity of design, placement, and maintenance of traffic control devices. (5)

5. Water crews trained to use their trucks as physical barriers to protect themselves from approaching vehicles should do so only when it does not add to traffic disruption.

6. Trucks parked at work sites should be equipped with flashing amber lights which should be in operation at all times. This should not be considered as a substitute for warning devices.

7. Headlights and floodlights used to illuminate work area should be placed in such a way that they do not shine on driver's eye.

8. There is need to protect crew members working in trenches or excavated ditches from cave-ins. Movable shoring units may be used to provide protection or a worker should be assigned to watch the banks of the ditch. Fill should be stored at a safe distance from the ditch. At night, excavations that are left open should be identified by barricade with flashing amber lights.

9. Crew members who are assigned the responsibility of traffic control and watching the banks of excavations should be alert, intelligent, courteous, posses good physical condition especially sight and hearing.

10. Safety programs should emphasize the need for safety glasses, hard hats, gloves, safety shoes, and adequate shoring techniques.

	Speed Limit		Distance from Site		Spacing Between Cones	
	km/h	mph	m	ft	m	ft
Warning Signs	40 or less	25 or less	7-15	25-50		
	56	35	45-60	150-200		
	72	45	76-106	250-350		
	88	55	106-152	350-500		
Traffic Cones	40 or less	25 or less			3	10
	56	35			10	35
	72	45			13	45
	88	55			16	55

Table 2
Placement of Traffic Warning Devices

CONCLUSION

This paper has provided an analysis of the activities of water distribution crews which have the primary responsibility for the miantenance and upkeep of water distribution systems. It also examined the working environment of work crews which is characterized by some hazards that include the danger posed by working in the middle of traffic, as well as working within excavated ditches. Safety guidelines which provide protection to work crews and the public, pedestrian and motorist, are identified. They include effective traffic control measures, proper shoring techniques and safety programs that lay emphasis on the use of appropriate safety devices.

REFERENCES

(1) AWAA Subcommittee Report; Water Pricing Research Needs, AWWA Journal 75(1983) 24-27.

(2) Damachi, N. A., A Minicomputer Based Information System for water utilities, Ph.D. Dissertation, Dept. of Mechanical and Industrial Engineering, Univ. of Cincinnati, (August 1981).

(3) Shell, R. L. and Damachi, N. A., Improving Water Utility Management, Water Engineering and Management 2(1984) 42-44).

(4) Shell, R. L., Damachi, N. A., Souder, H. R., Leigh, B., An analysis of Water Utility Operator Tasks, in: Mital, A. (ed.) Trends in Ergonomics/Human Factors, Vol. I (North-Holland, Amsterdam, 1984), 247-253.

(5) Uniform Traffic Control Devices for Streets and Highways. ANSI Manual D6. 1-1978. ANSI, New York (1978).

(6) Van Dyke, R. P., Hidden Costs of Accidents can Finance Safety Programs, AWWA Journal 75(1983) 73-77.

(7) Wantoch, H. J.; Traffic Control - Safety's First Line of Defense, AWAA Journal 75(1983) 68-72.

(8) Work Area Protection. NSC Publ. 129.06 - a Chicago (1979).

Trends in Ergonomics/Human Factors II
R.E. Eberts and C.G. Eberts (Editors)
Elsevier Science Publishers B.V. (North-Holland), 1985

A MODEL OF THE TRAINING PROCESS DERIVED FROM HUMAN FACTORS AND SAFETY LITERATURE

Roger C. Jensen, P.E.
Division of Safety Research
National Institute for Occupational Safety and Health
Morgantown, West Virginia

A block diagram model for characterizing adult training programs is proposed. The model concisely depicts the temporal relationships among activities and decisions involved in developing and evaluating training programs. The model may facilitate the systematic evaluation and management of an organization's current employee training programs as well as offer a structure for organizing training literature and identifying research needs.

INTRODUCTION

Training is a process of acquiring the skills and knowledge required for the performance of an activity. Examples of adult training programs include training for piloting an airplane, operating an industrial machine, performing a new job task, or performing a known task in a different manner. Clearly, training is important to organizations in terms of both resources expended on the training and the benefits resulting from subsequent performance. In the occupational safety field employee training for improving work practices is especially important because of a widespread reliance on worker behavior to minimize the risks associated with industrial processes. In order to clarify what research is needed to support occupational safety training activities, a search for literature concerning the process of employee training was undertaken.

Fortunately, the magnitude of this undertaking was eased by reference to literature summaries by authorities in the training field (Campbell, 1971; Goldstein, 1980; and Borman and Peterson, 1982). The study of these review articles suggested the need for a more concise and structured explanation of the training process which could be used for organizing literature and clarifying the gaps in knowledge. After developing a preliminary block diagram model depicting the basic components of training programs and their relationships, some associates recommended that it be shared with colleagues in ergonomics and industrial engineering. Consequently, this paper was developed with the intent of communicating these ideas in a manner which complements the contents of the review articles. This involves two important differences in approach.

First, unlike the review articles which focus on the process from the view of training specialists, this paper takes the view of the sponsoring organization. This difference in perspective is important because of the distinctly different roles played by the sponsor vis-a-vis the trainer. The distinction stems from the responsibilities of the sponsor in making decisions concerning the initiation, modification, and continuation of training programs as well as providing a management structure for implementing and monitoring training programs. Second, the review articles involve extended discussion of individual components of training programs which tends to obscure component interrelationships. In contrast, this very concise paper emphasizes the relationships among the components.

THE TRAINING PROGRAM MODEL

The proposed model of the training process is shown in Figure 1 in the form of a block diagram. Blocks represent activities, circles represent decisions, open arrows represent inputs from prior activities, and black arrows only show order. The blocks are arranged to depict the sequence of activities and decisions in a top to bottom order. Figure 1 shows the preferred sequence of steps as solid lines, and the non-preferred approach as dashed lines. The discussion of the model which follows includes numerous references to supporting literature from human factors, ergonomics, industrial engineering, and safety publications.

Assessment of Needs

Before undertaking a training program a determination should be made that training is the best method for achieving an organizational objective. For example, a company wishing to reduce their losses from employee disability due to low back pain should carefully consider their options involving task redesign before committing extensive resources to training programs of yet unproven effectiveness (Snook and White, 1984). If the decision is to undertake a training program, the first step is the assessment of needs, depicted in the top block of Figure 1. The literature suggests that a good needs assessment include: (1) sufficient communications within the sponsoring organization and between the sponsor and the trainers to clearly determine what the sponsor expects to achieve with the program (Goldstein, 1978), (2) an analysis of the job or activity to clearly understand the essential skills and knowledge requirements for successful performance (see Singleton, 1968; Bradbury, 1970; Tilley, 1969; Borman and Peterson, 1982), (3) an estimate of general trainee capabilities (e.g., reading abilities) or the current performance deficiencies which prompted the initiation of the training program, and (4) the identification of constraints on both the training program and the environment where the trainees will eventually be performing (Goldstein, 1978).

Development of the Program

The second block in Figure 1 is the development of the program. Walker (1965) recommended initiating the process with the preparation of a list of training objectives corresponding to the skill and knowledge requirements identified in the needs assessment. When the number of desirable skills exceeds the available time or resources of the training

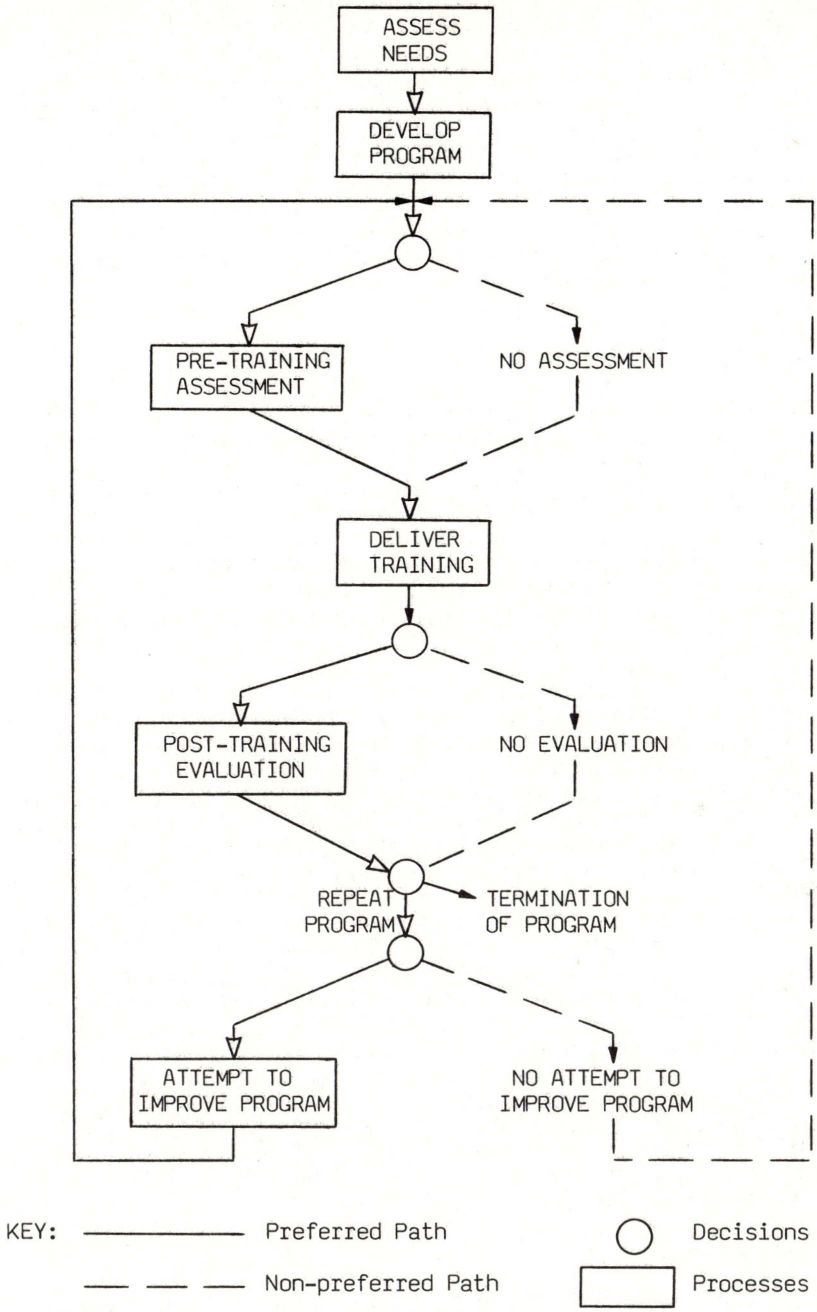

Figure 1
Block diagram model of the training process.

program, a priority list may be developed to help emphasize the most important skills (Swezey, 1979).

The second task in program development is the selection of the training techniques. A practical summary of the advantages and disadvantages of major training techniques is provided by Borman and Peterson (1982), although they comment that these selection principles are derived more from experience than from the results of objective research. Among the few studies are three which illustrate contributions from the human factors/ergonomics community. These studies concern manufacturing assembly instructions (Konz and Dickey, 1969), training the mildly mentally handicapped (Nettelbeck and Kirby, 1975), and developing typing skills (Baddeley, 1978).

A third task in the program development process includes developing the training media. Much of the literature on this subject is anecdotal, although considerable research concerning simulator development has been reported in the human factors/ergonomics literature (e.g., Valverde, 1973; Blaiwes, Puig, and Regan, 1973; Duncan and Shepherd, 1975; Johnson and Rouse, 1982; Johnson, 1981). Generally, these articles point to the conclusion that for effective transferrance of skills acquired with a simulator to on-the-job performance, the required degree of physical fidelity (the extent to which the simulator duplicates the functional characteristics of the real operational equipment) depends on the specific task being learned; the development of complex perceptual-motor skills requiring a much higher degree of physical fidelity than a strictly procedural task.

A fourth task in program development is the selection of evaluation criteria. Leading authorities indicate that a training program should have a set of specific objectives, each with criteria for assessing the extent to which the objective was met (Caro, 1973; Stringer, 1975; Goldstein, 1978; Jensen, 1982; Borman and Peterson, 1982). These objectives and their evaluation criteria may relate to several aspects of the program. A useful classification of four kinds of evaluations was provided by Kirkpatrick, 1967:

1. Reaction of the trainees, or what they thought of the program.
2. Learning of techniques, facts, principles, and skills.
3. Behavioral changes attributable to the training program.
4. Tangible results of the program such as reduced accident rates, or improved quality.

The first of these can be assessed by a post-training evaluation instrument. However, the other three types of evaluations require comparable measures obtained both before and after the training. Thus, the model in Figure 1 indicates that the evaluation criteria are developed in advance of the pre-training assessment phase.

The validity of performance measures is a critical concern. In this context validity refers to the extent to which the data used for evaluation are related to the result desired by the sponsor. For example, most States sponsor driver improvement programs aimed at getting habitually dangerous drivers to start driving more prudently. Measures of training effectiveness might be derived from pre- and post-training records such as the trainee's traffic violation citation record (a behavioral measure). From the sponsor's perspective the validity of this behavioral measure depends on how closely traffic citations reflect the individual's driving practices; validity being lower if all citations are included, and higher if it only includes those citations for offenses related to unsafe driving (Peck, 1976; Edwards and Ellis, 1976).

Another aspect of the validity issue occurs when the sponsor is looking for tangible results but the only practical source of information for evaluation is behavioral data. An example is the training program for industrial lift truck operators. The sponsor's objective is reducing accidents, but because accidents are relatively rare events in the statistical sense, accident rate is not a very sensitive criterion measure for evaluation. A more practical source of relevant data is obtained from the work behavior of the trainees. However, the sponsor needs to be concerned with the validity of the behavioral measure as a reflection of the extent to which the program achieves the objectives. No ideal methodology has been found for objectively establishing this connection between behavior and employee safety. Recent attempts to establish valid behavioral measures have used (1) prior accident reports at the establishment combined with suggestions from foremen, workers, and safety officers (Komaki, Barwick, and Scott, 1978; Komaki, Heinzmann, and Lawson, 1980), and (2) analysis of accident patterns from the establishment and from many other employers plus an analysis of the tasks (Cohen and Jensen, 1984).

The fifth and final task of training program development is a dry-run to improve the quality of the training program. A sponsoring organization should look for this step even though it is omitted from many training programs such as informal intra-organizational and on-the-job training .

Pre-training Assessment

After completing the program development, Figure 1 shows a decision to pursue one of two paths leading to the actual delivery of training. The preferred path includes pre-training skill/performance assessment, since three of the four kinds of program evaluations mentioned earlier require pre-training measures. In addition, the results of pre-training assessments may aid the trainers in adjusting the program to better fit the needs of the individual trainees.

Pre-training assessments should be based on the program objectives. If an objective of the program is to impart knowledge about a subject, an appropriate assessment may be based on a test of knowledge about the subject before and after the course (e.g., Kao, 1976) or at various stages during the training period (e.g., Jeantheau, 1969). If the objective is to change on-the-job behavior of workers, pre-training observations of the trainees performing the jobs before the training will be needed. This may be illustrated with a worker training program concerned with reducing employee back injuries by teaching workers "proper" lifting techniques. The proportion of lifts performed in the manner defined to be correct might be measured before the training using behavioral observations for subsequent comparison with comparable post-training data. In contrast, if the objective of this same training program were based on tangible results, the pre-training assessment would be based on the back injury experiences of the trainees.

Delivery of Training

The next block in the training model (Figure 1) is the actual training program. The human factors and safety literature includes numerous anecdotal articles, but only a few papers on training research. The value of active participation was shown by Czaja and Drury (1981). Studies of augmented feedback as a possible means of enhancing training have been reported by Lintern (1980) and Cote, Williges, and Williges (1981). Borman and Peterson (1982) indicate that program effectiveness is generally enhanced by practices such as dividing the overall content into smaller units and learning them one at a time in a logical order, active participation of the trainees to maintain vigilance and motivation, performance monitoring with feedback to the trainees, and having the trainees practice what they learned (although repetitive practice of tasks which vary each time may not lead to improvement according to Fisk, Scerbo, and Schneider, 1983).

On-the-job training is unique in that the training and the collection of performance data may occur concurrently; especially in manufacturing jobs using an incentive payment scheme (Hancock and Bayha, 1982). In order to appreciate how the model in Figure 1 accounts for such training, a distinction between collecting performance data and evaluating a training program should be recognized. The performance data (e.g., average time per cycle) is collected during the on-the-job training period. After the training period this data may be used as a criterion for evaluating the training program or perhaps comparing two or more different training programs.

Post-training Evaluation

Figure 1 indicates that after the training has been completed the program may include or not include a post-training evaluation. The preferred path includes the evaluation. Except for on-the-job training this normally involves the collection of post-training data and comparison of this data to pre-training measures to assess deficiencies as well as benefits from the program. The basic question is: To what extent have objectives been achieved?

Answers are not as easily obtained as one might think. Real world training programs include, in addition to the training program itself, numerous other factors which may also change during the evaluation period causing a confounding of the results. That is, if the evaluation shows an improvement over the pre-training status, the improvement may be objectively attributed to the training or to some other variable such as a change in production rates, supervisory personnel, or managerial practices Because these extra variables have the potential to confound the conclusions, they were called "threats to validity" by Cook and Campbell (1979). They described a range of methodologies for designing training programs and other field experiments so as to minimize these threats to validity. When there is some reason to suspect that the level of the dependent (performance) variable during a certain time period is affected by its value in prior time periods (i.e., serially dependent), then a time series statistical approach should be considered. Methods for such analyses are described by Cook and

Campbell (1979), Waller (1980), and Jones, Vaught, and Weinrott (1977). Thus, by utilizing appropriate designs one can increase the probability of being able to attribute improvements in trainee performance to the training programs rather than to some other factor.

Improvement of the Program

The next step in the model is a decision to either terminate or repeat the training program with another group of trainees. Since the decision may be influenced by the results of the post-training evaluation, an input arrow to the decision circle is shown on Figure 1. If the choice is to repeat, the program may be offered again with or without an attempt at improvement. Ordinarily one would expect that the program evaluation would identify one or more aspects which could be improved (e.g., Siegel and Federman, 1973). Consequently, Figure 1 shows the preferred path being the one which includes an attempt to improve the program before repeating it with another group of trainees.

SUMMARY

A model for characterizing adult training programs is proposed. Like other process models it may have several utilities. It provides a framework for comparing the characteristics of different training programs. It may serve as an outline for contracts with extra-organizational experts for providing in-house training. The model may provide a systematic structure for large organizations to critically examine their current employee training programs. Finally, the framework offered by the model could be useful for organizing training literature and identifying research needs concerning the training process.

REFERENCES

[1] Baddeley, A.D., The influence of length and frequency of training session on the rate of learning to type, Ergonomics 21 (1978) 627-635.

[2] Blaiwes, A.S., Puig, J.A. and Regan, J.J., Transfer of training and the measurement of training effectiveness, Human Factors 15 (1973) 523-533.

[3] Borman, W.C. and Peterson, N.G., Selection and Training of Personnel, Chapter 5.2, pgs. 5.2.17-5.2.27, in: Salvendy, G. (ed.), Handbook of Industrial Engineering (John Wiley and Sons, New York, 1982).

[4] Bradbury, R.J., Comment on 'some recent experiments on learning and their training implications', Ergonomics 13 (1970) 273-274.

[5] Brown, I.D., Some alternative methods of predicting performance among professional drivers in training, Ergonomics 11 (1968) 13-21.

[6] Campbell, J. P., Personnel training and development, Annual Reviews of Psychology 22 (1971) 565-602.

[7] Caro, P.W., Aircraft simulators and pilot training, Human Factors 15 (1973) 502-509.

[8] Cohen, H.H. and Jensen, R.C., Measuring the effectiveness of an industrial lift truck safety training program, Jrnl. of Safety Research, 15 (1984) 125-135.

[9] Cook, T.D. and Campbell, D.T., Quasi-Experimentation, Design & Analysis Issues for Field Settings (Houghton Mifflin Company, Boston, 1979).

[10] Cote, D.O., Williges, B.H. and Williges, R.C., Augmented feedback in adaptive motor skill training, Human Factors 23 (1981) 505-508.

[11] Czaja, S.J. and Drury, C.G., Training programs for inspection, Human Factors 23 (1981) 473-484.

[12] Duncan, K.D. and Shepherd, A., A simulator and training technique for diagnosing plant failures from control panels, Ergonomics 18 (1975) 627-641.

[13] Edwards, M.L. and Ellis, N.C., An evaluation of the Texas driver improvement training program, Human Factors 18 (1976) 327-334.

[14] Fisk, A.D., Scerbo, M.W. and Schneider, W., Issues in training for skilled performance, 392-396, Proceedings of the Human Factors Society 27th Annual Meeting (1983).

[15] Goldstein, I.L., Training in work organizations, Annual Reviews of Psychology 31 (1980) 229-272.

[16] Goldstein, I.L., The pursuit of validity in the evaluation of training programs, Human Factors 20 (1978) 131-144.

[17] Hancock, W.M. and Bayha, F.H., The Learning Curve, Chapter 4.3, in: Salvendy, G. (ed.), Handbook of Industrial Engineering (John Wiley and Sons, New York, 1982).

[18] Jeantheau, G.G., The use of multi-man system trainers, Ergonomics 12 (1969) 533-542.

[19] Jensen, R.S., Pilot judgment: Training and evaluation, Human Factors 24 (1982) 61-73.

[20] Johnson, S.L., Effect of training device on retention and transfer of a procedural task, Human Factors 23 (1981) 257-272.

[21] Johnson, W.B. and Rouse W.B., Training maintenance technicians for troubleshooting: Two experiments with computer simulations, Human Factors 24 (1982) 271-276.

[22] Jones, R.R., Vaught, R.S. and Weinrott, M., Time-series analysis in operant research, J. of Applied Behavior Anal. 10 (1977) 151-166.

[23] Kao, S.R., On educational ergonomics, Ergonomics 19 (1976) 667-681.

[24] Kirkpatrick, D.L., Evaluation of Training, Chapter 5, in: Craig and Lester (eds.), Training and Development Handbook, (McGraw-Hill, New York 1967).

[25] Komaki, J., Barwick, K.D. and Scott, L.R., A behavioral approach to occupational safety: pinpointing and reinforcing safe performance in a food manufacturing plant, Jrnl. of Applied Psych. 63 (1978) 434-445.

[26] Komaki, J., Heinzmann, A.T. and Lawson, L., Effect of training and feedback: component analysis of a behavioral safety program, Jrnl. of Applied Psych. 65 (1980) 261-270.

[27] Konz, S.A. and Dickey, G.L., Manufacturing assembly instructions: A summary, Ergonomics 12 (1969) 369-382.

[28] Lintern, G., Transfer of landing skill after training with supplementary visual cues, Human Factors 22 (1980) 81-88.

[29] Nettelbeck, T. and Kirby, N.H., Training the mildly mentally handicapped, Ergonomics 18 (1975) 517-522.

[30] Peck, R.C., Toward a dynamic system of driver improvement program evaluation, Human Factors 18 (1976) 493-506.

[31] Siegel, A.I. and Federman, P.J., Communications content training as an ingredient in effective performance, Ergonomics 16 (1973) 403-416.

[32] Singleton, W.T., Some recent experiments on learning and their training implications, Ergonomics 11 (1968) 53-59.

[33] Snook, S.H. and White, A.H., Education and Training, Chapter 12 in: Pope, Frymoyer, and Andersson (eds.), Occupational Low Back Pain (Praeger, New York, 1984).

[34] Stringer, P., Drawing training and spatial ability, Ergonomics 18 (1975) 101-108.

[35] Swezey, R.W., An application of a multi-attribute utilities model to training analysis, Human Factors 21 (1979) 187-189.

[36] Tilley, K.W., Developments in selection and training, Ergonomics 12 (1969) 585-597.

[37] Valverde, H.H., A review of flight simulator transfer of training studies, Human Factors 15 (1973) 510-523.

[38] Walker, R.W., An evaluation of training methods and their characteristics, Human Factors 7 (1965) 347-354.

[39] Waller, J.A., A systems model for safety program evaluation, Accid. Anal. & Prev. 12 (1980) 1-5.

VIII

ANTHROPOMETRY AND BIOMECHANICS

Trends in Ergonomics/Human Factors II
R.E. Eberts and C.G. Eberts (Editors)

DATA BASE SELECTION IN APPLIED ANTHROPOMETRY

Bruce Bradtmiller, Ph.D.

Anthropology Research Project, Inc.
503 Xenia Avenue
Yellow Springs, Ohio 45387
U.S.A.

In ergonomics and human factors research, anthropometry, the measurement of living men and women, has a place of importance in the design of workspaces and personal protective equipment. Because data bases differ in composition and focus, a lack of careful attention to data base selection can result in a poorly designed final product. Some of the factors which should influence the choice of anthropometric data are examined. Methods are presented for choosing the best data set, or altering existing data sets to improve their application to the design problem.

INTRODUCTION

Anthropometry, the measurement of the human body, has a place of acknowledged importance in the design of workspaces and personal protective equipment. This is evidenced by the considerable amount of space devoted to the topic in human factors manuals and handbooks. The backbone of a successful program of applying anthropometry to a design problem is the actual set of anthropometric data. Despite the awareness of a need for anthropometric data, scant consideration is often given to *which* anthropometric data to use, and what might be the consequences of the wrong choice. A brief survey of a few standard human factors handbooks (Diffrient *et al.*, 1974; Huchingson, 1981; McCormick and Sanders, 1982; Morgan *et al.*, 1963; Panero and Zelnik, 1979; Van Cott and Kinkade, 1972; Woodson, 1981) reveals a wide variety of data sets, ranging from children to adults, and from U.S. to foreign populations, both civilian and military. This paper will show, first, how anthropometric data sets can differ in important ways, and what the practical implications of those differences are, and second, suggest a series of steps one might use in selecting an appropriate anthropometric data set for a specific problem.

VARIABILITY AMONG ANTHROPOMETRIC DATA SETS

The simple approach, when designing to a mixed population might be simply to combine several surveys, as was done, for example, by Huchingson (1981) and Diffrient (1974). This is often done when single data sets do not represent the full range of the user population, or do not contain all the critical dimensions needed in the design. While such combining has a certain intuitive appeal, it often leads to serious problems, the most invisible of which is caused by the fact that certain dimensions, even when they have the same name, are often measured quite differently in any two surveys. Other pitfalls also beset the unsuspecting designer in attempting to combine certain survey populations.

The most fundamental distinction between anthropometric data sets is based on gender. The vast majority of anthropometric surveys have been conducted on either males only or females only. There is nothing inherently wrong with this, but most designers do not design for only one sex. There has been a tendency -- in the sizing and design of protective clothing and equipment, for example -- to combine height/weight data and treat women as though they were small men. This often results in a poor fit for women since there are marked differences not only in the sizes of men and women, but also in their proportions. Given equal height and weight, for example, women have larger hips than men, and considerably narrower shoulders. Facial dimensions, too, have been shown to vary in proportion as well as size -- differences which have been a source of problems for women who have been issued size "SMALL" respirators. This is not to say that items cannot be successfully designed for an integrated male/female population, but only that, as in other data set variances to be discussed, it requires a full understanding of the nature of the differences.

The second major division among data sets is civilian or military. Traditionally, the most thorough anthropometric surveys have been conducted by the military, both in the U.S. and abroad. Two reasons for this are the critical need for exact fit of equipment and workspaces in the military setting, and the relative ease with which one can assemble a large group of military personnel for lengthy measuring sessions. The major civilian surveys, in the U.S. at least, contain a large number of subjects, but relatively few dimensions. Anthropometrically, there are several important differences between the military and the civilian surveys. One of these differences is the range of values covered. Table 1 shows the range, expressed by the 1st and 99th percentiles of stature and weight, for several selected military and civilian surveys.

The largest differences in stature are at the lower end of the range, where the military distribution is truncated because of lower limit entrance requirements. The upper end of the distribution is roughly equivalent for civilian and military samples. In weight, however, although there is some difference at the lower end of the range (military personnel are slightly heavier), there are dramatic differences at the upper end of the range. Specifically, weight requirements in the military eliminate very heavy individuals. The difference is especially marked in the females.

VARIABILITY WITHIN DATA SETS

Samples may also differ in the proportion of various age and racial subgroups. To the extent that these factors influence anthropometric variables, they are critically important in determining the summary statistics of the entire sample. It is well known that head and face dimensions are strongly influenced by race. It is less well known that other body dimensions, such as leg length, are similarly affected. An analysis of variance, which shows the amount of variability in anthropometric dimensions due to race, was conducted on the 1977 survey of U.S. Army women (Churchill et al., 1977b). The null hypothesis that there is no

TABLE 1. FIRST AND 99TH PERCENTILE VALUES OF STATURE AND WEIGHT*

	1st%ile	STATURE (in cm)	99th%ile
Females	149.5	Air Force 1968	176.5
	148.2	Army 1977	178.4
	144.5	HES 1962 (civilian)	175.3
	145.8	HANES 1977 (civilian)	176.9
Males	163.2	Air Force 1967	191.8
	161.3	Air Force 1950	190.6
	156.5	HES 1962 (civilian)	189.1
	156.7	HANES 1977 (civilian)	190.7

	1st%ile	WEIGHT (in kg)	99th%ile
Females	43.7	Air Force 1968	79.5
	43.1	Army 1977	83.8
	41.0	HES 1962 (civilian)	107.3
	41.1	HANES 1977 (civilian)	114.8
Males	57.9	Air Force 1967	103.3
	55.7	Air Force 1950	98.6
	50.0	HES 1962 (civilian)	108.2
	49.3	HANES 1977 (civilian)	118.5

* Source: AFAMRL Anthropometric Data Book (1982)

difference in means of the racial subgroups was rejected for 58 of the 96 variables. Head and face dimensions are not part of the 96 variables. Thus the means for the racial groups differed, at the traditional 5% level, in over half the body measurements. Space limitations, unfortunately, do not permit an extended discussion of which variables differed by race, and the list would likely differ by sex. The point in the present context is only that race is an important variable which must be considered when selecting a data base for an anthropometrically based design.

Another often-overlooked variable in data base selection is age. Using the same group of Army women, measured in 1977 (Churchill et al., 1977b), an analysis of variance was conducted with the null hypothesis that the means of the various anthropometric dimensions are not different in seven age groups. The groups used were:

less than 20.5 years
20.5 to 25.5 years
25.5 to 30.5 years
30.5 to 35.5 years
35.5 to 40.5 years
40.5 to 45.5 years
greater than 45.5 years

The results of that analysis show that 35 of 69 anthropometric variables are significantly (at the 5% level) affected by age. As with racial categories, a discussion of specific dimensions is beyond the scope of this paper. Further, the list would be different for males, perhaps for civilian samples, and might depend to a certain extent on how the ages were grouped into categories. It is important to note, however, that age is a factor which can influence a large number of anthropometric variables.

Having examined age, race, and gender, it is worthwhile to return, briefly, to factors which distinguish some of the existing anthropometric data bases. None of the major military surveys are representative of the U.S. population in age because of the truncated distribution on both ends of the military samples. Even if one were designing, say, a workspace in which all users would be adults, the truncation at the upper end of the age distribution eliminates a large part of the age range. Even where older individuals exist in the military, those individuals do not usually become subjects in anthropometric surveys, because of the voluntary nature of the data collection process. Of the 1330 women in the Army survey mentioned above, only 14 were in the 45 and over category. There were only 11 subjects in the 40 to 45 year category. Thus the sample effectively ends at age 40. The same is true for all military surveys. In the case of the 1965 U.S. Air Force survey (Churchill et al., 1977a) nearly two-thirds of the subjects were basic trainees, in their first few months of military service. Because of the great number of those subjects, the overall mean for age in that survey is 22.68 (n=3869). There are other examples, of course. The civilian surveys are more likely to be representative of the civilian population, but each survey should be examined individually.

The racial or ethnic makeup of the various surveys is even more problematic because the information is often difficult to obtain. Unlike age, which might be published along with summary statistics of anthropometric variables, race is almost never summarized statistically. Often the only way to discern the racial representativeness of a given sample is to examine the raw data. Too often, this is not done.

SELECTING A DATA BASE

Having established that anthropometric data bases differ in important ways, how should one choose the most appropriate one for the design problem at hand? The first step is to define, very carefully, what dimensions are actually needed in the design project. Selecting the list of dimensions serves two purposes. First, it may reduce the number of possible data bases by eliminating those which do not contain the required dimensions. Second, an examination of the dimension list will show which other factors should be considered. For example, if one is designing a foothold for climbing a flat wall, foot breadth and foot circumference would be very important dimensions. Yet neither of these is significantly influenced by race or age, so these factors would not need to be considered in selecting the data base. In many other circumstances, however, a much longer list of dimensions will be required.

If age and race are important factors in the list of required dimensions, then the second step is an analysis of the age distribution and racial makeup of the intended user population. If the design is for broad consumer use, then the Bureau of the Census will have the required data. If a factory-specific workspace is to be created, then the plant personnel office should be able to provide the necessary figures.

When the age/race mix of the user population is determined, then it is a simple matter to select the data base whose age/race characteristics most closely match the user population. If none of the available data bases have a sufficiently close correspondence it is always possible to adjust the data base which comes closest to the target population. To "adjust", here, means to discard, in some random way, individuals from those age/race categories in which the prospective data base has a surplus. It is an unfortunate occurrence that many users of anthropometric data assume that because 5000 individuals were measured, all 5000 must be used in the design process. It must be added, of course, that adjusting should only be done where the initial sample size is sufficiently large to leave an adequate sample size when the adjusting is complete.

What if there is a data base where race and age match the user population perfectly, but the required measurements were not taken? There are three possible solutions to the problem. First, it is possible to calculate the values of the required dimension using other dimensions which are available. This can occasionally be done by substraction or addition, if two related dimensions are present. More often, regression equations, developed on a data base which contains the required dimension, can be applied to the desired data base to create the new dimension for each individual.

A second approach is to use the target data base, and match each subject on the basis of two or more common dimensions (usually stature and weight) to a data base which contains the required dimension. The value for that dimension can then be associated with each stature/weight pair in the target data base, in effect creating the required dimension where it did not previously exist. This approach has limitations, but when carefully applied, can be quite useful.

The third way to create a data base with the required dimensions is to actually have an anthropometric survey conducted on the target population. While this is the most costly of the three methods, it has the advantage of containing exactly the required dimensions and can have a perfect age/race/sex match with the user population. Further, in the case of a work station design for a factory, or series of factories, the sample size is small enough that such a survey would be less costly than many other aspects of the development process.

All three of these methods are more time-consuming than consulting a handbook, or published summary statistics of previous surveys. They have the advantage, however, of assuring that the design project will be directed toward the right group of users. Ultimately, such an approach is more cost- and time-effective than living with or correcting a product or work space designed for the wrong users.

REFERENCES

[1] AFAMRL Anthropometric Data Book, 3rd Edition. Manuscript on file, Anthropology Research Project, Inc. (1982).

[2] Churchill, E., T. Churchill and P. Kikta, The AMRL Anthropometric Data Bank Library: Volumes I - V, AMRL TR-77-1 (Aerospace Medical Research Laboratory, Wright-Patterson Air Force Base, Ohio, 1977).

[3] Churchill, E., T. Churchill, J.T. McConville and R.M. White, Anthropometry of Women of the U.S. Army - 1977: Report No. 2 - The Basic Univariate Statistics, NATICK/TR-77/024 (U.S. Army Natick Research and Development Command, Natick, Massachusetts, 1977).

[4] Diffrient, N., A.R. Tilley and J.C. Bardagjy, Humanscale 1/2/3 (The MIT Press, Massachusetts Institute of Technology, Cambridge, 1974).

[5] Huchingson, R.D., New Horizons for Human Factors in Design (McGraw-Hill, New York, 1981).

[6] McCormick, E.J. and M.S. Sanders, Human Factors in Engineering and Design (Fifth Edition) (McGraw-Hill, New York, 1982).

[7] Morgan, C.T., J.S. Cook, A. Chapanis and M.W. Lund, Human Engineering Guide to Equipment Design (McGraw-Hill, New York, 1963).

[8] Panero, J. and M. Zelnik, Human Dimension and Interior Space (Whitney Library of Design, and imprint of Watson-Guptill Publications, New York, 1979).

[9] Van Cott, H.P. and R.G. Kinkade, Human Engineering Guide to Equipment Design (Revised Edition) (American Institute for Research, Washington, 1972).

[10] Woodson, W.E., Human Factors Design Handbook (McGraw-Hill, New York, 1981).

Trends in Ergonomics/Human Factors II
R.E. Eberts and C.G. Eberts (Editors)

TEMPORAL CHARACTERISTICS OF HUMAN PERCEPTION OF CONTINUOUS ISOMETRIC FORCES

S. Deivanayagam, Ph.D., P.E.
and
Himat Tank, M.S.

Department of Industrial Engineering
The University of Texas at Arlington
Arlington, Texas 76019

In the absence of any external feedback mechanism, the human ability to maintain an isometric force constant, even for short durations, is quite limited. This paper reports an investigation conducted to study the human perception variability during the application of an isometric force. Several force levels, low to moderate high, were studied. The results indicate that the human perception is such that it tends to cause a decline in applied force, that can be described by a negative exponential expression. It is indicated that factors other than physiological fatigue may have major influences on the human perception in such cases.

INTRODUCTION

In spite of the large scale mechanization and automation that has been taking place in occupational fields, one can still notice a considerable number of occasions where application of manual forces is required. Often times the person is required to apply and maintain a constant isometric force for short durations of up to a minute or so. Human ability to maintain a constant isometric force depends, to a large extent, on external mechanisms such as displays or direct feedback from the object being manipulated. In the absence of any such external mechanisms, the human operator has to rely on his own internal mechanisms for the perception of the applied force. Past studies show that such internal mechanisms are of two general types, central to the organization or local to the muscle group/ skeletal joint. Pandlof (1978) has made an extensive literature study on the role of central and local factors in the rated perceived exertion during physical work. Table I presents the various parameters thought of having influence or some form of relationship to the perceived rate of exertion.

The role of local and central factors in the perception of exertion has not been proven conclusively one way or other. Considering the temporal effects on the perception of rated exertion, several researchers have found that a constant force will be perceived as increasing in intensity over time. Cain (1973) and Cain and Stevens (1970, 1971) in a series of experiments found that the actual force, applied during a supposedly constant static contraction, declined very rapidly initially and slowly

Table 1. Factors Suspected to Influence Human Perception of Exertion

Local Factors	Central Factors
Blood Lactate	Heart Rate
Anaerobic Matabilities	Tachycardia
Vanilmandelic Acid	Pulmon. Ventilation
Catecholomine Exertion	0_2 Consumption
Golgi Tendon Organ	Respiratory Rate
Muscle Temperature	Chemoreceptors
EMG	Baroreceptors
Chemoreceptors	
Mechanoreceptors	

thereafter. They attributed this effect to the subjective cost incurred, namely a build up of feelings of efforts and fatigue. The present research was undertaken to investigate the temporal variations in the isometric force that the subject perceives as constant during a continuous period of time lasting up to 60 seconds.

PRESENT STUDY

The primary objective of the present investigation was to determine the temporal characteristics in the perception of a continuously applied isometric force. This was accomplished by monitoring the actual force applied while the subject was required to maintain a perceived effort constant over the test duration of 60 seconds. Since there was no external feedback to the subject and the subject was considered to be well motivated, it was reasonable to expect that any change in the applied force was correlated to the internal perception of the applied force by the subject, who was all the time trying to maintain the applied force constant. In order to determine any effect of force level on the variation of the applied force, five levels of force were chosen. These were 12.5%. 25.0%. 37.5%, 50.0% and 62.5% of the maximum voluntary isometric force. The right elbow flexor muscles were chosen for this investigation.

The apparatus consisted of a chair with a padded elbow rest for the subject to be seated. An LVDT load cell anchored on the floor and connected to a handle at the other end through pieces of metallic chain was used to measure to force applied. The subject sat upright in the chair with elbow flexed at about 90° and forearm horizontal. Figure 1 shows the experimental set up. The handle was placed in the supinated palm of the subject which he clasped securely and comfortably. The chain length was adjusted tight and the subject was required to apply an upward isometric force by pulling up the handle by flexing at the elbow only. A digital display hidden from the view of the subject displayed the actual force to the experimentor. The experimentor then guided the subject orally to the desired level of force to start the trial. This was generally accomplished within about 5 seconds and then a clock was started. The subject then tried to maintain the same force on the handle, as he perceived it. At elapsed times of 5, 10, 20, 30, 40, 50, and 60 seconds after the starting of the clock, the experimentor recorded the actual force applied without providing any feedback to the subject. The subject was told to relax and rest at the end of 60 seconds.

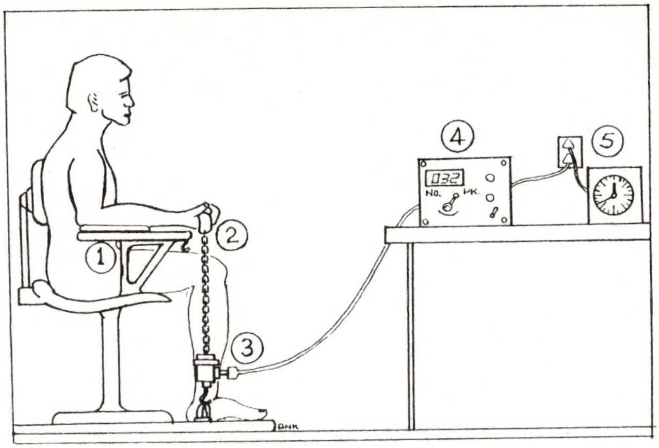

Figure 1. Experimental Set Up: 1. Padded Forearm Support, 2. Handle, 3. LVDT Load Cell, 4. Digital Display, and 5. Clock

The maximum voluntary isometric force of the subject was measured previously in the same apparatus in a similar manner. Two maximum force trials were done and their average was used for this study. All subjects performed all the test trials on two different days, first day in the order of increasing force levels and the second day in the order of decreasing force levels. There was a minimum of five minutes of rest between trials. A test session typically lasted about 90 minutes. At the end of the session the subject was asked about his feelings of pain, exhaustion, etc.

SUBJECTS

Fifteen university male students participated in the study. They were all either enrolled in the same class or friends of one of the experimentors and could be considered well motivated. They ranged in age from 22 to 28 years. None of them had any previous or current upper extremity injury or neurological disease.

RESULTS

It was observed that the applied force at the end of 60 seconds was much lower than the force at the start in all the trials. It was also observed that most often the force level fluctuated during the trial while steadily declining. Table II shows the average actual forces observed for the 15 subjects in the two trials during the 60 second duration. Cain (1973) has reported, based on his own as well as other investigations, that the decline follows a negative exponential relationship with time. The present results tend to confirm the earlier findings. Figure 2 presents the average actual force against the elapsed time. The relationship can be approximated as

$$f_t = Ae^{-kt} \quad \ldots .(1)$$

where f_t = applied force at time t

A = applied force at time 0

k = a rate constant, related to the subject's perception of his effort.

Table II. The Average Actual Forces (Lbs.) During the Trials

Force Level %	Elapased Time (Seconds)							
	0	5	10	20	30	40	50	60
12.5	4.37	3.97	3.87	3.63	3.30	3.13	2.77	2.60
25.0	8.70	7.67	7.30	6.77	6.07	5.60	5.17	4.83
37.5	12.10	11.27	10.23	9.80	8.80	8.57	7.63	7.10
50.0	17.20	14.87	13.23	11.80	10.63	9.77	9.33	8.93
62.5	21.67	18.47	16.80	15.13	13.37	12.47	11.20	10.73

One objective of the study was to determine whether the rate constant, "k", depends on the relative load level. If physiological fatigue is the major factor responsible for the differences in the perception of the applied force over time, the increasing level of fatigue due to higher load levels should result in higher rate constants when compared with low load levels. The rate constants for the individual subjects were computed for each load level as per expression 1. Table III shows the average ($\bar{X}_k$), and the

Table III. Statistics of Rate Constant

	Force Level (% of Maximum)				
	12.5	25	37.5	50.0	62.5
$\bar{X}_k$ $(X10^{-3})$	8.57	10.76	10.20	11.52	12.63
σ_k $(X10^{-3})$	4.8	4.7	4.7	4.6	6.1
R_{ave}	0.667	0.784	0.825	0.847	0.877

Note: All $\bar{X}_k$ values and R_{ave} values are negative.

standard deviation (σ_k) of the rate constants. Also presented in the table is the average linear correlation coefficient, R_{ave}, between the logarithm of the actual force and the elapsed time for each of the force levels.

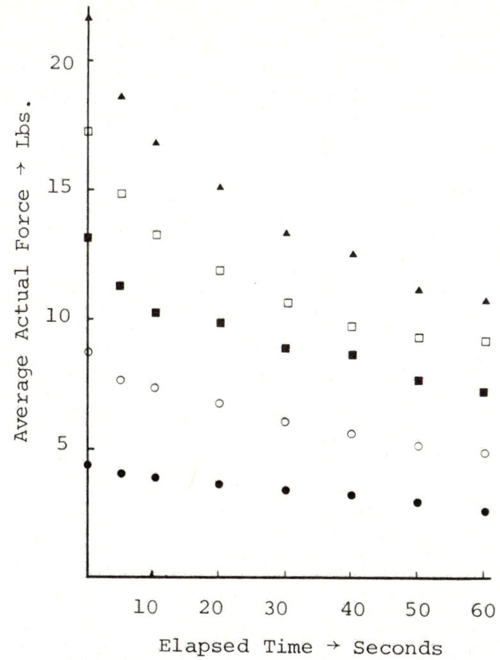

Figure 2. The Relationship Between the Average Actual Force and the Elapsed Time

● At 12.5% of max: $f_t = 1.406e^{-0.00857t}$; R = 0.667

○ At 25.0% of max: $f_t = 2.068e^{-0.01076t}$; R = 0.784

■ At 37.5% of max: $f_t = 2.445e^{-0.01020t}$; R = 0.825

□ At 50.0% of max: $f_t = 2.699e^{-0.01152t}$; R = 0.847

▲ At 62.5% of max: $f_t = 2.940e^{-0.01263t}$; R = 0.877

Figure 3 shows the relationship between $\bar{X}_k$ and the force levels. An increasing trend for $\bar{X}_k$ with increasing force levels can be noticed. However, the standard deviations are relatively large. An analysis of variance of the rate constant (k) values resulted in F-statistic = 1.407 while $F^*_{(0.95,\ 4,\ 70)} \simeq 2.5$. This indicates that the rate constants are not significantly different among the tested load levels. It must be mentioned that the correlation coefficient values indicate that the expression (1) is a reasonable descriptor of the process of declining force.

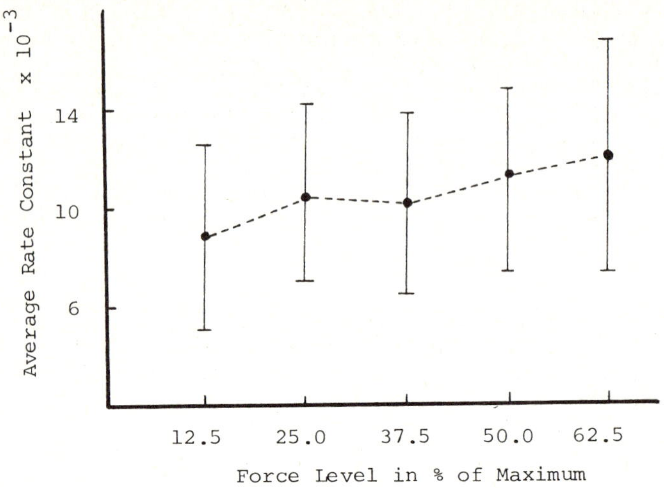

Figure 3. The Relationship Between Rate Constant and the Force Levels with Both the Average and Standard Deviation Values Shown

DISCUSSION

The present study was conducted to investigate the temporal characteristics of the human perception of voluntary isometric force. A time related process in the perception has been observed at low (12.5% of maximum) to moderate high (62.5% of maximum) levels of force. The actual force applied was found to decline continuously over time of up to 60 seconds.

An interesting observation was that all subjects exhibited this decline at all force levels. Thus a subject always perceived that the effort was getting harder no matter at what effort level he started. This perception of progressive difficulty resulted in him applying a declining force in order to maintain the same perception of his effort. It is also interesting to note that a subject starting at 62.5% of maximum would end up around 35% of maximum and the same subject starting at 12.5% of maximum would end up below 8.0%. If physiological fatigue was the major factor for the progressively harder perception, then very little decline should be observed at lower effort levels. The results of the study show that such is not the case. In all force levels, the ending force was about 50% to 60% of the starting force.

The actual force applied could be best described by a negative exponential function. Testing for the relationship between the rate of decline and the force levels showed inconclusive results. While the F test showed that there are no statistically significant differences among the rate constant values for the tested force levels, the power of the test was only 45%, rather poor. This leads us to suspect that the physiological fatigue process cannot be ruled out as not having an influence on the perception of the applied force. At the same time there probably are other factors such as the receptors in the joints, muscles, and tendons, the rate of accumulation of metabolics, etc. which may have more influence on the subjective perception.

The reporting of the subjective feelings at the end of test sessions indicated that 12 subjects felt light to moderate pain in the wrist area and 9 subjects felt the same in the forearm. Only a few subjects, less than 5 in each case, reported slight pain in the other parts of upper extremity or the body. All subjects reported feeling varying degrees of tiredness. Lack of significant pain in the elbow flexor muscles seems to indicate that the muscles were not excessively loaded and thereby reduces the possibility of local muscular fatigue influencing the decline in applied force.

ACKNOWLEDGEMENTS

The authors wish to thank Mr. D. N. Kulkarni, Mr. V. Ranginení, Mr. V. Reddy, Mr. R. G. Soni and the volunteer subjects for making this study possible. The efforts of Mrs. C. Murphy in the preparation of this manuscript are also deeply appreciated.

REFERENCES

Caferrelli, E., Peripheral and Central Inputs to the Effort Sense during Cycling Exercise, Europ. J. Appl. Physiol. 37 (1977) 181-189.

Cain, W. S., Nature of Perceived Effort and Fatigue: Roles of Strength and Blood Flow in Muscle Contractions, Jr. Mot. Behav. 5 (1973) 33-47.

Pandolf, K. B., Influence of Local and Central Factors in Dominating Rated Perceived Exertion during Physical Work, Percept. Mot. Skills 46 (1978) 683-698.

Sargeant, A. J., and Davies, C. T. M., Perceived Exertion during Rhythmic Exercise Involving Different Muscle Masses, J. Human Ergol. 2 (1973) 3-11.

Stein, R. B., Peripheral Control of Movement, Physiol. Rev. 54 (1974) 215-243.

Trends in Ergonomics/Human Factors II
R.E. Eberts and C.G. Eberts (Editors)

AN INTERACTIVE BIOMECHANICAL LIFTING MODEL

Chen Jen-Gwo
Brian Peacock

School of Industrial Engineering
University of Oklahoma
Norman, Oklahoma
U.S.A.

The purpose of this model is to investigate problems of equilibrium, work space dimensions, joint reaction forces, and muscle tension during saggital plane lifting. The model has been developed on a Tektronix 4054 microcomputer.* The interactive capability allows the investigator to set up parameters such as segment lengths, masses and centers of gravity, joint angles, object weight, lifting frequency and space restrictions. The outputs include analyses of moments, equilibrium, joint reaction and muscle forces as the load is moved to its destination and analyses of an extended lifting task based on the NIOSH guidelines.

1. INTRODUCTION

The ergonomist is making increasing use of computer simulation techniques in the evaluation and design of work situations. Particular applications have been in the areas of anthropometry and workplace design (Kingsley, 1981) and manual materials handling (Chaffin, 1970; Anil and Ayoub, 1980). Other mathematical models are presented in the NIOSH work practices guide for manual lifting (NIOSH, 1981) which are used to predict acceptable and maximum lifting limits given various postural, load and frequency conditions.

This paper describes an interactive computer model which relates anthropometric and task information to biomechanical data, obtained from the theory and literature, to produce detailed analyses of saggital plane lifting situations. A key element of this model is the interactive flexibility for modification of basic anthropometric and task data through the presentation of pictorial, numerical and graphical results. The model can be used, therefore, as a stimulating teaching tool, as a medium for professional analyses of manual materials handling situations and as a vehicle for research into the kinetics of musculo-skeletal injury.

*Copies of the basic program may be obtained from Brian Peacock.

2. PREVIOUS WORK

2.1 Biomechanical and Anthropometric Data

To develop a biomechanical model it is necessary to obtain data regarding masses of body segments, location of mass centers, segment lengths, centers of rotation, angles of pull of muscles, moments of inertia, and so on.

Segment lengths between each joint vary with age, body build, sex, and racial origin. Dempster (1967) summarized segment length and joint center location data relating to anatomical landmarks. Drillis and Contini (1963) expressed segment length as a percent of body height. These segment proportions serve as a good approximation in the absence of better individual data. Drillis and Contini (1963) also presented the ratios of segment mass to body mass. Hay (1973) indicated that there are basically two methods of approach which have been used in determining the location of the center of gravity of the human body—the direct (or whole body) and indirect (or segment) approach. Niels, et al. (1974) and Winters (1979) presented values for the location of the center of mass as a percentage of the segment length from either the distal or proximal end. The methods and data from these studies are invaluable in the development of biomechanical models.

2.2 Biomechanical Models

Pearson, et al. (1963) performed a dynamic analysis of the human upper extremity which presented the free body diagrams of arm segments and showed what forces are involved in planar motions. From graphical vector diagrams are derived equations for determining joint force and torque reactions to weight and inertia. Pearson (op cit) also showed how linear accelerations can be determined from angular accelerations, derived in turn by finite differences from displacement time data. Plagenhoef (1963) presented a kinematic and kinetic analysis of the human body in motion. The horizontal and vertical forces were obtained in any desired plane of any whole body action; however, most examples were limited to selected symmetrical motions in the saggital plane.

Slote and Stone (1963) investigated discrete voluntary ballistic forearm flexion in the saggital plane. Using mathematical representation of angular displacement as a function of time they developed predictions of work and power. Plagenhoef (1968) used a cinematographical technique to obtain body segment forces, moments of forces, total body centers of gravity and the contribution of each segment to the whole motion. He also developed two computer programs to obtain kinetic data on human movement.

Passerello and Huston (1971) developed equations of motion for a model of the human body. This human body model consists of a system of rigid segments attached to each other by hinge and ball and socket joints. They employed vector, tensor and matrix methods to provide a systematic organization of the geometry. In 1982, they extended their studies to present an analysis and a general procedure for studying the mechanics of human body motion using Lagrange's form of d'Alembert's principle to develop the governing equations. These dynamic equations are applicable in a wide variety of situations.

Chaffin (1969) developed a 7-link 2-dimensional static model to calculate joint forces and moments during material handling activities. The model also included an analysis of shear and compression forces in the lumbar spine. Chaffin and Barker (1970) extended this model into an eight-link one. By empirical tests, it was shown that the model can be used to predict the maximum strengths of persons asked to lift objects while assuming different postures. They modified this model later (Chaffin and Martin, 1972) to produce a series of strength prediction curves for typical industrial material handling tasks. They found that lift capacity is greatest when the hands are close to the knees and that lift capacity near the body is lowest when the load is waist-high as opposed to knee or shoulder-high. In 1975, they developed and validated a three-dimensional biomechanical model (Chaffin and Garg, 1975). The model calculates: (1) body position from articulation angles, (2) angular velocities and accelerations, (3) intertial moments and forces, and (4) reactive moments and forces at each articulation, including the L5/S1 joint.

An alternative approach to dynamic biomechanical modeling was developed by Ayoub and El-Bassoussi (1978). The model accepts operator, postural and task data. From these, the model simulates the lifting action and calculates the reactive forces and torques at the various joints. The model is mainly based on the acceleration patterns described by Slote and Stone (1963). The model also was used to determine the strain on the musculoskeletal system during the lifting action using "leg lifts" and "back lifts". The research data indicated that maximum compressive force is higher when a load is lifted with straight legs. These results were confirmed by Gray and Herrin (1979) who also noted the trade-off between biomechanical and metabolic stress. Gruver, et al. (1979) developed a 5-link 2-dimensional model of the human body to simulate manual lifting tasks. The task is described as a function space minimization problem with a nonlinear objective and subject to linear and nonlinear constraints. Inputs include anthropometric and load description, initial and final locations of the load, and performance time. The model indicates that the load should be kept close to the body throughout the lift, although additional constraints are needed to prevent the box from moving too close to the body.

2.3 Other Computer-aided Human Motion Models

The sports literature has also been a source of biomechanical models. Ramey and Yang (1981) developed a simulation procedure for human airborne motion. In these studies, the human body is considered to have nine body-segments (right and left lower leg, right and left thigh, right and left lower arm, right and left upper arm and main body). Somersault and hitch kick long jump motions were simulated. Dapena (1981) developed a method to permit the analysis of a real-life motion and subsequent simulation of a modified motion. The method was divided into two stages: (1) analysis of the actual motion of the subject, and (2) generation of the simulated motion of the subject.

Onyshko and Winters (1982) developed an interactive graphic package for simulation of human locomotion. The planar model was comprised of seven segments. The simulation package consists of four main subprograms. The first subprogram allows input data - the starting position, the angular velocities, the joint moments and the anthropometric details. The second subprogram allows the manipulation of the moments of the six joints. The third subprogram accepts the input, sets up the equations of motion and solves them numerically for a user defined period of time. The last subprogram plots the output as a stick figure in an overlaid or sequential form and has the option of superimposition of the driving joint moments.

3. THE BIOMECHANICAL LIFTING MODEL

The six segment saggital plane model described in this paper involves the following series of inputs, analyses and outputs. First, segment anthropometric, starting joint angle and load data is entered. This is followed by static equilibrium analysis around the foot support. The resulting picture is accompanied by an indication of whether or not equilibrium exists and, if not, what has to be done to obtain equilibrium. The investigator then has the opportunity of modifying the segment, joint angle and load data in an iterative fasion until equilibrium is obtained.

Next the investigator is invited to indicate a final position of the load and associated joint angles, together with the number of intermediate steps for analysis. At each step the forces at each joint, including muscle forces, are calculated. The step data are presented in stick diagram and digital form. Finally the investigator is invited to indicate the working time and frequency of the lift. The program then calculates and displays a comparison with an acceptable load based on the NIOSH guidelines.

3.1 Assumptions

a. Each segment is considered to be rigid with a constant relative location of its centre of mass.
b. The mass moment of inertia of each segment is constant during the movement.
c. All joints are considered to be simple hinge joints. The spinal movement is considered as a single joint located at the L5-S1 level.
d. Joints are considered to be frictionless.
e. Lifting is unrestricted and smooth.

3.2 Data Input

The model consists of the following six segments:
hand and forearm, upper arm, trunk, pelvis, thigh, lower leg and foot.

The investigator may choose to use default values for the segment lengths, masses and centres of mass based on the input of height and weight data. These proportional values and their sources are given in Table 1. The investigator must choose a load value and whether a stoop or a squat starting position is to be selected. Default joint angles for these alternative starting positions are assigned in the program but may be altered interactively.

Table 1. Default Value for Segments as Proportions of Body Height or Weight[1]

Segment	Weight	Segment Centre of Mass		Rad.GYR/ Seg. Length	Segment[5] Length
		Prox.	Dist.		
Shank & Foot	0.122	0.606	0.394	0.416	0.286
Thigh	0.200	0.433	0.567	0.323	0.243
Pelvis	0.064[2]	0.599[3]	0.401[3]	0.503[4]	0.0501
Trunk Neck Head	0.514	0.66	0.34	0.503	0.471
Upper Arm	0.056	0.436	0.564	0.322	0.161
Lower Arm & Hand	0.644	0.682	0.318	0.468	0.254

[1]From Winter (1979)
[2]From Plagenhoef (1963)
[3]From Hay (1973)
[4]Assume same as Trunk, Neck & Head
[5]From Niel, et al. (1974)

Alternatively the investigator may choose to input individual segment data, including segment lengths, masses, centres of mass as a proportion of the distance from the proximal end, and joint angles, measured counterclockwise from a horizontal plane passing through the joint. Joint angle range limits are included to guide investigators towards the choice of feasible postures (Figure 1).

As with the default value option the investigator has the opportunity to alter any of the segment, joint angle or load values until an acceptable starting position is obtained with static equilibrium.

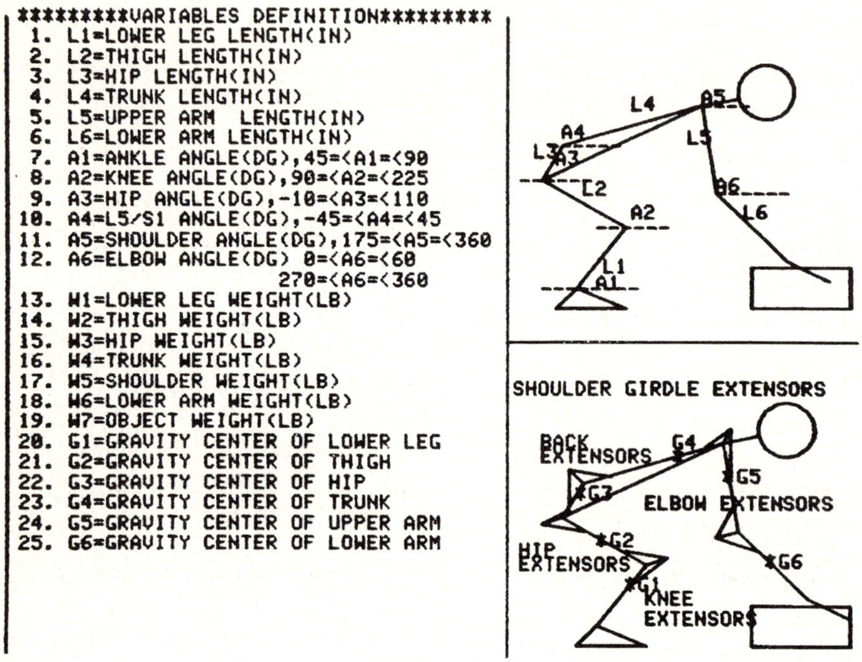

Figure 1. Initial Computer Screen Showing Input Requirements and Feasible Joint Ranges

3.3 Static Analysis

A list of input data is presented together with a stick diagram of the chosen posture (Figure 2). Moments are taken around the ankle and a statement is made regarding the equilibrium of the force system.

$$M = \sum_{i=1}^{6} W_i D_i + W_0 D_0$$

where M is the sum of the moments around the ankle joint
- W_i is the weight of the ith segment
- D_i is the horizontal distance of the centre of mass of the ith segment and the plane of the ankle joint
- W_0 is the weight of the load
- D_0 is the distance of the load from the plane of the ankle joint

```
IT IS TOO HEAVIY TO LIFT!
MY MAXIMUM LIFT CAPACITY
IS=47.6814536478
YOU CAN MOVE OBJECT NEAR TO FOOT
WITH DISTANCE=4.22979888669
OR CHANGE ANY INPUT VARIABLE.
PRESS<RETURN>, THEN TYPE <0> TO
CONTINUE.
19
19. PLEASE ENTER OBJECT WEIGHT
45
0


IT IS A EASY JOB!
I CAN LIFT OBJECT
WEIGHT=47.6814536478
PLEASE INPUT THE FOLLOWING
DATA TO GET BIOMECHANICS ANALYSIS
PLEASE ENTER FRAME INTERVAL(SEC)
.3
PLEASE ENTER TOTAL MOVEMENT TIME(SEC)
1.5
ENTER ANKLE'S MAX. DISPLACEMENT(DG)
10
ENTER KNEE JOINT'S MAX. DISPLACEMENT(DG)
25
ENTER HIP JOINT'S MAX. DISPLACEMENT(DG)
10
ENTER L5/S1 MAX. DISPLACEMENT(DG)
70
ENTER SHOULDER MAX. DISPLACEMENT(DG)
20
ENTER ELBOW MAX. DISPLACEMENT(DG)
40
```

```
********************************
*      INPUT DATA LISTING      *
********************************
 1. LOWER LEG LENGTH(IN)=17.16
 2. THIGH LENGTH(IN)=14.58
 3. HIP LENGTH(IN)=3.006
 4. TRUNK LENGTH(IN)=22.26
 5. UPPER ARM LENGTH(IN)=9.66
 6. LOWER ARM LENGTH(IN)=15.24
 7. ANKLE ANGLE(DG)=80
 8. KNEE ANGLE(DG)=115
 9. HIP ANGLE(DG)=50
10. L5/S1 ANGLE(DG)=-25
11. SHOULDER ANGLE(DG)=280
12. ELBOW ANGLE(DG)=270
13. LOWER LEG WEIGHT(LB)=18.3
14. THIGH WEIGHT(LB)=30
15. HIP WEIGHT(LB)=9.6
16. TRUNK WEIGHT(LB)=77.1
17. UPPER ARM WEIGHT(LB)=8.4
18. LOWER ARM WEIGHT(LB)=6.6
19. OBJECT WEIGHT(LB)=45
20. GRAVITY CENTER OF LOWER LEG=0.606
21. GRAVITY CENTER OF THIGH=0.433
22. GRAVITY CENTER OF HIP=0.599
23. GRAVITY CENTER OF TRUNK=0.66
24. GRAVITY CENTER OF UPPER ARM=0.436
25. GRAVITY CENTER OF LOWER ARM=0.672
```

Figure 2. Screen Showing Input Values, Results of Static Equilibrium Analysis and Prompts for Lifting Pattern

For equilibrium to exist it is necessary to adjust the load or the position of the load.

$$D_0' = \left(\sum_{i=1}^{6} W_i D_i\right)/W_0$$

where D_0' is the horizontal distance of the centre of mass of the load from the plane of the ankle joint given that equilibrium exists.

The investigator is instructed to move the load a distance d towards his feet to achieve equilibrium where $d = D_0 - D_0'$. To do this the investigator must alter segment, joint angle or load values. The process is interactive and any number of iterations may be carried out to test the various input data values and combinations. After each iteration a new analysis is presented and, if required, a hard copy obtained (Figure 2). It should be noted that this analysis does not take account of muscle tension or friction between the foot and floor.

4. KINEMATIC/KINETIC ANALYSIS OF A SINGLE LIFT

Given that an acceptable set of starting angles (θ_i') has been achieved the investigator is asked to indicate a final location of the load by inputting a final set of joint angles (θ_i'') or the change in each joint angle (θ_i)

where $\theta_i = \theta_i'' - \theta_i'$.

In addition the investigator is asked to input the duration of the lift (T) and a frame interval (Δt). Thus the number of frames (N) at which analysis is carried out is given by:

$$N = T/\Delta t.$$

4.1 Kinematic Analyses

The Slote and Stone (1963) equations are used for the following kinematic analyses:

4.1.1 Angular Motions

Angular Displacement (θ_{ij}) in time Δt

$$\theta_{ij} = \frac{\theta_i}{2\Pi}\left(\frac{2\Pi\Delta t}{T} - \sin\left(\frac{2\Pi\Delta t}{T}\right)\right)$$

Angular Velocity (ω_i)

$$\omega_i = \frac{d\theta_{ij}}{dt} = \frac{\theta_i}{T} * \left(1 - \cos\frac{2\Pi\Delta t}{T}\right)$$

Angular Acceleration

$$\alpha_i = \frac{d^2\theta_{ij}}{dt^2} = \frac{2\Pi\theta_i}{T^2} * \sin\left(\frac{2\Pi\Delta t}{T}\right)$$

The angular displacement over the duration Δt of each frame is

$$\theta_{ij}, \quad j = 1, N.$$

Thus, the total angular displacement following the nth frame in a movement is given by

$$\theta_i^0 = \theta_i^1 + \sum_{j=1}^{n} \theta_{ij}$$

These values, with the basic segment data, are used for the subsequent kinematic analyses.

4.1.2 Linear Motion of Joint Centres

These equations were taken from the work of Plagenhoef (1968). Linear Displacement at time t

$$X_k = \sum_{i=1}^{k} L_i \cos \theta_i^0$$

$$Y_k = \sum_{i=1}^{k} L_i \sin \theta_i^0$$

$$R_k = (X_k^2 + Y_k^2)^{\frac{1}{2}}$$

where X_k is the horizontal displacement of the kth joint
Y_k is the vertical displacement of the kth joint
R_k is the resultant displacement of the kth joint

<u>Linear Velocity</u>

$$\dot{X}_k = \sum_{i=1}^{k} (-L_i \omega_i \sin \theta_i^0)$$

$$\dot{Y}_k = \sum_{i=1}^{k} (L_i \omega_i \cos \theta_i^0)$$

$$\dot{R}_k = (\dot{X}_k^2 + \dot{Y}_k^2)^{\frac{1}{2}}$$

where ω_i is the angular velocity of the ith segment.

<u>Linear Acceleration</u>

$$\ddot{X}_k = \sum_{i=1}^{k} - L_i(\alpha_i \sin \theta_i^0 + \omega_i^2 \cos \theta_i^0)$$

$$\ddot{Y}_k = \sum_{i=1}^{k} L_i(\alpha_i \cos \theta_i^0 - \omega_i^2 \sin \theta_i^0)$$

$$\ddot{R}_k = (\ddot{X}_k^2 + \ddot{Y}_k^2)^{\frac{1}{2}}$$

Following each time frame interval of the chosen movement the new stick diagram (Figure 3) is shown on the screen, together with a digital display of the motion parameter values (Table 2).

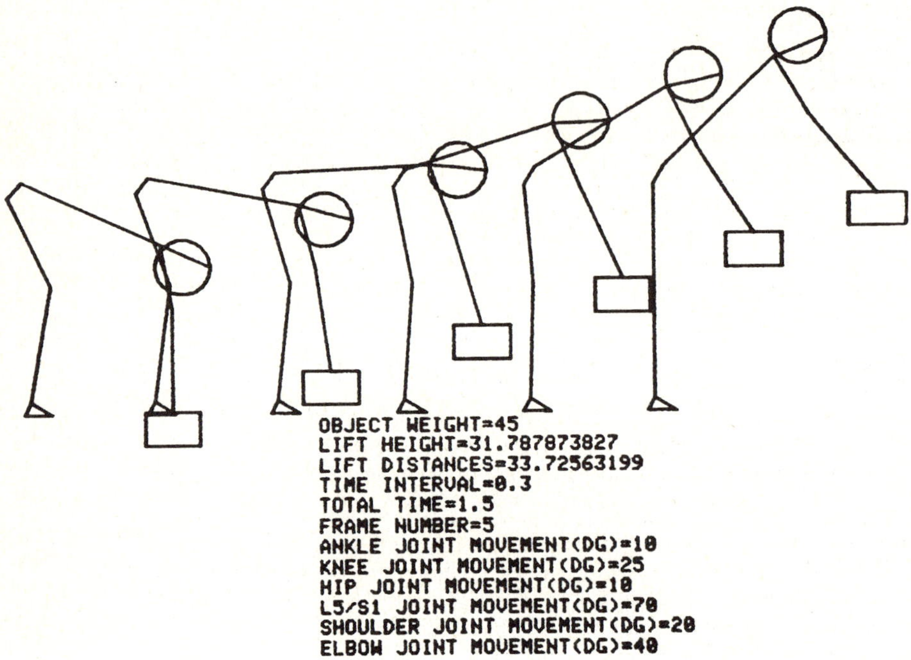

Figure 3. Intermediate Postures and Final Movement Ranges

Table 2. Example Frame Values for Kinematic and Kinetic Parameters

JOINT	1 Ankle	2 Knee	3 Hip	4 L5/S1	5 Shoulder	6 Elbow
Angular: Velocity ω_i Acceleration α_i	0.00066 0.25	0.0016 0.63	0.00065 0.25	0.0013 0.50	0.0013 0.50	0.0026 1.00
Linear: Velocity V_i Acceleration A_i	0.011 4.31	0.034 12.93	0.035 13.39	0.033 12.54	0.030 11.39	0.045 17.11
Joint Reaction Force	32.60	31.20	28.38	27.44	22.94	15.72
Joint Torque	370.23	-102.99	-388.36	-434.04	255.02	112.26
L5/S1: Compressive Force Shear Force					3941.95 4.76	
Extensor Muscle Force	0.134	0.106	0.104	0.386	6.36	-1.58

4.2 Kinetic Analyses

The purpose of kinetic analysis is to calculate the forces and moments acting around a joint as a movement takes place. It should be noted that the limb joints are considered to be frictionless and so only joint reaction forces are calculated. The spinal joints (L5/S1), however, are subject to both compressive and shear forces.

4.2.1 Segment Forces

Segment forces, at the segment centres of mass are calculated from the equations developed by Plagenhoef (1968) as follows:

$$F_{Y_i} = M_i\ddot{Y}_i + M_iL_iG_i\alpha_i \cos\theta_i{}^0 - M_iL_iG_i\omega_i{}^2 \sin\theta_i{}^0 + F_{Y_{i+1}}$$

$$F_{X_i} = M_i\ddot{X}_i - M_iL_iG_i\alpha_i \sin\theta_i{}^0 - M_iL_iG_i\omega_i{}^2 \cos\theta_i{}^0 + F_{X_{i+1}}$$

$$F_{R_i} = (F_{Y_i}{}^2 + F_{X_i}{}^2)$$

where F_{R_i} is the resultant force in the ith segment
F_{Y_i}, F_{X_i} are the vertical and horizontal components of joint reaction force
M_i is the mass of the ith segment
L_i is the length of the ith segment
G_i is the proportional distance of the centre of mass of the ith segment from the proximal end
α_i is the angular acceleration of the ith joint
ω_i is the angular velocity of the ith joint

θ_i^0 is the instantaneous angle of the segment to the horizontal.

It should be noted that these calculations do not take account of muscle tension.

4.2.2 Segment Moments

$$T_i = -W_iL_iG_i \cos\theta_i^0 + M_iK_i^2\alpha_i + M_iL_iG_i(\ddot{Y}_i\cos\theta_i^0 - \ddot{X}_i\sin\theta_i^0)$$
$$+ T_{i+1} + F_{Y_{i+1}} L_i\cos\theta_i^0 - F_{X_{i+1}} L_i\sin\theta_i^0$$

where T_i is the torque around the centre of gravity of the ith segment, and K_i is the radius of gyration of the ith segment.

4.2.3 Compressive and Shear Forces at L5/S1

Frievalds et al. (1984) discussed a number of assumptions that need to be met when calculating the forces acting at the L5/S1 joint:

1. the force acting on the L5/S1 disc is adjusted by 40° to account for the orientation of spine (Fisher, 1967).
2. in the vertical, erect posture the L5/S1 joint is assumed to be 19.5% of the distance between the hip joint and the shoulder (Lanier, 1939).
3. the erector spinae muscles exert a force parallel to the spine at a distance of 5 cm posterior to the centre of the discs (Bartelink, 1957; Perry, 1957).
4. the centre of gravity of the trunk is located approximately on a line between the shoulder and the hip throughout the lifting motion (Dempster, 1955).
5. forward pelvic tilt is determined by trunk flexion (Frievalds, 1984).
6. the contraction of the abdominal muscles during lifting causes an increase in intra abdominal pressure which exerts a force distributed over the area of the diaphram and pelvic floor. The area of the diaphram has been estimated by Morris et al. (1961) as 0.0465 m^2.

Calculation of the forces acting at the L5/S1 joint is as follows: (see Figure 4)

$$\text{Distance } HS^2 = L_3^2 + L_4^2 - 2L_3L_4\cos(180 - \theta_3 + \theta_4)$$

$$\sin\gamma = \frac{L_3\sin(180 - \theta_3 + \theta_4)}{HS}$$

$$\begin{aligned}\alpha &= 180 - \gamma - (180 - \theta_3 + \theta_4)\\ &= \theta_3 - \theta_4 - \gamma\\ \beta &= \theta_3 - \alpha\\ &= \theta_4 + \gamma\end{aligned}$$

where HS is the distance between the hip and shoulder.

The abdominal and L5/S1 forces are calculated from the equations described by Frievalds et al. (1984) and Chaffin (1969).

$$P_a = 0.01333(0.6516 - 0.005447(\theta_2 - \beta)(10.19\ T_3)^{1.8}$$
$$F_a = P_a(0.0465)$$
$$T_a = P_a(0.0465(0.067 + 0.082\ \sin(\theta_2 - \beta)))$$
$$F_e = (T_4 - T_a)/0.05$$
$$F_c = F_e - F_a + F_4\ \sin(\beta - 40)$$
$$F_s = F_4\ \sin(\beta - 40)$$

where

P_a is the intra abdominal pressure
F_a is the abdominal force on the diaphram
T_a is the torque exerted by the abdominal force
F_e is the tension in erector spinae
F_4 is the reactive force for the upper spinal segment
F_c is the compressive force at L5/S1
F_s is the shear force at L5/S1

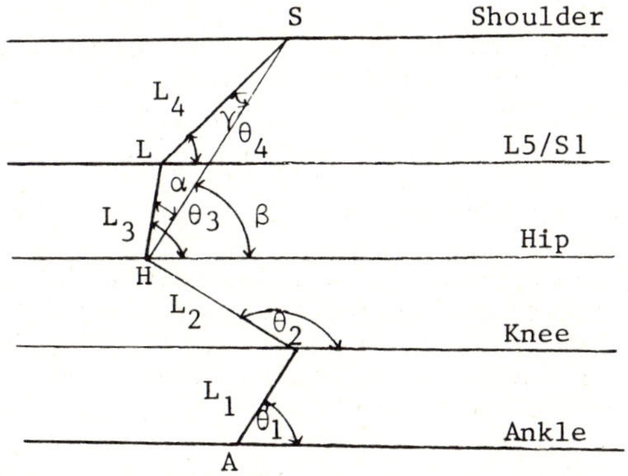

Figure 4. Segment Length and Joint Angle Summary

4.2.4 Muscle Forces

Muscle forces are calculated by static analysis following estimation of the moment arm (perpendicular distance) of the muscle with respect to the centre of rotation of each joint:

$$F_m = \frac{M}{D_m}$$

where F_m is the muscle force
D_m is the perpendicular distance of the muscle from the joint centre
M is the sum of the moments around the joint centre.

It should be noted that the estimations regarding muscle insertions and joint centres of rotation are limited by the availability of accurate data and the inherent difficulty of measurement.

5. LIFTING TASK GUIDELINES

The NIOSH guidelines (NIOSH, 1981, pp. 126, 127) indicated action and maximum permissible limits following extensive epidemiological, biomechanical physiological and psychophysical analyses:

$$AL = 90(6/H)(2 - 0.01|V - 30|)(0.7 + 3/D)(1 - F/F_{max})$$
$$MPL = 3(AL)$$

where AL = Action Limit
MPL = Maximum permissible limit
H = Horizontal distance of load from foot support
V = Vertical location of load at origin of lift
F = Average frequency of lift (lifts/minute)
F_{max} = Maximum frequency that can be sustained based on task duration and average vertical location.

The investigator is invited to input a lifting frequency and work duration. The program then calculates and displays the AL and MPL and the investigator is invited to modify the task, segment or load variables (Figure 5).

```
Please input lift frequency              0.2
     (lifts/minute)
Please input task duration               2.0
     (hours)
                  AL = 23.5
                 MPL = 70.5
          Load input = 50

    Task requires administrative control.
```

Figure 5. Lifting Task Guidelines (Screen Output)

Discussion

This model provides a convenient method of studying complex stress-strain relationships during saggital plane lifting with reference to the basic principles of mechanics. The interactive facility provides a medium for the assessment of tasks, workplaces and human characteristics. It is therefore a practical tool as well as a teaching tool for engineers and health scientists.

There are however a number of assumptions that must be made when using the model which give rise to the following limitations. First, the model is two dimensional and does not therefore account for the torsional movements that are of critical importance in the comprehensive analysis of lifting stress. Second, assumptions are made regarding segment and joint angle characteristics that may limit the validity of the model. However, the interactive characteristic of the program allows the investigator to test the sensitivity of the model to widely ranging segment, postural and task input values. Thus the investigator can account for the model assumptions.

An extension to this model will include the incorporation of alternative default sets of anthropometric and lifting pathway data. Thus, for example, it would be possible to select a 20th percentile man lifting an object to a high shelf at a chosen rate. Interaction with the model would be used to identify the particular aspects of biomechanical stress for this type of individual and task without extensive interactive data modification. Other extensions would be to include ground support friction, load size and shape and workplace variables.

6. REFERENCES

[1] Anil, M. and Ayoub, M. M., "Modeling of Strength and Lifting Capacity," Human Factors 22 (1980) 285-290.

[2] Ayoub, M. S. and El-Bassoussi, M. S. "Dynamic Biomechanical model for saggital plane lifting activities", in: Drury, C. G. (ed.), Safety in Manual Material Handling, 78-185 (NIOSH, Cincinnati, Ohio, 1978).

[3] Bartelink, D. L., "The role of abdominal pressure in relieving the pressuire on teh lumbar intervertebral discs", J. of Bone Joint Surgery 39B (1957) 718.

[4] Chaffin, D. B., "A Computerized Biomechanical Model--Development and use in studying gross body actions," J. Biomechanics 2 (1969) 429-441.

[5] Chaffin, D. B., "A Biomechanical Model for Analysis of Symmetric Sagittal Plane Lifting," AIIE Trans. 2 (1970) 16-27.

[6] Chaffin, D. B., "A Computer-Assisted Manual Work-Design Model," AIIE Trans. 2 (1970) 348-354.

[7] Chaffin, D. B. and Martin, J. B., "Biomechanical Computerized Simulation of Human Strength in Saggital-Plane Activities", AIIE Trans. 4 (1972) 19-28.

[8] Dapena, J., "Simulation of Modified Human Airborne Movement", J. Biomechanics 14 (1981) 81-89.

[9] Dempster, W. T. and Gaughran, G.R.L., "Properties of Body Segments Based on Size and Weight", American J. of Anatomy (1967) 120.

[10] Drillis, R. J., Contini, R. and Bluestein, M., "Determination of Body Segment Parameters, Human Factors 5 (1963) 493-504.

[11] Fisher, B. O., Jr., "Analysis of Spinal Stress During Lifting a Biomechanical Model", unpublished M.S. Thesis, Univ. of Michigan (1967).

[12] Frievalds, A. et al., "A Dynamic Biomechanical Evaluation of Lifting Maximum Acceptable Loads," J. Biomechanics 17 (1984) 251-262.

[13] Garg, A. and Chaffin, D. B., "A Biomechanical Computerized Simulation of Human Strength," AIIE Trans. 7 (1975) 1-10.

[14] Garg, A. and Herrin, G. D., "Stoop or Squat: A Biomechanical and Metabolic Evaluation", AIIE Trans. 11 (1979) 293-302.

[15] Gruver, W. A., Ayoub, M. A. and Muth, M. B., "A Model for Optimal Evaluation of Manual Lifting Task," J. of Safety Research 11 (1979) 61-71.

[16] Hay, J. G., "Moment of Inertia of the Human Body," Kinesiology IV (1973) 43-52.

[17] Huston, R. L., "On the Dynamics of a Human Body Model," J. Biomechanics 4 (1971) 369-378.

[18] Kingsley, E. C., Schofield, N. A. and Case, K., "A Computer Aid for Man Machine Modelling," Computer Graphics (1981) 163-169.

[19] Kroemer, K. H. E., "Human Strength: Terminology, Measurement, and Interpretation of Data," Human Factors 12 (1970) 297-313.

[20] Lanier, R. R., "Presacral Vertebrae of White and Negro Males, Am. J. Phys. Anthropol. 25 (1939) 343-420.

[21] McDaniel, J. W., "Computerized Biomechanical Man-Model," Human Factors 20 (1978) 384-389.

[22] Morris, J. M., Lucas, D. B. and Bressler, B., "Role of the trunk in stability of the spine", J. Bone Jt. Surg. 43A (1961) 327-351.

[23] Neils, D. et al. "Humanscale 1/2/3" (MIT, Massachusetts, 1981).

[24] Onyshko, S., Winter, D. A. and Well, R. P., "SIMU - An Interactive Computer Graphics Simulation of Human Gait," Biomechanics: Principles and Application (1982) 203-247.

[25] Passerello, C. E. and Huston, R. L., "On the Dynamics of a Human Body Model", J. of Biomechanics 4 (1971) 369-378.

[26] Passerello, C. E. and Huston, R. L., "The Mechanics of Human Body Motion," Human Dynamics Clarendon (1982) 203-247.

[27] Pearson, J. R., McGinley, D. R., and Butzel, "Dynamic Analysis of the Upper Extremity for Planar Motion," Human Factors 5 (1963) 59-70.

[28] Perey, D., "Fracture of the vertebral end plate in the lumbar spine, an experimental biomechanical investigation", Acta Orthop. Scandi Suppl. 25 (1957).

[29] Plagenhoef, F. J., Jr., "Methods for Obtaining Kinematic Data to Analyze Human Motion," Res. Q. 37 (1966) 103-112.

[30] Plagenhoef, F. J., Jr., "Computer Program for Obtaining Kinetic Data to Analyze Human Motion," Res. Q. 37 (1968) 1-15.

[31] Ramey, M. R. and Yang, A. T., "A simulation procedure for Human Motion Studies", J. Biomechanics 14 (1981) 203-213.

[32] Slote, L. and Stone, G., "Biomechanical Power Generated by Forearm Flexion," Human Factors 5 (1963) 443-452.

[33] Winter, D. A., "Biomechanics of Human Movement," (John Wiley & Sons, New York, 1979).

Trends in Ergonomics/Human Factors II
R.E. Eberts and C.G. Eberts (Editors)
© Elsevier Science Publishers B.V. (North-Holland), 1985

AN ANALYSIS OF ROTARY AND PULL STRENGTHS OF THE UPPER EXTREMITY

Sheik N. Imrhan

Dept. of Industrial Engineering, Louisiana Tech University, U.S.A.

Mohamed M. Ayoub

Dept. of Industrial Engineering, Texas Tech University, U.S.A.

Isokinetic and static rotary and linear pull strength of the upper extremity were compared from data on 10 males between the ages of 18 and 24 years. Maximum voluntary contractions of linear pull, elbow flexion and shoulder extension were tested in the saggital plane under various levels of 3 variables -- velocity of contraction, elbow angle and starting position of the arm. All 3 variables proved to be significant determinants of strength, with important similarities between pull and shoulder extension with respect to the starting position of arm.

INTRODUCTION

There are many working environments where an individual is expected to exert static or dynamic forces (or torques) of the upper extremity. Studies in upper extremity strength have been related predominantly to static contractions even though many tasks require dynamic contractions. In such studies the effects of important variables such as velocity of contraction and the position of the moving arm (or segment) remain unknown. The present study overcomes these deficiencies by incorporating velocity of contraction and starting position of the arm as variables under investigation. Many studies have also failed to distinguish between two types of dynamic contraction -- those with constant external load (often misnamed 'isotonic') and those performed at a constant velocity ('isokinetic'). Hislop and Perrine (1967) and Kroemer (1970) emphasize this point. There have been few studies on isokinetic contractions of the upper extremity, and none relating rotary contractions about the joints to linear pull. The present study focuses on this issue and deals specifically with the rotary contractions comprising a linear pull.

Past studies have established that the magnitude of static force or torque achieved by a pull depends on a multitude of variables, including body posture and limb configuration. Optimal hand locations for pulling tasks have been determined for the standing individual (Martin and Chaffin, 1972; and Ayoub and McDaniel, 1974). For a given body posture skeletal position and velocity of contraction have been found to have significant effects on maximum forces attained by the arm. For isometric contractions the 70-120 deg range of the included angle at the elbow have generated the greatest elbow forces (Clarke et alia, 1950; Elkins and Wakim, 1951; Provins and Salter, 1955; Liberson et alia, 1962; Doss and Karpovich, 1965; Singh and

Karpovich, 1965; Currier, 1972; and Petrofsky and Phillips, 1980). For isokinetic contractions maximal torques (and forces) about the elbow and shoulder have been found to decrease with increasing velocity, and were attained at a later stage in the contraction period for the higher velocities; that is, at higher velocities of contraction a greater degree of rotation was required to achieve maximal torque (Ayoub et alia, 1981 & 1982).

PURPOSE OF THE STUDY

The purpose of the present study was to analyze and compare rotary and linear pull strengths of the upper extremity. Both isokinetic and static contractions were studied, and rotary contractions were limited to those comprising linear pull, that is elbow flexion and shoulder extension.

METHODS

Theoretical considerations

When an individual pulls maximally towards his shoulder his elbow flexes and his shoulder extends simultaneously. Torque output gradually increases then decreases after reaching a maximum. For an isokinetic pull the angular velocities about the elbow and shoulder do not remain constant. At the point of maximum pull torque, for a given starting position of the arm and given velocity, the arm attains a configuration defined by the included angle at the elbow (θ) and angle at the shoulder (β). Figure 1 shows a typical configuration, where α represents angular rotation of the elbow from start to point of maximum torque ($\theta = 180 - \alpha$ deg) and β represents angular rotation of the shoulder. Also the angular velocities about the elbow ($\dot{\theta}$) and shoulder ($\dot{\beta}$) attain specific values depending on the velocity of pull. By measuring the distance of pull, and the lower and upper arm link lengths θ and β can easily be computed from the trigonometric relationships of the arm segments. From a knowledge of the velocity of pull (V) $\dot{\theta}$ and $\dot{\beta}$ can be obtained by applying differential calculus. The details of these methods can be found in Imrhan (1983).

Subjects

Ten male right-handed students of Texas Tech University, between the ages of 18 and 24 years were selected according a weight-stature plan (Roth et alia, 1976). All were in excellent health, free of musculo-skeletal injuries to the arm-shoulder-back region, and were not involved in any intensive muscle building/strengthening exercises. Each subject was paid $4.00 per hour for his effort. The experiment lasted 4.5 weeks.

Apparatus

The apparatus is described in detail in Imrhan(1983). It includes

1. a Cybex II isokinetic dynamometer for measuring torque (in ft-lb)
2. Cybex II accessories such as a speed selector, a torque and position angle recorder, metal arms and handgrips; an Integrated System for Automatic Acquisition and Control (ISAAC), and an Apple II computer
3. a stable adjustable chair
4. a plywood wheel of weight 10 kg and effective diameter 96.52 cm (38 in) fixed onto the dynamometer shaft. A steel cord attached at one point to the perimeter of the wheel allowed a linear pull to be converted to a rotation. The other end of the cord passed around a pulley on a steel

supporting frame and was attached to a rotatable handgrip
5. a large metal protracter for measuring angles at joints of the body
6. an Anthropometric kit manufactured by Pfister Import-Export Inc.

Procedure and experimental designs

Each subject was first asked to sign a consent form and to complete a personal data form. Anthropometric measurements, except for arm link lengths, were taken according to the methods described in 'NASA Reference Publication 1024: 1978.' Upper and lower arm link lengths were taken as the distance from the mid-hand/handgrip interface to the shoulder and elbow axes of rotation respectively. Data was then collected for linear pull, elbow flexion and shoulder extension, in that order. All contractions were maximal and voluntary and were maintained in the saggital plane only. For each measurement two repetitions were taken, at random, with torque values differing by less than 10%. Only the larger one was used for subsequent data analysis.

For linear pull, data was gathered at each combination of 5 starting positions of the outstretched arm and 4 velocities of pull. The positions were 10, 50, 90, 130 and 170 deg of the humerus from the long axis of the trunk of the seated subject (P10, P50, P90, P130 and P170). The velocities were 0, 18, 36 and 54 deg/s dynamometer velocity (0, 3, 6 and 9 rpm). A 5 x 4 factorial design was used to achieve this. It should be noted that 1 deg/s dynamometer velocity was equivalent to a linear hand velocity of 0.8467 cm/s (or 1 rpm was equivalent to 2 in/s). Elbow flexion data was gathered at 6 velocities -- 0, 24, 48, 72, 96 and 120 deg/s. Shoulder extension data was gathered at each combination of 3 starting positions of the humerus (P50, P90 and P130), 3 elbow angles (50, 85 and 120 deg) and 5 velocities (0, 18,36, 54 &72 deg/s). A 3 x 3 x 5 factorial design was used.

Because of the extensive time necessary to change starting position and elbow angle these two variables were not truly randomized. Velocities were randomized. Adequate rest times were given to the subjects between successive contractions to eliminate any fatigue effects.

Training

Before the final data gathering subjects were trained in exerting the contractions under investigation. This was to familiarize them with the tasks, develop the necessary skills and minimize or eliminate learning effects. When subjects were able to perform 3 successive contractions of the same type under specified conditions, with maximal torques differing by less than 10% they were considered ready for final testing.

Body posture

Subjects were seated strapped around the upper chest and pelvic regions to stabilize the body and minimize inputs from muscles extraneous to those under investigation. The seat pan was set at 5 deg to the horizontal and the back rest at 95 deg to the seat pan. Subjects' feet rested on the floor or foot rest with the knee angle at approximately 90 deg. They were instructed not to push with their legs to gain additional arm force. For rotary contractions the subject's arm was coupled to the dynamometer's arm by means of a handgrip on the latter. This ensured stabilization of the arm segments and hence rotation about the tested joints only. For linear pull subjects gripped the rotatable handgrip firmly, with the arm fully extended and pulled towards the mid-shoulder point as far as possible. The hand was maintained in the mid-supinated/pronated orientation for all contractions.

Data Analysis methods

SAS (Statistical Analysis System) was used to perform analyses of variance and multiple comparison of means by Duncan's test (wherever ANOVA showed a significant treatment effect at the 0.05 level of significance).

Since the moment arms of the three types of contractions were not equal all torque values were converted to corresponding forces for valid strength comparisons, by dividing by the corresponding moment arm. The resulting forces (in lb) were then converted into units of Newton (N) by multiplying by 4.448222.

RESULTS AND DISCUSSION

Selected anthropometric measurements are summarized in table 1 and strength measurements in tables 2, 3 and 4. The values given in tables 2, 3 and 4 are the means of n observations.

Though the magnitude of the three types of strength differed significantly there were comparable trends in the variation across the various levels of the independent variables, especially for shoulder extension and linear pull. Static pull force (and torque) was as great as the sum of elbow and shoulder forces (and torques). Static shoulder extension and pull strengths were greater than the corresponding isokinetic strengths. On the average the greatest isokinetic pull force was only 76.7% that of its static counterpart while the corresponding figure for pull force was 88.6%. Both pull and shoulder extension isokinetic forces (and torques) declined with increasing velocity but the former did so more rapidly. Duncan's multiple range test indicated that the forces of both pull and shoulder extension were significantly different at all levels of velocity tested. For elbow flexion, on the other hand, static force (and torque) was not significantly greater than the stronger, low velocity isokinetic forces (and torques). This confirmed the findings of Ayoub et alia (1981 & 1982). The reason was probably that the fully extended elbow, on which static contractions were measured, was at a lesser mechanical advantage than the flexed elbow on which isokinetic maximal strengths were obtained. Elbow flexion strength increased slightly from 24 deg/s to 48 deg/s then declined slowly with increasing velocity. This decline was not as sharp as that observed by Ayoub et alia but the difference was most likely due to the difference in postures adopted by the subjects in these two studies. In the study of Ayoub et alia subjects were tested while lying supine. In the seated posture of the present study subjects were better able to utilize the muscles of the upper body to aid in contraction and this effect seemed to be greater at the higher velocities. The result was a smaller decrement in strength at the various velocities tested.

Starting position of the arm had a highly significant effect ($p<0.0001$) on shoulder extension and linear pull strengths. Three distinct regions of strength were observed -- at the horizontal (P90), above and below it. For the isokinetic contractions the upper region generated greater torques (and forces) than the lower region but the mid region was highly superior to both. Static shoulder strength followed the same trend but static pull strength showed a slight difference. For static pull Duncan's test indicated that while the mid region was vastly superior to the others for torque generation the upper region (P130-P170) was not superior to the lower region (P10-P50), as illustrated by figure 2. This implies that as we move away from the horizontal on either side static strength drops very rapidly and almost symmetrically within the first 40 deg; thereafter it drops very slightly.

Shoulder strengthdepended also on the configuration of the arm ($p < 0.0001$). Torques increased (and forces decreased) non-linearly as the angle at the elbow increased, that is, with the arm in the more extended configuration. This was due to the larger moment arm of the torque at larger elbow angles.

For all contractions the time taken to reach maximum torque increased as the velocity of contraction increased. This translates into increased angular rotation as velocity increases. Similar findings were observed by Ayoub et alia (1981 & 1982) for a variety of rotary contractions of the upper extremity. One of the most important findings in this study relates to the configuration of the arm at the point of maximum pull torque. At this point, for any given velocity of pull, the angular configuration of the arm was very similar for all starting positions despite the fact that the magnitude of the torques differed significantly. The θ and also the β values were not significantly different at the 5 starting positions. Figure 3 shows these angular values plotted against starting position. Also of significance is the fact that the angular velocities, $\dot{\theta}$ and $\dot{\beta}$, about the elbow and shoulder respectively, are almost identical at the point of maximum torque, for all starting positions at a given velocity (figure 4).

CONCLUSION

In conclusion we state thatin equipment design and job design involving dynamic and static pulls or rotations about the joints of the upper extremity, we must consider not only the velocity of contraction but also the starting position and the configuration of the arm. For the seated operator velocity has only a slight effect on elbow flexion strength but a large negative effect on shoulder extension and linear pull strengths. This implies that, for tasks involving dynamic strength of the upper extremity, the operator must be given more time to reach his maximum torque (or force) as velocity of contraction increases. Provisions must be made for a greater range of motion of the arm for the higher velocities of contraction. For tasks requiring maximal strength handgrips should be located in such a position as to take advantage of the superiority of the horizontal starting position of the humerus, and positions above the horizontal should be preferred to positions below. Finally the results of this study allow designers to estimate the amount of arm space required for pulling tasks for individuals of all sizes. This is especially important in confined working spaces.

REFERENCES

Ayoub, M.M., Gidcumb, C.F., Reeder, M.J., Beshir, M.Y., Hafeez, H.A., Aghazadeh, F.A. and Bethea, N.S., Development of an Atlas of Strength and Establishment of an Appropriate Model Structure -- Males, Final Report, Texas Tech Institute of Biotechnology (1981).

Ayoub, M.M., Gidcumb, C.F., Reeder, M.J., Beshir, M.Y., Hafeez, H.A., Aghazadeh, F.A. and Bethea, N.S. Development of an Atlas of Strength and

Establishment of an Appropriate Model Structure -- Females, Final Report, Texas Tech Institute of Biotechnology (1982).

Clarke, H.H., Elkins, E.C., Martin, G.M. and Wakim, K.G., The relationship between body position and the application of muscle power to the movements of joints, Archives of Physical Medicine, Feb. 1950 81-89.

Currier, D.P., Maximal isometric tension of the elbow extensors at varied positions: Part 1. Assessment by cable tensiometer, Physical Therapy, 1972 52(10) 1043-1049.

Doss, W.F. and Karpovich, P.V., A comparison of concentric, eccentric and isometric strength of elbow flexors, Jrnl. of App. Physiol., 1965 20 351-53.

Elkins, E.C. and Wakim, W.G., Objective recording of the strength of normal muscles, Archives of Physical Medicine, October 1951 33 639-647.

Imrhan, S.N., Modelling Isokinetic Strength of the Upper Extremity, Ph.D. Dissertation, Dept. of Industrial Engineering, Texas Tech University (1983).

Kroemer, K.H.B., Human strength: terminology, measurement, and interpretation of data, Human Factors, 1970 12(3) 293-313.

Liberson, W.T., Dodney, M. and Asa, M.M., Brief repeated isometric maximal exercises, American Jrnl. of Physical Medicine, 1962 41 3-14.

Martin, J.B. and Chaffin, D.B., Biomechanical computerized simulation of human strength in saggital plane activities (AIIE Trans., Mar. 1972 293-313).

NASA reference publication, 1024: anthropometric source book: vol 1. Anthropometry for designers (July 1978).

Petrofsky, J.S. and Phillips, C.A., The effects of elbow angle on isometric strength and endurance of the elbow flexors in men and women, Jrnl. of Human Ergol., 1980 9 125-131

Provins, K. and Salter, N., Maximum torque about the elbow joint, Jrnl. of App. Physiol., January 1955 7 393-398.

Roth., J.T., Ayoub, M.M., Dixon, J.J. and Halcomb, C.J., Methodology and reach profiles for restrained and unrestrained seated operator, Improved Seat and Workplace Design (AIRTASK A 62736 N/W 55.525.403, Pacific Missile Center, California, U.S.A. 1976).

Singh, M. and Karpovich, P.V., Isotonic and isometric forces of the forearm flexors and extensors, Jrnl. of App. Physiol., 1966 21 1435-1437.

Williams, M., Tombelin, J.A. and Robertson, K.J., Muscle force curves of school children, Physical Therapy, 1965 45 539-549.

Hislop, A.J. and Perrine, J.J., The isokinetic concept of exercise, Physical Therapy, 1967 47(2) 114-117.

TABLE 1

SUMMARY OF SELECTED ANTHROPOMETRIC MEASUREMENTS

Variable	Mean	Standard Deviation	Range
Age (yr)	21.10	1.97	18.60-23.60
Body weight (kg)	77.38	11.57	58.86-95.91
Stature (cm)	176.70	6.96	166.40-189.90
Chest circumference (cm)	99.97	6.87	87.40-111.10
Lower arm link length (cm)	33.32	1.14	31.60-35.80
Upper arm link length	29.95	1.15	28.20-31.90

TABLE 2

MEANS OF ELBOW FLEXION FORCES AND ANGULAR ROTATIONS n = 10

	Velocity (deg/s)					
	0	24	48	72	96	120
Force (N)	148.3	157.4	158.4	153.8	141.8	133.4
(deg)	-	46.6	62.6	70.3	79.1	83.0
(deg)	-	133.4	117.4	109.7	100.9	97.0

TABLE 3

MEANS OF SHOULDER EXTENSION FORCES (N) n = 30

Position	Velocity (deg/s)				
	0	18	36	54	72
Shoulder					
P50	280.7	235.7	213.7	198.3	186.0
P90	289.6	260.2	247.8	234.3	219.8
P130	269.8	248.5	231.7	220.8	213.1
Elbow (deg)					
50	351.6	318.5	296.2	279.0	264.8
85	269.1	237.0	222.4	208.4	197.3
120	219.4	189.0	174.6	166.3	156.8

TABLE 4

MEANS OF LINEAR PULL FORCES (N) n = 10

Position	Velocity (deg/s)			
	0	18	36	54
P10	414.4	298.2	237.0	173.3
P50	434.7	309.0	241.5	180.9
P90	490.7	382.1	316.2	257.2
P130	418.5	325.0	262.9	202.4
P170	408.2	345.6	283.6	213.9

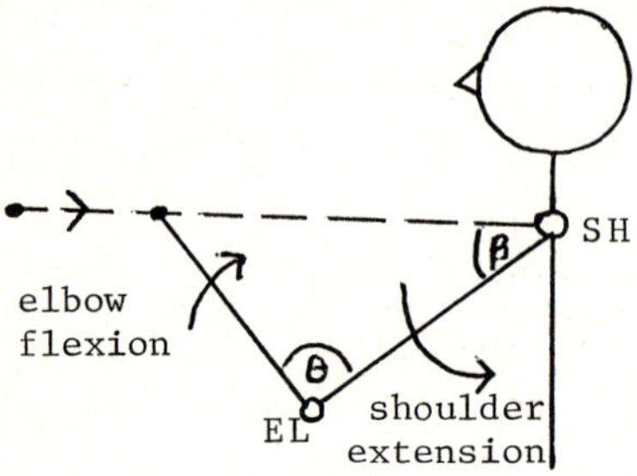

Figure 1

COMPONENTS OF LINEAR PULL

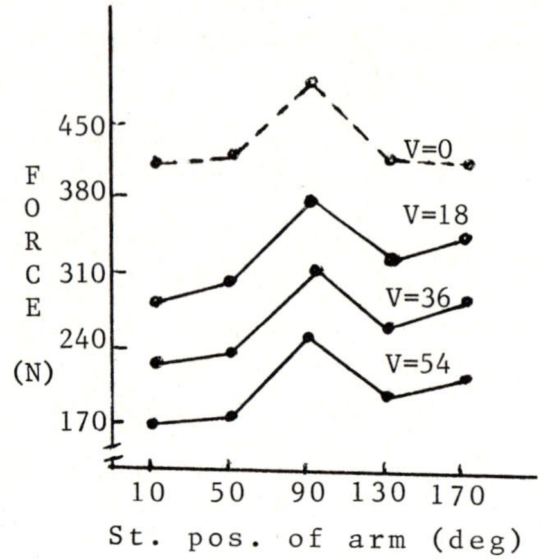

Figure 2

PULL FORCE VS ST. POS.
- BY VELOCITY

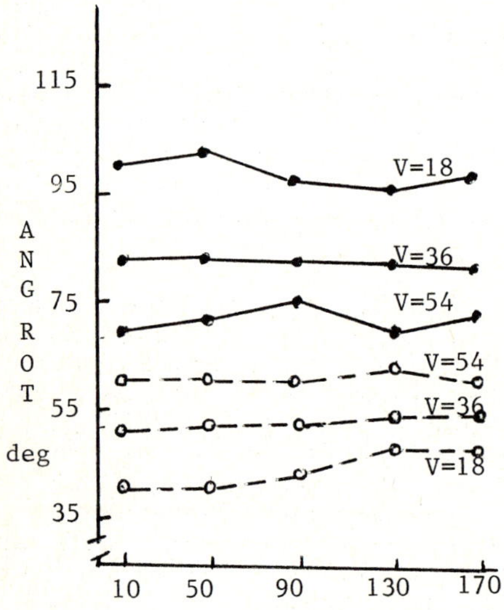

Figure 3

ANG. ROTATION VS ST. POSITION
- BY VELOCITY
(LINEAR PULL)

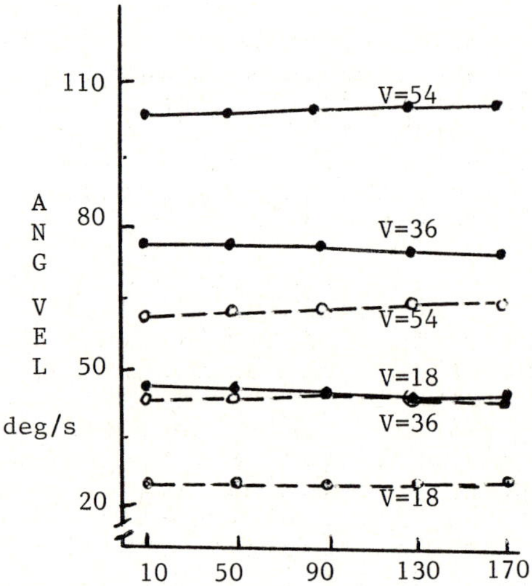

Figure 4

ANG. VELOCITY VS ST. POSITION
- BY VELOCITY
(LINEAR PULL)

Trends in Ergonomics/Human Factors II
R.E. Eberts and C.G. Eberts (Editors)

EFFECT OF GENDER ON PHYSICAL WORK CAPACITY AND STRENGTH

F. Aghazadeh
Industrial Engineering Department
Louisiana State University
Baton Rouge, Louisiana
and
Shriram R. Dharwadkar
School of Engineering Technology
South Carolina State College
Orangeburg, South Carolina

Determination of physical capacity of females becomes more important with the increase of females in industry and the armed forces. Previous studies indicate that an average woman's physical strength is a certain percentage of that of an average males's. This pilot study was conducted to investigate the physical performance of males and females in order to find out if the difference in physical performance is due to gender or to the anthropometric characteristics.

INTRODUCTION

It is generally accepted that the average physical work capacity and muscle strength of a woman are considerably less than her male counterpart. The relationship between the muscle strengths of men and women is a research topic that has been under investigation for several decades. The Equal Employment Opportunities Act of the United States government has opened new avenues for women to work and along with it, has rekindled renewed interest in research, concerning muscular strength of men and women.

There is a wide range of distribution of performance limits defined by the maximum oxygen intake as one of the most important criteria for work capacity. There is quite general agreement that the maximum oxygen intake (VO2 max.) is the best reference level for aerobic work capacity (Simonson, 1971). An individual's maximal aerobic power plays a decisive role in his work capacity. Some research has been done in the past to compare the maximal aerobic powers between men and women. It has been reported (Astrand and Rodahl, 1977) that before puberty there is no significant difference in maximal aerobic powers between boys and girls. Thereafter, the women's power is 70-75 percent of that of men. In both sexes there is a peak at 18-20 years of age followed by a gradual decline in the maximal oxygen uptake.

Two important variables which have an effect on human strength are age and sex. Various comparative techniques have been used in the past to report the effect of these two variables on strength. Overall, strength appears to increase with age up to approximately the age of 30-39. Then a leveling of strength occurs and finally the strength starts to decrease beginning approximately from the age of 50 (Ayoub et al., 1982). An average female has a lower muscular strength than a male, although some females may be stronger than males. According to Konz (1979), a female's strength is about sixty to sixty-five percent of the strength of a male of the same age. Konz also reports that a female of the same weight and age as a male will have about 80 percent of the male's strength. He attributes the 20 percent difference in strengths to physiological and cultural differences between men and women. Hosler and Morrow (1982) have indicated that although there is a significant difference between men and women in terms of arm and leg strengths, the impact of gender difference is rather small (about 2%), after allowing for body size and composition. In their study, relative strength was estimated by regression techniques after accounting for differences in body size and composition.

Many comparative strength studies usually involve absolute strengths. However, the difference in muscular strengths between sexes depends upon the part of the body being studied. Laubach (1976) has reviewed nine different studies reporting comparable static and dynamic muscle strength measurements between men and women. His observations are summarized in Table 1.

TABLE 1

FEMALE STRENGTH AS A PERCENTAGE OF MALE STRENGTH FOR DIFFERENT BODY PARTS

Part of the Body	Women's Strength as a Percentage of Men's	
	Range (%)	Mean (%)
(a) Upper Extremity	35 - 79	55.8
(b) Lower Extremity	57 - 86	71.9
(c) Trunk	37 - 70	63.8
(d) Dynamic Strengths	59 - 84	68.6

A literature review by Ayoub, Grasley et al. (1978) on sex differences involved anthropometry, biomechanics and physiological responses. They point out that the literature thus far available is unable to support any ruling based on sex alone and there is a need for further research in this area. The purpose of this study was to determine the effect of sex on static and dynamic strengths when anthropometrically similar pairs of males and females in approximately the same age group are considered.

SUBJECTS

The subjects were selected from a junior level student population. Participation by the subjects in this research was voluntary and no one was paid for the purpose. To begin with, anthropometric measurements were obtained from twenty-four subjects. From this population group, a sample of twelve subjects were selected who could meet the requirements of this study. The main criterion for the selection of the samples was that the subjects selected could be approximately paired with an anthropometrically similar subject of the other sex. The mean age of the male samples was 23.8 years and that of female samples was 20.6. A comparison of the average anthropometric dimensions of male and female subjects involved in this study is presented in Table 2.

TABLE 2

COMPARISON OF MALE AND FEMALE ANTHROPOMETRIC DATA

	Males		Females		
	$\overline{X}$	SD	$\overline{X}$	SD	M/F%
Weight (lbs)	145.3	12.3	144.0	14.3	99.1
Height (cms)	172.	4.6	171.0	1.9	98.9
Acromial Height (cms)	144.7	4.1	142.9	2.2	98.7
Iliac Crest Height (cms)	102.3	2.8	106.1	11.1	103.7
Knuckle Height (cms)	77.9	3.2	71.5	1.2	91.8
Knee Height (cms)	48.6	2.8	48.1	3.7	98.9
Forearm grip distance(cms)	37.0	5.8	33.0	1.4	89.2
Chest width (cms)	31.9	1.4	29.9	3.4	93.2
Circumferences (cms)					
Chest	91.6	4.6	89.3	4.5	97.5
Abdominal	79.5	3.2	70.2	6.7	88.3
Forearm	26.5	1.1	26.1	3.1	98.5
Biceps	25.3	8.1	28.7	3.1	113.4
Thigh	53.2	3.6	58.0	3.5	109.0
Calf	32.2	13.2	37.9	2.4	117.7

METHODS AND PROCEDURES

There is a linear relationship between pulse rate and oxygen uptake during submaximal work. The slope of the line changes with the state of physical training or physical fitness; a fit person is able to transport the same amount of oxygen at a lower heart rate than an unfit person.

On the basis of the oxygen uptake-heart rate relationship, it is possible to predict an individual's maximal oxygen uptake by the heart rate response to two submaximal loads, assuming a maximal heart rate of about 190 to 200 beats/min. Since the maximal heart rate declines with age after about age 20, it is necessary to make corrections for age. Hence a better approximation would be to consider maximum heart rate, as equal to 220 minus the age of the subject. Maximal oxygen uptake can be estimated on the basis of heart rate or pulse response to submaximal workload, using the monogram developed by Astrand and Astrand (1977). However, more accurate results can be obtained using the following relations (Bernard et al., 1979), where oxygen consumption may be calculated from the ventilation volume (V_E).

$$VO_2 = (0.69-0.002*Age)(V_E)^{0.44}-1.47 \text{ (for males)}$$

and

$$VO_2 = (1.64-0.002*Age)(V_E)^{0.25}-2.34 \text{ (for females)}$$

The physical work capacity of the subjects were predicted from submaximal techniques utilizing the above formulas. The apparatus included a bicycle ergometer and a respiration gas meter (Max-Planck-Institute). The subjects were instruted to refrain from energetic physical activities and smoking two hours preceding the test. In addition, the test was not performed earlier than about 1 hour after a light meal or after two to three hours if a heavier meal had been taken. Pulse rate and pulmonary ventilation volumes were recorded at the end of every minute. The load was increased in steps at the end of every four minutes. A total of three work levels were used. The results are presented in Table 3.

TABLE 3

COMPARISON OF PHYSICAL WORK CAPACITY OF MALES AND FEMALES (l/min)

Subject Pair	PWC Males	PWC Female	Difference PWCM-PWCF	PWCF/PWCM %
1	1.9	1.5	0.45	79
2	1.92	1.82	0.10	95
3	2.15	1.82	0.33	85
4	2.30	1.97	0.33	86
5	2.60	1.97	0.63	76
6	1.96	1.85	0.11	94

The procedure for obtaining anthropometric data was similar to the one used by Aghazadeh (1982). All strength measurements were made using a Mini-Gym Model 180 (Strength/Fitness Systems, Indep. Mo.). Both static and dynamic strengths could be measured by controlling a slow-fast knob on the mini-gym. Observations were recorded for dynamic strength from floor to shoulder height, and for static strengths of stooped back and composite. The testing procedures and subject's posture in each of the above observations were similar to the ones outlined by Aghazadeh (1982). Each test component consisted of a minimum of three maximum efforts with a duration of holding at maximum for about four seconds for static tests. In each case an average of three consistent observations was recorded. The subject's posture was closely watched every time to make sure that they were the same for each subject. The experimental results are presented in Table 4.

TABLE 4

STRENGTH MEASUREMENT DATA

Static Strengths (lbs.)						Dynamic Strength (lbs.)		
Stooped Back			Composite			Floor to Shoulder		
M	F	%	M	F	%	M	F	%
178	197	54	220	210	95	122	133	110
307	120	39	425	157	37	257	175	68
253	120	47	360	157	44	175	175	100
178	168	94	253	223	92	248	125	50
291	168	58	350	227	65	220	156	71
178	120	67	220	125	57	122	148	121

RESULTS AND CONCLUSIONS

It has been reported in the literature that the physical work capacities of women is generally seventy-five percent of their male counterpart (Astrand and Rodahl, 1977). However, when anthropometrically similar pairs are considered this number seems to be higher. From our results it was observed that female work capacities ranged from 76 to 95 percent of the work capacities of anthropometrically similar men, with a mean of 86 percent.

Konz (1979) has reported that a female of the same age as a man will have 80 percent of the man's strength. He attributes the 20 percent difference to physiological and cultural differences between men and women. However, Hosler and Morrow (1982) have concluded that the impact of gender difference is small (about 2%) after allowing for body size and composition. Their study was based however, on results of regression test, and not on data obtained by using actually anthropometrically similar paired sample. In this study we have considered six pairs of anthropometrically similar samples. Our observations were significantly different for dynamic strengths and static strengths. A comparison of static strengths indicated that the effect of gender difference may be substantial. Females strength ranged from 37 to 95 percent of that of males with an averge of 60 to 65 percent (Table 4). This number agrees with the findings of Konz (1979). In the case of dynamic strength however, the effect of gender difference was observed to be much lower than in static strength tests. The difference ranged from 50 to 121 percent with an average of 87 percent. This average agrees with the result of the PWC test. Even though this study concludes that females are stronger than the previous studies had indicated, it should be mentioned that this conclusion should be accepted with caution. This was a pilot study, and the number of subjects was very small. Future more elaborate studies are required to validate the findings of this investigation. Furthermore, the equipment used in this study to measure strength ,Mini-Gym 180, does not give very accurate readings. Future studies should incorporate a load-cell with a more accurate indicator to insure a better results.

REFERENCES

[1] Aghazadeh, F., Simulated Dynamic Lifting Strengths Models for Manual Lifting, Ph.D. Dissertation, Industrial Engineering Department, Texas Tech University, December 1982.

[2] Astrand, P. and Rodahl, K., Textbook of Work Physiology, McGraw-Hill, New York, 1977.

[3] Ayoub, M. M., Gidcumb, C.F., Reeder, M.J., Hafez, H.A., Beshir,M.Y., Aghazadeh, F., and Bethea, N.J., Development of a Female Atlas of Strengths. Final Report, McDonnell-Douglas Aircraft, Contract No. AS-20501-C, 1982.

[5] Bernard, T.E., Kamon,E., and Franklin, B.A., Estimation of Oxygen Consumption From Pulmonary Ventilation During Exercise. Human Factors, 21, 417-421, 1979.

[4] Grasley, C., Ayoub,M.M., and Bethea,N.J., Male-Female Differences in Variables Affecting Performance. Proceedings of the Human Factors Society, 22nd Annual Meeting, 416-420, 1978.

[6] Hosler, W. W. and Morrow, J.R., Arm and Leg Strength Compared Between Young Women and Men after Allowing for Differences in Bodysize and Composition, Ergonomics, 25, 309-313,1982.

[7] Konz, S., Work Design, Grid Publishing Inc., Ohio, 1979.

[8] Laubach, L.L., Comparative Muscular Strength of Mean and Women: A Review of the Literature, Aviation Space and Environmental Medicine, 534-542, May 1976.

[9] Simonson, E., Physiology of Work Capacity and Fatigue, Charles C. Thomas Publishing, Illinois,1971.

Trends in Ergonomics/Human Factors II
R.E. Eberts and C.G. Eberts (Editors)

A BIOMECHANICAL COMPARISON OF CURRENT MAILBAG DESIGNS

George B. Page B.S.E.

Center For Ergonomics
College of Engineering
The University of Michigan
Ann Arbor

This study analysed and compared the biomechanical attributes of five different mailbag designs. Using a three dimensional torso model developed by Dr. Albert Schultz, static and dynamic postures were analyzed at loads of 5, 20 and 35 pounds. In the worst cast scenario of walking at 120 paces per minute while carrying 35 lbs., the cross pack had the highest average peak compressive force of 1750 N. Also, in this scenario, the side and cross packs produced the largest differences between predicted muscle forces of the left and right erector spinaes. Of most importance, the cross pack demanded the highest percent of the subjects maximum voluntary contractive strength with a level of 58% of the MVC for lateral bending. Altogether, the compressive forces for all the mailbag designs are well below NIOSH's Action Limit of 3400 N. Therefore, the high "percent of the subjects" MVC is most likely the cause of fatigue and low back pain. Coupled with the asymmetrical loading of the spine, current side and cross pack designs seem most detrimental to the musculoskeletal system of the low back.

INTRODUCTION

Overexertion is a common cause of low back trauma, especially among those involved in manual materials handling occupations. Two-thirds of all compensable industrial overexertion injuries and illnesses involve the back.[1] Furthermore, these injuries compile to as high as $14 billion annually.[2]

Manual materials handling involves activities such as pushing, pulling, lifting, and carrying. Letter carriers are among this group of manual material handlers. Often overlooked, there are approximately 280,000 letter carriers in Canada and the U.S.[3] In addition, the mailbag is most often used by letter carriers and lends itself well to postural stress and strain.

The two most common types of mailbags include the side pack, and the cross the chest pack. When loaded on the shoulder, the side pack rests against the ipsilateral side of the pelvis, while the cross the chest pack rests against the contralateral side of the pelvis. Both of these loading postures result in asymmetrical loading of the back's musculoskeletal system.

A study conducted in May of 1980 showed that 70% of the letter carriers experienced low back discomfort.[4] A similar study was performed by Imrie and Crosbie.[5] Sixty-seven percent of the letter carriers they surveyed reported symptoms of back discomfort. This seems to suggest the nature of an unbalanced load resulting in a displaced center of gravity may be a detriment to the letter carrier.

Alternative mailbags have been designed to decrease unbalanced loading and alleviate low back muscular disharmony. One design, the trap strap, is a supplement adaptable to the side and cross the chest packs. The trap strap acts as a hip belt attached to the mailbag itself. This design supposedly reduces the stress placed on the shoulder and lower back by shifting it to the pelvis.

Another mailbag design is a derivative of the common backpack used by campers. This mailbag is reversed and carries its load in front of the body instead of on the back. This mailbag design lends itself well to symmetrical loading in the frontal plane. An increase in the compressive loads of the spine in this case is a major concern. With a seemingly large (in contrast to the other mailbags) moment arm, the front pack mailbag may cause greater muscular activity in the erector spinaes, therefore producing greater compressive forces on the lumbar spine. Also, with greater muscular activity in the low back muscles, fatigue may play a greater role in the development of low back pain and discomfort. Figure 1 below displays the five different bags.

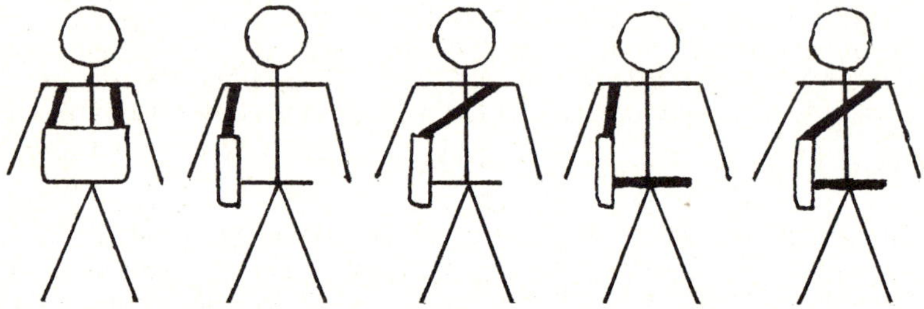

Figure 1
Packs (from left to right) Front Pack,
Side Pack, Cross Pack, Side Trap-Strap Pack,
Cross Trap-Strap Pack (Straps in bold print)

The purpose of this study was to compare the front pack, side pack, cross the chest pack, and the trap strap adaptors to the side and cross the chest pack mailbags. This study focused on biomechanically modelling the five mailbag types and their interactions upon the musculoskeletal system of the back. A three dimensional lateral torso model developed by Dr. Albert Schultz[6] was used in this analysis. With the biomechanical model, we were able to predict compressive and shear forces acting upon the L3 disc and muscle forces of 22 muscle equivalents of the back. Furthermore, an electromyographic study of the back was performed to aid in verifying our predicted data. With this data, the benefits and trade-offs between compressive forces upon the spine, asymmetrical loading, and overall loading in relation to muscle fatigue will be discussed.

METHODS

As previously mentioned, this study consists of two parts, First, a biomechanical analysis of the trunk muscle contraction forces and spine compressive and shear loads resulting from the external loads of the mailbags was performed. Second, an electromyographic analysis of the muscle activity at three different levels of the back was performed. This aided in validating the biomechanical analysis. In both sections of the study, the peak forces were recorded and computed in order to look at the worst case scenario.

First, the contractive forces of the trunk muscles and the compressive and shear loads acting upon the spine were estimated for each of the subjects. The theory that the load above and below an imaginary transverse plane at the L3 level must be in equilibrium is the basis of the analysis. Loads above the plane include internal loads: loads due to the body weight of each body segment above the L3 plane, and external loads due to external forces acting upon the body above the L3 plane. The moments created about the L3 disc by these loads are balanced by the equal and opposite moments created below the L3 plane by the contraction of the trunk muscles.

There were three main steps in the biomechanical analysis. First, the five mail bags had to be modelled using free body diagrams in order to help determine the magnitude and direction of external forces acting upon the body above the L3 plane. By measuring the tensile force imposed upon a supporting strap of the mailbag, these forces could be determined. Second, utilizing these reactive forces and the proper anthropometric dimensions of the body and mailbag, the three moments needed by the lower-trunk to equalize the load above the L3 plane were calculated. Third, the muscle contractive forces needed to produce the equalizing moments and the resultant compressive and shear forces were estimated using Schultz's[7] 22 muscle equivalent model.

Schultz uses a linear programming method for finding the muscle contractive forces. This model utilizes an optimization scheme which minimuzes the spinal compressive force while searching for a feasible solution with the smallest muscle intensities. "Use of this optimization scheme keeps antagonistic muscle activity to a minimum although it calls on nearly every muscle that can contribute to an activity to do so at nearly equal contraction intensities."[8]

The needed input for the biomechanical analysis included the weight of the subject, the lateral and anterior-posterior diameters of the trunk at the L3 level, the peak tensile force of a strap on the mailbag, directional angles of the supporting strap, and the location of the center of mass of each body segment above the L3 plane relative to the L3 disc. The mass of the arms, head and trunk above the L3 plane were estimated as 9, 5, and 36% based on data collected by Clauser et.al.[9]

In order to determine the potential level of fatigue experienced by the trunk muscles for the different packs, the muscle strength needed to maintain upright posture was compared to the predicted maximum voluntary contractive strength of the trunk muscles. Schanne's[10] strength equations and Stobbe's[11] strength data were used to predict the maximum voluntary contractive strength for back extension and lateral bending. These equations represent the muscle groups involved in maintaining upright posture while wearing the mailbags.

The second part of this study pertained to the analysis of the back using integrated EMGs for each of two subjects. Electromyographic activity of the left and right erector spinaes were measured at the L3, L1 and T9 levels of the spine for each test case. While these were measured, the tensile force of a strap of the mailbag was monitored using a force transducer.

The experiment consisted of a static stance, and dynamic test. For each of the two tests, the total (mailbag and contents) load was varied at 5, 20 and 35 pounds. Thirty-five pounds is the maximum load a letter carrier can carry in Canada. The static stance test measured the IEMGs and tensile forces for each mailbag at each of the 3 loads while walking at a slow (80 paces per minute) and fast (120 paces per minute) pace. Both dynamic and static tests monitored the muscle's electrical activity with electrodes spaced two inches apart on each side of the spine.

RESULTS

The results are broken into two sections. The first section includes the predicted compressive and shear forces of the lumbar spine for all of the static and dynamic test conditions. Also included are the predicted and observed differences in muscle activity between the left and right erector spinaes, and the contractive strengths of the erector spinaes as a percent of the predicted maximum voluntary contractive strength for back extension and lateral bending for each subject. The second section looks at the correlation between predicted and measured muscle activity at the L3 level of the back, and includes the observed results from the integrated EMGs at the T9 and L1 levels of the back.

SECTION 1

At a load of 35 lbs. the side pack and cross trap-strap mailbags have the highest average peak compressive forces; 1112N and 1086 N respectively. Furthermore, the side trap-strap bag has the lowest average peak compressive force at a 35 lbs. load with a value of 931 N. Also, the study revealed the cross pack as having the greatest average peak compressive forces of all mailbags at loads of 5, 20, and 35 lbs. while walking at a slow pace with values of 560, 992 and 1461 N respectively. Again, the side trap-strap bag had the lowest average peak compressive force at a 35 lbs. load with a value of 1119 N.

The cross pack also had the highest average peak compressive forces at 20 and 35 lbs. with values of 1089 and 1750 N while walking at a fast pace. Furthermore, the side trap-strap pack produces the lowest average compressive forces at a 35 lbs. load registering 1222 N.

Mean predicted peak shear forces at the L3 disc were at most 109 N in either the lateral or anterior-posterior direction. As anticipated, the anterior-posterior shear forces were greatest for the front pack. In addition, the shear forces in the lateral direction were greatest for the two side pack mailbags, while the two cross packs produced shear forces on average no higher than 20 N in either direction.

Next, this study looked at how load and mailbag type affect the difference in muscle activity levels between the left and right erector spinaes.

First, while statically standing, the side pack had the greatest difference at loads of 20 and 35 lbs. with the contralateral erector spinaes contracting 62 and 125 N more than the ipsilateral erector spinae. Meanwhile, the front pack had no predicted difference resulting from symmetrical loading in the frontal plane. This holds true for both walking paces also.

The study revealed the cross and side packs as having the greatest discrepancies between the left and right erector spinaes while walking at a pace of 80 paces per minute at the maximum load of 35 lbs. with values of 188 and 150 N respectively. While, the cross pack had the greatest predicted difference of 231 N while walking at a pace of 120 paces per minute.

The worst case situation of walking rapidly at 120 ppm while carrying a 35 lb. load shows relatively little difference in compressive forces between the front, side-trap, and cross-trap packs, while the side and cross packs produce relatively larger compressive forces. An analysis of variance of the compressive force data showed no significant difference at a .9 level of confidence between the five bags. This is most likely due to the small sample size of subjects tested.

The worst case situation of walking fast carrying a 35 lb. load shows the side and cross packs as having the greatest relative difference between predicted muscle forces of the left and right erector spinaes. Furthermore the front pack proves to have a significantly smaller difference in muscle activity at a .9 level of significance. Again, the small sample size limits performing an anlysis of variance on the gathered data.

The worst case situation also reveals the front pack as having a relatively smaller difference in observed muscle activity between the left and right erector spinaes. No significant difference was reported between the bags at a .9 level of significance. Once more, this is most likely due to the limited number of subjects tested.

Figure 2 displays the required lateral bending strength demands as a percent of the average predicted maximum voluntary lateral bending strength for the worst case scenario. As can be seen from this figure, the front pack requires a significantly smaller percent of the subjects maximum voluntary strength than the other four packs. Furthermore, the cross pack demands the highest percent of the subjects maximum voluntary contractive strength (MVC) with a level of 58% of the MVC. The other pack designs are not far behind with values ranging from 34% of the MVC for the side trap-strap pack to 45% of the MVC for the side pack.

Figure 3 shows the required back extension strength demands as a percent of the average predicted maximum voluntary back extension strength for the worst case scenario. This figure reveals the front pack requiring the highest level of the subjects MVC with a value of 7% of the MVC. The other four packs require an insignificant amount of the subjects MVC with values no higher than .48%.

SECTION 2

The correlation coefficients between the predicted and observed differences in muscle activity at the L3 level for each bag and test condition ranged between .791 and .999 with all but two conditions falling below .913. The lowest correlation coefficient of .791 occurs with the side pack while

PERCENT OF MAXIMUM VOLUNTARY CONTRACTIVE STRENGTH (LATERAL BENDING)

FOR THE WORST CASE SCENARIO OF 35 LBS AND 120 ppm

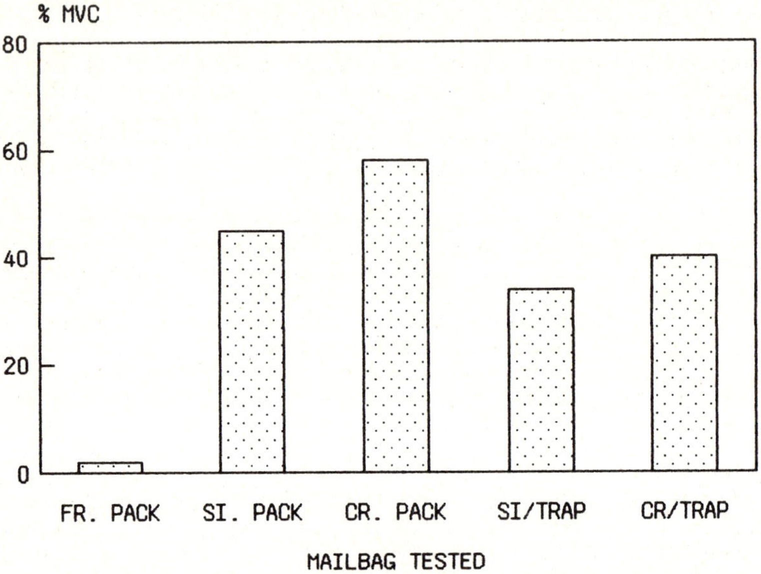

FIGURE 2

PERCENT OF MAXIMUM VOLUNTARY CONTRACTIVE STRENGTH (BACK EXTENSION)

FOR THE WORST CASE SCENARIO OF 35 LBS AND 120 ppm

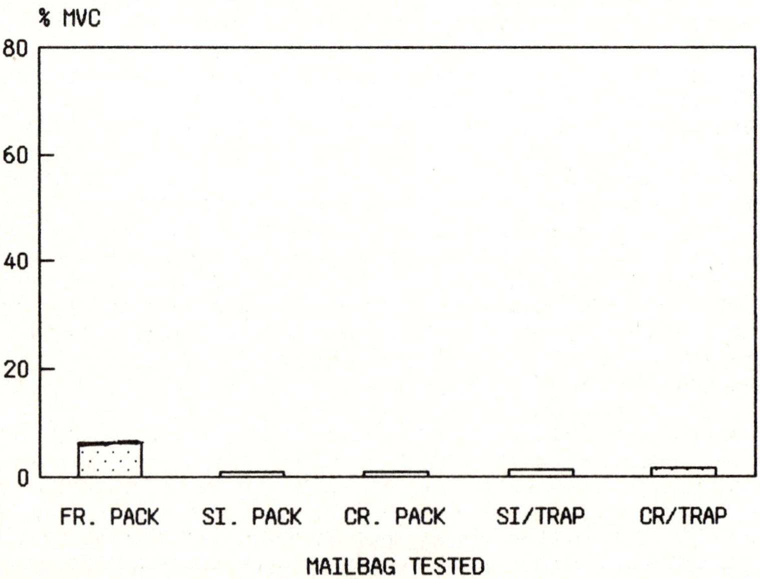

FIGURE 3

walking rapidly (120 paces per minute). This means the model explained 79% of the variance in the observed result in this case. It was not possible to determine correlation coefficients for the front pack data since it had predicted values of zero for the difference between the contractive forces of the left and right erector spinaes for all cases.

The observed difference in muscle activity at the T9 level of the spine proved significantly greater than the difference in muscle activity at the L3 and L1 level at a .9 level of significance for the front pack. No other mailbag proved to be significantly different at the different levels of the back. This also may be due to the small ample size of subjects tested.

CONCLUSIONS

The level of compressive forces being applied to the L3 are well below the NIOSH Action Limit of 3400 N.[12] The Action Limit is the level believed to represent nominal risk to most workers. This limit was developed from cadaver spine studies of Evans[13], Sonoda[14] and Perey.[15] Even though the Action Limit is set for compressive forces acting upon the L5/S1 disc and our data is based on the L3 disc, the relative difference is considered insignificant.

Relatively higher compressive forces were expected for the front pack as noted by Cook and Newmann.[16] On the contrary, high compressive forces did not develop for the front pack. This was most likely due to the hip belt which absorbed approximately 66% of the load in the worst case scenario and up to 88% of the load while standing still. This pelvic load bearing effect of the trap strap was not realized with the two side and cross pack conditions, where-in only approximately 9% of the total force was transferred to the pelvis. The front pack's hip belt can ride high on the pelvis, droop in front where the load is supported and still absorb a large portion of the load. While the side and cross packs' trap-strap can't comfortably be worn high enough to produce a similar effect. As mentioned by Vernon and Laufer,[17] the pelvis is a more rigid system than the back-shoulder region and better suited for carrying a load. Further research could include determining the effects of loading the pelvis upon the spine in terms of pelvic rotation and gait analysis.

Since the predicted compressive forces of all 5 mailbags seem to be well below dangerous levels, low back problems must be the result of causes from another arena. Also, the magnitude of shear forces the spine can withstand is not well documented.[18] Therefore, the results concerning lateral and anterior-posterior shear forces are in limbo until further progress is made in this area. The data looking at asymmetrical muscle contractive forces of the back and muscle loading as a percent of the muscle's maximum voluntary contractive strength may help explain the postural problem of current mailbag designs.

The large discrepancy between the different sides of the erector spinaes for the side, cross the chest, side trap-strap, and cross the chest trap-strap packs seems to show while one of the four bags may be better than the other, they all have a much greater difference in muscle activity than the front pack. These predicted values seem to be well supported by the high correlations observed with the difference in myoelectrical activity of the back for all the test conditions including the worst case scenario. In comparison to the other four mailbags, the symmetrical loading of the front pack is a definite benefit.

The worst case scenario for the required trunk muscle contraction strengths as percents of the subjects maximum voluntary contractive strengths are displayed in Figures 2 and 3. As can be seen from these Figures, the front pack requires the smallest overall percent of maximum voluntary trunk muscle contractive strength. This includes lateral bending and back extension. The other four packs require a significantly greater percent of the trunk muscle's maximum voluntary contractive strength. Even though these four packs require an insignificant amount of back extension strength, they do demand between 34 to 58 percent of the maximum voluntary contractive strength of the muscles involved with lateral bending. This may be due to the contralateral erector spinae's mechanical disadvantage in countering a load carried on one shoulder. Mainly, the load is countered by one set of erector spinae muscles vs. two for the front loaded pack. This seems to suggest that muscle fatigue may easily arise for the side, cross, and trap-strap packs. According to Rohmert[19], a person can maintain such loads for only 1 to 3 minutes before fatigue develops. Even though this is for the worst case scenario of carrying a 35 pound load while walking rapidly, it reveals a potential cause of low back discomfort among letter carriers. Also, since Rohmert determined exertion levels below 15% of the maximum voluntary contractive strength can be endured for an extended period of time, the front pack, with a value of 7.0% of the maximum voluntary contractive strength, seems better suited for extended wear.

Even though the exact cause of low back problems among letter carriers is not known, the presence of fatigue seems assured with such high levels of the subjects maximum voluntary contractive strength required. This is most likely compounded by the effects of asymmetrical loading of the musculoskeletal structure of the back.[20,21] Since compressive forces acting upon the back for all 5 mailbags seem insignificant to be the cause of low back problems among letter carriers, muscle fatigue and asymmetrical loading of the spine seem to be the more likely culprits. Among these last two factors, the front pack appears to be the least stressful to the musculoskeletal system of the low back. With reduced stress placed on the musculoskeletal system of the low back, a reduction in the severity and the amount of cases concerning low back pain among letter carriers should be expected. This translates further to a reduced burden in workers compensation or its equivalent for employers and healthier employees.

The front pack is further beneficial in splitting its load about 2 shoulders and the pelvis instead of one shoulder or one shoulder and the pelvis. Potential problems with the front pack mailbag include difficulties in getting into it, and more importantly, the reduced ability to see where you are placing your feet. Further research could also include investigating at what degree of loading asymmetrical muscle activity becomes a detriment to the spine.

A more complete analysis could be performed using a greater sample size of subjects. This would aid in lowering the overall variance in measured data allowing for a more meaningful analysis of variance between the different mailbags. Altogether, this would help determine how significantly different the 5 mailbags are for all the test conditions. Also, further research could include an investigation of fatigue in extended periods of muscle contraction.

REFERENCES

[1]Garg, A.,"Lifting and back injuries," Plant Engineering, October, 1982.
[2]Akeson, W. and Murphy, R., "Editorial Comment: Low back pain," Clinical Orthop., 129:2-3, 1977.
[3]Miller, M.G.W., "The postal postural problem," Proceedings of the 1984 International Conference on Occupational Ergonomics.
[4]Miller, M.G.W., "Postal Fatigue," A brief presented to the U.S. Postal Service, April, 1981.
[5]Imrie, J.D. and K.L. Crosbie, "The trap strap report," Unpublished paper, Industrial Medical Resources, Toronto, 1982.
[6]Schultz, A.B., Haderspeck, K., Warwick, D., and Portillo, D., "Use of trunk muscles in isometric performance of mechanically complex standing tasks," J. of Ortho. Research, 1:77-91, 1983.
[7]ibid 6
[8]ibid 6
[9]Clauser, D., McConville, J., Young, J., "Weight volume and center of mass of segments of the human body," AMRL-TR-69-70, Aerospace Medical Research Laboritory (USAF), Wright-Patterson Air Force Base, Ohio, 1969.
[10]Schanne, F.J., "A three-dimensional hand force capability model for a seated person," Unpublished Ph.D. Dissertation, University of Michigan, Ann Arbor, 1972.
[11]Stobbe, T.J., "The development of a practical strength testing program for industry," Unpublished Ph.D. Dissertation, University of Michigan, Ann Arbor, 1982.
[12]National Institute for Occupational Safety and Health, A Work Practices Guide for Manual Lifting; Tech. Report No. 81-122, U.S. Dept. of Health and Human Services (NIOSH), Cincinnati, Oh., 1981.
[13]Evans, F.G., and H.R. Lissner, "Studies on the energy absorbing capacity of human lumbar intervertibral discs," Proceedings of the Seventh STAPP Car Crash Conference, Springfield, Ill. 1965.
[14]Sonoda, T., "Studies of the strength for compression, tension and torsion of the human vertebral column," J. Kyota Prefect. Med. Univ., Vol. 71, 659-702, 1962.
[15]Perey, O., "Fractures of the vertebral endplate in the lumbar spine," Acta Ortho. Scand., Supp. 25, 1957.
[16]Cook, T.M., and D.A. Neuman, "An electromyographic analysis of the low back muscles during load carrying," Proceedings of the 1984 International Conference on Occupational Ergonomics, 1984.
[17]Vernon, H. and Laufer, S., "The trap strap: A study of its efficiency, the final report," Unpublished paper, C.M.C.C., 1982.
[18]Berkson, M.H., Nachemson A., Schultz, A.B., "Mechanical properties of human lumbar spine motion segments--Part 2: Responses in compression and shear; influence of gross morphology." J. Biomech. Eng. Trans. ASME 101, 46-32, 1979.
[19]Romert, W., Ermittlung von Erholung-Spausen for Statische Arbeit des Menchen, Int. Z. Angew. Physio., 18, 123-124, 1960.
[20]Noro, L., "Medical aspects of weight carrying," Industrial Medicine and Surgery, March, 1967.
[21]Grice, A., "Harmony of joint and muscle function," C.C.A., 1976.

A complete copy of this study may be attained by writing: George Page, Center for Ergonomics, The University of Michigan, IOE Bldg., 1205 Beal, Ann Arbor, MI 48109

IX

MANUAL MATERIALS HANDLING

Trends in Ergonomics/Human Factors II
R.E. Eberts and C.G. Eberts (Editors)

PHYSIOLOGIC ISSUES IN MANUAL MATERIALS HANDLING

Ashraf M. Genaidy, Shihab S. Asfour, Tarek M. Khalil and Sherif M. Waly

Departments of Biomedical and Industrial Engineering
University of Miami
Coral Gables, Florida
U.S.A.

The primary objective of the present paper is to highlight physiologic issues encountered in manual materials handling and to provide a solution to these issues. The issues discussed in this paper are : (1) the physiologic fatigue criterion, and (2) additivity of the physiological cost of simple tasks.

INTRODUCTION

Back injuries resulting from manual materials handling (MMH) activities are a major source of lost time and compensation claims. Reduction in the number of such injuries has therefore, become a priority concern of many industries, labour unions and researchers. Therefore, the main objective of this investigation is to highlight physiologic issues encountered in MMH and to provide a solution to these issues. The two topics discussed in this paper are: (1) physiologic fatigue criterion, and (2) additivity of the physiologic cost of simple tasks.

PHYSIOLOGIC FATIQUE CRITERION

The physiologic fatigue criterion (PFC) is the energy expenditure level that can be maintained by an individual throughout the working day without excessive fatigue. This fatigue is due to the accumulation of lactic acid in the blood at a rate faster than the rate of its removal. In doing work below the PCF, after a few minutes of adaptation, the energy expenditure is almost constant and is proportional to the external load. If work is performed above the PFC, the energy expenditure does not attain a constant level but continues to rise. The more the PFC is exceeded, the greater is the rise. At this stage, an oxygen debt or a certain amount of deficit is accumulated, and the subject will not be able to continue the task for a long period of time. PFC recommended by different researchers ranged between 33 and 50% of the maximal oxygen uptake (Garg and Ayoub (1980)).In most industrial applications the oxygen consumption value of 1.0 liter/min., which is equivalent to one-third of the maximal oxygen uptake, is usually recommended as the PFC for continuous work. The PFC of 1.0 liter/min was determined based on highly dynamic tasks (bicycle ergometer, motor-driven treadmill). Such a value was applied to manual tasks such as lifting without being justified and validated. Lifting is believed to be a combined static and dynamic task and utilizes muscle groups that are different from those used in cycling and running. Thus, this criterion is high when applied to lifting because the maximum oxygen uptake would be generally lower than that obtained in cycling or running tasks (Petrofsky and Lind (1978)). Therefore, there is a need to develop a physiologic fatigue criterion that is applicable to lifting under different task variables.

Two separate experiments were conducted to obtain PFC values that take into account both the psychophysical acceptability and physiologic endurance of workers for different task variables. In the first experiment, the oxygen consumption associated with the maximum acceptable load was recorded for different frequency-height combinations. In the second experiment, the physiologic endurance was considered as one-third of maximal oxygen uptake obtained from lifting for different frequency-height combinations. PFC was then computed as the minimum oxygen consumption value obtained from both experiments for each frequency-height combination.

Figure 1 depicts PFC as a function of the frequency and height of lift. PFC is expressed as a percentage of maximal oxygen uptake obtained from the bicycle ergometer. This figure clearly indicates that PFC increases with an increase in the frequency of lift across all heights as would be expected since lifting actions involve the lifting portions of the body weight and the task becomes more dynamic. The increase is most pronounced between frequencies 1 and 7 lifts/min. All PFC values reported show a tendency to level off beyond the frequency of 7 lifts/min. Across all frequencies, the 0-127 cm height recorded the highest PFC value, followed by the 0-76.2 cm and 76.2-127 cm heights respectively. This indicates that the muscle mass has an effect on PFC. The 76.2-127 cm range utilizes primarily the arm muscles (a comparatively small muscle group) and the 0-76.2 cm used mainly the leg muscles (larger, more powerful); the 0-127 cm range represents an amalgam of both muscle groups. It is to be noted that the PFC values for lifting were much lower than the 33% adopted in the literature based on the bicycle ergometer. Therefore, when designing lifting tasks a physiologic fatigue criterion based on a single value might not be appropriate over wide range of frequency and height of lift.

ADDITIVITY OF THE PHYSIOLOGICAL COST OF SIMPLE TASKS

Some researchers based their work on the assumption that the net total metabolic cost of a series of activities can be estimated by summing their net steady state individual metabolic costs as obtained from their separate performance (Garg, 1976). The net energy cost of a task is arrived at by subtracting the individual's standing oxygen consumption, at rest, from the amount of oxygen consumed while performing that task. This assumption was not verified experimentally and there is no literature in the field of MMH, that supports or contradicts this assumption. Based on that assumption, one would expect that the net oxygen consumed while lifting a box from the floor to 76.2 cm (VO_{f1}) added to the net oxygen consumed while lifting the same box from 76.2 cm to 127 cm above the floor (VO_{f2}) should be equal to the net oxygen consumed while lifting that box from the floor to 127 cm. above the floor (VO_{f3}). Thus based on that assumption, an estimate of the net oxygen consumed while lifting a box from the floor to 127 cm (EVO_{f3}) can be calculated from the following equation:

$$EVO_{f3} = VO_{f1} + VO_{f2}$$

For the purpose of our discussion, the actual net oxygen consumed while lifting a box from the floor to 127 cm is denoted by (AVO_{f3}) and the difference between EVO_{f3} and AVO_{f3} is called the "extra oxygen cost". For the additivity assumption to be valid, the "extra oxygen cost" should be equal to zero.

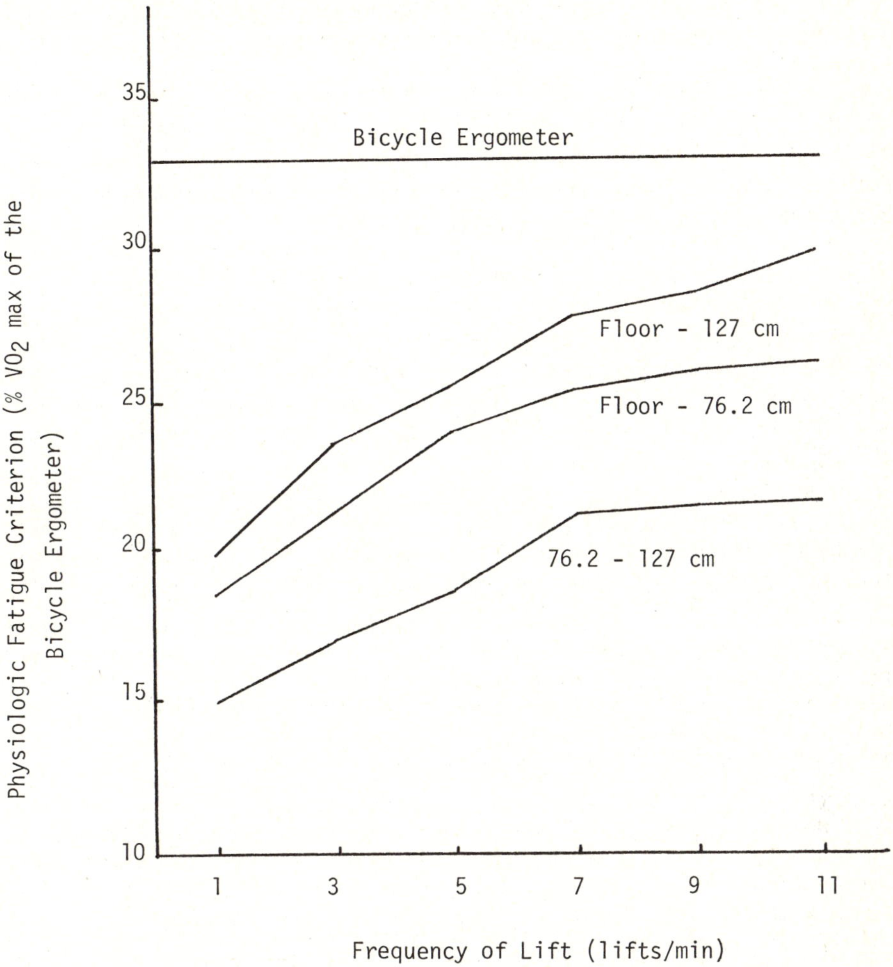

Figure 1

The Physiologic Fatigue Criterion as a Function of the Frequency and Height of Lift.

An experiment was conducted to verify the additivity assumption for different task variables: frequency of lift or lower (3, 6, 9 times/min), load lifted or lowered (6.8, 13.6, 20.5 kg), box width (38, 66 cm), box length (38, 66 cm), height of lift or lower (floor - 76.2 cm, floor - 127 cm, 76.2 - 127 cm) and angle of twist of the body (0°, 90°). Ten male subjects, 18 - 28 years of age, participated in this experiment. Five out of the subjects performed lifting tasks and five performed lowering tasks. Table 1 gives the mean "extra oxygen cost", for the data of this experiment. Performing student's paired t-tests showed that all the mean "extra oxygen cost" values reported in Table 1, were significantly different from zero at the 1% level. The mean "extra oxygen cost" values ranged between 60.5 ml/min and 267 ml/min. The average percentage error in estimating VO_{f3} from (VO_{f1} + VO_{f2}) ranged between 19% and 36.8% for lifting. The percentage error for lowering, however, ranged between 27.1% and 44.8%. Table 1 also shows that the "extra oxygen cost" increased with an increase in either the frequency of lift (lower) or the load lifted (lowered).

The multiple linear regression technique was used in developing two models that can help predict VO_{f3} (ml/min), from VO_{f1} (ml/min), VO_{f2} (ml/min), frequency of handling (times/min), and load handled (lb).

Model A: applicable to manual lifting tasks (R^2 = 0.910).

$$VO_{f3} = 36.33 + 0.5637 * VO_{f1} + 0.3893 * VO_{f2} + 1.1301 * \text{frequency} * \text{load}$$

Model B: applicable to manual lowering tasks (R^2 = 0.845).

$$VO_{f3} = 27.91 + 0.5137 * VO_{f1} + 0.6692 * VO_{f2} + 1.4234 * \text{frequency} * \text{frequency}$$

From the above discussion it can be easily seen that the net total energy cost of two activities cannot be estimated by simply summing their net steady state individual metabolic costs as obtained from their separate performance in the case of lifting or lowering tasks, at least for those tasks similar to the ones studied in this experiment.

REFERENCES

[1] Garg, A., A. metabolic prediction model for manual material handling jobs, Doctoral Dissertation, Dept. of Ind. Eng., Univ. of Michigan (1976).

[2] Garg, A. and Ayoub, M. M., What criteria exist for determining how much load can be lifted safely? Human Factors, 22 (1980) 475-486.

[3] Petrofsky, J. S., and Lind, A. R., Comparison of metabolic and ventilatory response of men to various lifting tasks and to bicycle ergometry, J. Appl. Physiol., 45 (1978) 60-63.

TABLE 1 THE MEAN 'EXTRA OXYGEN COST' AT DIFFERENT LEVELS OF TASK VARIABLES

Task	F	L	BL	BW	AT	Mean VO_{f3} (ml/min)	Mean VO_{f1} + VO_{f2} (ml/min)	Mean 'Extra O_2 Cost' (ml/min)	Error % in Estimation of VO_{f3}
	3					333	419	86	26
	6					610	762	151	25
	9					907	1096	189	21
		6.8				401	493	92	23
		13.6				608	751	143	24
		20.5				842	1033	19	23
L			38			557	669	112	20
			66			677	849	172	25
				38		590	720	129	22
				66		643	798	155	24
I					0	626	780	154	25
					90	608	738	130	21
	3	6.8				206	281	76	37
	3	13.6				337	422	85	25
F	3	20.5				458	556	98	21
	6	6.8				412	499	87	21
	6	13.6				589	749	161	27
	6	20.5				830	1038	207	25
T	9	6.8				584	669	115	20
	9	13.6				989	1083	184	21
	9	20.5				1238	1505	268	22
	3		38			370	370	61	21
I	3		66			469	469	112	31
	6		38			675	675	127	23
	6		66			849	849	176	26
	9		38			964	964	149	18
N	9		66			1228	1228	228	23
		6.8	38			446	446	71	19
		6.8	66			540	540	113	27
		13.6	38			660	660	121	22
G		13.6	66			843	843	166	25
		20.5	38			758	903	145	19
		20.5	66			927	1163	236	26
	3					213	303	90	42
L	6					388	542	154	40
	9					598	782	193	33
O		6.8				302	405	103	34
W		13.6				388	532	144	37
E		20.5				501	691	190	38
			38			350	479	129	37
R			66			443	606	163	37
I				38		372	508	136	37
N				66		422	578	155	37
G					0	402	559	157	39
					90	392	526	134	34

TABLE 1 (Cont'd.) THE MEAN 'EXTRA OXYGEN COST' AT DIFFERENT LEVELS OF TASK VARIABLES

Task	F	L	BL	BW	AT	Mean VO_{f3} (ml/min)	Mean VO_{f1} + VO_{f2} (ml/min)	Mean 'Extra O_2 Cost' (ml/min)	Error % in Estimation of VO_{f3}
	3	6.8				164	226	62	38
	3	13.6				218	301	88	41
L	3	20.5				263	382	119	45
	6	6.8				303	408	105	35
O	6	13.6				374	530	156	42
	6	20.5				488	689	201	41
W	9	6.8				439	582	143	33
	9	13.6				577	765	188	33
E	9	20.5				751	1001	295	33
	3		38			194	270	75	39
R	3		66			232	336	104	45
	6		38			342	488	146	43
I	6		38			434	596	162	37
	9		38			514	678	164	32
N	9		66			664	886	222	34
		6.8	38			279	355	76	27
G		6.8	66			324	455	131	40
		13.6	38			337	473	136	41
		13.6	66			440	491	151	34
		20.5	38			435	609	174	40
		20.5	66			466	772	206	37

Note: F = Frequency (times/min)
L = Load (kg)
BL = Box Length (cm)
BW = Box Width (cm)
AT = Angle of twist (deg.)

Trends in Ergonomics/Human Factors II
R.E. Eberts and C.G. Eberts (Editors)

THE COMPARISON OF STRESSFULNESS OF MANUAL MATERIALS HANDLING ACTIVITIES

Bernard C. Jiang, Industrial Engineering Department
Auburn University, Auburn University, Alabama 36849

James L. Smith, Industrial Engineering Department
Texas Tech University, Lubbock, Texas 79409

The psychophysical capacities for four individual MMH activities (lifting from floor to knuckle height, lifting from knuckle to shoulder height, lowering from knuckle to floor height, and carrying for 4.3 m) and three combined MMH activities, which were the combination of the above four individual activities, were determined under three frequency conditions (one time maximum, 1/minute, and 6/minute). Twelve male college students served as subjects for the study. In all experimental conditions, the psychophysically determined capacities were the maximum acceptable work load for a one hour work period. The capacities for all task condition were compared and the limiting capacity for a combined activity was identified. The results of the present study were also compared to the previous phychophysical data bases. Finally, a general recommendation for job design/redesign was given for a MMH job designer.

INTRODUCTION

Manual Materials Handling (MMH) activities produce the single largest percentage of compensable work injuries in U.S. industry, currently constituting one-fourth to one-third of all injuries (Kroemer 1983). In the U.S., 400,000 workers suffer disabling back injuries every year (Accident Facts 1978). The cost to U.S. industry is estimated to be $14 billion a year. As reported by Lahey (1984), a Canadian physician provides statistics showing that back pain is the second leading cause of absenteeism in industry, second only to the common cold.

Engineering control of reducing job stressfulness is obviously an effective way to decrease the risk of injury. To accomplish this, it is necessary to know the stressfulness of different types of MMH tasks so that the change of tasks will result in a reduction of physical stress. Some general recommendations for reducing job demand have been given by Ayoub (1982b), and Ayoub et al. (1983), such as changing a task from lifting to lowering, a change from carrying to pulling, and a change from pulling to pushing. In light of these recommendations, this paper compares the stressfulness of different types of MMH activities by using a psychophysical approach.

Several combined MMH activities were also examined. The capacities of these combined activities were compared with each other and the dominant element for certain combined activities was identified. Finally, general recommendations for a MMH job designer were given based on the present study and the previous data reported by Snook (1978), Ayoub et al. (1978, 1983), and Mital (1984).

PROBLEM IDENTIFICATION:

In order to minimize the injuries caused by MMH activities as well as to maximize job productivity, the job demands should match the capacities of individuals. This can be accomplished by either changing the type of work performed or by screening workers to fit the job requirements. Of these two methods, changing the job demands is more effective in terms of the risk of injury involved in the task.

The approaches that have been used to express the task stressfulness are physiological, biomechanical, and psychophysical. A physiological approach is used in MMH studies when the job being analyzed involves large muscle groups and a high frequency of handling. The physiological indices generally failed to represent the stressfulness of a task involving only small muscle groups or a low frequency of handling. A biomechanical approach is suitable for low frequency MMH tasks requiring maximal or near-maximal efforts. It is not surprising that discrepancies were reported when using physiological or biomechanical approaches alone (Asfour 1980, Ayoub et al. 1983) in the determination of MMH capacity. Because different muscle groups are involved in a MMH activity and there are different frequencies of handling, a psychophysical approach is an optimal approach to use in order to determine an operator's MMH capacity under various task conditions. Therefore, psychophysically determined capacity was adopted as an index for the stressfulness of a MMH task in this study.

Another problem of recent MMH studies was that most of the previous psychophysical studies were centered only on individual activities of MMH tasks, i.e., one of lifting, lowering, carrying, holding, pushing or pulling. The combination of two or more activities has never been examined. These kinds of combined activities are prevalent in industry and in our daily lives. The following cases of combined MMH activities were mentioned by Ayoub (1982a, 1982b):

> Packing operators--"In the inspection and packing area of a large manufacturing plant, packing operator must lift, transport, and lower product units weighing 5 kg and 6.4 kg for shipping," (Ayoub 1982a)
>
> Garbage collector--"Another illustration would be the lifting/carrying task associated with garbage collection. In this case, the behavior of a garbage collector is governed by the fact that he must remain close to the truck between collection points." (Ayoub 1982a)
>
> Handling power tools--"In a plant where bauxite is refined into alumina, the employees use airpowered jackhammer tools in cleaning the inside of tanks to clear clogged pipes and valves. The work is predominantly a combination of lifting and carrying heavy tools for periods from one to two hours." (Ayoub 1982b)

Assembly operations--"Transfer points on some conveyors were not located at the same height. Accordingly, employees were forced to manually lift and carry heavy components from one conveyor to the next." (Ayoub 1982b)

Currently, no interaction of task activities is assumed and the only way to analyze a combined activity is to analyze each component of the combined activity and use the most critical task as the limiting factor for the combined activity.

The objective of this study was to compare the stressfulness of four individual and three combined MMH activities. The results of this quantitative comparison provided recommendations to MMH job designers who are concerned with reducing the stressfulness of a task through job design/redesign.

REVIEW OF PAST PSYCHOPHYSICAL DATA

There were four major psychophysical MMH capacity data bases which were based on industrial populations or Air Force work force: Ayoub et al. (1978), Snook (1978), Ayoub et al. (1982), and Mital (1984). Ayoub et al. (1978) studied psychophysical capacities for 73 male and 73 female industrial workers. The tasks being studied consisted of six ranges of lifting - floor to knuckle (FK), floor to shoulder (FS), knuckle to shoulder (KS), knuckle to reach (KR), and shoulder to reach (SR); three box sizes - 30.5, 45.7, and 61 cm; and four frequencies - 2, 4, 6, and 8 lifts/min. Overall, the mean capacity for different range of lifting is in the following decreasing order: FK (27.80 kg), KS (26.12 kg), FS (23.28 kg), FR (22.33 kg), KR (24.34 kg), and SR (19.83 kg).

Based on the works done prior to 1978, Snook (1978) reported the psychophysical capacity data according to gender of operators, task type (lifting, lowering, carrying, pushing, or pulling), box size, distance of handling, and percent of industrial population. Graphical presentation of Snook's data for 50th percentile males and for box size of 36 cm is shown in Figure 1.

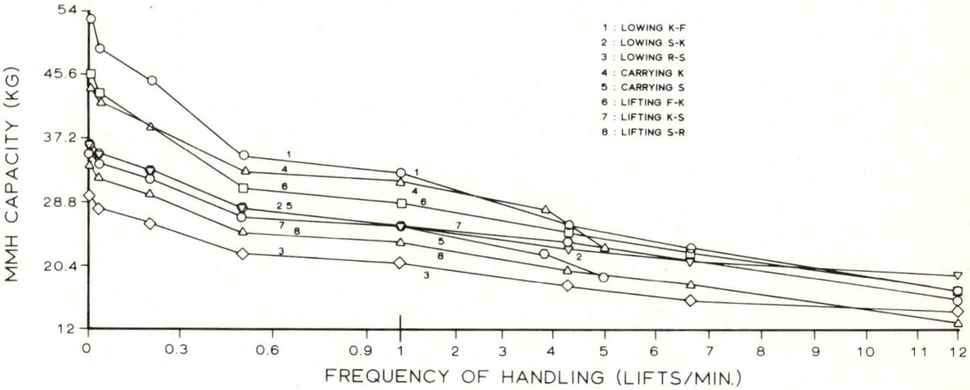

Figure 1
MMH Capacities for 50th Percentile Males
(Based on Snook 1978)

Based on the two above-mentioned psychophysically determined capacity data bases (Ayoub et al. 1978 and Snook 1978), Ayoub et al. (1983) developed lifting capacity prediction models for both males and females. The models can accommodate: 1) a wide range of frequencies (0.1 to 12 lifts/min); 2) various container sizes (30.5 to 76.2 cm in sagittal plane); 3) six ranges of lift; and 4) individuals and different percentiles of population. The models also allow for adjustment of lifting capacity based on presence or absence of handles and whether or not a twist of the body occurs during lifting. A graphic representation of lifting capacity for 50th percentile male, and box size of 45.7 cm is shown in Figure 2.

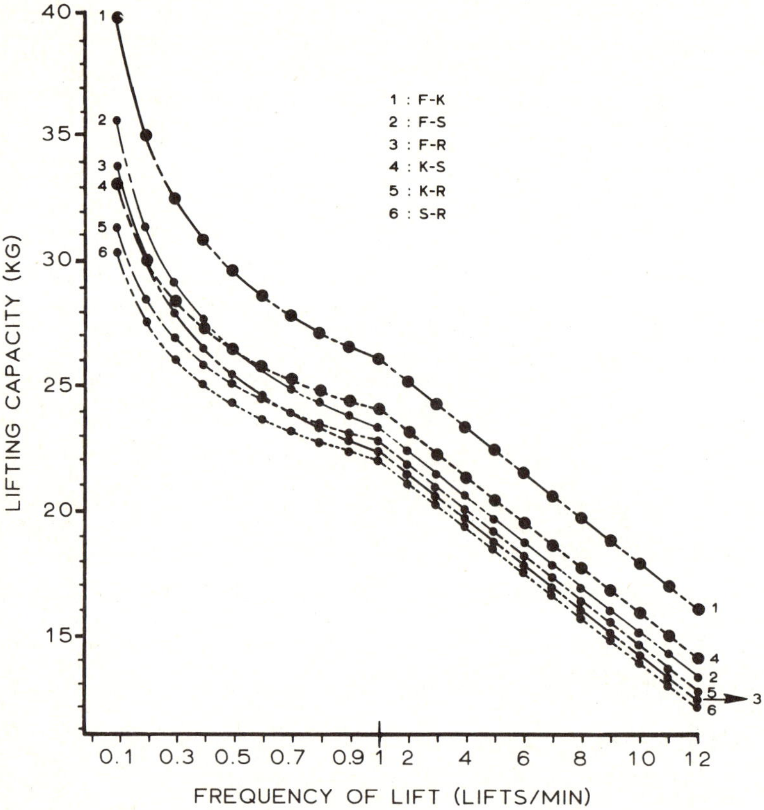

Figure 2
Lifting Capacities for 50th Percentile Males
(Based on Ayoub et al. 1983)

Ayoub et al. (1982) developed a criterion by which the Air Force could reliably evaluate the compatability of an individual's physical capacities with the physical demands of the various Air Force Specialty Codes (AFSCs). MMH capacities were determined psychophysically based on 571 airmen's test data and based on a one time maximum performance. The mean capacities based on this study were shown in the following decreasing order: lifting a regular box, FK (63.5 kg), lifting a regular box from floor to work bench height (57.2 kg), two-hand front carry (55.4 kg), two-hand side carry (49.3 kg), hold at shoulder height (47.0 kg), one-hand tool box carry (46.9 kg), hold at reach height (45.2 kg), low level push (43.4 kg), lifting a tool box from floor to work bench height (38 kg), low level push (36.0 kg), lifting a regular box, FS (32.5 kg), lifting a regular box from floor to 1.83 m height (26.4 kg), and high level push (25.2 kg).

Mital (1984) utilized a psychophysical approach to determine the lifting capacity for 37 male and 37 female industrial workers. Task conditions examined in this study were: four lifting frequencies (1, 4, 8, and 12 lifts per minute), three height levels (FK, KS, and SR), and three box sizes (30.5, 45,7 and 61.0 cm). Mean lifting capacity data for male subjects based on a box size of 45.7 cm are shown in Figure 3.

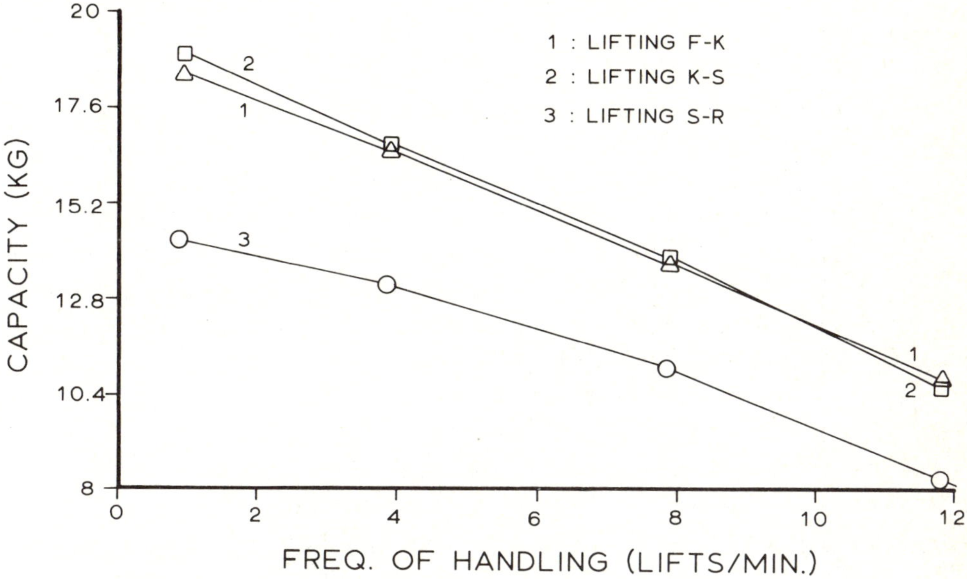

Figure 3
Mean Lifting Capacity Data (Based on Mital 1984)

EXPERIMENTAL PROCEDURES:

Subjects:

Twelve healthy male subjects who were free from back pain were recruited from a college student population. They were familiarized with the experimental procedures before the experimental data were collected. The subjects' ages varied between 18 and 24, with the average age being 20.6. The subjects' mean height was 175.7 cm (SD=4.91) and mean weight was 72.5 kg (SD=8.44). The subjects were paid volunteers who could be dismissed during the experiment for any of the following reasons: 1) subject's unwillingness; 2) physical problem observed after the experiment began; or 3) subject's inability to perform the experiment properly.

Familiarization session:

Although training was not a major variable in this study, the subjects were given seven one-hour sessions for familiarization. The familiarization sessions were not designed to increase the subjects' strength or endurance, but to acquaint them with the experimental procedure. During the seven one-hour familiarization sessions, each subject became familiarized with all strength testing procedures and performed six different combinations of individual and combined activities with different frequencies of handling. Normally, each subject conducted three one-hour sessions at a frequency of 1 handling/min, and three one-hour sessions at a frequency of 6 handlings/min. During these sessions, both individual and combined activities were conducted by each subject.

The purposes of the familiarization sessions were as follows:

1) The subjects could become familiar with the experimental environments and activities conducted.

2) The subjects could become familiar with the experimental procedures and psychophysical methodology.

3) The subjects could utilize their muscular groups that would be used during experimentation in order to have a natural response as the data collection began.

4) The experimenter could become familiar with the subjects in order to have better cooperation.

<u>Psychophysical MMH capacities</u>

Individual activities in this study refer to one of the following MMH activities: lifting, lowering, or carrying. The combined activities refer to the activities which combine lifting, lowering and/or carrying in one task.

Based on a survey conducted by Drury et al. (1982), and further discussed by Mital and Ilango (1983), the most common three MMH activities in industry are lifting, lowering and carrying, in that order. These three activities were examined in this study. The following individual activities were selected for study:

1) Lifting from floor to knuckle height (LFK):
   LFKM - LFK at the frequency of one time maximum
   LFK1 - LFK at the frequency of 1 handling/min
   LFK6 - LFK at the frequency of 6 handlings/min

2) Lifting from knuckle to shoulder height (LKS):
   LKSM - LKS at the frequency of one time maximum
   LKS1 - LKS at the frequency of 1 handling/min
   LKS6 - LKS at the frequency of 6 handlings/min

3) Lowering from knuckle to floor height (LOW):
   LOWM - LOW at the frequency of one time maximum
   LOW1 - LOW at the frequency of 1 handling/min
   LOW6 - LOW at the frequency of 6 handlings/min

4) Two hand front carrying for 4.3 m (C):
   CM - C at the frequency of one time maximum
   C1 - C at the frequency of 1 handling/min
   C6 - C at the frequency of 6 handlings/min.

The selection of two ranges of lift was made on the basis of different muscle groups involved in the lifting activity. The floor to knuckle lift required the work of a combination of arm, back, and leg muscles while the knuckle to shoulder lift required arm and back muscle involvement. The carrying distance of 3.4 m was selected according to the previous studies (Snook 1978, Ayoub et al., 1982, Mital and Ilango 1983), and it represented a medium carrying distance in industry.

Combined activities were chosen for two reasons: first, they simulated industrial situations and, second, they allowed for the examination of the relationship between individual and combined activities. The following combined activities were selected in this study:

1) Lifting FK + Carrying 3.4 m (LC):
   LCM - LC at the frequency of one time maximum
   LC1 - LC at the frequency of 1 handling/min
   LC6 - LC at the frequency of 6 handlings/min

2) Lifting FK + Carrying 3.4 m + Lifting KS (LCL):
   LCLM - LCL at the frequency of one time maximum
   LCL1 - LCL at the frequency of 1 handling/min
   LCL6 - LCL at the frequency of 6 handlings/min

3) Lifting FK + Carrying 3.4 m + Lowering KF (LCLO):
   LCLOM - LCLO at the frequency of one time maximum
   LCLO1 - LCLO at the frequency of 1 handling/min
   LCLO6 - LCLO at the frequency of 6 handlings/min.

The frequency of handling has been recognized as a critical task variable in MMH activities. It has been shown that a definite corresponding decrease in weight that can be lifted occurs with an increase in the frequency of lift (Snook and Irvine 1968, Jorgensen and Poulsen 1974, Ayoub et al. 1978, Snook 1978; Asfour 1980, Mital 1980 and Bakken 1983). The selected frequencies of MMH activities in this study were one and six handlings per minute. They represent an infrequent and a frequent handling in MMH activities. One time maximum capacities for all task conditions were also determined psychophysically. It provided a valuable reference for a subject's maximum voluntary capacity.

Two replications were conducted in order to compare the initial loads for each activity and to examine the repeatability of the psychophysical methodology. If two psychophysical measurements under a certain test condition were found to have a difference of 15% or more (Ciriello and Snook 1983, Griffin et al. 1984), another test was conducted until the difference between the two values was within 15%.

The average of these two measurements served as the capacity for that subject under a fixed task condition. The psychophysical capacity data for this study and the comparison with the data from previous researchers (Ciriello and Snook 1983, Snook 1978, Ayoub et al. 1978) is show in Table 1.

Table 1
Comparison of the MMH Capacities with Previous Studies (Male Subjects Only) *

	Present ** (N=12)	Ciriello and Snook, 1983 +	Snook 1978 ++	Ayoub et al. 1978
LFKM	73.14(18.86)	61.00(16.32)	46.00	
LKSM	46.09(6.95)	44.00(8.09)	36.00	
LOWM	80.77(21.05)	70.00(14.00)	53.00	
CM	79.09(19.55)	66.00(15.18)	44.00	
LFK1	40.32(10.18)	41.00(11.32)	29.00	27.82(7.68)
LKS1	30.59(4.50)	34.00(7.41)	26.00	26.27(6.68)
LOW1	51.68(11.18)	43.00(11.59)	33.00	
C1	43.64(7.27)	44.00(12.09)	32.00	
LFK6	26.23(5.68)	27.00(7.91)#	22.00#	25.05(7.05)
LKS6	23.18(4.86)	24.00(5.91)#	21.00#	23.50(6.95)
LOW6	38.82(7.55)	32.00(8.18)#	23.00#	
C6	28.73(5.95)	27.00(9.68)	23.00	

* all capacities units in kg, the values in parentheses are standard deviations

** box size was 29.2 cm in sagittal plane, capacities for one hour work duration, in all other studies capacities were estimated for 8 hrs work duration

\+ box size is 36.1 cm with handles

\# frequency is 6.7 handlings/min

++ box size is 36.1 cm and capacity for 50 percentile male

This experiment was different from other psychophysical experiments (Mital 1980, Asfour 1980, Aghazadeh 1982, Plott 1983, Hafez 1984) in that the subjects were told to work at their maximum acceptable level of performance for a one hour work bout instead of selecting an acceptable level of performance for an eight hour period. The one hour duration was selected because it more closely approximated an intermittent work schedule that would be encountered when heavy MMH activities occur only occasionally, such as unloading a truck of boxed material or participating in emergency services (Smith and Jiang 1984, Legg and Pateman 1984). For each simulated MMH activity, the subjects were allowed to adjust the weight of the load to the maximum amount that they could handle continuously for at least one hour, working as hard as they could, without becoming unusually tired, weakened, overheated, out of breath or straining themselves (Ayoub et al. 1978).

Heart rate (HR) was monitored by a small HR monitoring device called an "Exersentry." (Respironics Inc.) This device was described in detail by Smith and Wilson (1983). This HR monitoring device has been proven to be sufficiently reliable, convenient, inexpensive, self-constrained and to cause minimum interference with the operator (Smith and Wilson 1983, Plott 1983). "Exersentry" was especially suitable for the present study due to its light weight and lack of tethered wires, thereby minimizing interruption when the subject was conducting a carry activity.

The box used in the experiment was constructed using 1.3 cm plywood without handles. The size of the box was 44.5 cm (frontal) x 29.2 cm (sagittal) x 24.8 cm (vertical). There were two runners built at the bottom of the box 8.9 cm from both edges.

No information regarding performance was made available to the subject. During performance of the tests, the subject was isolated from other subjects in order to eliminate competition and other external influences.

Each data point was collected for approximately thirty minutes. A rest period of at least twenty minutes followed.

DISCUSSION:

Comparison of stressfulness for different MMH activities:

According to the survey conducted by Drury et al. (1982), the most common MMH activities in industry are lifting, lowering, and carrying, in that order. It was pointed out by Ayoub et al. (1983) that "lifting" is the most stressful activity among all MMH activities. However, a quantitative comparison for different MMH capacities has never been made. In the present study, four individual MMH activities--lifting FK, lifting KS, lowering KF, and carrying 3.4 m -- were examined under three frequencies, i.e., one time maximum, one handling per minute, and six handlings per minute. For all three frequencies, lifting KS happened to be the most stressful activity as indicated by the smallest capacity. The capacity for lifting KS was limited by the arm strength from knuckel to shoulder height, and it was assisted relatively less by other body parts when compared to other MMH activities in this study. It is concluded that, in order to reduce the risk of injury, lifting KS should be avoided as much as possible. It also means that if a task has to be conducted, such as lifting KS, the weight of the load should be less than that for other tasks, such as lifting FK, lowering KF, or carrying for 3.4 m.

As far as the effects of the frequency of handling, lifting KS was shown to have the smallest capacity compared to other activities tested. For the frequency of one time maximum, lowering capacity and carrying capacity are higher than that for lifting capacity; and they are closer to lifting FK capacity, while much lower than lifting for KS capacity. For the frequency of one handling per minute, the highest capacity occurred at the lowering activity, and lowest capacity occurred at lifting KS activity. For the frequency of six handlings per minute, the decreasing order of stressfulness of individual MMH activities was: lifting KS, lifting FK, carrying for 3.4 m and lowering KF.

The capacities for combined MMH activities have a close relationship with the limiting individual activity that composed the combined activity (Jiang 1984). Table 2 shows the capacity for combined activities, corresponding limiting activity, and the percent of reduction of a combined capacity from the limiting capacity. In general, the percent of reduction for a combined activity is between 0.3 and 26% depending on the activity and the frequency of handling.

Table 2
Capacities for Combined MMH Activities

Combined Activity	Capacity *	Limiting Activity	% of Reduction **
LCM	67.1(17.8)	LFKM	9.3%
LCLM	46.0(6.8)	LKSM	0.3%
LCLOM	64.4(20.4)	LFKM	11.9%
LC1	35.7(7.5)	LFK1	11.4%
LCL1	28.6(5.1)	LKS1	6.4%
LCLO1	32.5(6.1)	LFK1	19.3%
LC6	21.7(4.7)	LFK6	17.2%
LCL6	19.1(3.5)	LKS6	17.7%
LCLO6	19.4(3.4)	LFK6	26.0%

* all capacities units in kg
the values in parentheses are standard deviations

** $\% \text{ of reduction} = 1 - \dfrac{\text{capacity for combined activity}}{\text{capacity for individual activity}}$

In summary, it was shown that the lifting activity was the most stressful activity among the MMH activities examined in this study. It is noted here, for the purpose of job design/redesign, that the risk of injury can be reduced if a task can be changed from a more stressful activity to a less stressful activity, i.e., changing a task from carrying an object for 3.4 m to lowering the object from knuckle to floor level and providing a conveyor to carry the object for the 3.4 m distance. Furthermore, the stressfulness of a combined MMH activity can be reduced by redesigning the limiting activity involved in the combined activity.

Effect of heart rate:

Heart rate is a physiological index for physical work. It has been used with other physiological indices, such as oxygen consumption and blood pressure, to indicate the physiological cost for a certain task. Mean heart rates for each activity are shown in Table 3. In general, the mean heart rates in this study were higher than those for Ciriello and Snook (1983). The major reason was that a short work duration (one hour) was involved rather than an eight hour workday. Comparing the mean heart rate data with the physiological strain and work duration categories in the literature (Astrand et al. 1977, NIOSH 1981), the mean heart rate for activities at the frequencies of one handling per minute and six handlings per minute were categoried as moderate and heavy work, respectively. The mean heart rate values corresponding to each MMH activity for frequencies of 1 and 6/min are depicted also in Figure 4.

To conclude this study, the following MMH job design/redesign recommendations are made:

1. For individual MMH activities, the average capacity for the frequencies of one time max, 1/min, and 6/min are in the following decreasing order: lowering knuckle to floor height, carrying for 3.4 m, lifting FK, and lifting KS. Therefore, it is desirable to design a less stressful task rather than more stressful task, e.g., to design a low-

ering task rather than a lifting task, or to design a lifting FK task rather than a lifting KS task.

Table 3
Comparison of Mean Heart Rate Data with Data From Ciriello and Snook (1983)

	Present (N=12)	Ciriello and Snook (1983)
LFK1	98.3(12.4) *	84(6.6)
LKS1	90.5(11.5)	84(7.9)
LOW1	90.1(8.5)	84(7.7)
C1	93.5(13.3)	93(9.3)
LC1	100.2(10.4)	
LCLO1	101.7(11.1)	
LCL1	101.0(11.2)	
LFK6	126.3(11.9)	116(9.30) **
LKS6	108.7(11.3)	108(20.2) **
LOW6	115.7(11.6)	103(12.3) **
C6	115.7(14.3)	109(10.8)
LC6	133.9(14.2)	
LCLO6	133.7(13.5)	
LCL6	139.4(15.3)	

* the values in parentheses are standard deviations
** based on frequency of 6.7 handlings per minute

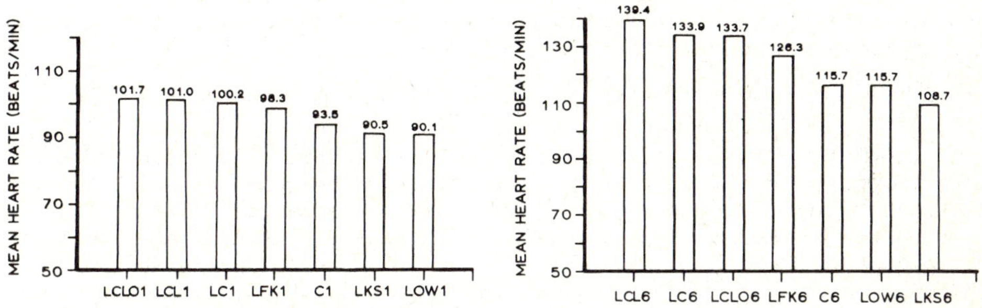

Figure 4
Mean Heart Rate for MMH Activities in Present Study

2. The capacity for a combined MMH task is less than the capacity for the limiting individual activity. The range of reduction is between 0.3 and 26% depending on the number of individual activities that composed the activity and the frequency of handling.

3. Lifting activity was identified as the critical activity among lifting, lowering, and carrying activities in this study. For a high frequency of handling (greater than 1/min), the mean capacities are in the following decreasing order: lifting FK, lifting KS, lifting FS, lifting KR, lifting FR, and lifting SR. For a low frequency of handling (less than 1/min), the highest capacity occured in lifting FK and the lowest capacity occured in lifting SR, however, the capacities for other

activities interacted with the frequency of handling.

4. High frequency of combined MMH tasks (e.g., lifting a box FK, carrying the box for 3.4 m, and then lifting the box KS) are not recommended because of their high physiological cost. This also means that the resting time is too short to recover for high frequency combined MMH activities. However, light work (such as walking) combined with a stressful MMH task is recommended, because a light exercise can accelerate the elimination of lactic acid based on physiological principles.

REFERENCES:

[1] Accident Facts, 1959-1981 Editions (National Safety Council, Chicago, Illinois).

[2] Aghazadeh, F., Simulated Dynamic Lifting Strength Models for Manual Lifting, Ph.D. Dissertation, Texas Tech University (1982).

[3] Asfour, S. S., Energy Cost Prediction Models for Manual Lifting and Lowering Tasks, Ph.D. Dissertation, Texas Tech University (1980).

[4] Astrand, P. and Rodahl, K., Textbook of Work Physiology (McGraw Hill, New York, 1977).

[5] Ayoub, M. A., Control of manual lifting hazards: I. training in safety handling, Journal of Occupational Medicine 24 (1982a) 573-577.

[6] Ayoub, M. A., Control of manual lifting hazards: II. job redesign, Journal of Occupational Medicine 24 (1982b) 668-676.

[7] Ayoub, M. M., Bethea, N. J., Deivanayagam, S., Asfour, S. S., Bakken, G. M., Liles, D., Mital, A., and Sherif, M., Determination and Modeling of Lifting Capacity, Final Report, HEW (NIOSH) Grant No. 5R01OH-000545-02 (1978).

[8] Ayoub, M. M., Denardo, J. D., Smith, J. L., Bethea, N. J., Lambert, B. K., Alley, L. R., Duran, B. S., Establishing Physical Criteria for Assigning Personnel to Air Force Jobs. Final Report, Air Force Office of Scientific Research, Contract No. F49620-79-C-006 (1982).

[9] Ayoub, M. M., Gidcumb, C. F., Hafez, H., Intaranont, K., Jiang, B. C., and Selan, J. L., A Design Guide for Manual Lifting, A Draft Report Submitted of OSHA (1983).

[10] Bakken, G. M., Lifting Capacity Determination as a Function of Task Variables, Ph.D. Dissertation, Texas Tech University (1983).

[11] Ciriello, V. M., and Snook, S. H., A study of size, distance, height, and frequency effects on manual handling task, Human Factors 25 (1983) 473-483.

[12] Drury, C. G., Law, C. H., and Pawenski, C. S., A survey of industrial box handling, Human Factors 24(1982) 553-565.

[13] Griffin, A. B., Troup, J. D. G., and Lloyd, C. D. E. F., Tests of lifting and handling capacity: their repeatability and relationship to back symptoms, Ergonomics 27 (1984) 305-320.

[14] Hafez, H. A. An Investigation of Biomechanical, Physiological and Environmental Heat Stresses Associated with Manual Lifting in Hot Environments, Ph.D. Dissertation, Texas Tech University (1984).

[15] Jiang, B. C., Psychophysical Capacity Modeling of Individual and Combined Manual Materials Handling Activities, Ph.D. Dissertation, Texas Tech University (1984).

[16] Jorgensen, K. and Poulsen, E., Physiological problems in repetitive lifting with special reference to tolerance limits to the maximum lifting capability, Human Factors 25 (1983) 493-506.

[17] Kroemer, K. H. E, An isoinertial technique to assess individual lifting capability, Human Factors 25 (1983) 493-506.

[18] Lahey, J. W., Bearing down on musculoskeletal disorders, National Safety News (1984) 37-39.

[19] Legg, S. J. and Pateman, C. M., A physiological study of the repetitive lifting capabilities of healthy young males, Ergonomics 27 (1984) 259-272.

[20] Mital, A., Effects of Task Variable Interactions in Lifting and Lowering, Ph.D. Dissertation, Texas Tech University (1980).

[21] Mital, A. and Ilango, M., Load characteristics and manual carrying capacities, in: Proceedings of the Human Factors Conference 27th (1983) 274-278.

[22] Mital, A., Revised lifting capability data base for the industrial workforce, in: Proceedings of the Human Factors Society 28th (1984) 581-585.

[23] NIOSH, U. S. Work Practice Guide for Manual Lifting (Publication No. 81-122, 1981).

[24] Plott, C. C., Alternative Strength Testing Methods for Employee Screening, Master Thesis, Texas Tech University (1983).

[25] Smith, J. L. and Jiang, B. C., A manual materials handling study of bag lifting, American Industrial Hygiene Association Journal 45 (1984) 505-508.

[26] Smith, L. A. and Wilson, D., An inexpensive system for remote monitoring of work heart rate, Ergonomics 26 (1983) 719-722.

[27] Snook, S. H. and Irvine, C. H., Maximum frequency of lifting acceptable to male industrial workers, American Industrial Hygiene Association Journal 29 (1968) 531-536.

[28] Snook, S. H., The design of manual handling tasks, Ergonomics 21 (1978) 963-985.

Trends in Ergonomics/Human Factors II
R.E. Eberts and C.G. Eberts (Editors)

NEW HORIZONS FOR ERGONOMICS RESEARCH IN LOW BACK PAIN

T. M. Khalil, S. S. Asfour, E. A. Moty, R. Steele and H. L. Rosomoff

College of Engineering and School of Medicine
University of Miami
Coral Gables, Florida 33124

Low back pain is one of the most prevalent and costly problems facing society today. Past research in ergonomics and safety has concentrated on manual materials handling and injuries associated with it. This paper reveals that problems of manual materials handling represent but a small fraction of the problems contributing to back pain. Statistics collected from hospital records on workers compensation low back pain cases over three years is used to illustrate areas of concern. Figures reveal that a much expanded horizon in ergonomics research and application is needed to effect a meaningful contribution to the solution of this problem.

The pioneering approach utilized by the University of Miami, in putting ergonomists in a hospital environment to work jointly with physicians and professionals from other disciplines, has provided an excellent forum for recognizing and solving many complex problems related to back pain. A case study is used to illustrate results obtained.

INTRODUCTION

Low back pain is one of the most prevalent and costly problems facing our society today. Many researchers and agencies have reported on the various aspects of low back pain. Exact figures are not available because there is no central reporting agency. Available statistics indicate that almost everyone (4 out of 5 people) at some time in his or her life, will experience back pain and will be incapacitated by it. (Nachemson, 1971; Eagle, 1979; Jensen, 1980; Locke, 1983).

Kelsey and White (1983), estimated that low back pain is the diagnosis in 10% of all chronic health conditions. White (1983), estimated the number of people in the United States alone complaining of low back pain to be 70 million people, and half of them are in the working group between 21 and 45 years of age.

Anderson et al. (1984) estimated that an average of 28.6 days/100 workers are lost each year because of back pain. Loeser (1979) put the losses to be equivalent to 170 million working days each year in the United States.

Back pain statistics reported in the literature has a great deal of variability. However, a common denominator that cannot be doubted is that low back pain problem is a major source of incapacitation, suffering, and cost to the world today. It significantly affects industrial productivity and it has a marked impact on the gross national product of a nation. Work and lost productivity represent but one aspect of the expensive back-pain problem. Other important aspects include humanitarian consideration of the injured individual, the billions of dollars spent each year on treatment and surgeries, and most importantly the unsuccessful restoration of functional abilities.

Low back pain represents a major economic problem both at the national and at the organizational levels. The staggering statistics (Khalil et al. 1983, Asfour et al. 1983) make the prevention and treatment of low back pain a compelling priority from a humane as well as from an economic point of view. A major challenge becomes the development of pragmatic means and methods dealing with the problem, i.e. the management of the low back pain problem.

THE CAUSES OF LOW BACK INJURY

The onset of low back pain has often been attributed to heavy work involving lifting (Keegan, 1953; Magora, 1973; Anderson, 1983, Troup, 1965). Past research in ergonomics and safety has concentrated on manual materials handling and injuries associated with it (Ayoub et al. 1978a,b, Chaffin 1975, Chaffin et al. 1973, Davis et al. 1977a, b, Snook 1978). More emphasis was placed on lifting research than any other type of research associated with the problem of manual materials handling or low back pain. The ergonomics research in manual lifting was very useful in establishing the work practices guide published by The National Institute for Occupational Safety and Health (NIOSH) in 1981. The guidelines established are useful and would hopefully lead to a reduction in injury statistics and in preventing low back pain. However, based on the data collected at the University of Miami Back Pain Center, cases of back pain resulting from manual lifting tasks represent but a small fraction of the total number of cases of low back pain.

There are many causes and conditions that could lead to back injury and/or result in low back pain. Kelsey (1975) showed that the risk of getting back pain was highest in sedentary occupations and driving motor vehicles. Those who drive for a living are 3 times more likely to suffer from back disorders than the rest of the population (Troup, 1978). Sitting for prolonged periods increases back symptomology (Magora, 1972; Nachemson and Elfstrom, 1970).

A review of one hundred cases of low back injury obtained from records of patients admitted over a one year period to the Comprehensive Pain Center at the University of Miami, was made. All cases reviewed were Worker's Compensation patients. A summary of the findings is given in Table (1). It can be seen from the statistics collected that back pain is caused by a variety of tasks, which may or may not involve load handling, spans all professions, affects both males and females and reoccurs in many victims, and at any time of the day. Sixty-eight percent of the sample studied was males compared to 32% females. This may be a reflection on the pattern of employments where male population still predominates the industrial environments. However, existing information and our own observations indicate equal incidents between males and females. Most female patients were involved in secretarial, nursing and housekeeping activities. In general most of the female population was involved in light work load whereas most of the male population such as truck drivers, mechanics and carpenters were involved in heavy workloads. The major cause of low back injury in our sample was due to slipping and falling (38%), followed by lifting (20%), car accidents (12%) and simple bending or reaching (10%).

1 -	Sample Size	100				
2 -	Sex	68% Males		32% Females		
3 -	History of prior injuries	32% Yes		65% No		
4 -	Occupation:	Truck drivers	12%	Mechanics	14%
		Carpenters	8%	Sedentary	18%
		Electricians	6%	Painters	4%
		Others	36%	No Work	2%
5 -	Cause of Injury:	Slip and Fall	38%	Lifting	20%
		Car Accident	12%	Bending	10%
		Struck by	9%	Pushing	4%
		Others	7%		
6 -	Time of Injury:	AM	40%		
		PM	38%		
		Do Not Remember	22%		

TABLE 1

Statistical Summary of 100 Cases of Low Back Pain

THE UNIVERSITY OF MIAMI PROGRAM

Figures cited above reveal that a much expanded horizon in Ergonomics research and application is needed to effect a solution to the low back pain problem. Research directed towards prevention of back injury as well as quick restoration of functional abilities should prove to be extremely useful.

A successful low back pain rehabilitation program that is capable of returning a low back pain patient to full-time employment would significantly increase productivity and reduce rising medical and disability costs. The University of Miami Comprehensive Pain Center, Back Pain Rehabilitation Program provides a good example for such a program. A detailed description of the program may be found in Rosomoff et al. (1981), Khalil et al. (1983). Several back pain programs and clinics have attempted various approaces similar to the one utilized by the University of Miami. It is however the integration of all the elements in an aggressive, comprehensive, multidisciplinary, workable program that is unique in the University of Miami's approach. More than 70 full-time professionals from various disciplines are applying their knowledge, simultaneously and in an integrated fashion to the solution of a patient's low back pain problem. Ergonomics research and application is an integral part of the program encompassing all its phases including diagnosis treatment and post injury management. In spite of the high health professional to patient ratio according to available cost figures, the Program proves to be a highly cost/effective approach (Asfour et al., 1983, Steele, 1983).

ROLE OF ERGONOMICS

Ergonomics is both a science and a technology and, when introduced into an organization, it has to be integrated into the formal organization and into the technical development projects within that organization. The task of the ergonomist is first to determine the capabilities, capacities, and limitations of human beings and then to attempt to build or modify the work environment according to these capabilities and limitations (fitting the environment to the man). Since ergonomics deals with human functional abilities and the efficiency of human performance, there is a good reason for applying ergonomics to problems of occupational health and rehabilitation.

The pioneering approach utilized by the University of Miami, in putting ergonomists in a hospital environment to work jointly with physicians and professionals from other disciplines, has provided an excellent forum for recognizing and solving many complex problems related to back pain. The Ergonomics Division of the University of Miami Comprehensive Pain Center is mainly concerned with the evaluation of the functional ability or performance of each patient. A performance profile is established for each patient upon admission to the program and progress is monitored throughout the program. Ergonomics evaluations of patients with low back pain are important and essential to optimize treatment and prognosis.

FUNCTIONAL ABILITIES RESEARCH AND APPLICATION

Back injuries will continue to occur to a large segment of the population. Most human performance techniques measures and data available in the literature pertain to healthy populations. These may not be directly applicable to injured, unhealthy or disabled populations. When an individual suffers an injury, particularly back injury a change occurs in the individual's characteristics resulting oftenly in a loss of functional abilities. Therefore restoration of the functional ability becomes the aim of most back pain rehabilitation programs. Ergonomics research in the areas of evaluation, methods and techniques of restoration of functional abilities is greatly needed.

At the University of Miami, a patient's evaluation includes the description of the following characteristics of the individual: anthropometry, strength, kinesiological and psychomotor abilities. These parameters help establish a human performance profile on the individual. Patients are followed throughout the program to determine their progress and evaluate the effectiveness of the rehabilitation programs in raising their level of performance. Measurements taken are intended to evaluate functional abilities rather than absolute or maximum levels. This is particularly pertinent to strength measurements where the Patients Acceptable Maximum Effort (AME) level is measured as opposed to their maximum strength capability. The latter measure is not realistic or practical in cases where injury and/or pain are present. AME defines the level of voluntary muscular effort byyond which a patient's level of pain becomes unacceptable. It determines the level at which a patient can perform tasks without excessive pain and discomfort.

The static lifting strength test system and postures described in Chaffin (1975) and Ayoub (1978) were adopted for the determination of the AME measures. The tests selected were those of the arm, shoulder, leg, back and composite strength test positions. These tests were selected because they utilized different muscle groups that are commonly used in daily activities. Acceptable Maximum Effort Measurements (AME) were taken for each worker's compensation case patient admitted into the program. AME was repeated up to three times, for every posture, and the two closest readings were averaged to be reported as a single AME value. The AME measurements were repeated for the same subjects two weeks after being admitted to the University of Miami program, then 4 weeks into the program.

In order to ascertain the repeatability and consistency of the AME measure, a series of paired t-tests were performed. The tests objective was to determine if there is a significant difference between values of the AME replications. For a sample size of 88 patients, results showed that at 1 percent level there was no significant difference between replications. This finding indicates that the patients were able to achieve their AME consistently. This finding was true throughout the program. Thus the AME measure proved to be both a valid and a reliable measure in the evaluation of the functional abilities of low back pain patients.

A CASE STUDY

Case: B.F.
Sex: Male
Age: 58 years
Occupation: Carpenter
Job Demands: Lifts, carries and installs doors. Cuts wood, frames buildings, mixes concrete, puts down metal drains, puts in metal or wooden studs, performs roughs and finish carpentry.
How Injury Occurred: In 1982, patient stepped in a hole while carrying a load of about 40 lbs.
Medical History: No prior back injuries.

ERGONOMICS EVALUATION

	Initial	Midterm	Final	% Change
1 - AME, lbs.				
Composite	66	117	265	+301%
Shoulder	59	74	110	+ 86%
Arm	39	62	122	+213%
Leg	128	260	288	+125%
Back	75	158	237	+216%
2 - Trunk flexion, degrees	40	40	125	+213%
3 - Walking pace, ft/sec.	2.54	1.7	1.54	+ 65%

Discussion: After his injury in 1982, patient suffered severe low back pain radiating to the right lower extremity which incapacitated him. Six months after his injury and being unable to return to gainful employment he was admitted to the University of Miami's program. At the end of a 4 week low back pain rehabilitation program patient returned to his initial job fully functional.

DISCUSSION AND CONCLUSIONS

The problem of low back pain is one of the most prevalent and costly problems in our society today. Many researchers and practioners have contributed their time and talent towards finding a prevention methodology or a cure to this problem. It seems however, that the occurring incidences and associated cost defies all effort undertaken so far. Many researchers, particularly from the disciplines of Ergonomics and Safety Engineering have concentrated their effort on prevention strategies of back injuries. This obviously is the most effective approach for avoiding the problem. Unfortunately it seems from the many cases reviewed by the authors, that injuries to the back will continue to occur at a very high rate. This is primarily due to the multitude of factors contributing to a back injury. Very few researchers in Ergonomics have devoted their effort to post injury management of low back pain. This was always thought of as the sole domain of the medical and health care professionals. Traditionally many physicians have opted to perform corrective surgeries to the back in an effort to relieve low back pain. Unfortunately surgical intervention has shown very little effectiveness in relieving the back pain problem particularly on a long term basis. Aggressive conservative pain management programs seem to be more effective in the treatment and restoration of functional abilities. The contribution of ergonomics in the treatment and evaluation of these programs is essential. As a matter of fact, Ergonomists hold the distinctive advantage of being able to participate in the effort to solve the low back pain problem at three phases of the problem: 1) pre-injury, through prevention; 2) injury, through evaluation and treatment techniques; and 3) post injury, through design and modification of the task and work environment. The University of Miami back rehabilitation program concentrates on that three pronged approach. Involvement in the injury evaluation and treatment program proved to be quite successful in establishing functional evaluation parameters and testing them on patient population. Ergonomic reasoning became an integral part of a large multidisciplinary involvement which proved to be quite effective in regaining functional capabilities to individuals with injured backs. It also permitted for the first time an objective, quantitative evaluation on patient's ability to perform specific tasks. This is quite significant in matching the tasks and jobs to individuals and results in safer and more appropriate job placement.

REFERENCES:

[1] Anderson, G. B. J., Pope, M. H. and Frymoyer, Epidemiology, in: Pope et al. (eds.), Occupational low back pain (Preager, N. Y., 1984) 101-113.

[2] Asfour, S. S., Khalil, T. M., Moty, E. A., Steele, R. and Rosomoff, H. L., Low back pain: a challenge to productivity, Proceedings of the 7th international conference on production research, (Windsor, Canada, 1983).

[3] Ayoub, M. M., Bethea, N. J., Deivanayagam, S., Asfour, S. S., Bakken, G. M., Liles, Mital, A. and Sherif, M., Determination and Modeling of lifting capacity, Final Report, DHEW (NIOSH) Grant 5 Rol OH 00545-02

(1978a).

[4] Ayoub, M. M., Dryden, R. D. and Knipfer, R. E., Psychophysical based models for the prediction of lifting capacity of the industrial worker, Automotive Engineering Congress and Exposition Conference (Detroit, 1976).

[5] Ayoub, M. M., Dryden, R. D., McDaniel, J. W., Knipfer, R. E., and Aghazadeh, F., Modeling of lifting capacity as a function of operator and task variables, In: Drury, C. G. (ed.), DHEW (NIOSH) Pub. No. 78-185 (1978b), 120-130.

[6] Chaffin, D. B., Ergonomics guide for the assessment of human static strength, Am. Ind. Hgy. Assoc. J. 36 (1975) 505.510.

[7] Chaffin, D. B., Manual Materials handling and low back pain, In: Zenz, C. (ed.), Occupational Medicine, Principles and Practical Applications (year book Medical PUblishers, Inc., Chicago, IL., 1975) Chapt. 19.

[8] Chaffin, D. B. and Park, K. S., A longitudinal study of low back pain as associated with occupational lifting factors, Am. Ind. Hgy. Assoc. J. 34 (1973) 513-525.

[9] Davis, P. R., Posture of the trunk during the lifting of weights, Br. Med. J. 1 (1959) 87-89.

[10] Davis, P. R. and Stubbs, D. A., Safe levels of manual forces for young males 1, Appl. Ergonomics 8 (1977a) 141-150.

[11] Davis, P. R. and Stubbs, D. A., Safe levels of manual forces for young males 2, Appl. ERgonomics 8 (1977b) 219-228.

[12] Eagle, R., A pain in the back, New Scientist (1979).

[13] Jensen, G. M., Biomechanics of the lumbar intervertebral disk; A review, Physical therapy 60 (1980) 765-773.

[14] Keegan, J., Alterations of the lumbar curve related to posture and seating, J. Bone Joint Surge Am. 35 (1953) 589-603.

[15] Kelsey, J. L., An epidemiological study of acute herniated lumbar intervertebral discs, Rheumatol. Rehab. 14 (1975) 144-159.

[16] Kelsey, J. L. and White, A. A., Epidemiology and impact on low back pain, Spine 5 (1980) 133-142.

[17] Khalil, T. M., Asfour, S. S., Moty, E. A., Rosomoff, H. C. and Steele, R., The management of low back pain: A comprehensive approach, In: Proceedings of the annual Ind. Eng. Conf. (Louisville, Ky., 1983).

[18] Locke, T. C., Stretching away from back pain injury, Occup. Health and Safety (1983) 8-14.

[19] Loeser, T. A., Low back pain: Introduction to planary session. In: Bonica, T. T. et al. (eds.) Advances in Pain Research 3 (Raven Press,

[20] Magora, A., Investigation of the relation between low back pain and occupation, Scand. J. Rehab. Med. 5 (1973) 186-190.

[21] Nachemson, A., Low back pain: its etiology and treatment, Clin. Med. Fam. Fl. (1971) 18-23.

[22] Nachemson, A. and Elfstrom, G., Intravital dynamic Pressure measurement in lumbar discs, Scand. J. Rehab. Med. Supp. (1970).

[23] Rosomoff, H. L., Green, C., Gilbert, M. and Steele, R., Pain and low back rehabilitation program at the University of Miami School of Medicine, In: Lorenz, K. Y. (ed.) New approaches to treatment of chronic pain, NIDA research monograph 36 (1981).

[24] Snooke, S. H., The design of manual handling tasks, Ergonomics 21 (1978) 963-985.

[25] Steele, R., Is the pain team cost effective. 1st National Conf. "Back Pain; Cutting the Cost of Employers", Chicago, Il. 1983.

[26] Troup, J. D. G., The relation of lumbar spine disorders to heavy manual work and lifting, Lancet 1 (1965) 857-861.

[27] Troup, J. D. G., Driver's back pain and its prevention, Applied Ergonomics (1978) 207-214.

[28] White, A. H., and Mattmiller, A. W. (1983 Backschool and other conservative approaches to low back pain, C. V. Mosby Co., St. Louis, 1983).

Trends in Ergonomics/Human Factors II
R.E. Eberts and C.G. Eberts (Editors)

HUMAN PERFORMANCE TIMES IN THE USE OF MANUAL HOIST SYSTEMS

Jeffrey C. Woldstad

Center for Ergonomics
University of Michigan
Ann Arbor, Michigan

Increasing use of material movement assist devices has highlighted the need for a better understanding of human interface with such systems. As a first step in this regard, this study reports standard movement times collected using a small overhead hoist under varying load, distance, and tolerance conditions. Performance times are modeled based upon a modification of Fitts' Index of Difficulty. Results are compared to those obtained for manual movement times under analogous conditions. Recommendations for the design and use of similar systems are proposed based upon the models developed.

INTRODUCTION

Recent emphasis on the danger of manually lifting heavy loads has prompted an increased interest in material handling assist devices. Use of a hoist or other mechanical handling device is often the most practical means of controlling operator risk in strenuous lifting situations. Unfortunately, methods work in the area of hoist use and design has been extremely limited. As such, operators often balk at the prospect of using assist devices that are more awkward and slower than manual methods. As a first step towards improving the interface between material handling assist devices and the worker, this report will document standard movement time values obtained using a small state-of-the-art pneumatic hoist. Performance times obtained will then be compared to manual carry times and to predetermined time system estimates.

In one of the first applications of Information Theory to human performance, Paul Fitts [3] developed the Index of Difficulty (ID) as a predictor of human movement time. The relationship, known as "Fitts Law" is of the form

$$MT = a + b(ID) = a + b \log_2 (2D/W),$$

where movement distance (D) is defined as the distance between two targets of fixed width (W) in a reciprocal tapping task. Fitts equation has been found to be an excellent predictor of mean movement time in a variety of situations, often accounting for over 90 percent of the total variation of average movement times. With slight variations in the regression parameters, Fitts index has been applied successfully to such diverse activities as foot motions [2], microscopic motions [6], underwater movements [5], and remote manipulators [7].

Several authors (Chung [1]; Welford [8]; and Woodworth [9]) have proposed that movements are composed of two distinct phases. Phase I, labeled "ballistic" movement is postulated to be preprogrammed before movement occurs, and is intended to bring the movement within an initial zone of radius W_0 around the target. This phase is followed by a slower feedback controlled phase labeled "homing" movement which brings the movement within the final tolerance W. This and similar models suggest that Fitts original equation should be expanded to the form :

$$MT = a + b(ID1) + c(ID2) = a + b \log_2 (D/W_0) + c \log_2 (W_0/W),$$

which reduces to :

$$MT = a + b \log_2 D - c \log_2 W$$

if W_0 is considered an unknown constant. In most instances prediction accuracy has been slightly improved when employing this modified model [1]. Although empirically very good, equations based on Fitts law are as of yet not well understood. However, the compact form of the equation and the robust relationships make them ideally suited to the development of standard time data [6].

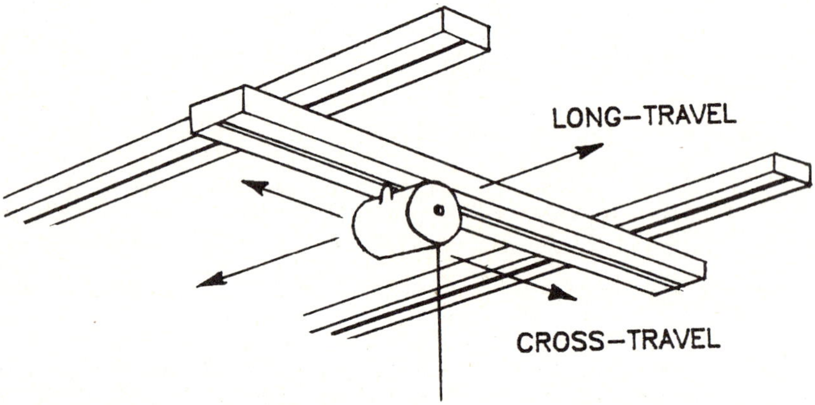

Figure 1: Diagram of hoist apparatus and directions of travel.

PILOT STUDY

A preliminary study was initially undertaken to investigate the effects of load mass transported, movement direction, positioning tolerance, and distance of move on movement times employing a hoist device. A single subject was employed in a balanced design involving three levels of each factor. The main question of interest in this investigation was whether the direction of hoist movement (long-travel, cross-travel, or mixed) had a large effect on movement time. Long-travel was defined as movement parallel to the main tracks of the carriage assembly, while cross travel moved perpendicular to these tracks (see Figure 1). It was hypothesized that the increased mass due to moving the entire cross-bridge assembly in long travel would substantially increase movement time. Results of the preliminary investigation indicated that direction of travel had very little effect on overall movement time. As such, movement direction will not be considered in the main investigation.

METHOD

SUBJECTS : One female and two male subjects were employed. The male subjects (S1 & S2) were current University students, while the female subject (S3) was the spouse of a student. Subject ages ranged from 22 to 25 years. All subjects were paid $5.00 dollars per hour for participation in the study. Subjects professed no problems that would prohibit potentially strenuous physical activity, and all were selected to be of unremarkable anthropometries. S1 was unique among the subjects in that he alone participated in the pilot study and the manual carrying trials.

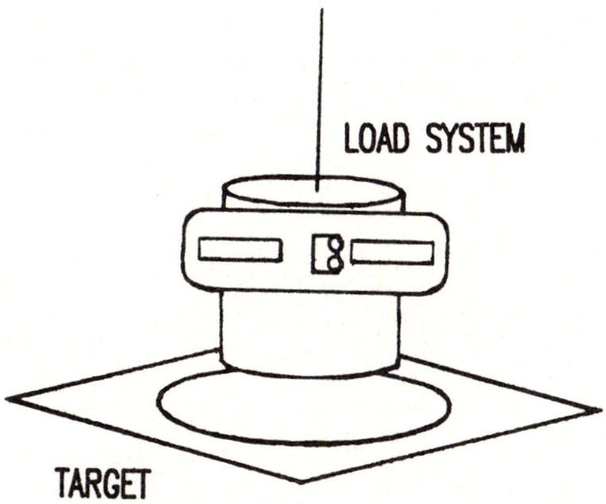

Figure 2 : Diagram of target and load configuration.

APPARATUS : Experimentation was conducted on a Zimmerman Manufacturing (Module-Air, series ZF, 150 lb capacity) pneumatic hoist installed in the laboratory (refer to Figure 1). The hoist apparatus was connected to a 12' x 6' overhead carriage system mounted to the ceiling. The lowest portion of the carriage was approximately 11 ft from the floor. Vertical movement of the hoist system was achieved by actuating two small buttons attached to the load. Horizontal and longitudinal movements were achieved by physically pushing the suspended load and overhead bridge assembly. Data acquisition was facilitated though the use of an IBM Personal Computer (PC) which measured and recorded all movement times on-line. The internal timer of the PC was employed to measure movement times which allowed an accuracy of approximately 0.025 seconds. The load consisted of a medium size (8 qt.) bucket filled with cast iron weights. Force sensitive handles were mounted to the bucket to provide easy hand grips. Handles were situated to allow easy actuation of the push button control (see Figure 2).

TASK : The experimental task was designed to resemble a large reciprocal tapping task. Subjects were asked to transport a load, either manually or employing the hoist, between circular target zones on two small tables (approx. 3 ft high) located in the middle of a large room. As part of the experimental manipulation the load transported, the target tolerance, and the distance traveled were systematically varied. On a typical trial the subject would start at a given target and move immediately to the next table, when the load was placed within the target zone a tone sounded and the subject moved back to the original table. Movement continued back and forth between tables for 10 to 15 movements before the trial was terminated. Movement time was considered the time spent traveling between targets, as well as, time spent on the target before movement began. Subjects were instructed to maintain a walking gait throughout the trials. It was suggested that the activity be performed at a pace they believed could be sustained for several hours of continuous work. In carry trials the subject was given additional instructions to keep the load as close to the body as possible, thus eliminating large swinging actions during movement. All trials using the hoist were conducted in the longitudinal (long-travel) direction of movement (refer to figure 1).

PROCEDURE : A full factorial experimental design was employed incorporating 3 subjects (S1, S2, S3), 3 weight conditions (20 lbs, 50 lbs, 80 lbs), 3 tolerance conditions (±.375 in, ±1.00 in, ±3.00 in), and 5 distance conditions (2 ft, 3 ft, 4 ft, 6 ft, 10 ft). S2 and S3 completed 15 movements per trial for the 45 different conditions while S1 completed only 10 movements per condition. The order of presentation was balanced across subjects for weight conditions, while distance was randomized within weight and tolerance was randomized within distance. Each subject was tested for approximately 2 hours on a single day. Subjects were asked repeatedly throughout the procedure whether they felt fatigued and were given rest accordingly. The manual carrying experiments followed an identical

TABLE 1 : Average manual and hoist movement time values (seconds).

		HOIST			MANUAL	
DIST(ft)	TOL(in)	WT=20lbs	WT=50lbs	WT=80lbs	WT=20lbs	WT=50lbs
2	.375	2.14	3.32	3.29	2.54	2.43
	1.00	1.67	2.92	3.01	2.18	2.14
	3.00	1.59	2.59	2.69	2.01	1.99
3	.375	2.30	3.43	3.79	2.38	3.61
	1.00	2.10	3.16	3.34	2.29	3.07
	3.00	1.73	3.09	3.19	2.35	2.82
4	.375	2.89	3.91	4.35	3.42	3.17
	1.00	2.63	3.49	3.49	3.27	3.13
	3.00	2.62	3.36	3.42	3.14	2.74
6	.375	4.52	4.47	4.75	3.60	3.87
	1.00	4.38	4.04	4.06	3.42	3.53
	3.00	4.24	3.78	3.92	3.32	3.56
10	.375	6.11	5.59	5.54	5.03	5.35
	1.00	6.01	5.23	5.03	4.68	4.72
	3.00	5.64	4.91	4.99	4.67	4.94

procedure except that only a single subject (S1) was employed and the heaviest weight condition was deleted.

As a measure of external validity a Predetermined Motion Time System (PMTS) was incorporated into the study. PMTS values are of interest because they represent the most widely used, highly tested standard times available. The Methods-Time Measurement-1 (MTM-1) system was employed to predict standard time values for the manual carry task. MTM-1 was chosen because it is one of the most popular systems currently in use, and in general has been very well studied in the past [4]. A qualified MTM-1 technician was present during much of the manual carry experimentation and completed his analysis independently.

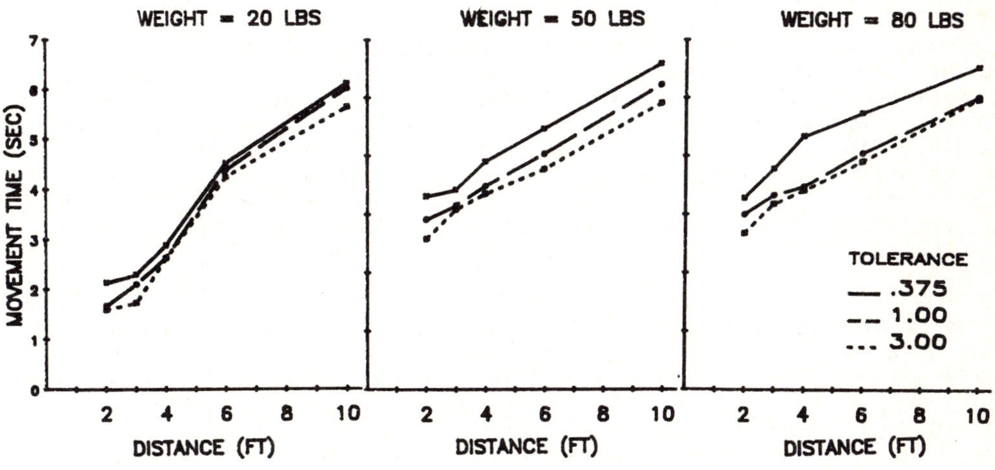

Figure 3 : Average movement times for hoist transport trials.

RESULTS

Movement times within each trial were checked for stability before analysis was begun. The results showed that mean times were sufficiently stable after the first movement, as such the first observation within each trial was excluded from further analysis. Mean movement times for all conditions in the both the hoist and the manual carry portions of the experiment are shown in Table 1. The same values are displayed graphically in figures 3 and 4. Least squares regression techniques were employed to quantify the relationships shown. Models under investigation included different forms of Fitts equation, as well as other simple linear and polynomial models. To simplify the relationships involved, each weight condition was modeled individually. The appropriate models with associated R values are shown in Table 2. All relations and coefficient values shown are highly significant ($p<.001$). Referring to the table, the variable D corresponds to the movement distance (in feet), while ID2 refers to the second index of difficulty in Fitts modified equation (ID2=log (1/tolerance), tolerance in feet). Regression coefficients for the hoist trials indicate that distance

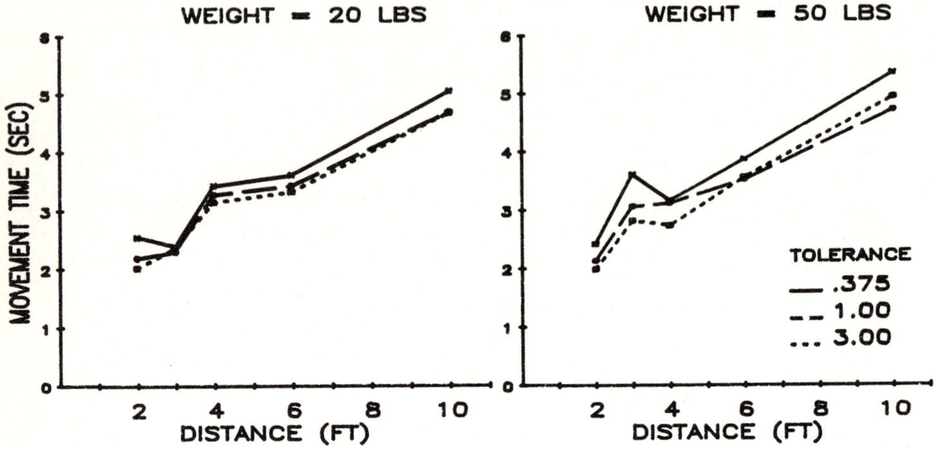

Figure 4. Average movement time for manual carry trials.

has the largest effect on movement time while weight has little effect except in the 20 pound condition. Tolerance is seen to have a relatively minor effect that becomes larger as the load becomes heavier and harder to handle. A somewhat surprising result is the large constant coefficient present in all equations. All relationships are seen to follow essentially

TABLE 2 : Regression equations for average and individual hoist and manual movement times.

WEIGHT	REGRESSION EQN	R2-MEAN	R2-IND
HOIST MOVEMENT :			
20	MT = 2.27 -1.13D +0.37D2 -0.02D3 +0.21ID2	.997	.838
50	MT = 1.69 +0.29D +0.29ID2	.991	.725
80	MT = 1.79 +0.26D +0.33ID2	.960	.633
MANUAL MOVEMENT :			
20	MT = 1.29 +0.32D +0.14ID2	.944	.838
50	MT = 1.26 +0.32D +0.23ID2	.929	.859

TABLE 3 : MTM-1 predicted manual movement times.

DIST(ft)	TOL(in)	WT=20 lbs	WT=50 lbs
2	.375	1.85	2.12
	1.00	1.91	2.18
	3.00	2.03	2.30
3	.375	1.85	2.12
	1.00	1.91	2.18
	3.00	2.03	2.30
4	.375	2.51	2.77
	1.00	2.56	2.83
	3.00	2.68	2.95
6	.375	2.51	2.77
	1.00	2.56	2.83
	3.00	2.68	2.95
10	.375	3.59	3.85
	1.00	3.63	3.91
	3.00	3.76	4.03

the same form except for hoist trials at the 20 lb weight condition. Unlike previous results, movement time is shown to be a function of distance instead of the logarithm of distance.

Results from the MTM-1 analysis are shown in Table 3. In some of the conditions distance was varied, however the subjects basic hand and foot motions remained unchanged. As a result the analyst was unable to distinguish between some of the different distance conditions. It should be noted that the MTM-1 estimates provided were based on observations of the subjects actual method and not upon preferred method considerations. To facilitate comparison, regression equations were calculated based on the MTM-1 predictions. The resulting relationships are shown in table 4. Comparison with the obtained manual movement equations (see table 2) reveals

TABLE 4 : Regression equations for MTM-1 prediction values.

WEIGHT	REGRESSION EQN
20	MT = 1.25 +0.22D +0.08ID2
50	MT = 1.51 +0.22D +0.09ID2

that the MTM-1 system placed smaller emphasis on distance and tolerance effects than actually observed. As such, on the whole MTM-1 estimates seem to be slightly smaller than manual movement values.

CONCLUSION

A comparison of hoist movement times, manual carry times, and MTM-1 prediction values indicates that use of the hoist assist device only slightly increases the time necessary to transport the load. Under equal conditions, the overall average movement time using the hoist device was 3.60 seconds compared to 3.31 seconds for manual handling, and 2.67 seconds for MTM-1 estimates. The majority of hoist movement time is the result of the distance traveled and a large constant time period. Weight is seen to have little effect except in the 20 lb condition, while tolerance has a small but significant effect. The consistency of the regression coefficient for distance suggests that subjects walk at a constant velocity regardless of the weight of the load. Accelerating and decelerating the load and carriage to this speed probably attributes for some of the constant time in each trial.

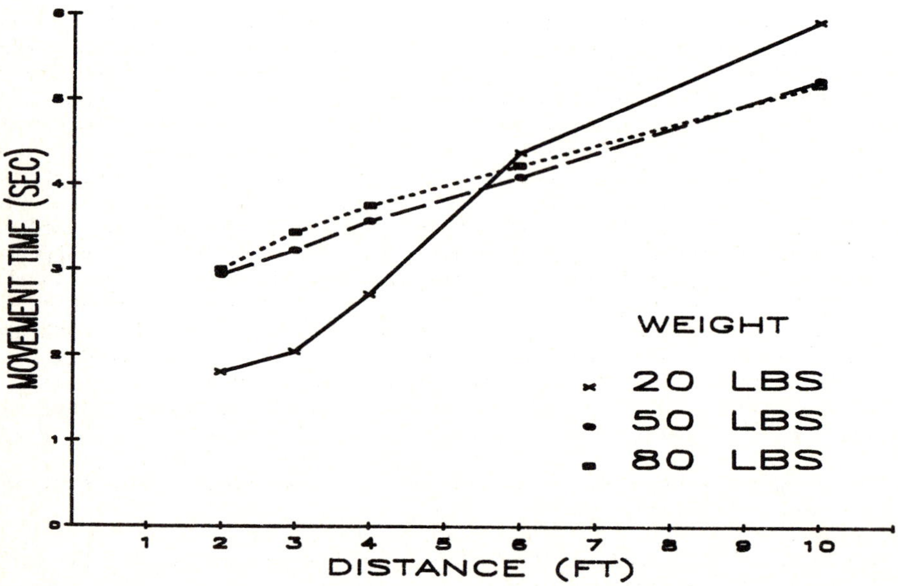

Figure 5. Average hoist movement time for the three weight conditions.

Figure 5 shows hoist movement time as a function distance for the three weight conditions. This figure makes apparent the substantially different results obtained in the 20 pound weight condition. It is suggested that under low weight, short distance conditions subjects were able to decrease transport times by taking advantage of the pendulum action of the hoist cable. Experimental observations suggest that under the best of conditions (2ft, 20lb) it was possible to swing the load from target to target without moving the carriage assembly at all. However, as the mass of the load increased the pendulum area significantly decreased, and the subjects were forced to drag the carriage along with their movements. Longer movement times for the low load, long distance movements can be explained in terms of the acceleration-deceleration effect mentioned earlier. The time necessary to accelerate and decelerate can be expected to increase slightly as the mass of the load increases due to effects of momentum. However, decreasing the load will increase the pendulum area and the time lag between stoppage of the load and the corresponding deceleration of the overhead carriage. Increasing the lag between deceleration of the load and deceleration of the carriage significantly increased placement control problems. In practice subjects where observed to either walk quickly to the target and wait for the carriage to stop before they placed the load, or significantly decrease the speed at which they moved to control carriage momentum. Increasing the load mass relative to the carriage mass decreased the pendulum area, and allowed the force of the load perpendicular to the line of movement to counteract the momentum of the carriage. This development suggests that that in many cases an optimum movement speed exist which is below the maximum or even preferred speed of the operator.

In summary, this research indicates several preliminary results that are of value in the design and use of hoist systems. First, it is apparent from this work that it is possible to accurately predict hoist movement times and that these times are only slightly larger than analogous manual times. Secondly, depending on the load and movement conditions in question, adjustment of the cable pendulum area can result in more efficient movement. If a relatively light load is to be moved a small distance, increasing the pendulum area through such measures as raising the carriage height will facilitate movement by minimizing carriage travel. In contrast, if a light load is to be moved a long distance, mounting carriage system low, or using a rigid mast instead of a cable is recommended. Third, once the hoist system has been installed, operators should be trained to move the load at a velocity that maximizes both speed and control. Excessive velocity will result in difficulty in decelerating the load mass and bridge assembly, while needlessly slow movement will increase the already large effects of distance on transport times. Finally, the results suggest that for fixed location movements, shock absorber or other braking systems that decelerate the carriage above the target position will prove effective.

REFERENCES

[1] Chung, C.S.,Arm Movements and P.M.T.S, Masters Thesis, University of Melbourne. (February 1983).

[2] Drury, C.G.,Application of Fitts' law to foot pedal design. Human Factors., 17(4) (1975) 368-373.

[3] Fitts, P.M.,The information capacity of human motor system in controlling the amplitude of movement, J. Exp. Psychol., 47 (1954).

[4] Karger, D.W., and Hancock, W.A. Advanced Work Measurement, (Industrial Press, New York, 1982)

[5] Kerr, R.,Movement time in an underwater environment, J. Motor Behav., 5(3) (1973) 175-178.

[6] Langolf, G.,Human Motor Performance in Precise Microscopic Work, Ph.D. Dissertation, the University of Michigan, Published by the MTM

Association, Fairlawn, N.J. (1973).

[7] McGovern, D.E.,Factors Affecting Control Allocation for Augmented Remote Manipulation, Ph.D. Dissertation, Stanford University. (1974)

[8] Welford, A.T.,Fundamentals of Skill (Methuen, London, 1968).

[9] Woodworth, R.S.,Accuracy of voluntary movement, Psychol. Rev. Monograph Supplement, 3(3) (1899).

ACKNOWLEDEMENTS

Partial support for this research was provided by the B. and A. O. Division of the Ford Motor Co. (Contract #NP43-467987). The author also received support on NIOSH Training Grant #5T15-OH07207. Hoist equipment was provided by D. W. Zimmerman Manufacturing of Detroit, Michigan. The author would like to thank Mr. James Foulke and Dr. Gary Langolf for their advice in this work.

Trends in Ergonomics/Human Factors II
R.E. Eberts and C.G. Eberts (Editors)

LIFTING CAPACITY NORMS BASED ON STRENGTH AND ENDURANCE OF WORKERS

Shihab S. Asfour, Ashraf M. Genaidy, Tarek M. Khalil
and Edward C. Greco
Departments of Industrial and Biomedical Engineering
University of Miami
Coral Gables, Florida
U.S.A.

The main objectives of this investigation are to establish lifting capacity norms based on a combination of the physiological and psychophysical approaches, and to compare the lifting capacity data from our study with past studies reported in the literature. The lifting capacity norms are developed for different percentages of the subject population.

INTRODUCTION

For repetitive lifting, the physiological and psychophysical approaches were adopted in the literature for the determination of lifting capacity. However, lifting capacity recommendations reported in the literature were based mainly on one approach at a time. Investigations conducted by the authors showed that a combination of the physiological and psychophysical approaches, rather than a single approach should be used in establishing lifting standards (Asfour et al. (1985)). Therefore, the main objectives of this study were: (1) to establish lifting capacity norms based on a combination of approaches, and (2) to compare the lifting capacity data from our study with past studies reported in the literature.

METHODS AND PROCEDURES

Ten male college students volunteered to participate in this study. The physical characteristics of the subjects, the details of the training program they participated in, and the experimental design and procedures are given in Asfour et al. (1985).

The lifting capacity data as obtained from the physiological and psychophysical approaches were computed as described elsewhere (Asfour et al. (1985)). Based on the means and standard deviations provided by the lifting capacity generated from both approaches, the lifting capacity norms were developed for different population percentages. These norms were computed for different population percentages. These norms were computed for different task variables (frequency of lift and height of lift).

RESULTS AND DISCUSSION

A comparison of the lifting capacity values as determined by the physiological criterion of the present and past studies is shown in Table 1. Figure 1 shows the maximum acceptable weights of lift based on the

TABLE 1 COMPARISON OF LIFTING CAPACITY DATA FROM THE PRESENT AND PAST STUDIES BASED ON THE PHYSIOLOGICAL APPROACH

Approach	Reference	Height of Lift (cm)	Frequency (lifts/min)	Box Size Length X Width (cm)	Lifting Capacity (kg) Mean (S.D.)
Metabolic (based on 1/3 VO_2 max of lifting tasks)	Present Study	Floor-76	1	38 x 38	26.4 (8.0)
Metabolic (based on 1.0 liter/min)	Genaidy (1983)	Floor-76	1	38 x 38	39.9 (9.4)
Metabolic (based on 0.86 liter/min)	Legg and Pateman (1983)	Floor Waist	1.6	50 x 26	66.3 (-)
Metabolic	Present Study	Floor-76	3	38 x 38	27.3 (8.3)
Metabolic (based on 1.0 liter/min)	Asfour (1980)	Floor-76	3	38 x 38	63.1 (-)
Metabolic	Genaidy (1983)	Floor-76	3	38 x 38	31.0 (7.8)
Metabolic	Legg and Pateman	Floor-waist	3.4	50 x 26	22.5 (-)
Metabolic	Present Study	Floor-76	9	38 x 38	10.7 (3.1)
Metabolic	Asfour (1980)	Floor-76	9	38 x 38	16.4 (-)
Metabolic	Present Study	76-127	3	38 x 38	25.1 (7.1)
Metabolic	Asfour (1980)	76-127	3	38 x 38	40.5 (-)
Metabolic	Present Study	76-127	9	38 x 38	4.4 (0.9)

TABLE 1 CONT'D. - COMPARISON OF LIFTING CAPACITY DATA FROM THE PRESENT AND PAST STUDIES BASED ON THE PHYSIOLOGIGAL APPROACH

Approach	Reference	Height of Lift (cm)	Frequency (lifts/min)	Box Size Length X Width (cm)	Lifting Capacity (kg) Mean (S.D.)
Metabolic	Asfour (1980)	76-127	9	38 x 38	26.8 (-)
Metabolic	Present Study	Floor - 127	3	38 x 38	27.6 (5.9)
Metabolic	Asfour (1980)	Floor - 127	3	38 x 38	36.4 (-)
Metabolic	Present Study	Floor - 127	9	38 x 38	9.1 (2.5)
Metabolic	Asfour (1980)	Floor - 127	9	38 x 38	10.5 (-)
Heart Rate (based on 110 bpm)	Legg and Pateman (1983)	Floor - Waist	1.7	50 x 26	66.3 (-)
Heart Rate (based on 115 bpm)	Genaidy (1983)	Floor - 76	2	38 x 38	41.3 (12.8)
Heart Rate	Genaidy (1983)	Floor - 76	3	38 x 38	31.4 (7.4)
Heart Rate	Legg and Pateman (1983)	Floor - Waist	3.4	50 x 26	22.5 (-)

Note: The subject population of all studies were trained college students except the study by Legg and Pateman which were trained soldiers.

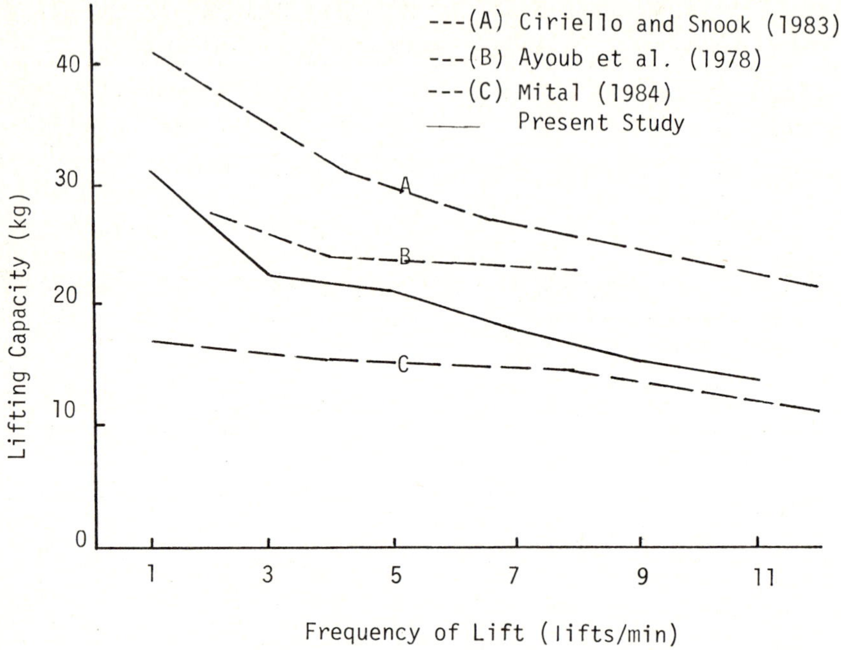

Figure 1

Comparison of Lifting Capacity Data from the Present and Past Studies Based on the Psychophysical Approach for the Floor to Knuckle Height of Lift

psychophysical methodology from the present and past studies for the floor to knuckle height. Table 1 and Figure 1 clearly indicate that there are differences in the lifting capacity data obtained from the different researchers using either similar or different criteria. Some of these differences could be attributed to differences in subject population. However, it is to be noted that the lifting capacity values of our study based on the psychophysical methodology are the closest to those of Mital (1984) who utilized industrial workers (Figure 1).

Table 2 shows the lifting capacity norms generated form both the physiological and psychophysical approaches. These norms were computed for different population percentages (5,10,25,50,75,90, and 95) under different frequencies and heights of lift. Table 2 also indicates that: (1) the lifting capacity decreases significantly between a frequency of 1 and 9 lifts/min, and (2) leg lift recorded the highest lifting capacity values.

Figure 2 shows a comparison between the 99% population limit developed in our study with the action limit defined by NIOSH (1981) for the different frequency-height combinations. This figure indicates that the lifting capacity data obtained from our study was generally lower than the action

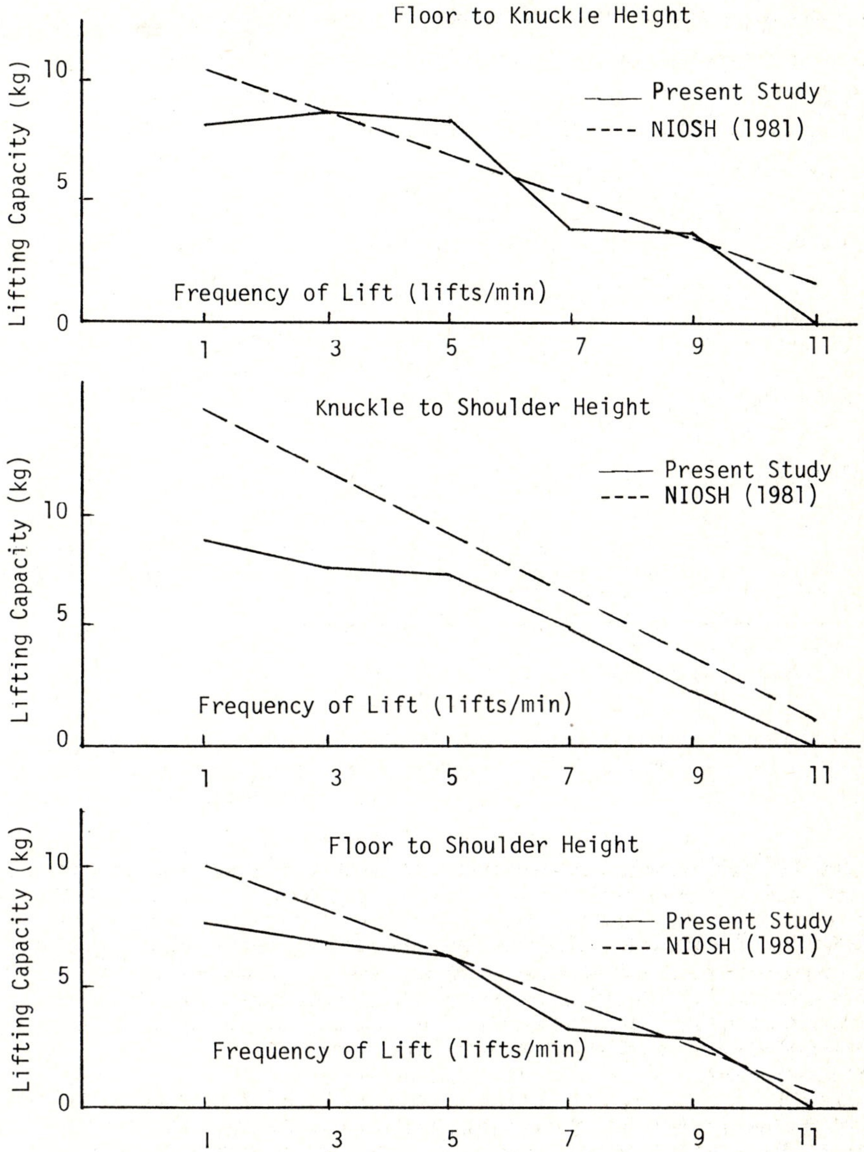

Figure 2

Comparison of Lifting Capacity between the Present Study and NIOSH (1981)

TABLE 2 RECOMMENDED LIFTING CAPACITY LIMITS FOR DIFFERENT PERCENTAGES OF THE SUBJECT POPULATION (MALE) USING COMBINATIONS OF PHYSIOLOGICAL AND PSYCHOPHYSICAL APPROACHES

Height of Lift	Frequency (lifts/min)	Lifting Capacity (kg) Percentages						
		5	10	25	50	70	90	95
Floor - 76.2 cm	1	46.2	43.0	37.6	31.7	25.7	20.3	17.1
	3	32.7	30.5	26.9	22.8	18.8	15.2	13.0
	5	30.4	28.5	25.2	21.6	17.9	14.7	12.7
	7	20.2	18.7	16.2	13.4	10.6	8.1	6.7
	9	15.8	14.7	12.8	10.7	8.6	6.7	5.6
Floor - 127 cm	1	36.5	33.9	29.5	24.5	19.6	15.2	12.6
	3	26.3	24.5	21.5	18.2	14.9	11.9	10.0
	5	23.5	21.9	19.3	16.3	13.4	10.7	9.1
	7	18.9	17.5	15.1	12.5	9.9	7.5	6.1
	9	13.2	12.3	10.8	9.1	7.4	5.9	5.0
76.2 - 127 cm	1	37.3	34.3	30.3	25.4	20.5	16.0	13.4
	3	27.1	25.3	22.3	18.9	15.5	12.5	11.1
	5	25.0	23.5	20.8	18.0	15.1	12.5	10.9
	7	13.5	12.8	11.5	10.0	8.6	7.2	6.5
	9	5.9	5.6	5.0	4.4	3.8	3.3	2.9

limit for the knuckle to shoulder height of lift. For the floor to knuckle and floor to shoulder heights of lift, the lifting capacity data from our study was close to the action limit. The only exception was for the floor to shoulder height. Our lifting capacity values were lower than those recommended by the NIOSH report at 1 and 3 lifts/min.

It is felt that the lifting capacity values provided by the present study could be useful for designers of lifting tasks. This is the only experimental data set available in the literature that utilizes both the physiological and psychophysical approaches on the same group of subjects. Most previous recommendations were based mainly on only one approach and NIOSH guidelines (1981) were based on the extrapolation of existing data.

REFERENCES

[1] Asfour, S. S., Energy cost prediction models for manual lifting and lowering tasks, Doctoral Dissertation, Dept. of Ind. Eng., Texas Tech Univ. (July 1980).

[2] Asfour, S. S., Genaidy, A. M., Khalil, T. M., and Greco, E. C., A combined approach for determination of lifting capacity, in: Eberts, R. (ed.), Trends in Ergonomics/Human Factors II (Elsevier Science Publishers B. V., North Holland, Amsterdam, 1985).

[3] Ayoub, M. M., Bethea, N. J., Deivanayagam, S., Asfour, S. S., Bakken, G. M., Liles, D., Mital, A., and Sherif, M., Determination and modeling of lifting capacity, DHEW (NIOSH) report, Grant #5-R01-OH-00545-02 (1978).

[4] Ciriello, C. M., and Snook, S. H., A Study of size, distance, height, and frequency effects on manual handling tasks, Human Factors, 25 (1983) 473-484.

[5] Genaidy, A. M., Physiological and psychophysical determination of lifting capacity at low frequency lifting tasks, Master thesis, Dept. of Biomed. Eng., Univ. of Miami (May 1983).

[6] Legg, S. J., and Pateman, C. M., Human capabilities in repetitive lifting, Proceedings of the Ergonomics Society Conference, 101-106 (1983).

[7] Mital, A., Comprehensive maximum acceptable weight of lift data base for regular 8-hour work shifts, Ergonomics, 27 (1984) 1127-1138.

[8] National Institute of Occupational Safety and Health (NIOSH), Work practices guide for manual lifting, Technical Report, DHHS (NIOSH) publication No. 81-122 (1981).

Trends in Ergonomics/Human Factors II
R.E. Eberts and C.G. Eberts (Editors)

A COMBINED APPROACH FOR DETERMINATION OF LIFTING CAPACITY

Shihab S. Asfour, Ashraf M. Genaidy, Tarek M. Khalil and Edward C. Greco
Departments of Industrial and Biomedical Engineering
University of Miami
Coral Gables, Florida
U.S.A.

The main objective of this study is to compare the physiological and psychophysical approaches over a wide range of load, frequency and height of lift. The main conclusion drawn from this study is that a combination of approaches rather than a single approach should be used for establishing lifting standards.

INTRODUCTION

In the last twenty years many studies have been made for the determination of the maximum permissible weight of lift. Different approaches have been adopted by different researchers for the development of criteria on which manual materials handling capacities, guidelines, and eventually, standards could be established. These approaches are (1) the physiological, (2) the psychophysical, and (3) the biomechanical. Although research has been conducted using each of these approaches separately, no studies, except Asfour et al. (1984b) have been conducted where any two or all three approaches were studied simultaneously using the same group of subjects. Consequently, neither the adequacy of one approach compared to another nor the possible combination of two or three approaches for the development of manual materials handling guidelines and/or standards can be properly determined.

Asfour et al. (1984b) studied simultaneously the physiological and psychophysical approaches using the same group of subjects. The major conclusion drawn from that study was that, at low frequencies of lift, the lifting capacity generated by the psychophysical approach is lower than that obtained by the physiological approach. However, this investigation was limited to low frequency of lift (1-4 lifts/min) and one height of lift (floor to knuckle height). A follow-up study was conducted to compare the physiological and psychophysical approaches over a wide range of weights of load, frequencies of lift, and different heights of lift requiring the use of different muscle groups.

METHODS AND PROCEDURES

Ten male subjects from the student population at the University of Miami, participated in this study. The subjects ranged in age from 22 to 28 years (mean 24 years). The mean weight of the sample was 79.34 kg (standard deviation 3.81). All subjects were informed of all experimental procedures, and were examined by a qualified physician prior to running this study. They had no history of medical problems. All subjects were trained

5 days for a period of 4 weeks in terms of flexibility, muscle strength, muscle endurance, and cardiovascular endurance. The details of the training program are given elsewhere (Asfour et al. (1984a)). All subjects were also trained for one week on the psychophysical methodology.

Two separate experiments were conducted, using the same group of subjects where the physiological methodology was used in Experiment (1) and the psychophysical methodology was applied in Experiment (2). A randomized complete block design with factorial treatment combinations was utilized in each of the two experiments. Each subject was considered as a block. Each subject performed all the possible treatment combinations in a random order.

For Experiment (1), the independent variables studied were the frequency of lift, load lifted, and height of lift. The dependent variables used were oxygen consumption and heart rate. Levels of frequency was set at 1,3,5,7,9 and 11 lifts/min in order to encompass low and high frequencies of lift encountered in industry. Levels of load were set at 6.82, 13.64, 20.45, 27.27, 34.09, and 40.91 kg to cover both light and heavy loads. The height ranges were: floor to knuckle height (floor level to 76.2 cm above the floor), knuckle to shoulder height (76.2 cm to 127 cm above the floor), and floor to shoulder height (floor level to 127 cm above the floor). These ranges were chosen because they represent typical lifting heights in industry requiring the use of different muscle groups. Experment (2) had the frequency of lift and height of lift as independent variables, and the maximum acceptable weight of lift as the dependent variable. The frequency and height levels used in Experiment (2) were the same as those used in Experiment (1). Each subject was required to lift a box (38.1x38.1x25.4 cm) using a free style technique. The box was provided with cutout handles, and was lowered by an assistant after each lift. The temperature and humidity were controlled (25°C and 70% respectively) in the laboratory where the experiments were conducted.

For Experiment (1), each treatment combination was performed by each subject for seven minutes. No recording of the physiological responses of any of the subjects was made until the first 4 minutes were over to insure that a steady state had been reached. Oxygen consumption and heart rate were then averaged during the last 3 minutes. Minute ventilation was measured using a Parkinson-Cowan ventilation meter. The percentage of oxygen in expired air was determined using a Beckman OM-11 O_2 analyzer. The percentage of carbon dioxide in expired air was determined using a Godard Capnograph infrared CO_2 analyzer. Heart rate for each subject was recorded on a 4 channel Narco-Biosystems physiograph. For Experiment (2), each subject was given twenty minutes during which he decided on the maximum acceptable weight of lift for a particular treatment combination. The maximum acceptable weight of lift was selected by each subject by instructing him to adjust the weight of load to the maximum amount that he could lift without strain or discomfort, and without becoming tired, weakened, over-heated or out of breath. At the end of the twenty minutes, the weight was recorded by the experimenter while the subject was then asked to lift that weight for seven minutes. Oxygen consumption, minute ventilation, and heart rate were averaged over the last 3 minutes of that 7 minute period.

Since the physiological work limit, based on the recommended oxygen consumption value of 1.0 liter/min or one-third of the maximal oxygen uptake, was determined based on highly dynamic tasks such as the standard bicycle ergometer and appears to be high when applied to lifting tasks, it was decided to use the physiological work limit based on lifting tasks (Petrofsky and Lind (1978)). This was achieved by estimating maximal oxygen uptake for lifting tasks using a progressive submaximal technique. The submaximal technique required performance of 3 progressive workloads (frequency x weight x height). For each frequency-height combination, three different weights were chosen to elicit heart rate values between 100 and 170 beats/min for each progressive workload. A line drawn through the heart rate and oxygen consumption values associated with the three progressive work loads, is extrapolated to the age expected maximal heart rate (220-age)(Astrand and Rodahl (1977)), from which maximal oxygen uptake can be estimated. The physiological work limit, was then computed for each subject for each frequency-height combination, as one-third of maximal oxygen uptake (VO_2 max).

RESULTS

The results of the experiments were analyzed using the analysis of variance technique. All the main effects of the independent variables studied were significant at the one percent level. All the interactions were not significant at the five percent level.

Table 1 shows the means and standard deviations of the lifting capacity determined using the physiological approach. These figures were computed as follows: (1) the physiological work limit was determined for each subject for each particular frequency-height combination as stated in the experimental procedures (one-third VO_2 max obtained from lifting tasks), (2) a relation was established between the amount of load lifted and the oxygen consumption, and the load that resulted in an oxygen consumption value equal to the physiological work limit was considered as the lifting capacity.

The means and standard deviations of the psychophysical determination of lifting capacity are also presented in Table 1. It is to be noted from Table 1 that the lifting capacity decreases with the increase of the frequency of lift across all heights as would be expected and the height of lift has a secondary effect on the lifting capacity as compared to the frequency of lift.

The lifting capacity values as determined by the physiological approach are compared to those obtained from the psychophysical approach in Figures 1 through 3 for the three heights of lift studied. The general trend in these figures shows that the physiological approach (based on 1/3 VO_2 max of lifting tasks) yields higher lifting capacity values than the psychophysical approach at low frequencies and lower values at higher frequencies. These conclusions support the notion that the physiological and psychophysical approaches are inherently different and would yield different lifting capacity values for different frequencies.

Figures 1 through 3 show the lifting capacity values based on the recommended physiological work limit of 1.0 liter/min. It is to be noted that the lifting capacity data based on 1.0 liter/min yields higher values than those based on 1/3 VO_2 max of lifting tasks, especially at low frequency lifts (1 and 3 lifts/min). This is attributed to the fact that VO_2 max for lifting tasks is lower than that for bicycling (Petrofsky and Lind

TABLE 1 MEANS (STANDARD DEVIATIONS) OF THE LIFTING CAPACITY (KG) BASED ON THE PHYSIOLOGICAL AND PSYCHOPHYSICAL APPROACHES

Approach	Height of Lift (cm)		Frequency of Lift (lifts/min)					
			1	3	5	7	9	11
Physio-logical	76-127	$\bar{X}$	32.5	25.1	17.5	10.0	4.4	0.0
		SD	9.2	7.1	4.3	2.2	0.9	-
	0-76	$\bar{X}$	26.4	27.3	23.2	13.4	10.7	0.0
		SD	8.0	8.3	6.8	4.1	3.1	-
	0-127	$\bar{X}$	25.45	21.6	18.8	12.5	9.1	0.0
		SD	6.0	5.9	4.6	3.9	2.5	-
Psycho-physical	76-127	$\bar{X}$	25.4	18.9	18.0	14.9	14.7	13.2
		SD	7.3	5.0	4.3	2.9	2.6	2.1
	0-76	$\bar{X}$	31.7	22.8	21.6	18.1	15.7	14.3
		SD	8.9	6.0	5.4	3.8	2.8	1.8
	0-127	$\bar{X}$	24.5	18.2	16.3	14.3	13.9	12.4
		SD	7.3	4.9	4.4	2.5	1.9	2.2

Note: $\bar{X}$: mean
SD: standard deviation

(1978)).

DISCUSSION

The results of the present study indicate that the physiological and psychophysical approaches are inherently different from each other. The physiological fatigue criterion of 1.0 liter/min, which is recommended by many researchers, seems to be high as a limit when applied to lifting tasks. This criterion is based on a percentage of the average maximal oxygen uptake obtained by a bicycle ergometer or a treadmill test. These tests are highly dynamic in nature. Since lifting tasks are comprised of both static and dynamic components and utilize muscle groups that are different from those used in cycling and running, the maximum oxygen uptake value obtained from a lifting task would be generally lower than that obtained in cycling or running tasks (Petrofsky and Lind (1978)). Thus, applying the physiological fatigue limit from one task to another would produce inaccurate results.

It is evident from the results obtained that at low frequencies of lift, the individual's muscular strength rather than physiological endurance becomes the limiting factor in determining the lifting capacity. At higher frequencies of lift the limiting factor is the individual's physiological endurance. It is to be noted that in the design of lifting tasks, the frequency should not exceed 10 lifts/min as the metabolic cost

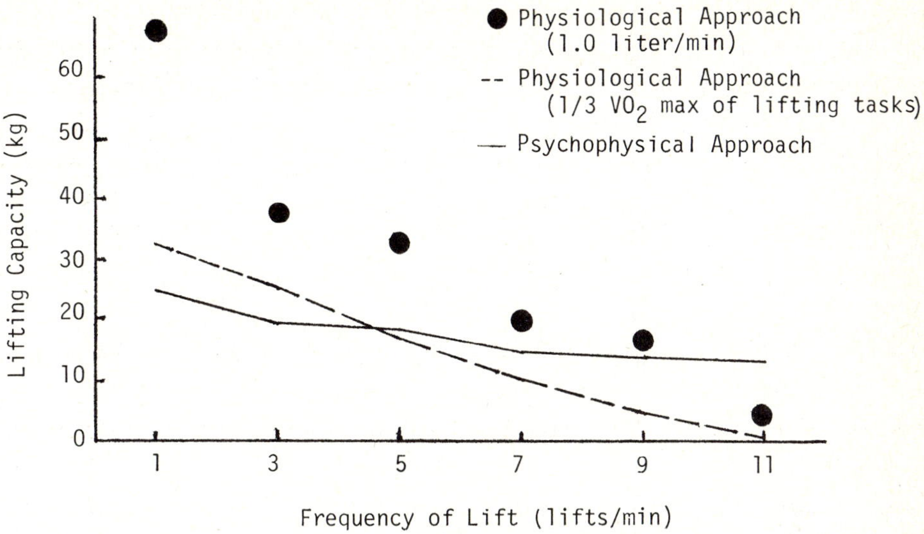

Figure 1

Comparison of Lifting Capacity Data between Physiological and Psychophysical Approaches (Height of Lift: 76 to 127 cm)

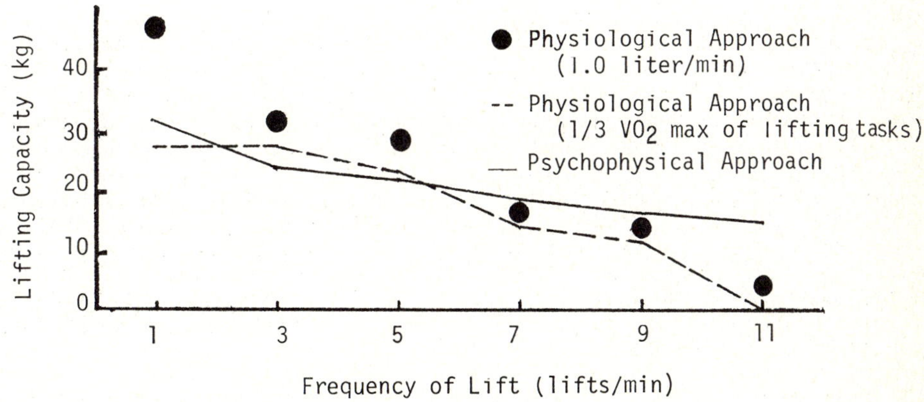

Figure 2

Comparison of Lifting Capacity Data between Physiological and Psychophysical Approaches (Height of Lift: 0 to 76 cm)

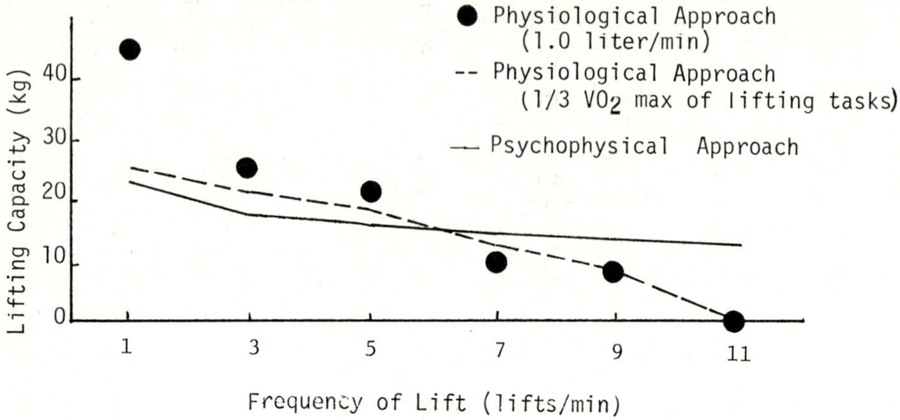

Figure 3
Comparison of Lifting Capacity Data between Physiological and Psychophysical Approaches (Height of Lift: 0 to 127 cm)

associated with lifting body weight alone may tax the individual at his physiological work limit.

CONCLUSIONS

The main conclusions drawn from the present study are:

(1) The frequency of lift is the crucial factor in designing lifting tasks.
(2) The height of lift has a secondary effect as compared to the frequency of lift.
(3) The lifting capacity data as determined based on 1/3 of maximal oxygen uptake for lifting tasks are considerably lower than those based on the 1.0 liter/min.
(4) The physiological approach is not recommended for setting standards for low frequency lifting tasks.
(5) The psychophysical approach is not recommended for setting standards for high frequency lifting tasks.
(6) A combination of approaches rather than a single approach should be used for establishing lifting standards.

REFERENCES

[1] Asfour, S. S., Ayoub, M. M., and Mital, A., Effects of an endurance and strength training programme on lifting capability of males, Ergonomics, 27 (1984a) 435-442.

[2] Asfour, S. S., Genaidy, A. M., Khalil, T. M., and Greco, E. C., Physiological and psychophysical determination of lifting capacity for low frequency tasks, in: Mital, A. (ed.), Trends in Ergonomics/ Human Factors I (Elsevier Science Publishers B.V., North-Holland, Amsterdam, 1984b), 149-153.

[3] Astrand, P. O., and Rodahl, K., Textbook of work physiology (McGraw-Hill, New York, 1977).

[4] Petrofsky, J. S., and Lind, A., Comparison of metabolic and ventilatory responses of men to various lifting tasks and bicycle ergometry, J. Appl. Physiol., 45 (1978) 60-63.

Trends in Ergonomics/Human Factors II
R.E. Eberts and C.G. Eberts (Editors)

STATIC STRENGTH AND MAXIMUM LIFTING CAPACITY FOR A TEAM OF TWO MALES

W. Karwowski
University of Louisville
Louisville, Kentucky
U.S.A.

A. Mital
University of Cincinnati
Cincinnati, Ohio
U.S.A.

and

N. Mulholland
University of Louisville
Louisville, Kentucky
U.S.A.

A hypothesis concerning additivity of static strength and maximum lifting capacity (MLC) for a team of two males was evaluated. Strength evaluations consisted of the arm, stooped back, leg and the composite measures. It was shown that the actual team strengths of the arm, stooped back and composite values were significantly lower than the respective sums of the individual strenghs. The MLC of the team was also lower than the sum of individual MLCs, and accounted on the average for about 87% of these sums.

INTRODUCTION

It has been shown in the literature (Poulsen and Jorgensen, 1971; Keyserling, Herrin and Chaffin, 1980) that isometric strength can be used for the pre-employment screening of individuals to reduce incidence of the low back overexertion injuries due to manual lifting.

Majority of studies in the area of manual materials handling were aimed at determination of the maximum acceptable or permissible loads to be moved by one worker only. On some occasions, however, the object to be lifted is too heavy for one person, and two workers may be needed to do the job, especially if mechanical aids are not available or their use would be prevented by the lack of space or awkward dimensions and shape of the object.

According to Davies (1972), it is unlikely that each worker in the team would exert his maximum force at exactly the same time, and therefore the total amount of load that can be pulled or lifted by a team is not the sum of each worker's capacity. Although the additivity of pulling and pushing forces was checked experimentally (Davies, 1972) the additivity of the static strengths has not yet been investigated.

The purpose of this study was to determine whether the static strength (i.e., the arm strength, stooped back strength, leg strength, and composite strength), as well as the maximum lifting capacity (MLC) for a team of two males are additive. The static strength has been successfully used in the past to predict the maximum lifting capacity of workers (Griffin, Troup and Lloyd, 1984) in order to prevent accidents and low-back injuries. Therefore the correlation matrix for the maximum lifting capacity and static strength measurements was also developed.

METHODS AND PROCEDURES

Six healthy males, college students who were trained in strength exertion and familiarized with the laboratory setup, voluntarily participated in the experiment. The subjects were inexperienced in manual lifting activities, and did not suffer in the past from any physical ailment.

Extensive anthropometric data for the subject population is given in Table 1.

TABLE 1

CHARACTERISTICS OF THE SUBJECTS

	Mean	Std. Dev.	Range
Age	24 years	1.79	21 - 16
Weight	79.7 Kg	19.40	58.9-115.66
Height	180.0 cm	6.78	171.6-189.0
Acromial Height	149.63 cm	6.46	142.15-159.95
Illiac Crest Height	105.03 cm	4.74	98.31-110.6
Arm Span Length	180.68 cm	7.67	171.5-192.4
Forearm Length	49.05 cm	3.65	43.5-53.2
Abdominal Circumference	89.96 cm	19.67	66.7-124.5
Grip Strength	46.87 Kg	6.80	38.0-56.5

The static strength evaluations consisted of four measures: arm strength, stooped back strength, leg strength, and composite strength. In addition to individual measures, the two men strengths (15 combinations for each of the four strengths) were investigated. A total of 120 data points were recorded for these static measures.

The order of all tests within the groups of one man, and the two man teams was randomized, and at least 4 minutes of rest time in between trials was provided. The measurements were taken over a period of 6 weeks.

Maximum lifting capacity was defined as the maximum load that the subject could lift in a single trial from the floor to a bench 89 cm above the floor. The subjects were asked to adjust the weight in the box to the maximum level they could lift without hurting themselves. Ten minute intervals were provided between successive trials to allow for recovery from physical exertion. It was observed that subjects would usually make about 10 adjustments to arrive at the maximum load they could lift.

A square box 30x30x15 cm was used for the single MLC determination, while a box of 30x60x15 cm was used to determine the maximum lifting capacity for a team of two subjects.

The procedure for measuring individual strengths was similar to that recommended by Chaffin (1975), with the exception that only peak forces were recorded. As shown by Kroemer and Marras (1981), the peak forces do not statistically differ from the level forces. To satisfy the Caldwell Regimen (Caldwell et al., 1974) the subjects were asked to build up the force to the maximum, and sustain it for about 3 seconds.

In order to evaluate the hypothesis about additivity of the static strength, the uniform measures of the forces exerted by the individuals and the teams had to be applied. In a situation where two persons are involved in a lifting action, it is unlikely that each member of the team would exert his maximum force at exactly the same time. Therefore, the peak exerted forces (Whitney, 1958) were recorded for the teamwork. Consequently, the same procedure was used in measurements of the individual strengths. The subjects were instructed to avoid any jerking. The peak maximum forces were measured using a load cell (Ametek) with a maximum capacity of 1000 lbs and accuracy ± 5%.

In general, two trials were performed for each type of strength, and the peak exerted forces were used in data analysis. However, if the difference between the first two trials was greater than 10%, a third trial was implemented, and two out of three values which did not differ more than 10% were used.

EQUIPMENT

In order to measure the stooped back and leg strengths of the two man team, a special handle, referred to here as an H-bar, was designed and constructed. In the stooped back measurement, each subject held onto the end of each bar; standing at the openings at either end of the H-bar. In the leg strength measurement, each subject held onto just one bar, similarly as in the individual measurement.

For the purpose of measuring the two man composite and arm strengths, the equipment used for the single measurements was modified. The long handle used for the one man arm measurement was padded on the sides, and used for the team composite strength. When measuring the team arm strength, the subjects faced each other and used the same long handle as they utilized in a single measurement of arm strength.

An important consideration in equipment design for the team measures was to provide for use of the same muscle groups as in the individual measurements. Therefore, the shorter subject was always accomodated to assure that the two members of the team maintained the posture similar (if not identical) to that used in case of individual strength exertions. Six 0.25 inch boards of wood were used to adjust the height of the shorter subject to the height of the taller subject.

RESULTS AND DISCUSSION

The values for the static strength measures for individuals and for a team of two mean are given in Tables 2 and 3, respectively. A paired T-test (Table 4) revealed that the actual team strengths for the arm, stooped back and composite measures were significantly lower than the respective sums of strengths of the individuals making a team.

TABLE 2

STRENGTH MEASURES AND MLC VALUES FOR INDIVIDUALS (Kg)

	N	Mean	S.D.	Range
Maximum Lifting Capacity (MLC)	12	59.04	5.54	52.6-66.2
Arm	12	44.71	2.18	41.2-47.6
Stooped Back	12	102.62	17.51	76.6-129.2
Leg	12	150.97	26.86	113.8-196.4
Composite	12	125.45	15.70	98.4-151.9

On the average, the actual team strengths were lower than the corresponding sums of individual strengths by 10.3%, 8.2% and 7.5% for the arm, stooped back and composite strength measures, respectively (see Table 5). Although not significant, there was a difference of 5.7% between the actual and sum of individual measures for the leg strength.

The maximum lifting capacity of a two men team was significantly lower than the sum of their individual capacities, and accounted on the average for about 87.5% of that sum.

The correlations between the arm and composite strengths, and MLC were significant for both the individual and the team measures. A correlation of 0.68 ($p \leq 0.05$) between the MLC for a team and a static arm strength of the weaker member of a team, indicates that the strength of the weaker subject is a limiting factor in the arm strength of the team.

TABLE 3

STRENGTH MEASURES AND MLC VALUES FOR A TEAM OF TWO MALES (Kg)

	N	Mean	S.D.	Range
Maximum Lifting Capacity (MLC)	30	105.7	13.5	94 -149
Arm	30	83.1	12.6	62 -103
Stooped Back	30	192.4	23.2	141 -240
Leg	30	287.4	38.3	217 -356
Composite	30	236.2	29.2	186 -299

TABLE 4

RESULTS OF THE 'T' TESTS

	df	T(df)	P
Arm Strength (Actual vs. Sum of Individual)	14	2.99	<0.01
Leg Strength (Actual vs. Sum of Individual)	14	1.43	NS
Stooped Back Strength (Actual vs. Sum of Individual)	14	2.44	<0.05
Composite Strength (Actual vs. Sum of Individual)	14	2.96	<0.01
Maximum Lifting Capacity (Actual vs. Sum of Individual)	14	7.04	<0.001

NA - not significant

TABLE 5

DIFFERENCES BETWEEN THE SUM OF TWO AND ACTUAL MEASURES OF STRENGTH

	Mean Differences (Kg) (Sum of individual - measures minus an actual 2-man value)	Mean Percentage Differences
Maximum Lifting Capacity (MLC)	12.37	12.6%
Arm Strength	6.52	10.3%
Stooped Back Strength	13.02	8.2%
Leg Strength	11.91	5.7%
Composite Strength	14.88	7.5%

The composite strength of a team correlated well with the MLC for a team ($r = 0.64$, $p \leq 0.05$). Also, it should be noticed that the grip strength of subjects significantly correlated with the individual measures of maximum weight lifted ($r = 0.55$).

CONCLUSIONS

The results of this study indicate that the maximum lifting capacity (MLC) for a team of two males is not additive, and is lower than the sum of individual MLCs by approximatley 13%. Also, with the exception of the isometric leg strength, the static strengths of a team are significantly lower than the corresponding sums. It should be emphasized that these results were obtained using a small sample of young, healthy college students, and further research is needed to validate these preliminary conclusions.

ACKNOWLEDGEMENT

This work was partially supported by a grant from the Graduate Research Council at the University of Louisville (UHSC 50-84).

TABLE 6

CORRELATIONS BETWEEN STATIC STRENGTHS AND MAXIMUM LIFTING CAPACITIES (MLC) FOR INDIVIDUALS AND A TEAM OF TWO MALES

STATIC STRENGTH MEASURES	Individual (MWL)	Team (MWL)
Grip Strength	0.549*	NA
Grip Strength of Stronger of Two	NA	NA
Grip Strength of Weaker of Two	NA	NA
Arm Strength	0.674**	0.312*
Arm Strength of Stronger of Two	NA	0.441**
Arm Strength of Weaker of Two	NA	0.681**
Stooped Back Strength	NS	NS
Stooped Back Strength of Stronger of Two	NA	NS
Stooped Back Strength of Weaker of Two	NA	NS
Leg Strength	NS	0.443**
Leg Strength of Stronger of Two	NA	NS
Leg Strength of Weaker of Two	NA	NS
Composite Strength	0.538*	0.639**
Composite Strength of Stronger of Two	NA	0.542**
Composite Strength of Weaker of Two	NA	0.352*

* - $p \leq 0.10$
** - $p \leq 0.05$
NS - not significant
NA - not applicable

REFERENCES

[1] Caldwell, L. S., Chaffin, D. B., Dukes-Dobos, F. N., Kroemer, K. H. E., Laubach, L. L., Snook, S. H., and Wasserman, D. E. A., A proposed standard procedure for static muscle strength testing. American Industrial Hygiene Association Journal, Vol. 35 (1974) 201-206.

[2] Carlson, S., A back and lift test, Applied Ergonomics, Vol. 11 (1980) 66-72.

[3] Chaffin, D. B., Ergonomics guide for the assessment of human static strength. American Industrial Hygiene Association Journal, Vol. 36 (1975) 505-511.

[4] Davies, B. T., Moving loads manually, Applied Ergonomics, Vol. 3 (1972) 190-194.

[5] Griffin, A. B., Troup, J.D.G. and Lloyd, D.C.E.F., Testing of lifting and handling capacity, their responsibility and relationship to back symptoms. Ergonomics, Vol. 27 (1984) 305-320.

[6] Keyserling, W. M., Herrrin, G. D. and Chaffin, D. B., Isometric strength testing as a means of controlling medical incidents on strenuous jobs, Journal of Occupational Medicine, Vol. 22 (1980) 332-336.

[7] Kroemer, K.H.E. and Marras, W. S., Evaluation of maximal and submaximal static muscle exertions. Human Factors, Vol. 23 (1981) 643-653.

[9] Mital, A. and Ayoub, M. M., Modeling of lifting capacity and isometric strength, Human Factors, Vol. 22 (1980) 285-290.

[10] Poulsen, E. and Jorgensen, K., Back muscle strength, lifting, and stooped working postures, Applied Ergonomics, Vol. 2 (1971) 133-137.

[11] Whitney, R. J., The strength of lifting action in man. Ergonomics, Vol. 1 (1958) 101-128.

Trends in Ergonomics/Human Factors II
R.E. Eberts and C.G. Eberts (Editors)

PREDICTING MALE AND FEMALE PHYSIOLOGICAL COSTS OF LOAD CARRIAGE

Ying-Hsin Liou and S. J. Morrissey

Department of Industrial Engineering
Auburn University
Auburn University, Alabama 36849
U.S.A.

Previously we presented two models for predicting physiological costs of load carriage; however, these models could not be used with females. This study was performed to examine the differences in male and female physiological responses to load carriage, and to develop regression models capable of predicting physiological costs of two-handed in front of the body (THIB) load carriage for males and females. Eight well-trained female subjects each carried 8 task loads with a variety of container widths, container weights and walking speeds on a level treadmill until steady-state heart rate values were achieved.

This data was then compared to data from males performing identical tasks. The results of this analysis showed a significant sex difference in both dependent variables ($p<0.001$). The male/female regression model for the task heart rate in beats/min accounted for 77 percent of the variability in the observed data ($R^2 = 0.773$); the metabolic rate model in watts accounted for over 94 percent of variability in the observed data ($R^2 = 0.943$).

BACKGROUND

Among the studies of load carriage, the majority of research has focused on backpack load carriage (Givoni and Goldman, 1971; Pandolf et al., 1977; Pimental et al., 1979). In industry, however, this type of carriage is rarely observed due to its inconvenience for carrying loads short distances, and Drury's survey (1982) indicates that 90.1 percent of industrial box handling activities are the "two handed-one box" case.

A few studies have developed regression models to predict the physiological costs of THIB load carriage (Garg et al., 1978; Morrissey and Liou, 1984; Liou, 1984), but none can be applied to females.

LITERATURE REVIEW

Givoni and Goldman (1971), and Pandolf et al. (1977) have extensively studied the metabolic rate of standing, walking and load carriage with the load carried on the back. However, the models they developed are not appropriate models to predict the metabolic rate of THIB load carriage (Morrissey and Liou 1984).

Recently, Morrissey and Liou (1984) and Morrissey (1984) developed a family of regression models for predicting the physiological costs of THIB load carriage, which took into account body weight, load carried, walking speed, container width in the sagittal plane, and task posture. These models offer a convenient and useful means of quantifying the physiological demand anticipated from this common type of industrial load carriage, but they can not be used with females.

Snook (1978) and Ciriello and Snook (1983) used psychophysical methods to determine the maximum acceptable weight that male and female industrial workers would carry under different combinations of distance and frequency. Their tabular results showed that maximum acceptable weights for females were significantly less than for males. While the subjects were performing at their respective maximum acceptable weights, it was found that the oxygen consumption of females was much lower than that of males while the males' and females' task heart rates were not significantly different. Ciriello and Snook's comparisons of males and females are questionable in that they did not take into consideration the effect of the subjects' body weight or the fact that the subjects were not carrying the same loads. Rather, the subjects were each performing at their maximum capabilities.

The purpose of this research was to investigate the differences in male and female physiological responses to load carriages on the basis of equal task weight, and to develop regression models capable of predicting physiological costs of THIB load carriage for males and females.

METHOD AND PROCEDURES

Subject selection and training

Eight female university students were used as experimental subjects, screened for general fitness and health, and then participated in the research. Subjects characteristics were age 21.4 ± 1.3 years; height 166.2 ± 6.2 cm; weight 60.6 ± 10.5 kg.

The training program designed for use in this research was 6 days long and consisted of 3 daily exercises to increase subjects' capacity to perform the task loadings; first, sit-ups and push-ups to improve upper body and abdominal strength; second, bicycle ergometer exercise for general cardiovascular fitness; and third, performing two different carrying tasks for up to 5 min per task to become familiar with the actual carrying tasks. The two tasks performed were randomly selected from the list of experimental task conditions in Table 1.

Independent variables

Three independent variables were studies: container width in the sagittal plane, container weight, and walking speed. A factorial design with two factors was used for this research (Montgomery, 1976). This design has two values for each independent variables, and a total of 8 independent combinations of these variables. Values for these independent variables were: container width, 15.3 and 22.9 cm; container weight, 4.5 and 13.6 kg; and walking speed, 1.12 and 1.56 m/s.

Table 1
Means and Standard Deviations of the Metabolic Rate and Heart Rate for the 8 Experimental Conditions

Container Width (cm)	Container Weight (kg)	Walking Speed (m/s)	Heart Rate (beats/min)	Metabolic Rate (watts)
15.3	4.5	1.12	101.5(11.9)	233.2(27.1)
	4.5	1.56	117.2(11.7)	337.6(45.2)
	13.6	1.12	109.9(9.4)	279.6(31.9)
	13.6	1.56	130.2(11.5)	400.6(45.9)
22.9	4.5	1.12	103.0(8.9)	241.4(32.4)
	4.5	1.56	116.7(9.3)	333.2(53.4)
	13.6	1.12	112.6(9.4)	299.6(34.1)
	13.6	1.56	135.5(13.3)	422.7(45.6)

Dependent variables and data collection

Four dependent variables were studied in the analysis of the physiological costs of load carriage. The first three, task heart rate, ventilation volume, and oxygen concentration in the expired air, were directly measured: the fourth, metabolic rate for task performance, was calculated by the Weir (1949) formula using the values of ventilation volume and oxygen concentration in the expired air.

During the data collection, each subject performed the 8 experimental conditions in a random order with 2 or 3 conditions being performed each day. For the carrying tasks subjects mounted the treadmill and were handed the container to carry. Subjects carried the load until their task heart had been at steady-state levels for at least 3 minutes. Before starting the next experimental condition the subject sat and rested until heart rates recovered to the pre-exertion level.

RESULTS AND DISCUSSION

Model development

The mean values and standard deviations of the task heart rate and metabolic rate for each of the 8 experimental conditions are given in Table 1. These data and the data from Liou's study (1984) of male load carriage, were analyzed using a computerized, stepwise regression package which generated several possible models from a set of candidate independent variables. The independent variables selected for consideration were those singular variables and their interactions shown in the previous research to be closely associated with the dependent variables. The models which resulted were evaluated in terms of the R-square value, mean square error (MSE), Cp statistic, and number of terms in the model.

The final male/female models selected for use in prediction of the metabolic rate and the task heart rate were:

$$HR(beats/min)=113.28-10.62G+21.45V^2+2.01(L+BW)(L/BW)^2 +0.67LV^2-0.56(L+BW)V \quad (1)$$

$$MR(watts)=25.4+24.1G+0.43ZV^2+(L+BW)(3.16+2.54V^2 +16(L/BW)^2 - 3.25V) \quad (2)$$

where BW=body weight, kg
L=load carried, kg
V=walking speed, m/s
Z=container width, cm
G=gender, male=1 and female=0

The task heart rate model accounts for 77 percent of the variability in the data ($R^2 = 0.773$) with a coefficient of variation of 0.073. The metabolic rate model accounts for over 94 percent of the variability ($R^2 = 0.943$) with a coefficient of variation of 0.073.

Sex difference

In the models above, it was of interest to notice that the sex variable was a factor for the heart rate model and the metabolic rate model. In order to examine the effects of sex differences in physiological responses to load carriage two methods were approached. First, the effects of the independent variables on the dependent variables were analyzed by analysis of variance (ANOVA). The results of ANOVA showed that the effects of sex, container weight and walking speed were significant for the task heart rate data ($p<0.001$). For the metabolic rate, the effects of sex, container weight, walking speed and their interaction were significant ($p<0.001$), and the effect of the interaction between sex and speed alone was also found to be significant at the 0.01 level. The second approach was to use equations reported by Morrissey and Liou (1984) for predicting male physiological costs. These predicted values were then compared to the corresponding observed values which can be found in Table 1. These differences were analyzed by the paired t-test. Table 2 shows that the predicted values of task heart rates and metabolic rates were significantly different from the observed values, with the heart rates of females higher than those of males but with lower metabolic rates. Similar results were found by Methany et al. (1942), Brouha (1960) and Morrissey et al. (1981).

Table 2
Results of Paired t-Test Which Show the Differences between the Female Observed Data and the Predicted Data from Morrissey and Liou's Models

Differences		t value	p
M & L HR Model*	- Observed HR Data	-9.83	0.0001<
M & L MR Model**	- Observed MR Data	5.94	0.0001<

* Morrissey and Liou's heart rate model for males
** Morrissey and Liou's metabolic rate for males

Model validation

Model validation was done by using equation (1) and equation (2) to predict the data given in the literature (Kamon and Belding, 1971; Morrissey and Liou, 1984). The carrying method in these previous studies was identical to the method used in this research. Table 3 shows the comparison between the model predicted metabolic rates and the observed metabolic rates from the study by Kamon and Belding. The observed values of Kamon's study are on the average 6.9% higher than the predicted values. Morrissey and Liou (1984) studied the metabolic costs of carrying load in erect and stooped postures. The observed heart rate and metabolic rate data of erect posture were used to compare with the model predicted data. Table 4 shows that the mean of the predicted heart rate, 100 beats/min, is not significantly different from the mean of the observed heart rate, 105 beats/min ($t=-1.81$, $p>0.09$, $n=21$), and the mean of the predicted metabolic rates, 368.4 watts, is also not significantly different from the mean of the observed metabolic rate, 351 watts ($t=1.46$, $p>0.16$, $n=21$).

Table 3
Comparison of the Model Predicted and the Observed Metabolic Rates from Kamon and Belding

Task Speed (m/s)	Task Load (kg)	Mean of Model Predicted MR (watts) A	Mean of Observed MR (watts) B	% of Difference $\frac{B-A}{A}$
1.11	0	271.9	288.1	6.0 %
	10	322.5	354.4	9.9 %
	15	368.9	400.5	8.6 %
	20	432.0	480.0	11.1 %
1.39	0	347.8	348.8	0.3 %
	10	407.1	466.0	14.5 %
	15	457.7	478.6	4.6 %
	20	523.8	524.0	0.0 %
			Average:	6.9 %

Table 4
Comparison of the Model Predicted and the Observed Heart Rates and Metabolic Rates from Morrissey and Liou

Speed (m/s)	Load (kg)		Mean of Model Predicted Data A	Mean of Observed Data B	A–B t value	A–B p
1.34	11.3	HR	100	105	-1.81	0.09
		MR	368.4	351.0	1.46	0.16

CONCLUSION

This research set out to describe and predict the physiological costs of THIB load carriages for males and females. Results found that when males and females performed equal task loads, females had higher task heart rates and lower metabolic rates. Two regression models, which offer a convenient and useful means of quantifying males and females physiological demand on THIB load carriage, were found to be accurate predictors of the data, accounting for 77 percent of the variance for task heart rate and 94 percent for metabolic rate.

REFERENCES

[1] Brouha, L., Physiology in Industry (Pergamon Press, New York, 1976).

[2] Ciriello, V. M. and Snook, S. H., A study of size, distance, height, and frequency effects on manual handling tasks, Human Factors 25(5) (1983) 473-483.

[3] Drury, C. G., Law, C. H. and Pawenski, C. S., A survey of industrial box handling, Human Factors 24(5) (1982) 553-565.

[4] Garg, A., Chaffin, D. B. and Herrin, G. D., Prediction of metabolic rates for manual material handling jobs, American Industrial Hygiene Association Journal 39 (1978) 661-674.

[5] Givoni, B. and Goldman, R. F., Predicting metabolic energy cost, Journal of Applied Physiology 30(3)(1971) 429-433.

[6] Kamon, E. and Belding, H. S., The physiological cost of carrying loads in temperate and hot environments, Human Factors 13(2) (1971) 153-161.

[7] Liou, Y. H., Development of Models for Predicting Physiological Costs of Load Carriage, Master's Thesis, Dept. of Industrial Engineering, Auburn University (August 1984).

[8] Methany, E., Johnson, R. E. and Forbes, W. H., Some physiological responses of women and men to moderate and strenuous exercises: a comparative study, American Journal of Physiology 137 (1942) 318-326.

[9] Montgomery, D. C., Design and Analysis of Experiments (John Wiley & Sons, New York, 1984).

[10] Morrissey, S. J., Prediction and evaluation of the metabolic costs of carrying loads in erect and stooped task postures, Proceedings of the 1984 International Conference on Occupational Ergonomics (1984) 564-568.

[11] Morrissey, S. J. and Liou, Y. H., Metabolic costs of load carriage with different container sizes, Ergonomics 27(8) (1984) 847-853.

[12] Morrissey, S. J. and Liou, Y. H., Development of models for predicting metabolic costs of carrying loads with a range of container sizes, in: Mital, A. (Editor), Trends in Ergonomics/Human Factors I (North-Holland, Amsterdam, 1984).

[13] Morrissey, S. J., and Ayoub, M. M., George, C. E. and Ramsey, J. D., Male and female responses to stoopwalking tasks, Proceedings of the 25th Human Factors Society (1981) 445-449.

[14] Pandolf, K. B., Givoni, B. and Goldman, R. F., Predicting energy expenditure with loads while standing or walking very slowly, Journal of Applied Physiology: Respiratory Environmental Exercises Physiology 43(4) (1977) 577-581.

[15] Pimental, N. A. and Pandolf, K. B., Energy expenditure while standing or walking slowly uphill or downhill with loads, Ergonomics 22(8) (1979) 963-973.

[16] Snook, S. H., The design of manual handling tasks, Ergonomics 21(12) (1978) 963-985.

[17] Weir, J. B. de V., New methods for calculating metabolic rate with special reference to protein metabolism, Journal of Physiology 109 (1949) 1-9.

Trends in Ergonomics/Human Factors II
R.E. Eberts and C.G. Eberts (Editors)

MODELLING OF HUMAN BODY VIBRATION - PART II

M.L. Amirouche

Mechanical Engineering Department
University of Illinois at Chicago
Chicago, Illinois 60680

This paper discusses an analysis procedure for studying vibration effects upon seated industrial workers. This includes vertical, horizontal, and torsional vibration. The procedure is based on Finite Segment Modelling (FSM) of the human body, thus treating it as a mechanical structure. The connective tissues are modelled by springs and dampers for which their coefficients are experimentally evaluated by previous investigators. In addition, the paper presents the transient response of different parts of the body due to a sinusoidal forcing function applied to the lower torso.

INTRODUCTION

Much research on the effects of mechanical vibration on humans is based upon statistical analysis of experimental data. Some results [9, 11] suggest that both neurological and physiological problems such as muscular weakness movements could be affected. This is especially the case in precision work.

The procedures developed in [1] could be useful in studying the effect of whole body vibration [2]. Indeed, this unique approach could be used to help establish minimal standards for vibration exposure. Some attempts have been made in modelling the human body as a series of lumped masses connected by springs and dampers, as described in references [1, 2, 5]. Most recently, Huston et al. [5, 7, 8] have shown that numerical procedures for studying human body vibration response could be studied by new techniques used for multibody systems. For low frequencies the governing dynamical equations could be obtained using Kane's equations as described in [1]. Experimental studies on transmission and tolerance of vertical vibrations, caused by tractors indicate that man is exposed to vibrations where he is vibrated (in the frequency range 0.5 - 11 Hz) at the level of intensity above the so-called intolerable [12, 10]. For rotational vibration it could be as low as 6 Hz [10, 12]. In this paper the whole body responses to a range of frequencies from 1 - 12 Hz are explored. The analysis presented herein is based upon integrating the equations of motions derived in [1] of the model presented by giving the vibrating chair a frequency range of 1 - 12 Hz. The paper is divided into three sections. Part I presents the model used and its characteristics, the second part discusses the vibration analysis and results. The final part consists of the conclusion and remarks.

DESCRIPTION OF THE HUMAN MODEL

Consider a seated human body as depicted by Figure 1.

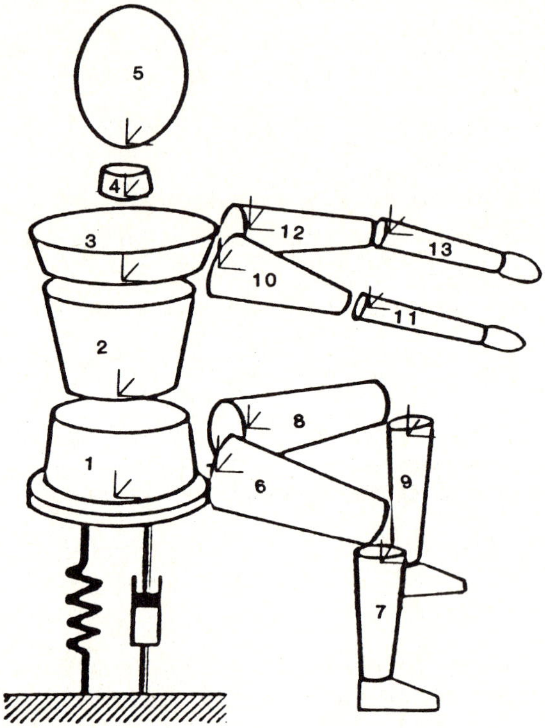

figure.1A SEATED HUMAN BODY MODEL

The model consists of 13 segments (bodies). The configuration of the structure is given by equations (1), (2), and (3) in reference [1] where the main body will be the trunk and the legs and arms, the branches. The dataset used here includes masses, moments of inertia, center of gravity location and distance between joints. This data is based on a 50th percentile sized part 572 dummy design. Table 1 presents the proposed model data collected from [13]. The connective tissues are modelled by springs and dampers. The values of the tissue springs and damping parameters are obtained from studies on the characteristics of specific subsystems (Musksian, 1974) as presented in reference [12]. Table 2 presents a listing of those values.

Since the data for modelling the connective tissues at the joints was not available for all individual parts of the human body, some of those linear coefficients, especially the legs and the arms, were approximated based on their respective cross-section areas. The stiffness and damping matrices are both diagonal matrices with coefficients given by Table 2.

Table 1: Model Dataset

Name	Mass kg	Moment of Inertia (Ix, Iy, Iz) kgm^2	Location Center of Gravity (m)	Location Origin of Joints (m)
Lower torso	11.76	(0.1297, 0.0817, 0.1393)	(0.026, 0., -0.079)	(0., 0., 0.)
Center torso	2.69	(0.0140, 0.0159, 0.0186)	(0.033, 0., 0.072)	(0., 0., 0.132)
Upper torso	17.36	(0.2352, 0.1896, 0.1508)	(0.029, 0., 0.162)	(0.065, 0., 0.318)
Neck	0.88	(0.0014, 0.0014, 0.0006)	(0., 0., 0.063)	(0., 0., 0.124)
Head	4.42	(0.0248), 0.0307, 0.0184)	(0.006, 0., 0028)	(0.03, 0.189, 0.260)
Upper Arm	2.22	(0.0161, 0.0156, 0.0016)	(0., 0., -0.122)	(0.08, ±0.2, -0.261)
Lower Arm	2.15	(0.0311, 0.301, 0.0014)	(0., 0., -0.167)	(0.042, 0.087, -0.07)
Upper Leg	9.68	(0.13, 0.1387, 0.0017)	(0., 0.006, -0.207)	(0., ±0.008, -0.405)
Lower Leg	4.42	(0.1315, 0.1271, 0.0046)	(0.016, 0., -0.272)	(0.46, 0., 0.)

Table 2: Characteristics of the Linear Coefficients of the Springs and Dampers

Damping Constant (c) (kgf/cm/sec)	Spring Constant (k) (kgf/cm)
$c_{11} = 0.378$	$k_{11} = 25.5$
$c_{22} = 0.298$	$k_{22} = 0.8941$
$c_{33} = 0.298$	$k_{33} = 0.8941$
$c_{44} = 3.651$	$k_{44} = 53.64$
$c_{55} = 3.651$	$k_{55} = 53.64$
$c_{66} = 0.3^*$	$k_{66} = 0.45^*$
$c_{77} = 0.3^*$	$k_{77} = 0.45^*$
$c_{88} = 0.3^*$	$k_{88} = 0.45^*$
$c_{99} = 0.3^*$	$k_{99} = 0.45^*$
$c_{10,10} = 0.3^*$	$k_{10,10} = 0.31^*$
$c_{11,11} = 0.3^*$	$k_{11,11} = 0.31^*$
$c_{12,12} = 0.3^*$	$k_{12,12} = 0.31^*$
$c_{13,13} = 0.3^*$	$k_{13,13} = 0.31^*$

* approximated values for legs and arms

VIBRATION ANALYSIS

In reference [1] part I of this paper, the equations of motion of the human body model may be written in the form of equation (16) as follows:

$$a_{\ell p}\ddot{x}_{\ell} = F_{\ell} \qquad (\ell, p = 1, \ldots, 6N) \tag{1}$$

where the $a_{\ell p}$ array is given by equation (17). Again it was shown in [2] that the generalized forces F_ℓ may be written as

$$F_\ell = -k_{\ell p}\, x_\ell - c_{\ell p}\, \dot{x}_\ell + f^*(t) \tag{2}$$

where

$k_{\ell p}$: stiffness matrix
$c_{\ell p}$: damping matrix
$f^*(t)$: forcing function

Combining (1) and (2) yields the following

$$a_{\ell p}\, \ddot{x}_\ell + c_{\ell p}\, \dot{x}_\ell + k_{\ell p}\, x_\ell = f^*_\ell(t) \tag{3}$$

For a linear representation of the connective tissues, the stiffness and damping matrices could be represented by diagonal matrices. Moreover, the values of the stiffness and damping for vertical vibration are given by Table 2. Let's model the chair oscillation by a sinusoidal function such that

$$f^*(t) = m\, x_m\, f^2 \sin \omega t \tag{4}$$

where the chair characteristics are given by

x_m: maximum amplitude
f: frequency
ω: angular frequency

Equations (3) and (4) are simulated in a computer program (currently in its final stage at the University of Illinois at Chicago), and integrated to solve for the displacement x_ℓ, the velocity $\dot{x}_\ell$ and acceleration $\ddot{x}_\ell$. The responses due to the sinusoidal input force to the lower torso are presented.

RESULTS

Figure 2 shows the response of the head at time t = 0.1s to vertical vibration, at frequency ranging from 1 to 12 Hz. The response is similar to a human body seating on a rigid chair. Response (experimental) of a man sitting on a rigid seat of a tractor (from Radke, 1957) shows that the peak response is approximately 4 Hz. Figure 1 shows that the peak response of the head is around 5 Hz.

Figure 3 represents the response of the head at time t = 0.25s and t = 0.3s for a frequency range of 1 - 12 Hz. The peak response is 2 and 3.6 Hz, respectively. Under the same condition and at t = 0.3 sec, the response of the lower torso is shown in Figure 4. It seems that at time t = 0.3s the peak response of the lower torso is the same as the peak response of the head. However, the amplitude of the lower torso is higher.

Again the response of the neck for a frequency ranging from 1 - 12 Hz at time t = 0.25s is presented in Figure 5. The neck has a higher amplitude of oscillation than the head or the lower torso (pelvis). Yet the peak response occurs at slightly lower frequency than the head.

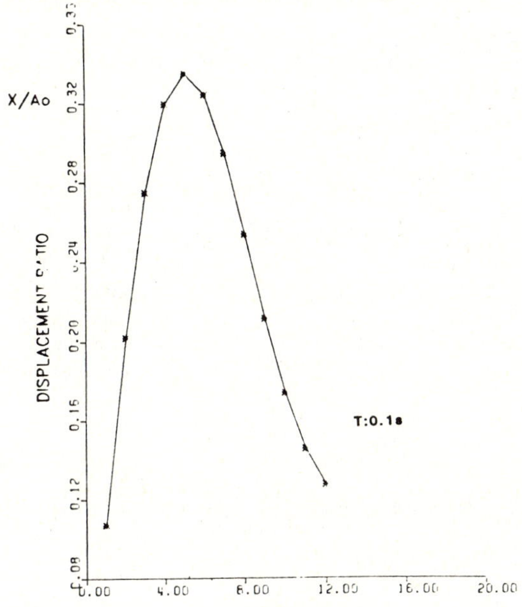

figure.2 RESPONSE OF THE HEAD

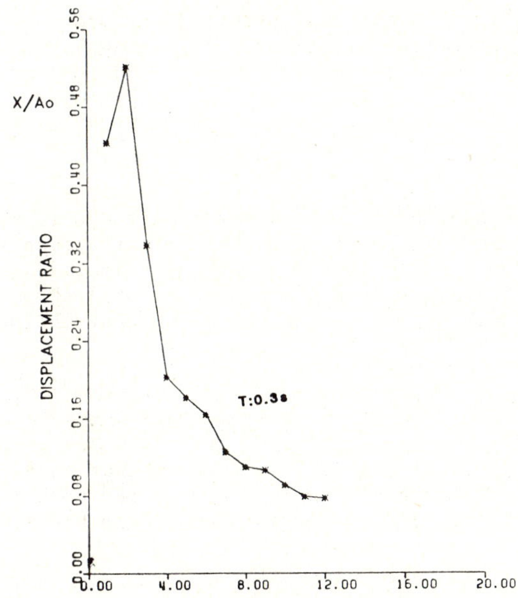

figure.3 RESPONSE OF THE HEAD

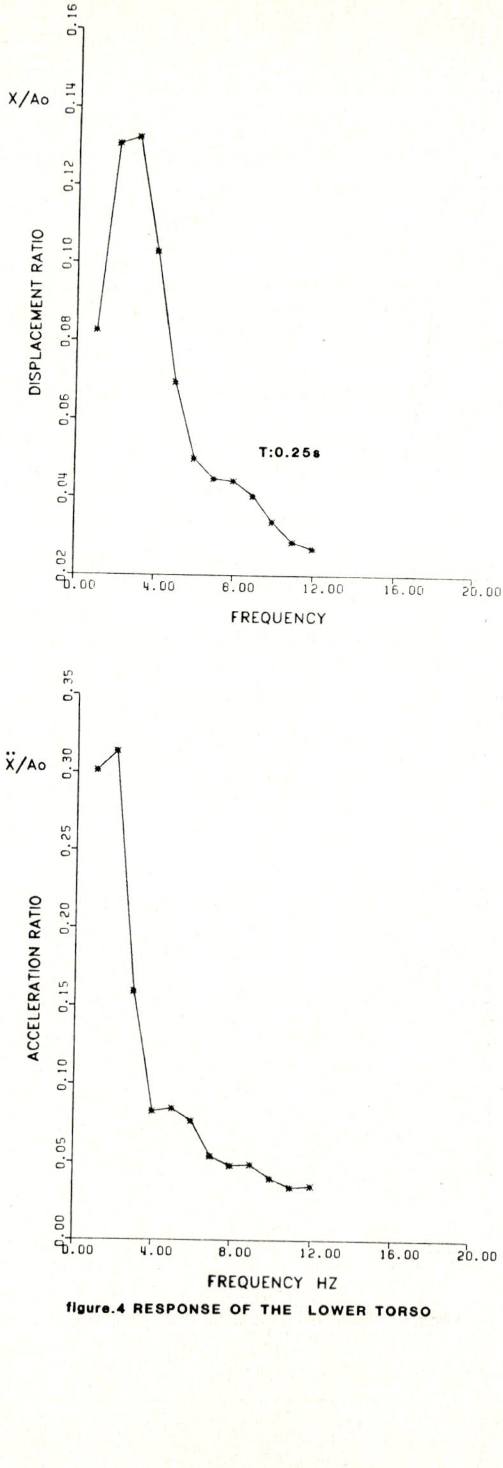

figure.4 RESPONSE OF THE LOWER TORSO

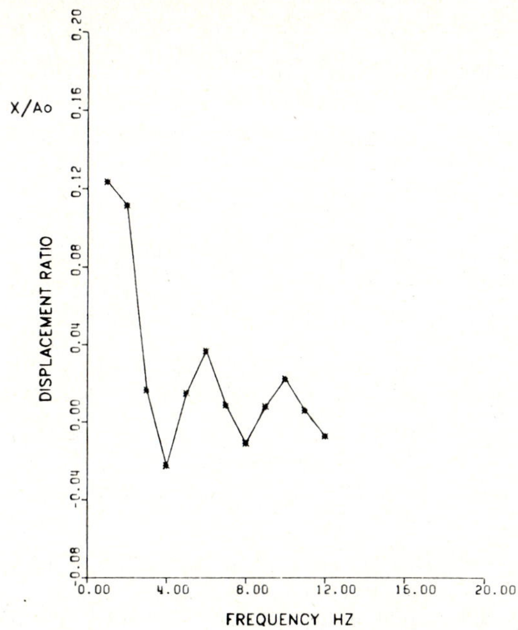

figure.5 RESPONSE OF THE NECK

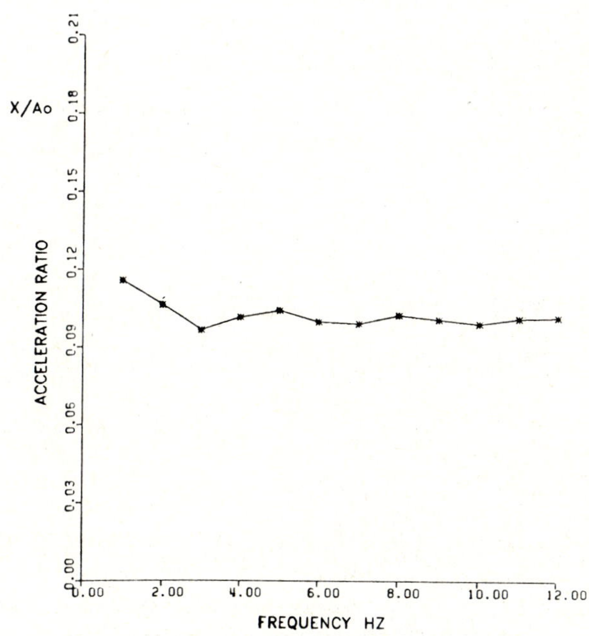

figure.6 ACCELERATION RATIO RESPONSE OF THE HEAD

From reference [12] the acceleration ratio of the head to the lower torso of the lumped mass model used is in a good agreement with experimental from Godman and Von Gierke (1960). In Figure 6 on the previous page the acceleration ratio of the head to the reference frame R taken at the seat is shown. The results prove to be very similar to those presented in reference [12], [10] and [14]. The difference in magnitude of the acceleration ratio is due to the scaling factors involved and to the characteristics of the seat.

CONCLUSIONS

The procedures developed in [1] and the analysis presented herein introduce an excellent alternative method to study the human body vibration. The method employed shows to be very effective in analyzing the responses of individual parts of the body. The computer code used can accommodate linear and non-linear modelling of the elasticity of the springs as well as the dampers. Indeed, the advantage of the use of the program developed is that it provides the responses due to rotation and translation of each individual part of the body along x, y, and z axis. To better select the appropriate parameters of a suspension like the vibrating chair such a study is needed. The characteristics of the oscillating chair used in the analysis were conveniently chosen to approximate the actual vibration of the chair.

It is suggested that more experimental data is needed for better formulation of the springs and dampers of the model presented. It is difficult to evaluate these constants because individuals react differently to various vibrating mediums. True mathematical relationships may become apparent from additional empirical data, therefore, reducing the amount of testing necessary to establish thresholds.

REFERENCES

[1] Amirouche, M.L., and Huston, R.L., Modelling of human body vibration - part I, First Mid-Central Ergonomics/Human Factors Conference, Cincinnati, Ohio, March 1984.

[2] Amirouche, M.L., Flexibility and transient effects in multi-body systems, Ph.D. Thesis, University of Cincinnati, Cincinnati, Ohio, June 1984.

[3] Amirouche, M.L., and Huston, R.L., Collaborative techniques in modal analysis, 26th AIAA SDM Conference, Orlando, Florida, April 1985.

[4] Amirouche, M.L., and Huston, R.L., Use of finite-segment modelling in computer-aided engineering CAE, International Conference, ASME, Minneapolis, Minnesota, August 1984.

[5] Huston, R.L., Multi-body dynamics including the effects of flexibility and compliance, Computers and Structures, 14 (1981) 443-451.

[6] Tien, Chein-Sheng, and Huston, R.L., BIODYNAMIC modelling of the head/neck system, Mid-Central Ergonomics/Human Factors Conference, Cincinnati, Ohio, March 1984.

[7] Huston, J.C., Passarello, C.E., and Huston, R.L., Numerical predict-tion of head/neck response to shock-impact, Measurement and Prediction of Structural and Biodynamic Crash-Impact Response, ASME, (1976) 137-150.

[8] Huston, R.L., Passarello, C.E., On Lagrange's form of d'Alembert's principle, The Matrix and Tensor Quarterly, 23 (1973) 109-112.

[9] Linder, G.S., Mechanical vibration effects on human beings, Aerospace Medicine, 33 (1962) 939-950.

[10] Muskian, R. and Nash, C.D. A model for the response of seated humans to sinusoidal displacements of the seat, J. Biomechanics, 7 (1974) 209-215.

[11] Radke, A.O., Vehicle vibration ... man's environment, ASME paper 57-A, (1957) 54.

[12] Patil, M.K., Palanichamu, M.S., and Ghista, D.N., Man-tractor system dynamics, towards a better suspension system for human ride comfort, J. Biomechanics, 11 (1978) 397-406.

[13] Wisman, J., Comparison of mass distribution data of the part 572 dummy, Ohio State University, 1983.

[14] Goldman, B.E. and Von Gierke, H.E., The effects of shock and vibration on man, Naval Medical Research Institute Report No. 60-3, Bethesda, Maryland (1960).

AUTHOR INDEX